CHEMICAL REACTOR DESIGN, OPTIMIZATION, AND SCALEUP

CHEMICAL REACTOR DESIGN, OPTIMIZATION, AND SCALEUP

E. Bruce Nauman

Rensselaer Polytechnic Institute
Troy, New York

McGRAW-HILL
New York Chicago San Francisco Lisbon London Madrid
Mexico City Milan New Delhi San Juan Seoul
Singapore Sydney Toronto

Library of Congress Cataloging-in-Publication Data

Nauman, E. B.
    Chemical reactor design, optimization, and scaleup / E. Bruce Nauman.
       p.   cm.
    Includes bibliographical references and index.
    ISBN 0-07-137753-0
    1. Chemical reactors.   I. Title.

TP157.N393   2001
660'.2832—dc21
                                                                                2001026699

McGraw-Hill
*A Division of The **McGraw·Hill** Companies*

Copyright © 2002 by The McGraw-Hill Companies, Inc. All rights reserved. Printed in the United States of America. Except as permitted under the United States Copyright Act of 1976, no part of this publication may be reproduced or distributed in any form or by any means, or stored in a data base or retrieval system, without the prior written permission of the publisher.

1 2 3 4 5 6 7 8 9 0 AGM/AGM 0 7 6 5 4 3 2 1

ISBN 0-07-137753-0

The sponsoring editor for this book was Kenneth P. McCombs and the production supervisor was Sherri Souffrance. It was set in Times Roman by Keyword Publishing Services, Ltd.

Printed and bound by Quebecor/Martinsburg.

McGraw-Hill books are available at special quantity discounts to use as premiums and sales promotions, or for use in corporate training programs. For more information, please write to the Director of Special Sales, McGraw-Hill Professional, Two Penn Plaza, New York, NY 10121-2298. Or contact your local bookstore.

 This book is printed on recycled, acid-free paper containing a minimum of 50% recycled, de-inked fiber.

Information contained in this work has been obtained by The McGraw-Hill Companies, Inc. ("McGraw-Hill") from sources believed to be reliable. However, neither McGraw-Hill nor its authors guarantee the accuracy or completeness of any information published herein and neither McGraw-Hill nor its authors shall be responsible for any errors, omissions, or damages arising out of use of this information. This work is published with the understanding that McGraw-Hill and its authors are supplying information but are not attempting to render engineering or other professional services. If such services are required, the assistance of an appropriate professional should be sought.

CONTENTS

TP
157
N393
2002
CHEM

Preface xiii
Notation xv

1. Elementary Reactions in Ideal Reactors — 1

1.1 Material Balances *1*
1.2 Elementary Reactions *4*
    1.2.1 First-Order, Unimolecular Reactions *6*
    1.2.2 Second-Order Reactions, One Reactant *7*
    1.2.3 Second-Order Reactions, Two Reactants *7*
    1.2.4 Third-Order Reactions *7*
1.3 Reaction Order and Mechanism *8*
1.4 Ideal, Isothermal Reactors *10*
    1.4.1 The Ideal Batch Reactor *10*
    1.4.2 Piston Flow Reactors *17*
    1.4.3 Continuous-Flow Stirred Tanks *22*
1.5 Mixing Times and Scaleup *25*
1.6 Batch versus Flow, and Tank versus Tube *28*
Problems *30*
References *33*
Suggestions for Further Reading *33*

2. Multiple Reactions in Batch Reactors — 35

2.1 Multiple and Nonelementary Reactions *35*
2.2 Component Reaction Rates for Multiple Reactions *37*
2.3 Multiple Reactions in Batch Reactors *38*
2.4 Numerical Solutions to Sets of First-Order ODEs *39*
2.5 Analytically Tractable Examples *46*
    2.5.1 The nth-Order Reaction *46*
    2.5.2 Consecutive First-Order Reactions, $A \to B \to C \to \cdots$ *47*
    2.5.3 The Quasi-Steady State Hypothesis *49*
    2.5.4 Autocatalytic Reactions *54*
2.6 Variable-Volume Batch Reactors *58*
    2.6.1 Systems with Constant Mass *58*
    2.6.2 Fed-Batch Reactors *64*
2.7 Scaleup of Batch Reactions *65*
2.8 Stoichiometry and Reaction Coordinates *66*
    2.8.1 Stoichiometry of Single Reactions *66*
    2.8.2 Stoichiometry of Multiple Reactions *67*
Problems *71*

Reference *76*
Suggestions for Further Reading *76*
Appendix 2: Numerical Solution of Ordinary Differential Equations *77*

3. Isothermal Piston Flow Reactors   81

3.1 Piston Flow with Constant Mass Flow *82*
    3.1.1 Gas-Phase Reactions *86*
    3.1.2 Liquid-Phase Reactions *95*
3.2 Scaleup of Tubular Reactions *99*
    3.2.1 Tubes in Parallel *100*
    3.2.2 Tubes in Series *101*
    3.2.3 Scaling with Geometric Similarity *106*
    3.2.4 Scaling with Constant Pressure Drop *108*
    3.2.5 Scaling Down *109*
3.3 Transpired-Wall Reactors *111*
Problems *113*
Reference *116*
Suggestions for Further Reading *116*

4. Stirred Tanks and Reactor Combinations   117

4.1 Continuous-Flow Stirred Tank Reactors *117*
4.2 The Method of False Transients *119*
4.3 CSTRs with Variable Density *123*
    4.3.1 Liquid-Phase CSTRs *123*
    4.3.2 Computation Scheme for Variable-Density CSTRs *125*
    4.3.3 Gas-Phase CSTRs *127*
4.4 Scaleup of Isothermal CSTRs *131*
4.5 Combinations of Reactors *133*
    4.5.1 Series and Parallel Connections *134*
    4.5.2 Tanks in Series *137*
    4.5.3 Recycle Loops *139*
Problems *142*
Suggestions for Further Reading *146*
Appendix 4: Solution of Simultaneous Algebraic Equations *146*
    A.4.1 Binary Searches *146*
    A.4.2 Multidimensional Newton's Method *147*

5. Thermal Effects and Energy Balances   151

5.1 Temperature Dependence of Reaction Rates *151*
    5.1.1 Arrhenius Temperature Dependence *151*
    5.1.2 Optimal Temperatures for Isothermal Reactors *154*
5.2 The Energy Balance *158*
    5.2.1 Nonisothermal Batch Reactors *160*
    5.2.2 Nonisothermal Piston Flow *163*
    5.2.3 Nonisothermal CSTRs *167*

5.3 Scaleup of Nonisothermal Reactors *173*
    5.3.1 Avoiding Scaleup Problems *174*
    5.3.2 Scaling Up Stirred Tanks *176*
    5.3.3 Scaling Up Tubular Reactors *179*
Problems *183*
References *186*
Suggestions for Further Reading *186*

6. Design and Optimization Studies   187

6.1 A Consecutive Reaction Sequence *187*
6.2 A Competitive Reaction Sequence *202*
Problems *203*
Suggestions for Further Reading *205*
Appendix 6: Numerical Optimization Techniques *205*
    A.6.1 Random Searches *206*
    A.6.2 Golden Section Search *207*
    A.6.3 Sophisticated Methods for Parameter Optimization *207*
    A.6.4 Functional Optimization *207*

7. Fitting Rate Data and Using Thermodynamics   209

7.1 Analysis of Rate Data *209*
    7.1.1 Least-Squares Analysis *210*
    7.1.2 Stirred Tanks and Differential Reactors *212*
    7.1.3 Batch and Piston Flow Reactors *218*
    7.1.4 Confounded Reactors *224*
7.2 Thermodynamics of Chemical Reactions *226*
    7.2.1 Terms in the Energy Balance *227*
    7.2.2 Reaction Equilibria *234*
Problems *250*
References *254*
Suggestions for Further Reading *255*
Appendix 7.1: Linear Regression Analysis *255*
Appendix 7.2: Code for Example 7.16 *258*

8. Real Tubular Reactors in Laminar Flow   263

8.1 Isothermal Laminar Flow with Negligible Diffusion *264*
    8.1.1 A Criterion for Neglecting Diffusion *265*
    8.1.2 Mixing-Cup Averages *265*
    8.1.3 A Preview of Residence Time Theory *268*
8.2 Convective Diffusion of Mass *269*
8.3 Numerical Solution Techniques *272*
    8.3.1 The Method of Lines *273*
    8.3.2 Euler's Method *275*
    8.3.3 Accuracy and Stability *276*
    8.3.4 The Trapezoidal Rule *277*
    8.3.5 Use of Dimensionless Variables *282*

- 8.4 Slit Flow and Rectangular Coordinates *285*
- 8.5 Special Velocity Profiles *287*
  - 8.5.1 Flat Velocity Profiles *287*
  - 8.5.2 Flow Between Moving Flat Plates *289*
  - 8.5.3 Motionless Mixers *290*
- 8.6 Convective Diffusion of Heat *291*
  - 8.6.1 Dimensionless Equations for Heat Transfer *293*
  - 8.6.2 Optimal Wall Temperatures *296*
- 8.7 Radial Variations in Viscosity *297*
- 8.8 Radial Velocities *301*
- 8.9 Variable Physical Properties *303*
- 8.10 Scaleup of Laminar Flow Reactors *304*
  - 8.10.1 Isothermal Laminar Flow *304*
  - 8.10.2 Nonisothermal Laminar Flow *305*

Problems *306*
References *309*
Suggestions for Further Reading *309*
Appendix 8.1: The Convective Diffusion Equation *310*
Appendix 8.2: Finite Difference Approximations *311*
Appendix 8.3: Implicit Differencing Schemes *314*

9. Real Tubular Reactors in Turbulent Flow   317

- 9.1 Packed-Bed Reactors *318*
- 9.2 Turbulent Flow in Tubes *327*
- 9.3 The Axial Dispersion Model *329*
  - 9.3.1 The Danckwerts Boundary Conditions *330*
  - 9.3.2 First-Order Reactions *332*
  - 9.3.3 Utility of the Axial Dispersion Model *334*
- 9.4 Nonisothermal Axial Dispersion *336*
- 9.5 Numerical Solutions to Two-Point Boundary Value Problems *337*
- 9.6 Scaleup and Modeling Considerations *344*

Problems *345*
References *347*
Suggestions for Further Reading *347*

10. Heterogeneous Catalysis   349

- 10.1 Overview of Transport and Reaction Steps *351*
- 10.2 Governing Equations for Transport and Reaction *352*
- 10.3 Intrinsic Kinetics *354*
  - 10.3.1 Intrinsic Rate Expressions from Equality of Rates *355*
  - 10.3.2 Models Based on a Rate-Controlling Step *358*
  - 10.3.3 Recommended Models *361*
- 10.4 Effectiveness Factors *362*
  - 10.4.1 Pore Diffusion *363*
  - 10.4.2 Film Mass Transfer *366*
  - 10.4.3 Nonisothermal Effectiveness *367*
  - 10.4.4 Deactivation *369*

10.5 Experimental Determination of Intrinsic Kinetics *371*
10.6 Unsteady Operation and Surface Inventories *375*
Problems *376*
References *380*
Suggestions for Further Reading *380*

11. Multiphase Reactors   381

11.1 Gas–Liquid and Liquid–Liquid Reactors *381*
    11.1.1 Two-Phase Stirred Tank Reactors *382*
    11.1.2 Measurement of Mass Transfer Coefficients *397*
    11.1.3 Fluid–Fluid Contacting in Piston Flow *401*
    11.1.4 Other Mixing Combinations *406*
    11.1.5 Prediction of Mass Transfer Coefficients *409*
11.2 Three-Phase Reactors *412*
    11.2.1 Trickle-Bed Reactors *412*
    11.2.2 Gas-Fed Slurry Reactors *413*
11.3 Moving Solids Reactors *413*
    11.3.1 Bubbling Fluidization *416*
    11.3.2 Fast Fluidization *417*
    11.3.3 Spouted Beds *417*
11.4 Noncatalytic Fluid–Solid Reactions *418*
11.5 Reaction Engineering for Nanotechnology *424*
    11.5.1 Microelectronics *424*
    11.5.2 Chemical Vapor Deposition *426*
    11.5.3 Self-Assembly *427*
11.6 Scaleup of Multiphase Reactors *427*
    11.6.1 Gas–Liquid Reactors *427*
    11.6.2 Gas–Moving-Solids Reactors *430*
Problems *430*
References *432*
Suggestions for Further Reading *432*

12. Biochemical Reaction Engineering   435

12.1 Enzyme Catalysis *436*
    12.1.1 Michaelis-Menten and Similar Kinetics *436*
    12.1.2 Inhibition, Activation, and Deactivation *440*
    12.1.3 Immobilized Enzymes *441*
    12.1.4 Reactor Design for Enzyme Catalysis *443*
12.2 Cell Culture *446*
    12.2.1 Growth Dynamics *448*
    12.2.2 Reactors for Freely Suspended Cells *452*
    12.2.3 Immobilized Cells *459*
Problems *459*
References *461*
Suggestions for Further Reading *461*

13. Polymer Reaction Engineering   463

13.1 Polymerization Reactions *463*
    13.1.1 Step-Growth Polymerizations *464*
    13.1.2 Chain-Growth Polymerizations *467*
13.2 Molecular Weight Distributions *470*
    13.2.1 Distribution Functions and Moments *470*
    13.2.2 Addition Rules for Molecular Weight *472*
    13.2.3 Molecular Weight Measurements *472*
13.3 Kinetics of Condensation Polymerizations *473*
    13.3.1 Conversion *473*
    13.3.2 Number and Weight Average Chain Lengths *474*
    13.3.3 Molecular Weight Distribution Functions *475*
13.4 Kinetics of Addition Polymerizations *478*
    13.4.1 Living Polymers *479*
    13.4.2 Free-Radical Polymerizations *482*
    13.4.3 Transition Metal Catalysis *487*
    13.4.4 Vinyl Copolymerizations *487*
13.5 Polymerization Reactors *492*
    13.5.1 Stirred Tanks with a Continuous Polymer Phase *492*
    13.5.2 Tubular Reactors with a Continuous Polymer Phase *496*
    13.5.3 Suspending-Phase Polymerizations *501*
13.6 Scaleup Considerations *503*
    13.6.1 Binary Polycondensations *504*
    13.6.2 Self-Condensing Polycondensations *504*
    13.6.3 Living Addition Polymerizations *504*
    13.6.4 Vinyl Addition Polymerizations *505*
Problems *505*
Reference *507*
Suggestions for Further Reading *507*
Appendix 13.1: Lumped Parameter Model of a Tubular Polymerizer *508*
Appendix 13.2: Variable-Viscosity Model for a Polycondensation in a Tubular Reactor *512*

14. Unsteady Reactors   517

14.1 Unsteady Stirred Tanks *517*
    14.1.1 Transients in Isothermal CSTRs *519*
    14.1.2 Nonisothermal Stirred Tank Reactors *527*
14.2 Unsteady Piston Flow *531*
14.3 Unsteady Convective Diffusion *534*
Problems *534*
References *538*
Suggestions for Further Reading *538*

15. Residence Time Distributions   539

15.1 Residence Time Theory *540*

15.1.1 Inert Tracer Experiments *540*
  15.1.2 Means and Moments *543*
15.2 Residence Time Models *545*
  15.2.1 Ideal Reactors and Reactor Combinations *545*
  15.2.2 Hydrodynamic Models *555*
15.3 Reaction Yields *561*
  15.3.1 First-Order Reactions *562*
  15.3.2 Other Reactions *564*
15.4 Extensions of Residence Time Theory *574*
  15.4.1 Unsteady Flow Systems *574*
  15.4.2 Contact Time Distributions *575*
  15.4.3 Thermal Times *575*
15.5 Scaleup Considerations *576*
Problems *577*
References *580*
Suggestions for Further Reading *580*

Index   581

PREFACE

This book is an outgrowth of an earlier book, *Chemical Reactor Design*, John Wiley & Sons, 1987. The title is different and reflects a new emphasis on optimization and particularly on scaleup, a topic rarely covered in undergraduate or graduate education but of paramount importance to many practicing engineers. The treatment of biochemical and polymer reaction engineering is also more extensive than normal.

Practitioners are the primary audience for the new book. Here, in one spot, you will find a reasonably comprehensive treatment of reactor design, optimization and scaleup. Spend a few minutes becoming comfortable with the notation (anyone bothering to read a preface obviously has the inclination), and you will find practical answers to many design problems.

The book is also useful for undergraduate and graduate courses in chemical engineering. Some faults of the old book have been eliminated. One fault was its level of difficulty. It was too hard for undergraduates at most U.S. universities. The new book is better. Known rough spots have been smoothed, and it is easier to skip advanced material without loss of continuity. However, the new book remains terse and somewhat more advanced in its level of treatment than is the current U.S. standard. Its goal as a text is not to train students in the application of existing solutions but to educate them for the solution of new problems. Thus, the reader should be prepared to work out the details of some examples rather than expect a complete solution.

There is a continuing emphasis on numerical solutions. Numerical solutions are needed for most practical problems in chemical reactor design, but sophisticated numerical techniques are rarely necessary given the speed of modern computers. The goal is to make the techniques understandable and easily accessible and to allow continued focus on the chemistry and physics of the problem. Computational elegance and efficiency are gladly sacrificed for simplicity.

Too many engineers are completely in the dark when faced with variable physical properties, and tend to assume them away without full knowledge of whether the effects are important. They are often unimportant, but a real design problem—as opposed to an undergraduate exercise or preliminary process synthesis—deserves careful assembly of data and a rigorous solution. Thus, the book gives simple but general techniques for dealing with varying physical properties in CSTRs and PFRs. Random searches are used for optimization and least-squares analysis. These are appallingly inefficient but marvelously robust and easy to implement. The method of lines is used for solving the partial differential equations that govern real tubular reactors and packed beds. This technique is adequate for most problems in reactor design.

No CD ROM is supplied with the book. Many of the numerical problems can be solved with canned ODE and PDE solvers, but most of the solutions are quite simple to code. Creative engineers must occasionally write their own code to solve engineering problems. Due to their varied nature, the solutions require use of a general-purpose language rather than a specific program. Computational examples in the book are illustrated using Basic. This choice was made because Basic is indeed basic enough that it can be sight-read by anyone already familiar with another general-purpose language and because the ubiquitous spreadsheet, Excel, uses Basic macros. Excel provides input/output, plotting, and formatting routines as part of its structure so that coding efforts can be concentrated on the actual calculations. This makes it particularly well suited for students who have not yet become comfortable with another language. Those who prefer another language such as C or Fortran or a mathematical programming system such as Mathematica, Maple, Mathcad, or Matlab should be able to translate quite easily.

I continue with a few eccentricities in notation, using $a, b, c, \ldots$ to denote molar concentrations of components A, B, C, ... I have tried to avoid acronyms and other abbreviations unless the usage is common and there is a true economy of syllables. Equations are numbered when the results are referenced or the equations are important enough to deserve some emphasis. The problems at the back of each chapter are generally arranged to follow the flow of the text rather than level of difficulty. Thus, some low-numbered problems can be fairly difficult.

Bruce Nauman
Troy, New York

NOTATION

Roman Characters

Symbol	Description	Equation where used
a	Concentration of component A	1.6
a	Vector of component concentrations ($N \times 1$)	2.38
$a(0-)$	Concentration just before the entrance of an open reactor	Exam. 9.3
$a(0+)$	Concentration just after the entrance of an open reactor	Exam. 9.3
$a(L-)$	Concentration just before the exit of an open reactor	Exam. 9.3
$a(L+)$	Concentration just after the exit of an open reactor	Exam. 9.3
$a(t, z)$	Concentration of component A in an unsteady tubular reactor	14.14
a'	Auxiliary variable, da/dz, used to convert second-order ODEs to first order	Exam. 9.6
a^*	Dimensionless concentration	Exam. 2.5
a^*	Concentration of component A at the interface	11.4
a_0	Initial concentration of component A	1.23
a_b	Concentration of component A in the bubble phase	11.46
$a_{batch}(t)$	Concentration in a batch reactor at time t	8.9
a_c	Catalyst surface area per mass of catalyst	10.38
a_e	Gas-phase concentration of component A in the emulsion phase	11.45
a_{equil}	Concentration of component A at equilibrium	Prob. 1.13
a_{full}	Concentration of component A when reactor becomes full during a startup	Exam. 14.3
a_g	Concentration of component A in the gas phase	11.1
a_g^i	Concentration of component A at the interface in the gas phase	11.4
a_{in}	Inlet concentration of component A	1.6
$a_{in}(t)$	Time-dependent inlet concentration of component A	Sec. 14.1
a_j	Amine concentration on jth tray	Exam. 11.7
a_l	Concentration of component A in the liquid phase	11.1
$a_l(l)$	Concentration at position l within a pore that is located at point (r, z)	Sec. 10.4.1
$a_l(l, r, z)$	Concentration at location l in a pore, the mouth of which is located at point (r, z)	Sec. 10.1
a_l^i	Concentration of component A at the interface in the liquid phase	11.4
a_{mix}	Concentration at mixing point	4.19

Roman Characters—Continued

Symbol	Description	Equation where used
a_{mix}	Mixing-cup average concentration	8.4
a_{new}	Concentration at new axial position	8.25
a_{out}	Outlet concentration of component A	1.48
$a_{out}(t)$	Time-dependent outlet concentration of component A	Exam. 14.1
a_s	Gas-phase concentration adjacent to surface	Exam. 11.13
$a_s(r, z)$	Concentration on surface of catalyst at location (r, z) in the reactor	Sec. 10.1
a_{trans}	Concentration of transpired component	3.46
a_{wall}	Concentration of component A at the wall	Sec. 8.2
A	Denotes an A-type endgroup in a condensation polymerization	Sec. 13.1
A	Denotes component A	1.12
A	Amount of injected tracer	Exam. 15.1
[A]	Concentration of component A	1.8
A, B, C	Constants in finite difference approximation	8.20
A, B, C	Constants in quadratic equation	App. 8.2
A, B, C, D	Constants in enthalpy equation	7.19
A_b	Cross-sectional area associated with the bubble phase	11.46
A_c	Cross-sectional area of tubular reactor	Sec. 1.4.2
A_e	Cross-sectional area of the emulsion phase	11.45
A_{ext}	External surface area	5.14
A'_{ext}	External surface per unit length of reactor	5.22
A_g	Cross-sectional area of the gas phase	11.28
A_i	Interfacial area	11.1
A'_i	Interfacial area per unit height of reactor	11.27
A_{inlet}	Cross-sectional area at reactor inlet	Prob. 3.6
A_l	Cross-sectional area of the liquid phase	11.27
A_s	Cross-sectional area associated with the solid phase	11.44
A_s	External surface area of the catalyst per unit volume of gas phase	10.2
[AS]	Concentration of A in the adsorbed state	10.5
Av	Avogadro's number	1.9
b	Concentration of component B	1.8
b_0	Initial concentration of component B	1.33
b_{in}	Inlet concentration of component B	Exam. 1.6
b_l	Liquid-phase concentration of component B	Exam. 11.6
b_{out}	Outlet concentration of component B	1.48
B	Denotes component B	1.12
B	Denotes a B-type endgroup in a condensation polymerization	Sec. 13.1

NOTATION xvii

Roman Characters—Continued

Symbol	Description	Equation where used
[B]	Concentration of component B	1.8
[BS]	Concentration of B in the adsorbed state	10.5
c	Concentration of component C	1.19
$c(l)$	Concentration of polymer chains of length l	Sec. 13.2.1
c_j	Carbon dioxide concentration in the gas phase on the jth tray	Exam. 11.7
c_J	Outlet concentration of gaseous carbon dioxide	Exam. 11.8
c_l	Concentration of polymer chains having length l	13.33
$c_{polymer}$	Summed concentration of all polymer chains	13.7
C	Denotes component C	1.19
C	Constant in various equations	1.28
C	Concentration of a nonreactive component	Prob. 1.1
C	Scaling exponent	Prob. 4.18
C	Concentration of inert tracer	15.1
C	Concentration of inert tracer in main tank of the side capacity model	Exam. 15.7
$C(t, z)$	Concentration of inert tracer in an unsteady tubular reactor	Exam. 15.4
C_0	Initial value for tracer concentration	15.1
C_0, C_1	Constants	Sec. 5.2.3
C_1, C_2	Constants of integration	9.18
C_1, C_2	Parameter groupings	Exam. 11.2
C_A	Capacity of ion-exchange resin for component A	11.49
C_{AB}	Collision rate between A and B molecules per volume	1.10
C_h	Constant in heat transfer correlation	5.34
$C_{in}(t)$	Inlet concentration of inert tracer	Exam. 15.4
$C_{out}(t)$	Outlet concentration of inert tracer	15.1
C_P	Heat capacity	5.15
C_R	Specific heat of the agitator	Exam. 14.9
CSTR	Acronym for continuous-flow stirred tank reactor	Sec. 1.4
d	Concentration of component D	2.1
data	Refers to set of experimental data	Sec. 7.1.1
d_j	Concentration of dissolved carbon dioxide in the liquid on the jth tray	Exam. 11.7
d_p	Diameter of a catalyst particle	Exam. 10.8
d_p	Diameter of particle	3.17
d_{pore}	Diameter of a pore	Sec. 10.4.1
d_t	Tube diameter	9.6
dw	Incremental mass of polymer being formed	Exam. 13.9
D	Denotes component D	2.20
D	Axial dispersion coefficient	9.14

Roman Characters—Continued

Symbol	Description	Equation where used
$\mathscr{D}_A$	Diffusion coefficient for component A	8.3
D_e	Axial dispersion coefficient in the emulsion phase	11.45
$\mathscr{D}_{eff}$	Effective diffusivity	10.27
D_g	Axial dispersion coefficient for gas phase	11.34
D_I	Diameter of impeller	5.34
D_{in}	Axial dispersion coefficient in entrance region of an open reactor	Fig. 9.9
$\mathscr{D}_K$	Knudsen diffusivity	10.26
D_l	Axial dispersion coefficient for liquid phase	11.33
D_{out}	Axial dispersion coefficient in exit region of an open reactor	Fig. 9.9
$\mathscr{D}_P$	Diffusivity of product P	10.7
D_r	Radial dispersion coefficient	9.1
D_z	Axial dispersion coefficient for concentration in PDE model	Sec. 9.1
e	Concentration of component E	Exam. 2.2
e	Epoxy concentration	Exam. 14.9
E	Denotes component E	2.1
E	Activation energy	5.1
E	Axial dispersion coefficient for heat	9.24
$\mathscr{E}$	Enhancement factor	11.41
E_0	Concentration of active sites	12.1
E_f	Activation energy for forward reaction	Sec. 5.1.2
E_r	Activation energy for reverse reaction	Sec. 5.1.2
E_r	Radial dispersion coefficient for heat in a packed-bed	9.3
E_z	Axial dispersion coefficient for temperature in PDE model	Sec. 9.1
f	Refers to forward reaction	1.14
f	Arbitrary function	App. 8.2
f	Initiator efficiency factor	13.39
$f(t)$	Differential distribution function for residence times	8.10
$f(t)$	Differential distribution function for exposure times	Sec. 11.1.5
$f(l)$	Number fraction of polymer chains having length l	13.8
$f-$	Value of function at backward point	App. 8.2
$f+$	Value of function at forward point	App. 8.2
f_0	Value of function at central point	App. 8.2
f_A°	Fugacity of pure component A	7.29
$\hat{f}_A$	Fugacity of component A in the mixture	7.29
$f_c(t_c)$	Differential distribution of contact times	Sec. 15.4.2
$f_{dead}(l)$	Number fraction of terminated polymer chains having length l	Sec. 13.4.2

Roman Characters—Continued

Symbol	Description	Equation where used
$f_T(t_T)$	Differential distribution function for thermal times	15.54
f_{in}, f_{out}	Material balance adjustment factors	7.12
f_R	Reaction efficiency factor	1.9
F	Arbitrary function	App. 4
F	Constant value for F_j	Exam. 11.8
$F(t)$	Cumulative distribution function	15.4
$F(r)$	Cumulative distribution function expressed in terms of tube radius for a monotonic velocity profile	15.29
Fa	Fanning friction factor	3.16
F_j	Volumetric flow of gas from the jth tray	Exam. 11.7
g	Grass supply	Sec. 2.5.4
g	Acceleration due to gravity	Exam. 4.7
$g(l)$	Weight fraction of polymer chains having length l	13.11
$g(t)$	Impulse response function for an open system	15.41
$g(t)_{rescaled}$	Impulse response function for an open system after rescaling so that the mean is $\bar{t}$	15.41
G	Arbitrary function	App. 4
G_1, G_2	Growth limitation factors for substrates 1 and 2	12.10
G_1, G_2	Viscosity integrals	Exam. 8.10
G_A	Discretization constant for concentration	Exam. 9.1
G_P	Growth limitation factor for product	12.13
G_S	Growth limitation factor for substrate	12.13
G_T	Discretization constant for temperature equation	Exam. 9.1
Gz	Graetz number	Sec. 5.3.3
h	Concentration of component H	2.41
h	Heat transfer coefficient on the jacket-side	5.34
h	Hydrogen ion concentration	Exam. 14.9
h_i	Interfacial heat transfer coefficient	11.18
h_r	Coefficient for heat transfer to the wall of a packed-bed	9.4
H	Denotes possibly hypothetical component with a stoichiometric coefficient of $+1$	2.41
H	Enthalpy	5.14
H	Enthalpy per mole of reaction mixture	7.42
H	Distance between moving plates	8.51
H_A, H_B, H_I	Component enthalpies	7.20
i	Concentration of component I	3.12
i	Index variable in radial direction	8.21
i	Concentration of adsorbable inerts in the gas phase	10.14

Roman Characters—Continued

Symbol	Description	Equation where used
I	Refers to inert component I	3.13
I	System inventory	1.2
I	Number of radial increments	Sec. 8.3.1
I–IV	Refers to reactions I–IV (Roman numerals)	Sec. 2.1
I_0	Initiator concentration at $t = 0$	13.31
[I]	Concentration of inactive sites	Prob. 12.1
$[IX_n]$	Concentration of growing polymer chains of length n that end with an X group	Sec. 13.4.4
$[IY_n]$	Concentration of growing polymer chains of length n that end with a Y group	Sec. 13.4.4
j	Index variable for axial direction	Exam. 3.3
j	Index variable for data	5.2
J	Number of iterations	Sec. A.4.1
J	Number of experimental data	5.2
J	Number of axial increments	Sec. 8.3.1
J	Number of trays	Exam. 11.8
J_{min}	Minimum number of axial increments, $L/\Delta z_{max}$	Exam. 8.4
J_r	Diffusive flux in radial direction	Sec. 8.2
J_{used}	Number of axial step actually used	Exam. 8.4
J_z	Diffusive flux in axial direction	Sec. 8.2
k	Reaction rate constant	1.8
k'	Pseudo-first-order rate constant	Sec. 1.3
k'	Rate constant with units of reciprocal time	Exam. 2.9
k''	Linear burn rate	11.51
k_0	Pre-exponential rate constant	5.1
k_a	Adsorption rate constant	10.4
k_a^+	Forward rate constant for reversible adsorption step	Exam. 10.2
k_a^-	Reverse rate constant for reversible adsorption step	Exam. 10.2
k_A	Denominator rate constant for component A	10.12
k_A, k_B, k_C	Rate constants for consecutive reactions	2.20
k_{AB}	Denominator constant	12.5
k_B	Denominator rate constant for component B	7.5
k_c	Rate constant for termination by combination	13.39
k_C	Denominator rate constant for component C	Exam. 4.5
k_d	Rate constant for cell death	12.17
k_d	Rate constant for termination by disproportionation	13.39
k_d	Desorption rate constant	10.6
k_d^+	Forward rate constant for reversible desorption step	Exam. 10.2
k_d^-	Reverse rate constant for reversible desorption step	Exam. 10.2
k_D	Rate constant for catalyst deactivation	10.35
k_f	Rate constant for forward reaction	1.14

Roman Characters—Continued

Symbol	Description	Equation where used
k_g	Mass transfer coefficient based on gas-phase driving force	11.5
k_i	Rate constant for chemical initiation	13.39
k_I	Rate constant for reaction I	2.1
k_I	Denominator rate constant for inerts I	10.14
k_{-I}, k_{-II}	Rate constants for reverse reactions I, II	Prob. 4.6
k_l	Mass transfer coefficient based on liquid-phase driving force	11.5
k_P	Denominator rate constant for product P	10.13
k_p	Propagation rate constant	13.31
k_r	Rate constant for reverse reaction	1.14
k_R	Reaction rate constant in denominator	7.5
k_R	Rate constant for surface reaction	10.5
k_R^+	Forward rate constant for reversible surface reaction	Exam. 10.2
k_R^-	Reverse rate constant for reversible surface reaction	Exam. 10.2
k_s	Mass transfer coefficient for a catalyst particle	10.2
k_S	Rate constant for catalyst deactivation	Sec. 10.4.4
k_S	Reaction rate constant in denominator	7.5
k_{SI}	Denominator constant for noncompetitive inhibition	12.6
k_{XX}	Rate constant for monomer X reacting with a polymer chain ending with an X unit	Sec. 13.4.4
k_{XY}	Rate constant for monomer Y reacting with a polymer chain ending with an X unit	Sec. 13.4.4
k_{YX}	Rate constant for monomer X reacting with a polymer chain ending with a Y unit	Sec. 13.4.4
k_{YY}	Rate constant for monomer Y reacting with a polymer chain ending with a Y unit	Sec. 13.4.4
K	Equilibrium constant	1.15
K^*	Dimensionless rate constant	1.29
K_0, K_1, K_2, K_3	Factors for the thermodynamic equilibrium constant	7.35
K_1	Equilibrium constant	Exam. 14.9
K_2	Constant	12.3
K_2	Equilibrium constant	Exam. 14.9
K_a	Kinetic equilibrium constant for adsorption	Exam. 10.4
K_d	Kinetic equilibrium constant for desorption	Exam. 10.3
K_{equil}	Kinetic equilibrium constant	Prob. 3.7
K_g	Mass transfer coefficient based on overall gas-phase driving force	11.1
K_H	Henry's law constant	11.1
K_H^*	Liquid–gas equilibrium constant at the interface	11.4
$K_{kinetic}$	Kinetic equilibrium constant	7.28
K_l	Mass transfer coefficient based on overall liquid-phase driving force	11.2

Roman Characters—Continued

Symbol	Description	Equation where used
K_m	Mass transfer coefficient between the emulsion and bubble phases in a gas fluidized bed	11.45
K_M	Michaelis constant	12.2
K_R	Kinetic equilibrium constant for surface reaction	Exam. 10.3
K_{thermo}	Thermodynamic equilibrium constant	7.29
l	Lynx population	Sec. 2.5.4
l	Position within a pore	10.3
l	Chain length of polymer	13.1
l, m, p, q	Chain lengths for termination by combination	Sec. 13.4.2
l_N	Number average chain length	13.4
l_W	Weight average molecular weight	13.12
L	Length of tubular reactor	1.38
$\mathscr{L}$	Length of a pore	Sec. 10.4.1
$L-$	Location just before reactor outlet	Exam. 9.3
$L+$	Location just after reactor outlet	Exam. 9.3
m	Reaction order exponent	1.20
m	Monomer concentration	Exam. 4.3
m	Exponent in Arrhenius equation	5.1
m	Exponent on product limitation factor	12.13
m	Denotes chain length of polymer	13.2
m, n, r, s	Parameters to be determined in regression analysis	7.48
m_A	Mass of an A molecule	1.10
mix	Refers to a property of the mixture	7.44
m_R	Mass of agitator	Exam. 14.9
M	Denotes monomer	4.6
M	Denotes any molecule that serves as an energy source	Prob. 7.7
M	Denotes a middle group in a condensation polymerization	Sec. 13.1
M	Number of simultaneous reactions	2.9
M_0	Monomer charged to system prior to initiation	13.31
M_A	Molecular weight of component A	Exam. 2.9
M_O	Maintenance coefficient for oxygen	Table 12.1
M_S	Maintenance coefficient, mass of substrate per dry cell mass per time	12.15
n	Reaction order exponent	1.20
n	Index variable for number of tanks	4.16
n	Zone number	Exam. 6.5
n	Index variable for moments of the molecular weight distribution	13.9
n	Number of moment	15.11

Roman Characters—Continued

Symbol	Description	Equation where used
N	Vector of component moles ($N \times 1$)	2.39
N	Denotes a middle group in a condensation polymerization	Sec. 13.1
N	Number of chemical components	2.9
N	Number of tanks in series	4.18
$\overline{N}_A$	Molar flow rate of component A	3.3
N_0	Moles initially present	Exam. 7.13
N_A	Moles of component A	2.30
N_{data}	Number of experimental data	Exam. 7.4
N_I	Rotational velocity of impeller	1.60
N_{total}	Total moles in the system	Sec. 7.2.1
Nu	Nusselt number	5.34
N_{zones}	Number of zones used for temperature optimization	Exam. 6.5
O	Operator indicating order of magnitude	Exam. 2.4
p	Concentration of product P	10.8
p	Parameter in analytical solution	9.19
p_1, p_2	Optimization parameters	App. 6
p_l	Concentration of product P at location l within a pore	10.7
p_{max}	Growth-limiting value for product concentration	12.13
p_{old}	Old or current value for optimization parameter	App. 6
p_s	Concentration of product P at the external surface of the catalyst	10.8
p_{trial}	Trial value for optimization parameter	App. 6
P	Denotes product P	Exam. 2.5
P	Pressure	3.12
P·	Concentration of growing chains summed over all lengths	13.39
P_0	Standard pressure	Sec. 7.2
Pe	Peclet number, $\bar{u}_s d_p / D_r$, for PDE model	Sec. 9.1
Pe	Peclet number for axial dispersion model, $\bar{u}L/D$	9.15
P_g	Partial pressure of oxygen in the gas phase	Exam. 11.9
P_l	Partial pressure of oxygen that would be in equilibrium with the oxygen dissolved in the liquid phase	Exam. 11.9
P_l	Denotes polymer of chain length l	13.1
Power	Agitator power	1.61
Pr	Prandlt number	Sec. 5.3.3
P_R	Probability that a molecule will react	Sec. 15.3.1
Products	Denotes summation over all products	12.14
[PS]	Concentration of P in the adsorbed state	10.6

Roman Characters—Continued

Symbol	Description	Equation where used
q	Transpiration volumetric flow per unit length	3.46
q	Recycle rate	Sec. 4.5.3
q	Energy input by agitator	Exam. 14.9
q	Volumetric flow rate into side tank of side capacity model	Exam. 15.7
$q_{generated}$	Rate of heat generation	5.32
$q_{removed}$	Rate of heat removal	5.33
Q	Volumetric flow rate	1.3
Q_0	Volumetric flow at initial steady state	14.9
Q_{full}	Volumetric flow rate at steady state	Exam. 14.4
Q_g	Gas volumetric flow rate	11.12
Q_{in}	Input volumetric flow rate	1.3
Q_l	Liquid volumetric flow rate	11.11
Q_{mass}	Mass flow rate	1.2
Q_{out}	Discharge volumetric flow rate	1.3
r	Radial coordinate	Sec. 1.4.2
r	Rabbit population	Sec. 2.5.4
$\hat{r}$	Dimensionless radius, r/R	8.5
r'	Dummy variable of integration	13.50
r_1	Dummy variable of integration	8.64
r_A	Radius of an A molecule	1.10
r_B	Radius of a B molecule	1.10
r_p	Radial coordinate for a catalyst particle	10.32
r_X, r_Y	Copolymer reactivity ratio	13.41
R	Refers to component R	1.12
R	Radius of tubular reactor	3.14
$\mathcal{R}$	Vector of reaction rates ($M \times 1$)	2.38
$\bar{R}$	Average radius of surviving particles	11.55
$\mathcal{R}'$	Multicomponent, vector form of $\mathcal{R}'_A$	3.9
R_0	Initial particle radius	11.52
$\mathcal{R}_0$	Initial reaction rate	Prob. 7.9
$\mathcal{R}_A$	Rate of formation of component A	1.6
$(\mathcal{R}_A)_0$	Reaction rate at of component A at the centerline	8.22
$(\mathcal{R}_A)_g$	Rate of formation of component A in the gas phase	11.12
$(\mathcal{R}_A)_l$	Rate of formation of component A in the liquid phase	11.11
$\mathcal{R}_{data}$	Experimental rate data	Sec. 7.1.1
$\mathcal{R}'$	Effective reaction rate for a tubular reactor with variable cross section	3.8
Re	Reynolds number	3.16
$(\mathbf{Re})_{impeller}$	Reynolds number based on impeller diameter	4.11
$(\mathbf{Re})_p$	Reynolds number based on particle diameter	3.17

Roman Characters—Continued

Symbol	Description	Equation where used
R_g	Gas constant	1.10
R_h	Radius of central hole in a cylindrical catalyst particle	Prob. 10.14
$\mathscr{R}_I$	Rate of reaction I, $I = 1$ to M	2.8
$\mathscr{R}_{max}$	Maximum growth rate	12.2
$\mathscr{R}_{model}$	Reaction rate as predicted by model	Sec. 7.1.1
$\mathscr{R}_P$	Rate of formation of product P	10.7
R_p	Radius of a catalyst particle	Exam. 10.6
$\mathscr{R}_r$	Rate of reverse reaction	Exam. 7.11
$\mathscr{R}_S$	Reaction rate for solid	Exam. 11.16
$\mathscr{R}_S$	Reaction rate for substrate	12.15
$\mathscr{R}_X$	Rate of formation of dry cell mass	12.10
s	Substrate concentration	12.1
s	Transform parameter	Sec. 13.4.2
s	Sulfate concentration	Exam. 14.9
s	Laplace transform parameter	Exam. 15.2
s_0	Initial substrate concentration	Exam. 12.5
S	Refers to component S	1.12
S	Refers to the substrate in a biological system	Sec. 12.1
S	Scaling factor for throughput	1.57
S	Concentration of inert tracer in the side tank of the capacity model	Exam. 15.7
[S]	Concentration of vacant sites	10.4
S^2	Sum-of-squares errors	5.2
S'	Scaleup factor per tube, S/S_{tubes}	Sec. 3.2.1
S_0	Total concentration of sites, both occupied and vacant	Exam. 10.1
S_1, S_2	Roots of quadratic equation	2.24
S_{AB}	Stoichiometric ratio of A endgroups to B endgroups at onset of polymerization	13.3
S_A^2, S_B^2, S_C^2	Sum-of-squares for individual components	7.16
Sc	Schmidt number, $\mu/(\rho \mathscr{D}_A)$	Sec. 9.1
$S_{Inventory}$	Scaling factor based on inventory	1.58
S_L	Scaleup factor for tube length	3.31
S_R	Scaleup factor for tube radius	3.31
$S^2_{residual}$	Sum of squares after data fit	Sec. 7.1.1
S_{tubes}	Scaleup factor for the number of tubes	3.31
t	Time	1.2
t	Residence time associated with a streamline, $L/V_z(r)$	Sec. 8.1.3
$\bar{t}$	Mean residence time	1.41
$\bar{t}_{loop}$	Mean residence time for a single pass through the loop	5.35

Roman Characters—Continued

Symbol	Description	Equation where used
$\bar{t}_n$	Residence time in the nth zone	Exam. 6.5
t'	Dummy variable of integration	11.49
t'	Time that molecules entered the reactor	Sec. 14.2
t^*	Dimensionless time	Exam. 2.5
t_0	Initial time	12.9
$t_{1/2}$	Reaction half-life	1.27
t_b	Residence time for a segregated group of molecules	Sec. 15.3.2
t_c	Contact time in a heterogeneous reactor	15.52
t_{empty}	Refers to time when reactor becomes empty	Exam. 14.10
t_{first}	First appearance time when $W(t)$ first goes below 1	Sec. 15.2.1
t_{full}	Time to fill reactor	Exam. 14.3
t_{hold}	Holding time following a fast fill	14.6
t_{max}	Time required to burn a particle	11.54
t_{mix}	Mixing time	Sec. 1.5
t_s	Time constant in a packed-bed, $L/\bar{u}_s$	9.9
t_T	Thermal time	15.53
$\mathcal{T}$	Dimensionless temperature	8.61
T_{ext}	External temperature	5.14
T_g	Temperature in the gas phase	Sec. 11.1.1
T_l	Temperature in the liquid phase	Sec. 11.1.1
T_{max}	Maximum temperature in the reactor	Exam. 9.2
T_n	Temperature in the nth zone	Exam. 6.5
T_{ref}	Reference temperature for enthalpy calculations	5.15
T_s	Temperature at external surface of a catalyst particle	10.4.3
T_{set}	Temperature setpoint	Exam. 14.8
$\bar{u}$	Average axial velocity	1.35
u_b	Gas velocity in the bubble phase	11.46
u_e	Gas velocity in the emulsion phase	11.45
$\bar{u}_g$	Average velocity of the gas phase	11.28
$\bar{u}_l$	Average velocity of the liquid phase	11.27
u_{min}	Minimum fluidization velocity	Sec. 11.3
$\bar{u}_s$	Superficial velocity in a packed-bed	3.17
$(\bar{u}_s)_g$	Superficial gas velocity, Q/A_c	Exam. 11.18
U	Overall heat transfer coefficient	5.14
U'	Heat transfer group	Exam. 7.6
$\mathbf{v}$	Velocity vector in turbulent flow	9.12
V	Volume	1.3
$\mathbf{V}$	Time average velocity vector	9.13
V_0	Velocity at centerline	Prob. 8.2
V_A	Molar volume of component A	7.32

Roman Characters—Continued

Symbol	Description	Equation where used
V_{full}	Full volume of reactor	Exam. 14.3
V_g	Volume of the gas phase	Sec. 11.1.1
V_l	Volume of the liquid phase	Sec. 11.1.1
V_m	Volume of the main tank in the side capacity model	Exam. 15.7
$\mathscr{V}_r$	Dimensionless velocity component in the axial direction, $V_r/\bar{u}$	13.49
V_S	Volumetric consumption rate for solid	11.50
V_S	Volume of side tank in side capacity model	Exam. 15.7
V_y	Velocity in the y-direction	Sec. 8.8
V_z	Axial component of velocity	8.1
$\mathscr{V}_z$	Dimensionless velocity profile, $V_z/\bar{u}$	8.34
$V_z(r)$	Axial component of velocity as a function of radius	8.1
$V_z(y)$	Axial velocity profile in slit flow	8.37
V_θ	Tangential velocity component	Sec. 8.7
w_1, w_2	Weight of polymer aliquots	13.14
w_A, w_B, w_C	Weighting factors for individual components	7.1.3
W	Mass flow rate	Exam. 6.1
$W(t)$	Washout function	15.2
$W(\theta, t)$	Washout function for an unsteady system	Sec. 15.4.1
W_1, W_2	Randomly selected values for the washout function	Exam. 15.6
x	Concentration of comonomer X	13.41
x_i	Mole fraction of component I	Sec. 7.2
x_p	Concentration of X monomer units in the copolymer	13.41
X	Denotes nonreactive or chain-stopping endgroup	Sec. 13.1
X	Denotes monomer X in a copolymerization	Sec. 13.4.4
X	Dry cell mass per unit volume	12.8
X_0	Initial cell mass per unit volume	12.9
X_1, X_2, X_3	Independent variables in regression analysis	7.49
X_A	Molar conversion of component A	1.26
X_A	Conversion of limiting endgroup A	13.16
X_M	Conversion of monomer	4.11
y	Slit or flat-plate coordinate in cross-flow direction	8.37
$\mathscr{y}$	Dimensionless coordinate, y/Y	8.45
y_A	Mole fraction of component A	7.30
y_p	Concentration of Y monomer units in the copolymer	13.41
Y	Denotes monomer Y in a copolymerization	Sec. 13.4.4
Y	Half-height of rectangular channel	Sec. 8.4
Y'	Fraction unreacted if the density did not change	Exam. 2.10
Y_A	Molar fraction of component A that has not reacted	1.25
Y_M	Fraction unreacted for monomer	4.10

Roman Characters—Continued

Symbol	Description	Equation where used
$Y_{P/S}$	Product mass produced per substrate mass	12.14
$Y_{X/S}$	Dry cell mass produced per substrate mass	12.14
$\widehat{Y}_{X/S}$	Theoretical yield of dry cell mass per mass of substrate	12.16
z	Axial coordinate	1.35
$\bar{z}$	Dimensionless axial coordinate, z/L	8.34
z_R	Location of reaction front	11.49
Z	Denotes a middle group in a random condensation polymerization	Sec. 13.1

Greek Characters

Symbol	Description	Equation where used
α	Time constant for lag phase	Prob. 12.7
α_T	Thermal diffusivity, $\lambda/(\rho C_P)$	8.52
α_{12}	Interaction parameter for dual substrate limitations	12.12
β	Constant in pressure drop equation	3.22 and 3.23
β	Volumetric coefficient of thermal expansion	7.18
β	Heat generation number for nonisothermal effectiveness model	Sec. 10.4.3
γ_A	Activity coefficient of component A	7.32
δ	Thickness of stagnant film in the film model	11.36
δ	Fractional increment in flow rate	14.10
$\delta(t)$	Delta function	15.9
Δ	Change in result upon changing step time	Exam. 2.15
ΔA_j	Interfacial area per tray	Exam. 11.7
ΔC_P	Specific heat difference for reaction	7.24
ΔH_F°	Standard free energy of formation	Sec. 7.2.2
ΔG_R	Free energy of reaction	Exam. 7.10
ΔG_R°	Standard free energy of reaction	7.29
ΔH_F°	Standard heat of formation	Sec. 7.2.1
ΔH_R	Heat of reaction	5.17
ΔH_R°	Standard heat of reaction	7.35
$(\Delta H_R)_I$	Heat of reaction for reaction I	5.17
$\Delta H_R \mathscr{R}$	Implied summation of heats of reaction	5.17
ΔP	Pressure drop	3.34
Δ_p	Range of random change	App. 6
ΔP_i	Difference in partial pressures across the interface	Exam. 11.9
Δr	Radial step size, R/I	Sec. 8.3.1
ΔS_{mix}	Entropy of mixing	Sec. 7.2

Greek Characters—Continued

Symbol	Description	Equation where used
Δt	Time step for numerical integration	Sec. 2.4
ΔT	Temperature driving force	Table 5.1
$\Delta T_{adiabatic}$	Adiabatic temperature change	5.20
ΔV_j	Volume of liquid on the jth tray	Exam. 11.7
ΔV_{act}	Activation volume	Prob. 5.4
Δx	Thickness of membrane	11.9
Δz	Axial step size, L/J	Sec. 8.3.1
Δz_{max}	Maximum axial increment for discretization stability	8.29
$\Delta \bar{z}_{max}$	Maximum axial increment in dimensionless form	8.35
$\Delta \rho$	Density difference between polymer and monomer	4.7
ε	Void fraction in a packed-bed	3.17
ε	Reaction coordinate for a single reaction	7.39
$\boldsymbol{\varepsilon}$	Reaction coordinate vector ($M \times 1$)	2.39
ε_I	Reaction coordinate for reaction I	2.40
ε_{total}	Void fraction including internal voids in the catalyst particles	10.38
η	Effectiveness factor	10.23
η	non-Newtonian flow index	Prob. 8.2
η	Effectiveness factor relative to enzyme in its native state	12.7
η_{fresh}	Effective factor for fresh catalyst	10.35
η_0	Effectiveness diffusivity ignoring film resistance	10.31
η_0	0th moment of normalized molecular weight distribution, equal to 1	Sec. 13.2.1
η_n	nth moment of the normalized molecular weight distribution with a 0th moment of 1	13.9
θ	Tangential coordinate	Sec. 1.4.2
θ	Time catalytic or ion-exchange reactor has been on stream	Sec. 10.4.4
θ	Time variable for an unsteady reactor	15.51
κ	Ratio of monomer to polymer density	4.7
λ	Thermal conductivity	Sec. 8.6
λ_{eff}	Effective thermal conductivity for a catalyst particle	10.32
λ_r	Effective thermal conductivity in the radial direction	9.4
μ	Viscosity	3.14
μ	Growth rate for cell mass, cell mass formed per cell mass present per unit time	12.8
μ_0	Viscosity before polymerization	Prob. 8.16
μ_0	0th moment of the molecular weight distribution prior to normalization, equal to $c_{polymer}$	13.10
μ_1	First moment of molecular weight distribution	13.10
μ_2	Second moment of the molecular weight distribution	13.12
μ_∞	Long time value for maximum growth rate	Prob. 12.7
μ_{bulk}	Viscosity in the main flow	5.36

Greek Characters—Continued

Symbol	Description	Equation where used
μ_{max}	Maximum growth rate for cell mass	12.8
μ_n	nth moment of the molecular weight distribution when the 0th moment is $c_{polymer}$	13.9
μ_{wall}	Viscosity at the reactor wall	5.34
$\boldsymbol{\nu}$	$N \times M$ matrix of stoichiometric coefficients	2.37
ν	Change in number of moles upon reaction	7.30
ν	Local rate of power dissipation per unit mass of fluid	15.51
ν_A	Stoichiometric coefficient for component A	1.12
$\nu_{A,I}$	Stoichiometric coefficient for component A in reaction I	2.8
ν_I	Change in number of moles upon reaction for reaction I	7.47
ρ	Mass density	1.3
$\bar{\rho}$	Density averaged with respect to flow rate	8.65
ρ_c	Catalyst mass per total reactor volume	10.38
ρ_∞	Mass density for complete reaction	Exam. 2.10
ρ_{molar}	Molar density of reacting mixture	3.12
σ^2	Dimensionless variance of residence time distribution	15.17
σ_t^2	Variance of residence time distribution	15.15
$\sigma_A, \sigma_B, \sigma_T$	Standard deviations for individual variables	Exam. 7.6
$\sigma_{residual}$	Standard deviation after data fit	Sec. 7.1.1
τ	Dimensionless reaction time	Exam. 2.10
τ	Dimensionless time	5.30
τ	Mean exposure time	11.39
τ	Transformed time for an anionic polymerization	13.35
$\hat{\phi}$	Denotes a benzene ring in chemical formulas	Prob. 2.5
$\hat{\phi}_A$	Fugacity coefficient of component A	7.30
Φ_A	Molar flux of component A	3.6
$\boldsymbol{\Phi}$	Vector form of Φ_A	3.9
$\boldsymbol{\Psi}$	Fluctuating velocity vector in turbulent flow	9.12
ω	Proportionality factor relating concentrations of consecutive chain lengths	Sec. 13.4.2

Special Characters

Symbol	Description	Equation where used
$\mathbf{0}$	Zero matrix	2.42
$0-$	Location just before reactor entrance	Exam. 9.3
$0+$	Location just after reactor entrance	Exam. 9.3
0_M	Zero molecule	1.12
$\sim$	Denotes an arbitrary segment of a polymer chain	Sec. 13.1.1

Special Characters—Continued

Symbol	Description	Equation where used
•	Denotes free radical	13.39
£	Laplace transformation operator	Exam. 15.8

Subscripts

Symbol	Description	Equation where used
A	Refers to component A	1.6
act	Refers to activation of a chemical reaction	5.1
actual	Refers to actual operating conditions in the reactor	3.25
adiabatic	Refers to adiabatic operation	5.20
azeotrope	Refers to conditions where the monomers and polymer have the same composition	13.43
b	Refers to the bubble phase in a fluidized bed	11.46
B	Refers to component B	1.12
batch	Denotes final time or concentration in a batch reactor	Exam. 2.5
C	Refers to component C	1.19
catalyst mass	Refers to reaction rate based on catalyst mass	Exam. 10.9
data	Refers to set of experimental of data	5.2
dead	Refers to terminated polymer chains	Sec. 13.4.2
e	Refers to the emulsion phase in a gas fluidized bed	11.45
E	Refers to component E	Sec. 2.2
empty	Refers to condition when reactor becomes empty	14.10
experiment	Refers to experimental data	7.16
ext	Refers to external conditions	5.14
first	Refers to first appearance time	Sec. 15.2.1
full	Refers to conditions when reactor becomes full	Exam. 14.3
full-scale	Refers to production facility	1.57
hold	Refers to holding time without flow during a startup	14.6
i	Denotes discretized radial position	8.21
I–IV	Refers to reactions I–IV (Roman numerals)	Sec. 2.1
in	Refers to condition at reactor inlet	1.2
inflect	Refers to a value at the inflection point	11.24
inlet	Denotes condition at reactor inlet	3.25
instant	Refers to properties of polymer being made at a particular time in a batch reactor or location in a tubular reactor	Exam. 13.9
j	Refers to jth data point	5.2
j	Denotes discretized axial position	9.27
l	Denotes chain length of polymer	13.1
L	Refers to scaleup of the tube length	3.31

Subscripts—Continued

Symbol	Description	Equation where used
large	Refers to large vessel	1.60
live	Refers to population of growing polymer chains	13.40
mass	Refers to variables that have mass rather than molar units	1.3
max	Refers to maximum value	Exam. 2.4
min	Denotes minimum value	App. 4
mix	Denotes mixing-cup average	Sec. 8.1.2
mix	Refers to property of a mixture	Sec. 7.2.1
monomer	Denotes monomer property	Exam. 4.3
new	Refers to value at new time or point	2.12
old	Refers to value at the old time or point	2.12
open	Denotes characteristics of an open system	15.41
optimal	Refers to optimal value	5.5
out	Refers to condition at reactor outlet	1.2
pilot-scale	Refers to pilot facility	1.57
polymer	Denotes polymer property	4.7
r	Refers to reverse reaction	1.14
r	Refers to radial direction	8.1
R	Refers to component R	1.12
R	Refers to scaleup of the tube radius	3.31
Reactions	Denotes summation over all reactions	2.8
ref	Refers to reference or standard conditions	5.15
rescaled	Denotes property of a system that has been rescaled to have the correct value for $\bar{t}$	15.42
S	Refers to component S	1.12
set	Refers to controller setpoint	Exam. 14.8
small	Refers to small vessel	1.60
spatial	Refers to a spatial average	Prob. 8.5
species	Refers to the collection of chemical components	7.20
surface area	Refers to reaction rate based on catalyst surface area	Exam. 10.9
trans	Refers to transpired component	3.46
tubes	Refers to number of tubes in parallel	3.31
wall	Refers to conditions at the wall	8.55
z	Refers to axial direction	Sec. 8.2
∞	Refers to complete conversion	Exam. 2.10
∞	Refers to a large ratio of tube to packing diameter	9.6
0	Denotes an initial value	1.23
1, 2	Refers to reactors 1 and 2 in a composite system	4.14
1, 2	Refers to polymer aliquots being mixed	13.14

Superscripts

Symbol	Description	Equation where used
m	Reaction rate parameter	7.3
n	Reaction rate parameter	7.3
$order$	Reaction order	1.29
r	Reaction rate parameter	7.3
s	Reaction rate parameter	7.3
*	Denotes dimensionless variable	1.29
*	Denotes value at interface	11.4
^	Denotes average with respect to volume	1.3
′	Denotes external heating area per unit length of tube	5.22
′	Denotes interfacial contact area per unit length of reactor	11.27

Abbreviations

CSTR	Continuous-flow stirred tank reactor
CVD	Chemical vapor deposition
MWD	Molecular weight distribution
NEMS	Nanoelectromechanical system
NPV	Net present value
ODE	Ordinary differential equation
PD	Polydispersity
PDE	Partial differential equation
PFR	Piston flow reactor
RND	Random number with range 0 to 1
RTD	Residence time distribution

Basic Language Code

Program segments and occasional variables within the text are set in a fixed-width font to indicate that they represent computer code.

The Basic language does not allow continuation statements. Instead, long statements give long lines of code. Margin requirements for printing require continuations. These are denoted by a plus sign, +, in the first column of the code. To run a program, delete the + and move the code to the end of the previous line. See Appendix 7 for examples.

CHAPTER 1
ELEMENTARY REACTIONS IN IDEAL REACTORS

Material and energy balances are the heart of chemical engineering. Combine them with chemical kinetics and they are the heart of chemical reaction engineering. Add transport phenomena and you have the intellectual basis for chemical reactor design. This chapter begins the study of chemical reactor design by combining material balances with kinetic expressions for elementary chemical reactions. The resulting equations are then solved for several simple but important types of chemical reactors. More complicated reactions and more complicated reactors are treated in subsequent chapters, but the real core of chemical reactor design is here in Chapter 1. Master it, and the rest will be easy.

1.1 MATERIAL BALANCES

Consider any region of space that has a finite volume and prescribed boundaries that unambiguously separate the region from the rest of the universe. Such a region is called a *control volume*, and the laws of conservation of mass and energy may be applied to it. We ignore nuclear processes so that there are separate conservation laws for mass and energy. For mass,

$$\text{Rate at which mass enters the volume} = \text{Rate at which mass leaves the volume} + \text{Rate at which mass accumulates within the volume} \tag{1.1}$$

where "entering" and "leaving" apply to the flow of material across the boundaries. See Figure 1.1. Equation (1.1) is an *overall mass balance* that applies to the total mass within the control volume, as measured in kilograms or pounds. It can be written as

$$(Q_{mass})_{in} = (Q_{mass})_{out} + \frac{dI}{dt} \tag{1.2}$$

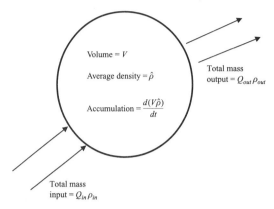

FIGURE 1.1 Control volume for total mass balance.

where Q_{mass} is the mass flow rate and I is the mass inventory in the system. We often write this equation using volumetric flow rates and volumes rather than mass flow rates and mass inventories:

$$Q_{in}\rho_{in} = Q_{out}\rho_{out} + \frac{d(\hat{\rho}V)}{dt} \qquad (1.3)$$

where Q is the volumetric flow rate (volume/time) and ρ is the mass density (mass/volume). Note that $\hat{\rho}$ is the average mass density in the control volume so that $\hat{\rho}V = I$.

Equations (1.1) to (1.3) are different ways of expressing the overall mass balance for a flow system with variable inventory. In steady-state flow, the derivatives vanish, the total mass in the system is constant, and the overall mass balance simply states that input equals output. In batch systems, the flow terms are zero, the time derivative is zero, and the total mass in the system remains constant. We will return to the general form of Equation (1.3) when unsteady reactors are treated in Chapter 14. Until then, the overall mass balance merely serves as a consistency check on more detailed *component balances* that apply to individual substances.

In reactor design, we are interested in chemical reactions that transform one kind of mass into another. A material balance can be written for each component; however, since chemical reactions are possible, the rate of formation of the component within the control volume must now be considered. The *component balance* for some substance A is

Rate at which component A enters the volume
+ net rate at which component A is formed by reaction
= rate at which component A leaves the volume
+ rate at which component A accumulates within the volume (1.4)

or, more briefly,

$$\text{Input} + \text{formation} = \text{output} + \text{accumulation} \tag{1.5}$$

See Figure 1.2. A component balance can be expressed in mass units, and this is done for materials such as polymers that have ill-defined molecular weights. Usually, however, component A will be a distinct molecular species, and it is more convenient to use molar units:

$$Q_{in}a_{in} + \hat{\mathscr{R}}_A V = Q_{out}a_{out} + \frac{d(V\hat{a})}{dt} \tag{1.6}$$

where a is the concentration or molar density of component A in moles per volume, and $\hat{\mathscr{R}}_A$ is the net rate of formation of component A in moles per volume per time. There may be several chemical reactions occurring simultaneously, some of which generate A while others consume it. $\hat{\mathscr{R}}_A$ is the net rate and will be positive if there is net production of component A and negative if there is net consumption. Unless the system is very well mixed, concentrations and reaction rates will vary from point to point within the control volume. The component balance applies to the entire control volume so that $\hat{a}$ and $\hat{\mathscr{R}}_A$ denote spatial averages.

A version of Equation (1.4) can be written for each component, A, B, C, ... If these equations are written in terms of mass and then summed over all components, the sum must equal Equation (1.1) since the net rate of mass formation must be zero. When written in molar units as in Equation (1.6), the sum need not be zero since chemical reactions can cause a net increase or decrease in the number of moles.

To design a chemical reactor, the average concentrations, $\hat{a}, \hat{b}, \hat{c}, \ldots$, or at least the spatial distribution of concentrations, must be found. Doing this is simple for a few special cases of elementary reactions and ideal reactors that

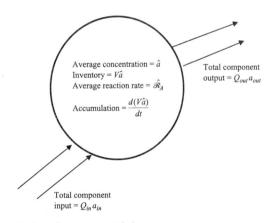

FIGURE 1.2 Control volume for component balance.

are considered here in Chapter 1. We begin by discussing elementary reactions of which there are just a few basic types.

1.2 ELEMENTARY REACTIONS

Consider the reaction of two chemical species according to the stoichiometric equation

$$A + B \rightarrow P \tag{1.7}$$

This reaction is said to be *homogeneous* if it occurs within a single phase. For the time being, we are concerned only with reactions that take place in the gas phase or in a single liquid phase. These reactions are said to be *elementary* if they result from a single interaction (i.e., a collision) between the molecules appearing on the left-hand side of Equation (1.7). The rate at which collisions occur between A and B molecules should be proportional to their concentrations, a and b. Not all collisions cause a reaction, but at constant environmental conditions (e.g., temperature) some definite fraction should react. Thus, we expect

$$\mathscr{R} = k[A][B] = kab \tag{1.8}$$

where k is a constant of proportionality known as the *rate constant*.

Example 1.1: Use the kinetic theory of gases to rationalize the functional form of Equation (1.8).

Solution: We suppose that a collision between an A and a B molecule is necessary but not sufficient for reaction to occur. Thus, we expect

$$\mathscr{R} = \frac{C_{AB} f_R}{\mathrm{Av}} \tag{1.9}$$

where C_{AB} is the collision rate (collisions per volume per time) and f_R is the reaction efficiency. Avogadro's number, Av, has been included in Equation (1.9) so that $\mathscr{R}$ will have normal units, mol/(m$^3 \cdot$s), rather than units of molecules/(m$^3 \cdot$s). By hypothesis, $0 < f_R < 1$.

The molecules are treated as rigid spheres having radii r_A and r_B. They collide if they approach each other within a distance $r_A + r_B$. A result from kinetic theory is

$$C_{AB} = \left[\frac{8\pi R_g T(m_A + m_B)}{\mathrm{Av} \, m_A m_B} \right]^{1/2} (r_A + r_B)^2 \mathrm{Av}^2 ab \tag{1.10}$$

where R_g is the gas constant, T is the absolute temperature, and m_A and m_B are the molecular masses in kilograms per molecule. The collision rate is proportional to the product of the concentrations as postulated in Equation (1.8). The reaction rate constant is

$$k = \left[\frac{8\pi R_g T(m_A + m_B)}{\text{Av}m_A m_B}\right]^{1/2} (r_A + r_B)^2 \text{Av} f_R \qquad (1.11)$$

Collision theory is mute about the value of f_R. Typically, $f_R \ll 1$, so that the number of molecules colliding is much greater than the number reacting. See Problem 1.2. Not all collisions have enough energy to produce a reaction. Steric effects may also be important. As will be discussed in Chapter 5, f_R is strongly dependent on temperature. This dependence usually overwhelms the $T^{1/2}$ dependence predicted for the collision rate.

Note that the rate constant k is positive so that $\mathscr{R}$ is positive. $\mathscr{R}$ is *the rate of the reaction*, not the rate at which a particular component reacts. Components A and B are consumed by the reaction of Equation (1.7) and thus are "formed" at a negative rate:

$$\mathscr{R}_A = \mathscr{R}_B = -kab$$

while P is formed at a positive rate:

$$\mathscr{R}_P = +kab$$

The sign convention we have adopted is that the rate of a reaction is always positive. The *rate of formation of a component* is positive when the component is formed by the reaction and is negative when the component is consumed.

A general expression for any single reaction is

$$0_M \to \nu_A A + \nu_B B + \cdots + \nu_R R + \nu_S S + \cdots \qquad (1.12)$$

As an example, the reaction $2H_2 + O_2 \to 2H_2O$ can be written as

$$0_M \to -2H_2 - O_2 + 2H_2O$$

This form is obtained by setting all participating species, whether products or reactants, on the right-hand side of the stoichiometric equation. The remaining term on the left is the *zero molecule*, which is denoted by 0_M to avoid confusion with atomic oxygen. The $\nu_A, \nu_B, \ldots$ terms are the *stoichiometric coefficients* for the reaction. They are positive for products and negative for reactants. Using them, the general relationship between the rate of the reaction and the rate of formation of component A is given by

$$\mathscr{R}_A = \nu_A \mathscr{R} \qquad (1.13)$$

The stoichiometric coefficients can be fractions. However, for elementary reactions, they must be small integers, of magnitude 2, 1, or 0. If the reaction of

Equation (1.12) were reversible and elementary, its rate would be

$$\mathscr{R} = k_f[A]^{-\nu_A}[B]^{-\nu_B} \cdots - k_r[R]^{\nu_R}[S]^{\nu_S} \qquad (1.14)$$

and it would have an equilibrium constant

$$K = \frac{k_f}{k_r} = [A]^{\nu_A}[B]^{\nu_B} \cdots [R]^{\nu_R}[S]^{\nu_S} = \frac{[R]^{\nu_R}[S]^{\nu_S} \cdots}{[A]^{-\nu_A}[B]^{-\nu_B} \cdots} \qquad (1.15)$$

where $A, B, \ldots$ are reactants; $R, S, \ldots$ are products; k_f is the rate constant for the forward reaction; and k_r is the rate constant for the reverse reaction.

The functional form of the reaction rate in Equation (1.14) is dictated by the reaction stoichiometry, Equation (1.12). Only the constants k_f and k_r can be adjusted to fit the specific reaction. This is the hallmark of an elementary reaction; its rate is consistent with the reaction stoichiometry. However, reactions can have the form of Equation (1.14) without being elementary.

As a shorthand notation for indicating that a reaction is elementary, we shall include the rate constants in the stoichiometric equation. Thus, the reaction

$$A + B \underset{k_r}{\overset{k_f}{\rightleftharpoons}} 2C$$

is elementary, reversible, and has the following rate expression:

$$\mathscr{R} = k_f ab - k_r c^2$$

We deal with many reactions that are not elementary. Most industrially important reactions go through a complex kinetic mechanism before the final products are reached. The mechanism may give a rate expression far different than Equation (1.14), even though it involves only short-lived intermediates that never appear in conventional chemical analyses. Elementary reactions are generally limited to the following types.

1.2.1 First-Order, Unimolecular Reactions

$$A \xrightarrow{k} \text{Products} \qquad \mathscr{R} = ka \qquad (1.16)$$

Since $\mathscr{R}$ has units of moles per volume per time and a has units of moles per volume, the rate constant for a first-order reaction has units of reciprocal time: e.g., s^{-1}. The best example of a truly first-order reaction is radioactive decay; for example,

$$U^{238} \rightarrow Th^{234} + He^4$$

since it occurs spontaneously as a single-body event. Among strictly chemical reactions, thermal decompositions such as

$$CH_3OCH_3 \rightarrow CH_4 + CO + H_2$$

follow first-order kinetics at normal gas densities. The student of chemistry will recognize that the complete decomposition of dimethyl ether into methane, carbon monoxide, and hydrogen is unlikely to occur in a single step. Short-lived intermediates will exist; however, since the reaction is irreversible, they will not affect the rate of the forward reaction, which is first order and has the form of Equation (1.16). The decomposition does require energy, and collisions between the reactant and other molecules are the usual mechanism for acquiring this energy. Thus, a second-order dependence may be observed for the pure gas at very low densities since reactant molecules must collide with themselves to acquire energy.

1.2.2 Second-Order Reactions, One Reactant

$$2A \xrightarrow{k} \text{Products} \qquad \mathscr{R} = ka^2 \qquad (1.17)$$

where k has units of $m^3\,mol^{-1}\,s^{-1}$. It is important to note that $\mathscr{R}_A = -2ka^2$ according to the convention of Equation (1.13).

A gas-phase reaction believed to be elementary and second order is

$$2HI \rightarrow H_2 + I_2$$

Here, collisions between two HI molecules supply energy and also supply the reactants needed to satisfy the observed stoichiometry.

1.2.3 Second-Order Reactions, Two Reactants

$$A + B \xrightarrow{k} \text{Products} \qquad \mathscr{R} = kab \qquad (1.18)$$

Liquid-phase esterifications such as

$$C_2H_5OH + CH_3\overset{\overset{O}{\|}}{C}OH \rightarrow C_2H_5O\overset{\overset{O}{\|}}{C}CH_3 + H_2O$$

typically follow second-order kinetics.

1.2.4 Third-Order Reactions

Elementary third-order reactions are vanishingly rare because they require a statistically improbable three-way collision. In principle, there are three types of third-order reactions:

$$\begin{aligned} 3A \xrightarrow{k} \text{Products} & \qquad \mathscr{R} = ka^3 \\ 2A + B \xrightarrow{k} \text{Products} & \qquad \mathscr{R} = ka^2 b \\ A + B + C \xrightarrow{k} \text{Products} & \qquad \mathscr{R} = kabc \end{aligned} \qquad (1.19)$$

A homogeneous gas-phase reaction that follows a third-order kinetic scheme is

$$2\text{NO} + \text{O}_2 \rightarrow 2\text{NO}_2 \qquad \mathscr{R} = k[\text{NO}]^2[\text{O}_2]$$

although the mechanism is believed to involve two steps[1] and thus is not elementary.

1.3 REACTION ORDER AND MECHANISM

As suggested by these examples, the *order* of a reaction is the sum of the exponents $m, n, \ldots$ in

$$\mathscr{R} = k a^m b^n \ldots \qquad \text{Reaction order} = m + n + \cdots \qquad (1.20)$$

This definition for reaction order is directly meaningful only for irreversible or forward reactions that have rate expressions in the form of Equation (1.20). Components A, B, ... are consumed by the reaction and have negative stoichiometric coefficients so that $m = -\nu_A$, $n = -\nu_B, \ldots$ are positive. For elementary reactions, m and n must be integers of 2 or less and must sum to 2 or less.

Equation (1.20) is frequently used to correlate data from complex reactions. Complex reactions can give rise to rate expressions that have the form of Equation (1.20), but with fractional or even negative exponents. Complex reactions with observed orders of 1/2 or 3/2 can be explained theoretically based on mechanisms discussed in Chapter 2. Negative orders arise when a compound retards a reaction—say, by competing for active sites in a heterogeneously catalyzed reaction—or when the reaction is reversible. Observed reaction orders above 3 are occasionally reported. An example is the reaction of styrene with nitric acid, where an overall order of 4 has been observed.[2] The likely explanation is that the acid serves both as a catalyst and as a reactant. The reaction is far from elementary.

Complex reactions can be broken into a number of series and parallel elementary steps, possibly involving short-lived intermediates such as free radicals. These individual reactions collectively constitute the *mechanism* of the complex reaction. The individual reactions are usually second order, and the number of reactions needed to explain an observed, complex reaction can be surprisingly large. For example, a good model for

$$\text{CH}_4 + 2\text{O}_2 \rightarrow \text{CO}_2 + 2\text{H}_2\text{O}$$

will involve 20 or more elementary reactions, even assuming that the indicated products are the only ones formed in significant quantities. A detailed model for the oxidation of toluene involves 141 chemical species in 743 elementary reactions.[3]

As a simpler example of a complex reaction, consider (abstractly, not experimentally) the nitration of toluene to give trinitrotoluene:

$$\text{Toluene} + 3\text{HNO}_3 \rightarrow \text{TNT} + 3\text{H}_2\text{O}$$

or, in shorthand,

$$A + 3B \to C + 3D$$

This reaction cannot be elementary. We can hardly expect three nitric acid molecules to react at all three toluene sites (these are the ortho and para sites; meta substitution is not favored) in a glorious, four-body collision. Thus, the fourth-order rate expression $\mathcal{R} = kab^3$ is implausible. Instead, the mechanism of the TNT reaction involves at least seven steps (two reactions leading to *ortho*- or *para*-nitrotoluene, three reactions leading to 2,4- or 2,6-dinitrotoluene, and two reactions leading to 2,4,6-trinitrotoluene). Each step would require only a two-body collision, could be elementary, and could be governed by a second-order rate equation. Chapter 2 shows how the component balance equations can be solved for multiple reactions so that an assumed mechanism can be tested experimentally. For the toluene nitration, even the set of seven series and parallel reactions may not constitute an adequate mechanism since an experimental study[4] found the reaction to be 1.3 order in toluene and 1.2 order in nitric acid for an overall order of 2.5 rather than the expected value of 2.

An irreversible, elementary reaction must have Equation (1.20) as its rate expression. A complex reaction may have an empirical rate equation with the form of Equation (1.20) and with integral values for n and m, without being elementary. The classic example of this statement is a second-order reaction where one of the reactants is present in great excess. Consider the slow hydrolysis of an organic compound in water. A rate expression of the form

$$\mathcal{R} = k[\text{water}][\text{organic}]$$

is plausible, at least for the first step of a possibly complex mechanism. Suppose [organic] $\ll$ [water] so that the concentration of water does not change appreciably during the course of the reaction. Then the water concentration can be combined with k to give a composite rate constant that is approximately constant. The rate expression appears to be first order in [organic]:

$$\mathcal{R} = k[\text{water}][\text{organic}] = k'[\text{organic}] = k'a$$

where $k' = k[\text{water}]$ is a *pseudo-first-order rate constant*. From an experimental viewpoint, the reaction cannot be distinguished from first order even though the actual mechanism is second order. Gas-phase reactions also appear first order when one reactant is dilute. Kinetic theory still predicts the collision rates of Equation (1.10), but the concentration of one species, call it B, remains approximately constant. The observed rate constant is

$$k' = \left[\frac{8\pi R_g T(m_A + m_B)}{\mathbf{A}\mathrm{v} m_A m_B}\right]^{1/2} (r_A + r_B)^2 \mathbf{A}\mathrm{v} f_R b$$

which differs by a factor of b from Equation (1.11).

The only reactions that are strictly first order are radioactive decay reactions. Among chemical reactions, thermal decompositions may seem first order, but an external energy source is generally required to excite the reaction. As noted earlier, this energy is usually acquired by intermolecular collisions. Thus, the reaction rate could be written as

$$\mathscr{R} = k[\text{reactant molecules}][\text{all molecules}]$$

The concentration of all molecules is normally much higher than the concentration of reactant molecules, so that it remains essentially constant during the course of the reaction. Thus, what is truly a second-order reaction appears to be first order.

1.4 IDEAL, ISOTHERMAL REACTORS

There are four kinds of ideal reactors:

1. The batch reactor
2. The piston flow reactor (PFR)
3. The perfectly mixed, continuous-flow stirred tank reactor (CSTR)
4. The completely segregated, continuous-flow stirred tank reactor

This chapter discusses the first three types, which are overwhelmingly the most important. The fourth type is interesting theoretically, but has limited practical importance. It is discussed in Chapter 15.

1.4.1 The Ideal Batch Reactor

This is the classic reactor used by organic chemists. The typical volume in glassware is a few hundred milliliters. Reactants are charged to the system, rapidly mixed, and rapidly brought up to temperature so that reaction conditions are well defined. Heating is carried out with an oil bath or an electric heating mantle. Mixing is carried out with a magnetic stirrer or a small mechanical agitator. Temperature is controlled by regulating the bath temperature or by allowing a solvent to reflux.

Batch reactors are the most common type of industrial reactor and may have volumes well in excess of 100,000 liters. They tend to be used for small-volume specialty products (e.g., an organic dye) rather than large-volume commodity chemicals (e.g., ethylene oxide) that are normally reacted in continuous-flow equipment. Industrial-scale batch reactors can be heated or cooled by external coils or a jacket, by internal coils, or by an external heat exchanger in a pump-around loop. Reactants are often preheated by passing them through heat exchangers as they are charged to the vessel. Heat generation due to the reaction can be significant in large vessels. Refluxing is one means for controlling the exotherm. Mixing in large batch vessels is usually carried out with a mechanical agitator, but is occasionally carried out with an external pump-around loop where the momentum of the returning fluid causes the mixing.

Heat and mass transfer limitations are rarely important in the laboratory but may emerge upon scaleup. Batch reactors with internal variations in temperature or composition are difficult to analyze and remain a challenge to the chemical reaction engineer. Tests for such problems are considered in Section 1.5. For now, assume an *ideal batch reactor* with the following characteristics:

1. Reactants are quickly charged, mixed, and brought to temperature at the beginning of the reaction cycle.
2. Mixing and heat transfer are sufficient to assure that the batch remains completely uniform throughout the reaction cycle.

A batch reactor has no input or output of mass after the initial charging. The amounts of individual components may change due to reaction but not due to flow into or out of the system. The component balance for component A, Equation (1.6), reduces to

$$\frac{d(Va)}{dt} = \mathscr{R}_A V \tag{1.21}$$

Together with similar equations for the other reactive components, Equation (1.21) constitutes the *reactor design equation* for an ideal batch reactor. Note that $\hat{a}$ and $\hat{\mathscr{R}}_A$ have been replaced with a and $\mathscr{R}_A$ because of the assumption of good mixing. An ideal batch reactor has no temperature or concentration gradients within the system volume. The concentration will change with time because of the reaction, but at any time it is everywhere uniform. The temperature may also change with time, but this complication will be deferred until Chapter 5. The reaction rate will vary with time but is always uniform throughout the vessel. Here in Chapter 1, we make the additional assumption that the volume is constant. In a liquid-phase reaction, this corresponds to assuming constant fluid density, an assumption that is usually reasonable for preliminary calculations. Industrial gas-phase reactions are normally conducted in flow systems rather than batch systems. When batch reactors are used, they are normally constant-volume devices so that the system pressure can vary during the batch cycle. Constant-pressure devices were used in early kinetic studies and are occasionally found in industry. The constant pressure at which they operate is usually atmospheric pressure.

The ideal, constant-volume batch reactor satisfies the following component balance:

$$\frac{da}{dt} = \mathscr{R}_A \tag{1.22}$$

Equation (1.22) is an ordinary differential equation or ODE. Its solution requires an initial condition:

$$a = a_0 \quad \text{at} \quad t = 0 \tag{1.23}$$

When $\mathscr{R}_A$ depends on a alone, the ODE is variable-separable and can usually be solved analytically. If $\mathscr{R}_A$ depends on the concentration of several components (e.g., a second-order reaction of the two reactants variety, $\mathscr{R}_A = -kab$), versions of Equations (1.22) and (1.23) are written for each component and the resulting equations are solved simultaneously.

First-Order Batch Reactions. The reaction is

$$A \xrightarrow{k} \text{Products}$$

The rate constant over the reaction arrow indicates that the reaction is elementary, so that

$$\mathscr{R} = ka$$
$$\mathscr{R}_A = v_A \mathscr{R} = -ka$$

which agrees with Equation (1.16). Substituting into Equation (1.22) gives

$$\frac{da}{dt} + ka = 0$$

Solving this ordinary differential equation and applying the initial condition of Equation (1.23) gives

$$a = a_0 e^{-kt} \tag{1.24}$$

Equation (1.24) is arguably the most important result in chemical reaction engineering. It shows that the concentration of a reactant being consumed by a first-order batch reaction decreases exponentially. Dividing through by a_0 gives the *fraction unreacted*,

$$Y_A = \frac{a}{a_0} = e^{-kt} \tag{1.25}$$

and

$$X_A = 1 - \frac{a}{a_0} = 1 - e^{-kt} \tag{1.26}$$

gives the *conversion*. The *half-life* of the reaction is defined as the time necessary for a to fall to half its initial value:

$$t_{1/2} = 0.693/k \tag{1.27}$$

The half-life of a first-order reaction is independent of the initial concentration. Thus, the time required for the reactant concentration to decrease from a_0 to $a_0/2$ is the same as the time required to decrease from $a_0/2$ to $a_0/4$. This is not true for reactions other than first order.

Second-Order Batch Reactions with One Reactant. We choose to write the stoichiometric equation as

$$2A \xrightarrow{k/2} \text{Products}$$

Compare this with Equation (1.17) and note the difference in rate constants. For the current formulation,

$$\mathscr{R} = (k/2)a^2$$

$$\mathscr{R}_A = \nu_A \mathscr{R} = -2\mathscr{R} = -ka^2$$

Substituting into Equation (1.21) gives

$$\frac{da}{dt} + ka^2 = 0$$

Solution gives

$$-a^{-1} + C = -kt$$

where C is a constant. Applying the initial condition gives $C = (a_0)^{-1}$ and

$$\frac{a}{a_0} = \frac{1}{1 + a_0 k t} \tag{1.28}$$

Observe that $a_0 k$ has units of reciprocal time so that $a_0 k t$ is dimensionless. The grouping $a_0 k t$ is the *dimensionless rate constant* for a second-order reaction, just as kt is the dimensionless rate constant for a first-order reaction. Equivalently, they can be considered as *dimensionless reaction times*. For reaction rates governed by Equation (1.20),

$$\text{Dimensionless rate constant} = K^* = a_0^{order-1} kt \tag{1.29}$$

With this notation, all first-order reactions behave as

$$\frac{a}{a_0} = e^{-K^*} \tag{1.30}$$

and all second-order reactions of the one-reactant type behave as

$$\frac{a}{a_0} = \frac{1}{1 + K^*} \tag{1.31}$$

For the same value of K^*, first-order reactions proceed much more rapidly than second-order reactions. The reaction rate for a first-order reaction will decrease to half its original value when the concentration has decreased to half the original concentration. For a second-order reaction, the reaction rate will decrease to a quarter the original rate when the concentration has decreased to half the original concentration; compare Equations (1.16) and (1.17).

The initial half-life of a second-order reaction corresponds to a decrease from a_0 to $a_0/2$ and is given by

$$t_{1/2} = \frac{1}{a_0 k} \tag{1.32}$$

The second half-life, corresponding to a decrease from $a_0/2$ to $a_0/4$, is twice the initial half-life.

Second-Order Batch Reactions with Two Reactants. The batch reaction is now

$$A + B \xrightarrow{k} \text{Products}$$

$$\mathscr{R} = kab$$

$$\mathscr{R}_A = \nu_A \mathscr{R} = -\mathscr{R} = -kab$$

Substituting into Equation (1.22) gives

$$\frac{da}{dt} + kab = 0$$

A similar equation can be written for component B:

$$\frac{db}{dt} + kab = 0$$

The pair of equations can be solved simultaneously. A simple way to proceed is to note that

$$\frac{da}{dt} = \frac{db}{dt}$$

which is solved to give

$$a = b + C$$

where C is a constant of integration that can be determined from the initial conditions for a and b. The result is

$$a - a_0 = b - b_0 \tag{1.33}$$

which states that A and B are consumed in equal molar amounts as required by the reaction stoichiometry. Applying this result to the ODE for component A gives

$$\frac{da}{dt} + ka(a - a_0 + b_0) = 0$$

The equation is variable-separable. Integrating and applying the initial condition gives

$$\frac{a}{a_0} = \frac{b_0 - a_0}{b_0 \exp[(b_0 - a_0)kt] - a_0} \qquad (1.34)$$

This is the general result for a second-order batch reaction. The mathematical form of the equation presents a problem when the initial stoichiometry is perfect, $a_0 = b_0$. Such problems are common with analytical solutions to ODEs. Special formulas are needed for special cases.

One way of treating a special case is to carry out a separate derivation. For the current problem, perfect initial stoichiometry means $b = a$ throughout the reaction. Substituting this into the ODE for component A gives

$$\frac{da}{dt} + ka^2 = 0$$

which is the same as that for the one-reactant case of a second-order reaction, and the solution is Equation (1.28).

An alternative way to find a special formula for a special case is to apply L'Hospital's rule to the general case. When $b_0 \to a_0$, Equation (1.34) has an indeterminate form of the 0/0 type. Differentiating the numerator and denominator with respect to b_0 and then taking the limit gives

$$\frac{a}{a_0} = \lim_{b_0 \to a_0} \left[\frac{1}{\exp[(b_0 - a_0)kt] + b_0 kt \exp[(b_0 - a_0)kt]} \right] = \frac{1}{1 + a_0 kt}$$

which is again identical to Equation (1.28).

Reactor Performance Measures. There are four common measures of reactor performance: fraction unreacted, conversion, yield, and selectivity. The fraction unreacted is the simplest and is usually found directly when solving the component balance equations. It is $a(t)/a_0$ for a batch reaction and a_{out}/a_{in} for a flow reactor. The conversion is just 1 minus the fraction unreacted. The terms conversion and fraction unreacted refer to a specific reactant. It is usually the stoichiometrically limiting reactant. See Equation (1.26) for the first-order case.

Batch reactors give the lowest possible fraction unreacted and the highest possible conversion for most reactions. Batch reactors also give the best yields and selectivities. These terms refer to the desired product. The *molar yield* is the number of moles of a specified product that are made per mole of reactant charged. There is also a *mass yield*. Either of these yields can be larger than 1. The *theoretical yield* is the amount of product that would be formed if all of the reactant were converted to the desired product. This too can be expressed on either a molar or mass basis and can be larger than 1. *Selectivity* is defined as the fractional amount of the converted portion of a reactant that is converted to the desired product. The selectivity will always be 100% when there is only one reaction, even though the conversion may be less than 100%. Selectivity is a trivial concept when there is only one reaction, but becomes an important consideration when there are multiple reactions. The following example illustrates a reaction with high conversion but low selectivity.

Example 1.2: Suppose it is desired to make 1,4-dimethyl-2,3-dichlorobenzene by the direct chlorination of *para*-xylene. The desired reaction is

$$p\text{-xylene} + Cl_2 \rightarrow \text{desired product} + 2HCl$$

A feed stream containing 40 mole percent *p*-xylene and 60 mole percent chlorine was fed to the reactor. The results of one experiment in a batch reactor gave the following results on a molar basis:

Component	Moles Output per mole of mixed feed
p-xylene	0.001
Chlorine	0.210
Monochloroxylene	0.032
1,4-dimethyl-2,3-dichlorobenzene	0.131
Other dichloroxylenes	0.227
Trichloroxylene	0.009
Tetrachloroxylenes	0.001
Total	0.611

Compute various measures of reactor performance.

Solution: Some measures of performance based on xylene as the limiting component are

Fraction unreacted $= 0.001/0.4 = 0.0025$

Conversion $= 1 - 0.0025 = 0.9975$

Yield $= 0.131/0.40 = 0.3275$ moles of product per mole of xylene charged

Percent of theoretical yield $= 0.131/0.4\ (100) = 32.8\%$

Selectivity $= 0.131/[0.9975(0.40)]\ (100) = 32.83\%$

This example expresses all the performance measures on a molar basis. The mass yield of 1,4-dimethyl-2,3-dichlorobenzene sounds a bit better. It is 0.541 lb of the desired product per pound of xylene charged.

Note that the performance measures and definitions given here are the typical ones, but other terms and other definitions are sometimes used. Be sure to ask for the definition if there is any ambiguity.

1.4.2 Piston Flow Reactors

Continuous-flow reactors are usually preferred for long production runs of high-volume chemicals. They tend to be easier to scaleup, they are easier to control, the product is more uniform, materials handling problems are lessened, and the capital cost for the same annual capacity is lower.

There are two important types of ideal, continuous-flow reactors: the piston flow reactor or PFR, and the continuous-flow stirred tank reactor or CSTR. They behave very differently with respect to conversion and selectivity. The piston flow reactor behaves exactly like a batch reactor. It is usually visualized as a long tube as illustrated in Figure 1.3. Suppose a small clump of material enters the reactor at time $t=0$ and flows from the inlet to the outlet. We suppose that there is no mixing between this particular clump and other clumps that entered at different times. The clump stays together and ages and reacts as it flows down the tube. After it has been in the piston flow reactor for t seconds, the clump will have the same composition as if it had been in a batch reactor for t seconds. The composition of a batch reactor varies with time. The composition of a small clump flowing through a piston flow reactor varies with time in the same way. It also varies with position down the tube. The relationship between time and position is

$$t = z/\bar{u} \tag{1.35}$$

where z denotes distance measured from the inlet of the tube and $\bar{u}$ is the velocity of the fluid. Chapter 1 assumes steady-state operation so that the composition at point z is always the same. It also assumes constant fluid density and constant reactor cross section so that $\bar{u}$ is constant. The age of material at point z is t, and the composition at this point is given by the constant-volume version of the component balance for a batch reaction, Equation (1.22). All that has to be done is to substitute $t = z/\bar{u}$. The result is

$$\bar{u}\frac{da}{dz} = \mathscr{R}_A \tag{1.36}$$

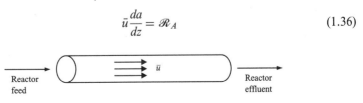

FIGURE 1.3 Piston flow reactor.

The initial condition is that

$$a = a_{in} \quad \text{at} \quad z = 0 \qquad (1.37)$$

Only the notation is different from the initial condition used for batch reactors. The subscripts *in* and *out* are used for flow reactors. The outlet concentration is found by setting $z = L$.

Example 1.3: Find the outlet concentration of component A from a piston flow reactor assuming that A is consumed by a first-order reaction.

Solution: Equation (1.36) becomes

$$\bar{u}\frac{da}{dz} = -ka$$

Integrating, applying the initial condition of Equation (1.37), and evaluating the result at $z = L$ gives

$$a_{out} = a_{in} \exp(-kL/\bar{u}) \qquad (1.38)$$

The quantity $L/\bar{u}$ has units of time and is the mean residence time, $\bar{t}$. Thus, we can write Equation (1.38) as

$$a_{out} = a_{in} \exp(-k\bar{t}) \qquad (1.39)$$

where

$$\bar{t} = L/\bar{u} \qquad (1.40)$$

Equation (1.40) is a special case of a far more general result. The *mean residence time* is the average amount of time that material spends in a flow system. For a system at steady state, it is equal to the mass inventory of fluid in the system divided by the mass flow rate through the system:

$$\bar{t} = \frac{\text{Mass inventory}}{\text{Mass throughput}} = \frac{\bar{\rho}V}{\rho Q} \qquad (1.41)$$

where $\rho Q = \rho_{out}Q_{out} = \rho_{in}Q_{in}$ is a consequence of steady-state operation. For the special case of a constant-density fluid,

$$\bar{t} = V/Q \qquad (1.42)$$

where $Q = Q_{in} = Q_{out}$ when the system is at steady-state and the mass density is constant. This reduces to

$$\bar{t} = L/\bar{u} \qquad (1.43)$$

for a tubular reactor with constant fluid density and constant cross-sectional area. Piston flow is a still more special case where all molecules have the same velocity and the same residence time. We could write $\bar{t} = L/u$ for piston flow since the velocity is uniform across the tube, but we prefer to use Equation (1.43) for this case as well.

We now formalize the definition of piston flow. Denote position in the reactor using a cylindrical coordinate system (r, θ, z) so that the concentration at a point is denoted as $a(r, \theta, z)$ For the reactor to be a *piston flow reactor* (also called plug flow reactor, slug flow reactor, or ideal tubular reactor), three conditions must be satisfied:

1. The axial velocity is independent of r and θ but may be a function of z, $V_z(r, \theta, z) = \bar{u}(z)$.
2. There is complete mixing across the reactor so that concentration is a function of z alone; i.e., $a(r, \theta, z) = a(z)$.
3. There is no mixing in the axial direction.

Here in Chapter 1 we make the additional assumptions that the fluid has constant density, that the cross-sectional area of the tube is constant, and that the walls of the tube are impenetrable (i.e., no transpiration through the walls), but these assumptions are not required in the general definition of piston flow. In the general case, it is possible for $\bar{u}$, temperature, and pressure to vary as a function of z. The axis of the tube need not be straight. Helically coiled tubes sometimes approximate piston flow more closely than straight tubes. Reactors with square or triangular cross sections are occasionally used. However, in most of this book, we will assume that PFRs are circular tubes of length L and constant radius R.

Application of the general component balance, Equation (1.6), to a steady-state flow system gives

$$Q_{in}a_{in} + \hat{\mathscr{R}}_A V = Q_{out}a_{out}$$

While true, this result is not helpful. The derivation of Equation (1.6) used the entire reactor as the control volume and produced a result containing the average reaction rate, $\hat{\mathscr{R}}_A$. In piston flow, a varies with z so that the local reaction rate also varies with z, and there is no simple way of calculating $\hat{\mathscr{R}}_A$. Equation (1.6) is an overall balance applicable to the entire system. It is also called an *integral balance*. It just states that if more of a component leaves the reactor than entered it, then the difference had to have been formed inside the reactor.

A *differential balance* written for a vanishingly small control volume, within which $\mathscr{R}_A$ is approximately constant, is needed to analyze a piston flow reactor. See Figure 1.4. The differential volume element has volume ΔV, cross-sectional area A_c, and length Δz. The general component balance now gives

Moles in + moles formed = moles out

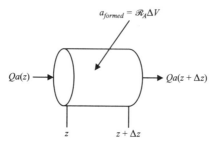

FIGURE 1.4 Differential element in a piston flow reactor.

or

$$Qa(z) + \mathcal{R}_A \Delta V = Qa(z + \Delta z)$$

Note that $Q = \bar{u}A_c$ and $\Delta V = A_c \Delta z$. Then

$$Q\frac{a(z + \Delta z) - a(z)}{\Delta V} = \bar{u}\frac{a(z + \Delta z) - a(z)}{\Delta z} = \mathcal{R}_A$$

Recall the definition of a derivative and take the limit as $\Delta z \to 0$:

$$\lim_{\Delta z \to 0}\left[\bar{u}\frac{a(z + \Delta z) - a(z)}{\Delta z}\right] = \bar{u}\frac{da}{dz} = \mathcal{R}_A \quad (1.44)$$

which agrees with Equation (1.36). Equation (1.36) was derived by applying a variable transformation to an unsteady, batch reactor. Equation (1.44) was derived by applying a steady-state component balance to a differential flow system. Both methods work for this problem, but differential balances are the more general approach and can be extended to multiple dimensions. However, the strong correspondence between time in a batch reactor and position in a piston flow reactor is very important. The composition at time t in a batch reactor is identical to the composition at position $z = \bar{u}t$ in a piston flow reactor. This correspondence—which extends beyond the isothermal, constant-density case—is detailed in Table 1.1.

Example 1.4: Determine the reactor design equations for the various elementary reactions in a piston flow reactor. Assume constant temperature, constant density, and constant reactor cross section. (Whether or not all these assumptions are needed will be explored in subsequent chapters.)

Solution: This can be done by substituting the various rate equations into Equation (1.36), integrating, and applying the initial condition of Equation (1.37). Two versions of these equations can be used for a second-order reaction with two reactants. Another way is to use the previous results for

ELEMENTARY REACTIONS IN IDEAL REACTORS

TABLE 1.1 Relationships between Batch and Piston Flow Reactors

Batch reactors	Piston flow reactors
Concentrations vary with time	Concentrations vary with axial position
The composition is uniform at any time t	The composition is uniform at any position z
Governing equation, (1.22)	Governing equation, (1.44)
Initial condition, a_0	Initial condition, a_{in}
Final condition, $a(t)$	Final condition, $a(L)$
Variable density, $\rho(t)$	Variable density, $\rho(z)$
Time equivalent to position in a piston flow reactor, $t = z/\bar{u}$	Position equivalent to time in a batch reactor, $z = \bar{u}t$
Variable temperature, $T(t)$	Variable temperature, $T(z)$
Heat transfer to wall, $dq_{removed} = hA_{wall}(T - T_{wall})dt$	Heat transfer to wall, $dq_{removed} = h(2\pi R)(T - T_{wall})dz$
Variable wall temperature, $T_{wall}(t)$	Variable wall temperature, $T_{wall}(z)$
Variable pressure, $P(t)$	Pressure drop, $P(z)$
Variable volume (e.g., a constant-pressure reactor), $V(t)$	Variable cross section, $A_c(z)$
Fed batch reactors, $Q_{in} \neq 0$	Transpired wall reactors
Nonideal batch reactors may have spatial variations in concentration	Nonideal tubular reactors may have concentrations that vary in the r and θ directions

a batch reactor. Replace t with $z/\bar{u}$ and a_0 with a_{in}. The result is $a(z)$ for the various reaction types.

For a first-order reaction,

$$\frac{a(z)}{a_{in}} = \exp(-kz/\bar{u}) \tag{1.45}$$

For a second-order reaction with one reactant,

$$\frac{a(z)}{a_{in}} = \frac{1}{1 + a_{in}kz/\bar{u}} \tag{1.46}$$

For a second-order reaction with two reactants,

$$\frac{a(z)}{a_{in}} = \frac{b_{in} - a_{in}}{b_{in}\exp[(b_{in} - a_{in})kz/\bar{u}] - a_{in}} \tag{1.47}$$

The outlet concentration is found by setting $z = L$.

Piston flow reactors and most other flow reactors have spatial variations in concentration such as $a = a(z)$. Such systems are called *distributed*. Their behavior is governed by an ordinary differential equation when there is only one spatial variable and by a partial differential equation (PDE) when there are two or three spatial variables or when the system has a spatial variation and also varies with time. We turn now to a special type of flow reactor where the entire reactor volume is well mixed and has the same concentration, temperature, pressure, and so forth. There are no spatial variations in these parameters. Such systems are called *lumped* and their behavior is governed by an algebraic equation when the system is at steady state and by an ordinary differential equation when the system varies with time. The continuous-flow stirred tank reactor or CSTR is the chemical engineer's favorite example of a lumped system. It has one lump, the entire reactor volume.

1.4.3 Continuous-Flow Stirred Tanks

Figure 1.5 illustrates a flow reactor in which the contents are mechanically agitated. If mixing caused by the agitator is sufficiently fast, the entering feed will be quickly dispersed throughout the vessel and the composition at any point will approximate the average composition. Thus, the reaction rate at any point will be approximately the same. Also, the outlet concentration will be identical to the internal composition, $a_{out} = \hat{a}$.

There are only two possible values for concentration in a CSTR. The inlet stream has concentration a_{in} and everywhere else has concentration a_{out}. The reaction rate will be the same throughout the vessel and is evaluated at the outlet concentration, $\hat{\mathscr{R}}_A = \mathscr{R}_A(a_{out}, b_{out}, \ldots)$. For the single reactions considered in this chapter, $\mathscr{R}_A$ continues to be related to $\mathscr{R}$ by the stoichiometric coefficient and Equation (1.13). With $\mathscr{R}_A$ known, the integral component balance, Equation (1.6), now gives useful information. For component A,

$$Qa_{in} + \mathscr{R}_A(a_{out}, b_{out}, \ldots)V = Qa_{out} \tag{1.48}$$

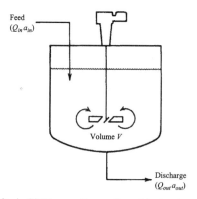

FIGURE 1.5 The classic CSTR: a continuous-flow stirred tank reactor with mechanical agitation.

Note that we have assumed steady-state operation and set $Q = Q_{in} = Q_{out}$, which assumes constant density. Dividing through by Q and setting $\bar{t} = V/Q$ gives

$$a_{in} + \mathscr{R}_A(a_{out}, b_{out}, \ldots)\bar{t} = a_{out} \qquad (1.49)$$

In the usual case, $\bar{t}$ and a_{in} will be known. Equation (1.49) is an algebraic equation that can be solved for a_{out}. If the reaction rate depends on the concentration of more than one component, versions of Equation (1.49) are written for each component and the resulting set of equations is solved simultaneously for the various outlet concentrations. Concentrations of components that do not affect the reaction rate can be found by writing versions of Equation (1.49) for them. As for batch and piston flow reactors, stoichiometry is used to relate the rate of formation of a component, say $\mathscr{R}_C$, to the rate of the reaction $\mathscr{R}$, using the stoichiometric coefficient ν_C, and Equation (1.13). After doing this, the stoichiometry takes care of itself.

A reactor with performance governed by Equation (1.49) is a steady-state, constant-density, perfectly mixed, continuous flow reactor. This mouthful is usually shortened in the chemical engineering literature to CSTR (for Continuous-flow Stirred Tank Reactor). In subsequent chapters, we will relax the assumptions of steady state and constant density, but will still call it a CSTR. It is also called an ideal mixer, a continuous-flow perfect mixer, or a mixed flow reactor. This terminology is ambiguous in light of micromixing theory, discussed in Chapter 15, but is well entrenched. Unless otherwise qualified, we accept all these terms to mean that the reactor is perfectly mixed. Such a reactor is sometimes called a *perfect mixer*. The term denotes instantaneous and complete mixing on the molecular scale. Obviously, no real reactor can achieve this ideal state, just as no tubular reactor can achieve true piston flow. However, it is often possible to design reactors that very closely approach these limits.

Example 1.5: Determine the reactor design equations for elementary reactions in a CSTR.

Solution: The various rate equations for the elementary reactions are substituted into Equation (1.49), which is then solved for a_{out}.

For a first-order reaction, $\mathscr{R}_A = -ka$. Set $a = a_{out}$, substitute $\mathscr{R}_A$ into Equation (1.49), and solve for a_{out} to obtain

$$\frac{a_{out}}{a_{in}} = \frac{1}{1 + k\bar{t}} \qquad (1.50)$$

For a second-order reaction with one reactant, $\mathscr{R}_A = -ka^2$. Equation (1.49) becomes a quadratic in a_{out}. The solution is

$$\frac{a_{out}}{a_{in}} = \frac{-1 + \sqrt{1 + 4a_{in}k\bar{t}}}{2a_{in}k\bar{t}} \qquad (1.51)$$

The negative root was rejected since $a_{out} \geq 0$.

For a second-order reaction with two reactants, $\mathcal{R}_A = \mathcal{R}_B = -kab$. Write two versions of Equation (1.49), one for a_{out} and one for b_{out}. Solving them simultaneously gives

$$\frac{a_{out}}{a_{in}} = \frac{-1 - (b_{in} - a_{in})k\bar{t} + \sqrt{1 + (b_{in} - a_{in})k\bar{t}^2 + 4a_{in}k\bar{t}}}{2a_{in}k\bar{t}} \qquad (1.52)$$

Again, a negative root was rejected. The simultaneous solution also produces the stoichiometric relationship

$$b_{in} - b_{out} = a_{in} - a_{out} \qquad (1.53)$$

The above examples have assumed that a_{in} and $\bar{t}$ are known. The solution then gives a_{out}. The case where a_{in} is known and a desired value for a_{out} is specified can be easier to solve. The solution for $\bar{t}$ is

$$\bar{t} = \frac{a_{out} - a_{in}}{\mathcal{R}_A(a_{out}, b_{out}, \ldots)} \qquad (1.54)$$

This result assumes constant density and is most useful when the reaction rate depends on a single concentration, $\mathcal{R}_A = \mathcal{R}_A(a_{out})$.

Example 1.6: Apply Equation (1.54) to calculate the mean residence time needed to achieve 90% conversion in a CSTR for (a) a first-order reaction, (b) a second-order reaction of the type $A + B \rightarrow$ Products. The rate constant for the first-order reaction is $k = 0.1\,s^{-1}$. For the second-order reaction, $ka_{in} = 0.1\,s^{-1}$.

Solution: For the first-order reaction, $\mathcal{R}_A = -ka_{out} = -k(0.1a_{in})$. Equation (1.54) gives

$$\bar{t} = \frac{a_{out} - a_{in}}{-ka_{out}} = \frac{0.1a_{in} - a_{in}}{-k(0.1a_{in})} = \frac{9}{k} = 90\,s$$

For the second-order case, $\mathcal{R}_A = -ka_{out}b_{out}$. To use Equation (1.54), stoichiometry is needed to find the value for b_{out} that corresponds to a_{out}. Suppose for example that B is in 50% excess so that $b_{in} = 1.5a_{in}$. Then $b_{out} = 0.6a_{in}$ if $a_{out} = 0.1a_{in}$. Equation (1.54) gives

$$\bar{t} = \frac{a_{out} - a_{in}}{-ka_{out}b_{out}} = \frac{0.1a_{in} - a_{in}}{-k(0.1a_{in})(0.6a_{in})} = \frac{15}{ka_{in}} = 150\,s$$

1.5 MIXING TIMES AND SCALEUP

Suppose a homogeneous reaction is conducted in a pilot plant reactor that is equipped with a variable speed agitator. Does changing the agitator speed (say by ± 20%) change the outcome of the reaction? Does varying the addition rate of reactants change the selectivity? If so, there is a potential scaleup problem. The reaction is sensitive to the *mixing time*, t_{mix}.

The mixing time in a batch vessel is easily measured. To do this, add unmixed ingredients and determine how long it takes for the contents of the vessel to become uniform. For example, fill a vessel with plain water and start the agitator. At time $t = 0$, add a small quantity of a salt solution. Measure the concentration of salt at various points inside the vessel until it is constant within measurement error or by some other standard of near equality. Record the result as t_{mix}. A popular alternative is to start with a weak acid solution that contains an indicator so that the solution is initially colored. A small excess of concentrated base is added quickly at one point in the system. The mixing time, t_{mix}, corresponds to the disappearance of the last bit of color. The acid–base titration is very fast so that the color will disappear just as soon as the base is distributed throughout the vessel. This is an example where the reaction in the vessel is limited strictly by mixing. There is no kinetic limitation. For very fast reactions such as combustion or acid–base neutralization, no vessel will be perfectly mixed. The components must be transported from point to point in the vessel by fluid flow and diffusion, and these transport processes will be slower than the reaction. Whether a reactor can be considered to be perfectly mixed depends on the speed of the reaction. What is effectively perfect mixing is easy to achieve when the reaction is an esterification with a half-life of several hours. It is impossible to achieve in an acid–base neutralization with a half-life of microseconds. The requirement for perfect mixing in a batch vessel is just that

$$t_{mix} \ll t_{1/2} \qquad (1.55)$$

When this relation is satisfied, the conversion will be limited by the reaction kinetics, not by the mixing rate. As a practical matter, the assumption of perfect mixing is probably reasonable when $t_{1/2}$ is eight times larger than t_{mix}.

Mixing times in mechanically agitated vessels typically range from a few seconds in laboratory glassware to a few minutes in large industrial reactors. The classic correlation by Norwood and Metzner[5] for turbine impellers in baffled vessels can be used for order of magnitude estimates of t_{mix}.

In a batch vessel, the question of good mixing will arise at the start of the batch and whenever an ingredient is added to the batch. The component balance, Equation (1.21), assumes that uniform mixing is achieved before any appreciable reaction occurs. This will be true if Equation (1.55) is satisfied. Consider the same vessel being used as a flow reactor. Now, the mixing time must be short compared with the mean residence time, else newly charged material could flow out of the reactor before being thoroughly mixed with the contents. A new condition to be satisfied is

$$t_{mix} \ll \bar{t} \tag{1.56}$$

In practice, Equation (1.56) will be satisfied if Equation (1.55) is satisfied since a CSTR will normally operate with $t_{1/2} \ll \bar{t}$.

The net flow though the reactor will be small compared with the circulating flow caused by the agitator. The existence of the throughput has little influence on the mixing time so that mixing time correlations for batch vessels can be used for CSTRs as well.

In summary, we have considered three characteristic times associated with a CSTR: t_{mix}, $t_{1/2}$, and $\bar{t}$. Treating the CSTR as a perfect mixer is reasonable provided that t_{mix} is substantially shorter than the other characteristic times.

Example 1.7: Suppose a pilot-scale reactor behaves as a perfectly mixed CSTR so that Equation (1.49) governs the conversion. Will the assumption of perfect mixing remain valid upon scaleup?

Solution: Define the *throughput scaleup factor* as

$$S = \frac{\text{Mass flow through full-scale unit}}{\text{Mass flow through pilot unit}} = \frac{(\rho Q)_{full\text{-}scale}}{(\rho Q)_{pilot\text{-}scale}} \tag{1.57}$$

Assume that the pilot-scale and full-scale vessels operate with the same inlet density. Then ρ cancels in Equation (1.57) and

$$S = \frac{Q_{full\text{-}scale}}{Q_{pilot\text{-}scale}} \quad \text{(constant density)}$$

Also assume that the pilot- and full-scale vessels will operate at the same temperature. This means that $\mathcal{R}_A(a_{out}, b_{out}, \ldots)$ and $t_{1/2}$ will be the same for the two vessels and that Equation (1.49) will have the same solution for a_{out} provided that $\bar{t}$ is held constant during scaleup. Scaling with a constant value for the mean residence time is standard practice for reactors. If the scaleup succeeds in maintaining the CSTR-like environment, the large and small reactors will behave identically with respect to the reaction. Constant residence time means that the system inventory, $\hat{\rho}V$, should also scale as S. The *inventory scaleup factor* is defined as

$$S_{Inventory} = \frac{\text{Mass inventory in the full-scale unit}}{\text{Mass inventory in the pilot unit}} = \frac{(\hat{\rho}V)_{full\text{-}scale}}{(\hat{\rho}V)_{pilot\text{-}scale}} \tag{1.58}$$

and

$$S_{Inventory} = \frac{V_{full\text{-}scale}}{V_{pilot\text{-}scale}} \quad \text{(constant density)}$$

So, in the constant-density case, the inventory scaleup factor is the same as the volumetric scaleup factor.

Unless explicitly stated otherwise, the throughput and inventory scaleup factors will be identical since this means that the mean residence time will be constant upon scaleup:

$$S_{Inventory} = S \quad \text{(constant } \bar{t}\text{)} \tag{1.59}$$

These usually identical scaleup factors will be denoted as S.

It is common practice to use *geometric similarity* in the scaleup of stirred tanks (but not tubular reactors). This means that the production-scale reactor will have the same shape as the pilot-scale reactor. All linear dimensions such as reactor diameter, impeller diameter, and liquid height will change by the same factor, $S^{1/3}$. Surface areas will scale as $S^{2/3}$. Now, what happens to t_{mix} upon scaleup?

The correlation of Norwood and Metzner shows t_{mix} to be a complex function of the Reynolds number, the Froude number, the ratio of tank-to-impeller diameter, and the ratio of tank diameter to liquid level. However, to a reasonable first approximation for geometrically similar vessels operating at high Reynolds numbers,

$$(N_I t_{mix})_{Large} = \text{constant} = (N_I t_{mix})_{Small} \tag{1.60}$$

where N_I is the rotational velocity of the impeller. This means that scaleup with constant agitator speed will, to a reasonable approximation, give constant t_{mix}. The rub is that the power requirements for the agitator will increase sharply in the larger vessel. Again, to a reasonable first approximation for geometrically similar vessels operating at high Reynolds numbers,

$$\left(\frac{\text{Power}}{\rho N_I^3 D_I^5}\right)_{Large} = \left(\frac{\text{Power}}{\rho N_I^3 D_I^5}\right)_{Small} \tag{1.61}$$

where D_I is the impeller diameter and will scale as $S^{1/3}$. If N_I is held constant, power will increase as $D_I^5 = S^{5/3}$. A factor of 10 increase in the linear dimensions allows a factor of 1000 increase in throughput but requires a factor of 100,000 increase in agitator power! The horsepower per unit volume must increase by a factor of 100 to maintain a constant t_{mix}. Let us hope that there is some latitude before the constraints of Equations (1.55) and (1.56) are seriously violated. Most scaleups are carried out with approximately constant power per unit volume and this causes N_I to decrease and t_{mix} to increase upon scaleup. See Problem 1.15.

The primary lesson from this example is that no process is infinitely scalable. Sooner or later, additional scaleup becomes impossible, and further increases in production cannot be *single-train* but must add units in parallel. Fortunately for the economics of the chemical industry, the limit is seldom reached.

1.6 BATCH VERSUS FLOW, AND TANK VERSUS TUBE

Some questions that arise early in a design are: Should the reactor be batch or continuous; and, if continuous, is the goal to approach piston flow or perfect mixing?

For producing high-volume chemicals, flow reactors are usually preferred. The ideal piston flow reactor exactly duplicates the kinetic behavior of the ideal batch reactor, and the reasons for preferring one over the other involve secondary considerations such as heat and mass transfer, ease of scaleup, and the logistics of materials handling. For small-volume chemicals, the economics usually favor batch reactors. This is particularly true when general-purpose equipment can be shared between several products. Batch reactors are used for the greater number of products, but flow reactors produce the overwhelmingly larger volume as measured in tons.

Flow reactors are operated continuously; that is, at steady state with reactants continuously entering the vessel and with products continuously leaving. Batch reactors are operated discontinuously. A batch reaction cycle has periods for charging, reaction, and discharging. The continuous nature of a flow reactor lends itself to larger productivities and greater economies of scale than the cyclic operation of a batch reactor. The volume productivity (moles of product per unit volume of reactor) for batch systems is identical to that of piston flow reactors and is higher than most real flow reactors. However, this volume productivity is achieved only when the reaction is actually occurring and not when the reactor is being charged or discharged, being cleaned, and so on. Within the class of flow reactors, piston flow is usually desired for reasons of productivity and selectivity. However, there are instances where a close approach to piston flow is infeasible or where a superior product results from the special reaction environment possible in stirred tanks.

Although they are both flow reactors, there are large differences in the behavior of PFRs and CSTRs. The reaction rate decreases as the reactants are consumed. In piston flow, the reactant concentration gradually declines with increasing axial position. The local rate is higher at the reactor inlet than at the outlet, and the average rate for the entire reactor will correspond to some average composition that is between a_{in} and a_{out}. In contrast, the entire volume of a CSTR is at concentration a_{out}, and the reaction rate throughout the reactor is lower than that at any point in a piston flow reactor going to the same conversion.

Figures 1.6 and 1.7 display the conversion behavior for first-and second-order reactions in a CSTR and contrast the behavior to that of a piston flow reactor. It is apparent that piston flow is substantially better than the CSTR for obtaining high conversions. The comparison is even more dramatic when made in terms of the volume needed to achieve a given conversion; see Figure 1.8. The generalization that

$$\text{Conversion in a PFR} > \text{conversion in a CSTR}$$

is true for most kinetic schemes. The important exceptions to this rule, autocatalytic reactions, are discussed in Chapter 2. A second generalization is

$$\text{Selectivity in a PFR} > \text{selectivity in a CSTR}$$

which also has exceptions.

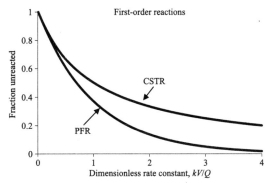

FIGURE 1.6 Relative performance of piston flow and continuous-flow stirred tank reactors for first-order reactions.

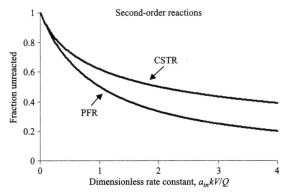

FIGURE 1.7 Relative performance of piston flow and continuous-flow stirred tank reactors for second-order reactions.

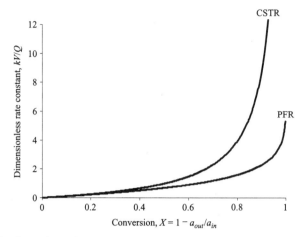

FIGURE 1.8 Comparison of reactor volume required for a given conversion for a first-order reaction in a PFR and a CSTR.

PROBLEMS

1.1. (a) Write the overall and component mass balances for an unsteady, perfectly mixed, continuous flow reactor.
(b) Simplify for the case of constant reactor volume and for constant-density, time-independent flow streams.
(c) Suppose there is no reaction but that the input concentration of some key component varies with time according to $C_{in} = C_0$, $t < 0$; $C_{in} = 0$, $t > 0$. Find $C_{out}(t)$.
(d) Repeat (c) for the case where the key component is consumed by a first-order reaction with rate constant k.

1.2. The homogeneous gas-phase reaction

$$NO + NO_2Cl \rightarrow NO_2 + NOCl$$

is believed to be elementary with rate $\mathcal{R} = k[NO][NO_2Cl]$. Use the kinetic theory of gases to estimate f_R at 300 K. Assume $r_A + r_B = 3.5 \times 10^{-10}$ m. The experimentally observed rate constant at 300 K is $k = 8$ m^3/(mol·s).

1.3. The data in Example 1.2 are in moles of the given component per mole of mixed feed. These are obviously calculated values. Check their consistency by using them to calculate the feed composition given that the feed contained only *para*-xylene and chlorine. Is your result consistent with the stated molar composition of 40% xylene and 60% chlorine?

1.4. Suppose that the following reactions are elementary. Write rate equations for the reaction and for each of the components:

(a) $2A \underset{k_r}{\overset{k_f}{\rightleftarrows}} B + C$

(b) $2A \underset{k_r}{\overset{k_f/2}{\rightleftharpoons}} B+C$

(c) $B+C \underset{k_r}{\overset{k_f}{\rightleftharpoons}} 2A$

(d) $2A \xrightarrow{k_I} B+C$

$B+C \xrightarrow{k_{II}} 2A$

1.5. Determine $a(t)$ for a first-order, reversible reaction, $A \underset{k_r}{\overset{k_f}{\rightleftharpoons}} B$, in a batch reactor.

1.6. Compare $a(z)$ for first- and second-order reactions in a PFR. Plot the profiles on the same graph and arrange the rate constants so that the initial and final concentrations are the same for the two reactions.

1.7. Equation (1.45) gives the spatial distribution of concentration, $a(z)$, in a piston flow reactor for a component that is consumed by a first-order reaction. The local concentration can be used to determine the local reaction rate, $\mathcal{R}_A(z)$.

(a) Integrate the local reaction rate over the length of the reactor to determine $\hat{\mathcal{R}}_A$.

(b) Show that this $\hat{\mathcal{R}}_A$ is consistent with the general component balance, Equation (1.6).

(c) To what value of a does $\hat{\mathcal{R}}_A$ correspond?

(d) At what axial position does this average value occur?

(e) Now integrate a down the length of the tube. Is this spatial average the same as the average found in part (c)?

1.8. Consider the reaction

$$A+B \xrightarrow{k} P$$

with $k=1\,\mathrm{m^3/(mol \cdot s)}$. Suppose $b_{in}=10\,\mathrm{mol/m^3}$. It is desired to achieve $b_{out}=0.01\,\mathrm{mol/m^3}$.

(a) Find the mean residence time needed to achieve this value, assuming piston flow and $a_{in}=b_{in}$.

(b) Repeat (a) assuming that the reaction occurs in a CSTR.

(c) Repeat (a) and (b) assuming $a_{in}=10 b_{in}$.

1.9. The esterification reaction

$$\mathrm{RCOOH+R'OH} \underset{k_r}{\overset{k_f}{\rightleftharpoons}} \mathrm{RCOOR'+H_2O}$$

can be driven to completion by removing the water of condensation. This might be done continuously in a stirred tank reactor or in a horizontally compartmented, progressive flow reactor. This type of reactor gives a reasonable approximation to piston flow in the liquid phase while providing a vapor space for the removal of the by-product water. Suppose it is desired to obtain an ester product containing not more than 1% (by mole) residual alcohol and 0.01% residual acid.

(a) What are the limits on initial stoichiometry if the product specifications are to be achieved?
(b) What value of $a_{out}k\bar{t}$ is needed in a CSTR?
(c) What value of $a_{out}k\bar{t}$ is needed in the progressive reactor?
(d) Discuss the suitability of a batch reactor for this situation.

1.10. Can an irreversible elementary reaction go to completion in a batch reactor in finite time?

1.11. Write a plausible reaction mechanism, including appropriate rate expressions, for the toluene nitration example in Section 1.3.

1.12. The reaction of trimethylamine with n-propyl bromide gives a quaternary ammonium salt:

$$N(CH_3)_3 + C_3H_7Br \rightarrow (CH_3)_3(C_3H_7)NBr$$

Suppose laboratory results at 110°C using toluene as a solvent show the reaction to be second order with rate constant $k = 5.6 \times 10^{-7}$ m³/(mol · s). Suppose $[N(CH_3)_3]_0 = [C_3H_7Br]_0 = 80$ mol/m³.

(a) Estimate the time required to achieve 99% conversion in a batch reactor.
(b) Estimate the volume required in a CSTR to achieve 99% conversion if a production rate of 100 kg/h of the salt is desired.
(c) Suggest means for increasing the productivity; that is, reducing the batch reaction time or the volume of the CSTR.

1.13. Ethyl acetate can be formed from dilute solutions of ethanol and acetic acid according to the reversible reaction

$$C_2H_5OH + CH_3COOH \rightarrow C_2H_5OOCCH_3 + H_2O$$

Ethyl acetate is somewhat easier to separate from water than either ethanol or acetic acid. For example, the relatively large acetate molecule has much lower permeability through a membrane ultrafilter. Thus, esterification is sometimes proposed as an economical approach for recovering dilute fermentation products. Suppose fermentation effluents are available as separate streams containing 3% by weight acetic acid and 5% by weight ethanol. Devise a reaction scheme for generating ethyl acetate using the reactants in stoichiometric ratio. After reaction, the ethyl acetate concentration is increased first to 25% by weight using ultrafiltration and then to 99% by weight using distillation. The reactants must ultimately be heated for the distillation step. Thus, we can suppose both the esterification and membrane separation to be conducted at 100°C. At this temperature,

$$k_f = 8.0 \times 10^{-9} \text{ m}^3/(\text{mol} \cdot \text{s})$$

$$k_r = 2.7 \times 10^{-9}\, m^3/(mol \cdot s)$$

Determine $\bar{t}$ and a_{out} for a CSTR that approaches equilibrium within 5%; that is,

$$\frac{a_{out} - a_{equil}}{a_{in} - a_{equil}} = 0.05$$

1.14. Rate expressions for gas-phase reactions are sometimes based on partial pressures. A literature source[5] gives $k = 1.1 \times 10^{-3}\, mol/(cm^3 \cdot atm^2 \cdot h)$ for the reaction of gaseous sulfur with methane at 873 K.

$$CH_4 + 2S_2 \rightarrow CS_2 + 2H_2S$$

where $\mathscr{R} = k P_{CH_4} P_{S_2}\, mol/(cm^3 \cdot h)$. Determine k when the rate is based on concentrations: $\mathscr{R} = k[CH_4][S_2]$. Give k in SI units.

1.15. Example 1.7 predicted that power per unit volume would have to increase by a factor of 100 in order to maintain the same mixing time for a 1000-fold scaleup in volume. This can properly be called absurd. A more reasonable scaleup rule is to maintain constant power per unit volume so that a 1000-fold increase in reactor volume requires a 1000-fold increase in power. Use the logic of Example 1.7 to determine the increase in mixing time for a 1000-fold scaleup at constant power per unit volume.

REFERENCES

1. Tsukahara, H., Ishida, T., and Mitsufumi, M., "Gas-phase oxidation of nitric oxide: chemical kinetics and rate constant," *Nitric Oxide*, **3**, 191–198 (1999).
2. Lewis, R. J. and Moodie, R. B., "The nitration of styrenes by nitric acid in dichloromethane," *J. Chem. Soc., Perkin Trans.*, **2**, 563–567 (1997).
3. Lindstedt, R. P. and Maurice, L. Q., "Detailed kinetic modeling of toluene combustion," *Combust. Sci. Technol.*, **120**, 119–167 (1996).
4. Chen, C.Y., Wu, C.W., and Hu, K. H., "Thermal hazard analysis of batch processes for toluene mononitration," *Zhongguo Huanjing Gongcheng Xuekan*, **6**, 301–309 (1996).
5. Norwood, K. W. and Metzner, A. B., "Flow patterns and mixing rates in agitated vessels," *AIChE J.*, **6**, 432–437 (1960).
6. Smith, J. M., *Chemical Engineering Kinetics*, 1st ed., McGraw-Hill, New York, 1956, p. 131.

SUGGESTIONS FOR FURTHER READING

There are many good texts on chemical engineering kinetics, and the reader may wish to browse through several of them to see how they introduce the subject.

A few recent books are

Levenspiel, O., *Chemical Reaction Engineering*, 3rd ed., Wiley, New York, 1998.

Schmidt, L. D., *The Engineering of Chemical Reactions*, Oxford University Press, New York, 1998.

King, M. B. and Winterbottom, M. B., *Reactor Design for Chemical Engineers*, Chapman & Hall, London, 1998.

A relatively advanced treatment is given in

Froment, F. and Bischoff, K. B., *Chemical Reactor Analysis and Design*, 2nd ed., Wiley, New York, 1990.

An extended treatment of material balance equations, with substantial emphasis on component balances in reacting systems, is given in

Reklaitis, G. V. and Schneider, D. R., *Introduction to Material and Energy Balances*, Wiley, New York, 1983.

See also

Felder, R. M. and Rousseau, R. W., *Elementary Principles of Chemical Processes*, 3rd ed., Wiley, New York, 2000.

CHAPTER 2
MULTIPLE REACTIONS IN BATCH REACTORS

Chapter 1 treated single, elementary reactions in ideal reactors. Chapter 2 broadens the kinetics to include multiple and nonelementary reactions. Attention is restricted to batch reactors, but the method for formulating the kinetics of complex reactions will also be used for the flow reactors of Chapters 3 and 4 and for the nonisothermal reactors of Chapter 5.

The most important characteristic of an ideal batch reactor is that the contents are perfectly mixed. Corresponding to this assumption, the component balances are ordinary differential equations. The reactor operates at constant mass between filling and discharge steps that are assumed to be fast compared with reaction half-lives and the batch reaction times. Chapter 1 made the further assumption of constant mass density, so that the working volume of the reactor was constant, but Chapter 2 relaxes this assumption.

2.1 MULTIPLE AND NONELEMENTARY REACTIONS

Multiple reactions involve two or more stoichiometric equations, each with its own rate expression. They are often classified as *consecutive* as in

$$A + B \xrightarrow{k_I} C \quad \mathscr{R}_I = k_I ab$$
$$C + D \xrightarrow{k_{II}} E \quad \mathscr{R}_{II} = k_{II} cd \tag{2.1}$$

or *competitive* as in

$$A + B \xrightarrow{k_I} C \quad \mathscr{R}_I = k_I ab$$
$$A + D \xrightarrow{k_{II}} E \quad \mathscr{R}_{II} = k_{II} ad \tag{2.2}$$

or *completely independent* as in

$$A \xrightarrow{k_I} B \qquad \mathscr{R}_I = k_I a$$
$$C + D \xrightarrow{k_{II}} E \qquad \mathscr{R}_{II} = k_{II} cd \qquad (2.3)$$

Even *reversible* reactions can be regarded as multiple:

$$A + B \xrightarrow{k_I} C \qquad \mathscr{R}_I = k_I ab$$
$$C \xrightarrow{k_{II}} A + B \qquad \mathscr{R}_{II} = k_{II} c \qquad (2.4)$$

Note that the Roman numeral subscripts refer to numbered reactions and have nothing to do with iodine. All these examples have involved elementary reactions. Multiple reactions and apparently single but nonelementary reactions are called *complex*. Complex reactions, even when apparently single, consist of a number of elementary steps. These steps, some of which may be quite fast, constitute the *mechanism* of the observed, complex reaction. As an example, suppose that

$$A \xrightarrow{k_I} B + C \qquad \mathscr{R}_I = k_I a$$
$$B \xrightarrow{k_{II}} D \qquad \mathscr{R}_{II} = k_{II} b \qquad (2.5)$$

where $k_{II} \gg k_I$. Then the observed reaction will be

$$A \rightarrow C + D \qquad \mathscr{R} = ka \qquad (2.6)$$

This reaction is complex even though it has a stoichiometric equation and rate expression that could correspond to an elementary reaction. Recall the convention used in this text: when a rate constant is written above the reaction arrow, the reaction is assumed to be elementary with a rate that is consistent with the stoichiometry according to Equation (1.14). The reactions in Equations (2.5) are examples. When the rate constant is missing, the reaction rate must be explicitly specified. The reaction in Equation (2.6) is an example. This reaction is complex since the mechanism involves a short-lived intermediate, B.

To solve a problem in reactor design, knowledge of the reaction mechanism may not be critical to success but it is always desirable. Two reasons are:

1. Knowledge of the mechanism will allow fitting experimental data to a theoretical rate expression. This will presumably be more reliable on extrapolation or scaleup than an empirical fit.

2. Knowing the mechanism may suggest chemical modifications and optimization possibilities for the final design that would otherwise be missed.

The best way to find a reaction mechanism is to find a good chemist. Chemical insight can be used to hypothesize a mechanism, and the hypothesized mechanism can then be tested against experimental data. If inconsistent, the mechanism must be rejected. This is seldom the case. More typically, there are several mechanisms that will fit the data equally well. A truly definitive study of reaction mechanisms requires direct observation of all chemical species, including intermediates that may have low concentrations and short lives. Such studies are not always feasible. Working hypotheses for the reaction mechanisms must then be selected based on general chemical principles and on analogous systems that have been studied in detail. There is no substitute for understanding the chemistry or at least for having faith in the chemist.

2.2 COMPONENT REACTION RATES FOR MULTIPLE REACTIONS

The component balance for a batch reactor, Equation (1.21), still holds when there are multiple reactions. However, the net rate of formation of the component may be due to several different reactions. Thus,

$$\mathcal{R}_A = \nu_{A,I}\mathcal{R}_I + \nu_{A,II}\mathcal{R}_{II} + \nu_{A,III}\mathcal{R}_{III} + \cdots \quad . \quad (2.7)$$

Here, we envision component A being formed by Reactions I, II, III, ..., each of which has a stoichiometric coefficient with respect to the component. Equivalent to Equation (2.7) we can write

$$\mathcal{R}_A = \sum_{Reactions} \nu_{A,I}\mathcal{R}_I = \sum_I \nu_{A,I}\mathcal{R}_I \quad (2.8)$$

Obviously, $\nu_{A,I} = 0$ if component A does not participate in Reaction I.

Example 2.1: Determine the overall reaction rate for each component in the following set of reactions:

$$A + B \xrightarrow{k_I} C$$
$$C \xrightarrow{k_{II}} 2E$$
$$2A \xrightarrow{k_{III}/2} D$$

Solution: We begin with the stoichiometric coefficients for each component for each reaction:

$$\begin{array}{lll} \nu_{A,I} = -1 & \nu_{A,II} = 0 & \nu_{A,III} = -2 \\ \nu_{B,I} = -1 & \nu_{B,II} = 0 & \nu_{B,III} = 0 \\ \nu_{C,I} = +1 & \nu_{C,II} = -1 & \nu_{C,III} = 0 \end{array}$$

$$\nu_{D,I} = 0 \quad \nu_{D,II} = 0 \quad \nu_{D,III} = +1$$
$$\nu_{E,I} = 0 \quad \nu_{E,II} = +2 \quad \nu_{E,III} = 0$$

The various reactions are all elementary (witness the rate constants over the arrows) so the rates are

$$\mathscr{R}_I = k_I ab$$
$$\mathscr{R}_{II} = k_{II} c$$
$$\mathscr{R}_{III} = (k_{III}/2)a^2$$

Now apply Equations (2.7) or (2.8) to obtain

$$\mathscr{R}_A = -k_I ab - k_{III} a^2$$
$$\mathscr{R}_B = -k_I ab$$
$$\mathscr{R}_C = +k_I ab - k_{II} c$$
$$\mathscr{R}_D = (k_{III}/2)a^2$$
$$\mathscr{R}_E = +2k_{II} c$$

2.3 MULTIPLE REACTIONS IN BATCH REACTORS

Suppose there are N components involved in a set of M reactions. Then Equation (1.21) can be written for each component using the rate expressions of Equations (2.7) or (2.8). The component balances for a batch reactor are

$$\frac{d(Va)}{dt} = V\mathscr{R}_A = V(\nu_{A,I}\mathscr{R}_I + \nu_{A,II}\mathscr{R}_{II} + \nu_{A,III}\mathscr{R}_{III} + \cdots + M \text{ terms})$$
$$\frac{d(Vb)}{dt} = V\mathscr{R}_B = V(\nu_{B,I}\mathscr{R}_I + \nu_{B,II}\mathscr{R}_{II} + \nu_{B,III}\mathscr{R}_{III} + \cdots) \quad (2.9)$$
$$\frac{d(Vc)}{dt} = V\mathscr{R}_C = V(\nu_{C,I}\mathscr{R}_I + \nu_{C,II}\mathscr{R}_{II} + \nu_{C,III}\mathscr{R}_{III} + \cdots)$$

This is a set of N ordinary differential equations, one for each component. The component reaction rates will have M terms, one for each reaction, although many of the terms may be zero. Equations (2.9) are subject to a set of N initial conditions of the form

$$a = a_0 \quad \text{at} \quad t = 0 \quad (2.10)$$

The number of simultaneous equations can usually be reduced to fewer than N using the methodology of Section 2.8. However, this reduction is typically more trouble than it is worth.

Example 2.2: Derive the batch reactor design equations for the reaction set in Example 2.1. Assume a liquid-phase system with constant density.

Solution: The real work has already been done in Example 2.1, where $\mathscr{R}_A, \mathscr{R}_B, \mathscr{R}_C, \ldots$ were found. When density is constant, volume is constant, and the V terms in Equations (2.9) cancel. Substituting the reaction rates from Example 2.1 gives

$$\frac{da}{dt} = -k_I ab - k_{III} a^2 \qquad a = a_0 \quad \text{at} \quad t = 0$$

$$\frac{db}{dt} = -k_I ab \qquad b = b_0 \quad \text{at} \quad t = 0$$

$$\frac{dc}{dt} = +k_I ab - k_{II} c \qquad c = c_0 \quad \text{at} \quad t = 0$$

$$\frac{dd}{dt} = (k_{III}/2) a^2 \qquad d = d_0 \quad \text{at} \quad t = 0$$

$$\frac{de}{dt} = +2 k_{II} c \qquad e = e_0 \quad \text{at} \quad t = 0$$

This is a fairly simple set of first-order ODEs. The set is difficult to solve analytically, but numerical solutions are easy.

2.4 NUMERICAL SOLUTIONS TO SETS OF FIRST-ORDER ODEs

The design equations for multiple reactions in batch reactors can sometimes be solved analytically. Important examples are given in Section 2.5. However, for realistic and industrially important kinetic schemes, the component balances soon become intractable from the viewpoint of obtaining analytical solutions. Fortunately, sets of first-order ODEs are easily solved numerically. Sophisticated and computationally efficient methods have been developed for solving such sets of equations. One popular method, called Runge-Kutta, is described in Appendix 2. This or even more sophisticated techniques should be used if the cost of computation becomes significant. However, computer costs will usually be inconsequential compared with the costs of the engineer's personal time. In this usual case, the use of a simple technique can save time and money by allowing the engineer to focus on the physics and chemistry of the problem rather than on the numerical mathematics. Another possible way to save engineering time is to use higher-order mathematical programming systems such as Mathematica®, Matlab®, or Maple® rather than the more fundamental programming languages such as Fortran, Basic, or C. There is some risk to this approach in that the engineer may not know when either he or the system has made a mistake. This book adopts the conservative approach of illustrating numerical methods by showing programming fragments in the general-purpose language known as Basic. Basic was chosen because it can be sight-read by anyone familiar with computer programming, because it is widely available on personal computers, and because it is used as the programming component for the popular spreadsheet Excel®.

The simplest possible method for solving a set of first-order ODEs—subject to given initial values—is called *marching ahead*. It is also known as *Euler's method*. We suppose that all concentrations are known at time $t=0$. This allows the initial reaction rates to be calculated, one for each component. Choose some time increment, Δt, that is so small that, given the calculated reaction rates, the concentrations will change very little during the time increment. Calculate these small changes in concentration, assuming that the reaction rates are constant. Use the new concentrations to revise the reaction rates. Pick another time increment and repeat the calculations. Continue until the specified reaction time has been reached. This is the tentative solution. It is tentative because you do not yet know whether the numerical solution has *converged* to the true solution with sufficient accuracy. Test for convergence by reducing Δt and repeating the calculation. Do you get the same results to say four decimal places? If so, you probably have an adequate solution. If not, reduce Δt again. Computers are so fast that this brute force method of solving and testing for convergence will take only a few seconds for most of the problems in this book.

Euler's method can be illustrated by the simultaneous solution of

$$\frac{da}{dt} = \mathscr{R}_A(a,b)$$
$$\frac{db}{dt} = \mathscr{R}_B(a,b)$$
(2.11)

subject to the usual initial conditions. The marching equations can be written as

$$a_{new} = a_{old} + \mathscr{R}_A(a_{old}, b_{old}) \Delta t$$
$$b_{new} = b_{old} + \mathscr{R}_B(a_{old}, b_{old}) \Delta t$$
$$t_{new} = t_{old} + \Delta t$$
(2.12)

The computation is begun by setting $a_{old} = a_0$, $b_{old} = b_0$, and $t_{old} = 0$. Rates are computed using the old concentrations and the marching equations are used to calculate the new concentrations. Old is then replaced by new and the march takes another step.

The marching-ahead technique systematically overestimates $\mathscr{R}_A$ when component A is a reactant since the rate is evaluated at the old concentrations where a and $\mathscr{R}_A$ are higher. This creates a systematic error similar to the numerical integration error shown in Figure 2.1. The error can be dramatically reduced by the use of more sophisticated numerical techniques. It can also be reduced by the simple expedient of reducing Δt and repeating the calculation.

MULTIPLE REACTIONS IN BATCH REACTORS

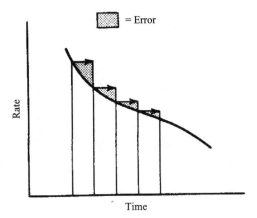

FIGURE 2.1 Systematic error of Euler integration.

Example 2.3: Solve the batch design equations for the reaction of Example 2.2. Use $k_I = 0.1\,\text{mol}/(\text{m}^3 \cdot \text{h})$, $k_{II} = 1.2\,\text{h}^{-1}$, $k_{III} = 0.6\,\text{mol}/(\text{m}^3 \cdot \text{h})$. The initial conditions are $a_0 = b_0 = 20\,\text{mol}/\text{m}^3$. The reaction time is 1 h.

Solution: The following is a complete program for performing the calculations. It is written in Basic as an Excel macro. The rather arcane statements needed to display the results on the Excel spreadsheet are shown at the end. They need to be replaced with PRINT statements given a Basic compiler that can write directly to the screen. The programming examples in this text will normally show only the computational algorithm and will leave input and output to the reader.

```
DefDbl A-Z
Sub Exp2_3()

k1 = 0.1
k2 = 1.2
k3 = 0.06
tmax = 1

dt = 2

For N = 1 To 10

aold = 20
bold = 20
cold = 0
dold = 0
eold = 0

t = 0
```

```
dt = dt/4

Do

    RA = -k1 * aold * Bold - k3 * aold ^2
    RB = -k1 * aold * Bold
    RC = k1 * aold * Bold - k2 * cold
    RD = k3/2 * aold ^2
    RE = 2 * k2 * cold
    anew = aold + dt * RA
    bnew = bold + dt * RB
    cnew = cold + dt * RC
    dnew = dold + dt * RD
    enew = eold + dt * RE

    t = t + dt

    aold = anew
    bold = bnew
    cold = cnew
    dold = dnew
    eold = enew

Loop While t < tmax

Sum = aold + bold + cold + dold + eold

'The following statements output the results to the
'Excel spreadsheet
Range("A"& CStr(N)).Select
ActiveCell.FormulaR1C1 = dt
Range("B"& CStr(N)).Select
ActiveCell.FormulaR1C1 = aold
Range("C"& CStr(N)).Select
ActiveCell.FormulaR1C1 = bold
Range("D"& CStr(N)).Select
ActiveCell.FormulaR1C1 = cold
Range("E"& CStr(N)).Select
ActiveCell.FormulaR1C1 = dold
Range("F"& CStr(N)).Select
ActiveCell.FormulaR1C1 = eold
Range("G"& CStr(N)).Select
ActiveCell.FormulaR1C1 = Sum

Next N

End Sub
```

The results from this program (with added headers) are shown below:

Δt	$a(t_{max})$	$b(t_{max})$	$c(t_{max})$	$d(t_{max})$	$e(t_{max})$	Sum
0.5000000	−16.3200	0.0000	8.0000	8.1600	24.0000	23.8400
0.1250000	2.8245	8.3687	5.1177	2.7721	13.0271	32.1102
0.0312500	3.4367	8.6637	5.1313	2.6135	12.4101	32.2552
0.0078125	3.5766	8.7381	5.1208	2.5808	12.2821	32.2984
0.0019531	3.6110	8.7567	5.1176	2.5729	12.2513	32.3095
0.0004883	3.6195	8.7614	5.1168	2.5709	12.2437	32.3123
0.0001221	3.6217	8.7625	5.1166	2.5704	12.2418	32.3130
0.0000305	3.6222	8.7628	5.1165	2.5703	12.2413	32.3131
0.0000076	3.6223	8.7629	5.1165	2.5703	12.2412	32.3132
0.0000019	3.6224	8.7629	5.1165	2.5703	12.2411	32.3132

These results have converged to four decimal places. The output required about 2 s on what will undoubtedly be a slow PC by the time you read this.

Example 2.4: Determine how the errors in the numerical solutions in Example (2.3) depend on the size of the time increment, Δt.

Solution: Consider the values of $a(t_{max})$ versus Δt as shown below. The indicated errors are relative to the fully converged answer of 3.6224.

Δt	$a(t_{max})$	Error
0.5000000	−16.3200	19.9424
0.1250000	2.8245	−0.7972
0.0312500	3.4367	−0.1853
0.0078125	3.5766	−0.0458
0.0019531	3.6110	−0.0114
0.0004883	3.6195	−0.0029
0.0001221	3.6217	−0.0007
0.0000305	3.6222	−0.0002

The first result, for $\Delta t = 0.5$, shows a negative result for $a(t_{max})$ due to the very large value for Δt. For smaller values of Δt, the calculated values for $a(t_{max})$ are physically realistic and the errors decrease by roughly a factor of 4 as the time step decreases by a factor of 4. Thus, the error is proportional to Δt. Euler's method is said to *converge order* Δt, denoted $\mathbf{O}(\Delta t)$.

Convergence order Δt for Euler's method is based on more than the empirical observation in Example 2.4. The order of convergence springs directly from the way in which the derivatives in Equations (2.11) are calculated. The simplest approximation of a first derivative is

$$\frac{da}{dt} \approx \frac{a_{new} - a_{old}}{\Delta t} \qquad (2.13)$$

Substitution of this approximation into Equations (2.11) gives Equations (2.12). The limit of Equation (2.13) as $\Delta t \to 0$ is the usual definition of a derivative. It assumes that, locally, the function $a(t)$ is a straight line. A straight line is a first-order equation and convergence $O(\Delta t)$ follows from this fact. Knowledge of the convergence order allows extrapolation and acceleration of convergence. This and an improved integration technique, Runge-Kutta, are discussed in Appendix 2. The Runge-Kutta technique converges $O(\Delta t^5)$. Other things being equal, it is better to use a numerical method with a high order of convergence. However, such methods are usually harder to implement. Also, convergence is an *asymptotic* property. This means that it becomes true only as Δt approaches zero. It may well be that the solution has already converged with adequate accuracy by the time the theoretical convergence order is reached.

The convergence of Euler's method to the true analytical solution is assured for sets of linear ODEs. Just keep decreasing Δt. Occasionally, the word length of a computer becomes limiting. This text contains a few problems that cannot be solved in single precision (e.g., about seven decimal digits), and it is good practice to run double precision as a matter of course. This was done in the Basic program in Example 2.3. Most of the complex kinetic schemes give rise to nonlinear equations, and there is no absolute assurance of convergence. Fortunately, the marching-ahead method behaves quite well for most nonlinear systems of engineering importance. Practical problems do arise in *stiff* sets of differential equations where some members of the set have characteristic times much smaller than other members. This occurs in reaction kinetics when some reactions have half-lives much shorter than others. In free-radical kinetics, reaction rates may differ by 3 orders of magnitude. The allowable time step, Δt, must be set to accommodate the fastest reaction and may be too small to follow the overall course of the reaction, even for modern computers. Special numerical methods have been devised to deal with stiff sets of differential equations. In free-radical processes, it is also possible to avoid stiff sets of equations through use of the *quasi-steady-state hypothesis*, which is discussed in Section 2.5.3.

The need to use specific numerical values for the rate constants and initial conditions is a weakness of numerical solutions. If they change, then the numerical solution must be repeated. Analytical solutions usually apply to all values of the input parameters, but special cases are sometimes needed. Recall the special case needed for $a_0 = b_0$ in Example 1.4. Numerical solution techniques do not have this problem, and the problem of specificity with respect to numerical values can be minimized or overcome through the judicious use of *dimensionless variables*. Concentrations can be converted to dimensionless concentrations by dividing by an initial value; e.g. $a^* = a/a_0$, $b^* = b/a_0$, and so on. The normal choice is to normalize using the initial concentration of a stoichiometrically limiting component. Time can be converted to a dimensionless variable by dividing by some characteristic time for the system. The mean residence time is often used as the characteristic time of a flow system. In a batch system, we could use the batch reaction time, t_{batch}, so that $t^* = t/t_{batch}$ is one possibility for a dimensionless time. Another possibility, applicable to both flow and batch systems, is to base the characteristic time on the reciprocal of a rate constant. The quantity k_1^{-1} has units of time when k_1 is a first-order rate constant and $(a_0 k_2)^{-1}$ has units of time when k_2 is a second-order rate constant. More generally, $(a_0^{order-1} k_{order})^{-1}$ will have units of time when k_{order} is the rate constant for a reaction of arbitrary order.

Example 2.5: Consider the following competitive reactions in a constant-density batch reactor:

$$A + B \rightarrow P \quad \text{(Desired product)} \quad \mathcal{R}_I = k_I ab$$
$$2A \rightarrow D \quad \text{(Undesired dimer)} \quad \mathcal{R}_{II} = k_{II} a^2$$

The selectivity based on component A is

$$\text{Selectivity} = \frac{\text{Moles P produced}}{\text{Moles A reacted}} = \frac{p}{a_0 - a} = \frac{p^*}{1 - a^*}$$

which ranges from 1 when only the desired product is made to 0 when only the undesired dimer is made. Components A and B have initial values a_0 and b_0 respectively. The other components have zero initial concentration. On how many parameters does the selectivity depend?

Solution: On first inspection, the selectivity appears to depend on five parameters: a_0, b_0, k_I, k_{II}, and t_{batch}. However, the governing equations can be cast into dimensionless form as

$$\frac{da}{dt} = -k_I ab - 2k_{II} a^2 \quad \text{becomes} \quad \frac{da^*}{dt^*} = -a^* b^* - 2K_{II}^*(a^*)^2$$

$$\frac{db}{dt} = -k_I ab \quad \text{becomes} \quad \frac{db^*}{dt^*} = -a^* b^*$$

$$\frac{dp}{dt} = k_I ab \quad \text{becomes} \quad \frac{dp^*}{dt^*} = a^* b^*$$

$$\frac{dd}{dt} = k_{II} a^2 \quad \text{becomes} \quad \frac{dd^*}{dt^*} = K_{II}^*(a^*)^2$$

where the dimensionless time is $t^* = k_{II} a_0 t$. The initial conditions are $a^* = 1, b^* = b_0/a_0, p^* = 0, d^* = 0$ at $t^* = 0$. The solution is evaluated at $t^* = k_{II} a_0 t_{batch}$. Aside from the endpoint, the numerical solution depends on just two dimensionless parameters. These are b_0/a_0 and $K_{II}^* = k_{II}/k_I$. There are still too many parameters to conveniently plot the whole solution on a single graph, but partial results can easily be plotted: e.g. a plot for a fixed value of $K_{II}^* = k_{II}/k_I$ of selectivity versus t^* with b_0/a_0 as the parameter identifying various curves.

2.5 ANALYTICALLY TRACTABLE EXAMPLES

Relatively few kinetic schemes admit analytical solutions. This section is concerned with those special cases that do, and also with some cases where preliminary analytical work will ease the subsequent numerical studies. We begin with the *nth-order reaction*.

2.5.1 The nth-Order Reaction

$$A \rightarrow \text{Products} \qquad \mathscr{R} = ka^n \qquad (2.14)$$

This reaction can be elementary if $n = 1$ or 2. More generally, it is complex. Noninteger values for n are often found when fitting rate data empirically, sometimes for sound kinetic reasons, as will be seen in Section 2.5.3. For an isothermal, constant-volume batch reactor,

$$\frac{da}{dt} = -ka^n \qquad a = a_0 \quad \text{at} \quad t = 0 \qquad (2.15)$$

The first-order reaction is a special case mathematically. For $n = 1$, the solution has the exponential form of Equation (1.24):

$$\frac{a}{a_0} = e^{-kt} \qquad (2.16)$$

For $n \neq 1$, the solution looks very different:

$$\frac{a}{a_0} = \left[1 + (n-1)a_0^{n-1}kt\right]^{1/(1-n)} \qquad (2.17)$$

but see Problem 2.7. If $n > 1$, the term in square brackets is positive and the concentration gradually declines toward zero as the batch reaction time increases. Reactions with an order of 1 or greater never quite go to completion. In contrast, reactions with an order less than 1 do go to completion, at least mathematically. When $n < 1$, Equation (2.17) predicts $a = 0$ when

$$t = t_{max} = \frac{a_0^{1-n}}{(1-n)k} \qquad (2.18)$$

If the reaction order does not change, reactions with $n < 1$ will go to completion in finite time. This is sometimes observed. Solid rocket propellants or fuses used to detonate explosives can burn at an essentially constant rate (a zero-order reaction) until all reactants are consumed. These are multiphase reactions limited by heat transfer and are discussed in Chapter 11. For single phase systems, a zero-order reaction can be expected to slow and become first or second order in the limit of low concentration.

For $n < 1$, the reaction rate of Equation (2.14) should be supplemented by the condition that
$$\mathscr{R} = 0 \quad \text{if} \quad a \leq 0 \tag{2.19}$$
Otherwise, both the mathematics and the physics become unrealistic.

2.5.2 Consecutive First-Order Reactions, A → B → C → ···

Consider the following reaction sequence

$$A \xrightarrow{k_A} B \xrightarrow{k_B} C \xrightarrow{k_C} D \xrightarrow{k_D} \cdots \tag{2.20}$$

These reactions could be elementary, first order, and without by-products as indicated. For example, they could represent a sequence of isomerizations. More likely, there will be by-products that do not influence the subsequent reaction steps and which were omitted in the shorthand notation of Equation (2.20). Thus, the first member of the set could actually be

$$A \xrightarrow{k_A} B + P$$

Radioactive decay provides splendid examples of first-order sequences of this type. The naturally occurring sequence beginning with ^{238}U and ending with ^{206}Pb has 14 consecutive reactions that generate α or β particles as by-products. The half-lives in Table 2.1—and the corresponding first-order rate constants, see Equation (1.27)—differ by 21 orders of magnitude.

Within the strictly chemical realm, sequences of pseudo-first-order reactions are quite common. The usually cited examples are hydrations carried out in water and slow oxidations carried out in air, where one of the reactants

TABLE 2.1 Radioactive Decay Series for ^{238}U

Nuclear Species	Half-Life
^{238}U	4.5 billion years
^{234}Th	24 days
^{234}Pa	1.2 min
^{234}U	250,000 years
^{230}Th	80,000 years
^{226}Ra	1600 years
^{222}Rn	3.8 days
^{218}Po	3 min
^{214}Pb	27 min
^{214}Bi	20 min
^{214}Po	160 μs
^{210}Pb	22 years
^{210}Bi	5 days
^{210}Po	138 days
^{206}Pb	Stable

(e.g., water or oxygen) is present in great excess and hence does not change appreciably in concentration during the course of the reaction. These reactions behave identically to those in Equation (2.20), although the rate constants over the arrows should be removed as a formality since the reactions are not elementary.

Any sequence of first-order reactions can be solved analytically, although the algebra can become tedious if the number of reactions is large. The ODEs that correspond to Equation (2.20) are

$$\frac{da}{dt} = -k_A a$$
$$\frac{db}{dt} = -k_B b + k_A a$$
$$\frac{dc}{dt} = -k_C c + k_B b$$
$$\frac{dd}{dt} = -k_D d + k_C c$$
(2.21)

Just as the reactions are consecutive, solutions to this set can be carried out consecutively. The equation for component A depends only on a and can be solved directly. The result is substituted into the equation for component B, which then depends only on b and t and can be solved. This procedure is repeated until the last, stable component is reached. Assuming component D is stable, the solutions to Equations (2.21) are

$$a = a_0 e^{-k_A t}$$
$$b = \left[b_0 - \frac{a_0 k_A}{k_B - k_A}\right] e^{-k_B t} + \left[\frac{a_0 k_A}{k_B - k_A}\right] e^{-k_A t}$$
$$c = \left[c_0 - \frac{b_0 k_B}{k_C - k_B} + \frac{a_0 k_A k_B}{(k_C - k_A)(k_C - k_B)}\right] e^{-k_C t}$$
$$+ \left[\frac{b_0 k_B}{k_C - k_B} - \frac{a_0 k_A k_B}{(k_C - k_B)(k_B - k_A)}\right] e^{-k_B t} + \left[\frac{a_0 k_A k_B}{(k_C - k_A)(k_B - k_A)}\right] e^{-k_A t}$$
$$d = d_0 + (a_0 - a) + (b_0 - b) + (c_0 - c)$$
(2.22)

These results assume that all the rate constants are different. Special forms apply when some of the k values are identical, but the qualitative behavior of the solution remains the same. Figure 2.2 illustrates this behavior for the case of $b_0 = c_0 = d_0 = 0$. The concentrations of B and C start at zero, increase to maximums, and then decline back to zero. Typically, component B or C is the desired product whereas the others are undesired. If, say, B is desired, the batch reaction time can be picked to maximize its concentration. Setting $db/dt = 0$ and $b_0 = 0$ gives

$$t_{max} = \frac{\ln(k_B/k_A)}{k_B - k_A}$$
(2.23)

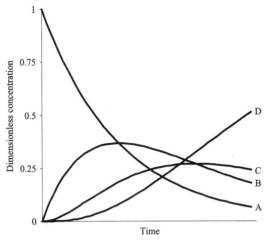

FIGURE 2.2 Consecutive reaction sequence, A → B → C → D.

Selection of the optimal time for the production of C requires a numerical solution but remains conceptually straightforward.

Equations (2.22) and (2.23) become indeterminate if $k_B = k_A$. Special forms are needed for the analytical solution of a set of consecutive, first-order reactions whenever a rate constant is repeated. The derivation of the solution can be repeated for the special case or L'Hospital's rule can be applied to the general solution. As a practical matter, identical rate constants are rare, except for multifunctional molecules where reactions at physically different but chemically similar sites can have the same rate constant. Polymerizations are an important example. Numerical solutions to the governing set of simultaneous ODEs have no difficulty with repeated rate constants, but such solutions can become computationally challenging when the rate constants differ greatly in magnitude. Table 2.1 provides a dramatic example of reactions that lead to stiff equations. A method for finding analytical approximations to stiff equations is described in the next section.

2.5.3 The Quasi-Steady State Hypothesis

Many reactions involve short-lived intermediates that are so reactive that they never accumulate in large quantities and are difficult to detect. Their presence is important in the reaction mechanism and may dictate the functional form of the rate equation. Consider the following reaction:

$$A \underset{k_r}{\overset{k_f}{\rightleftarrows}} B \overset{k_B}{\rightarrow} C$$

This system contains only first-order steps. An exact but somewhat cumbersome analytical solution is available.

The governing ODEs are

$$\frac{da}{dt} = -k_f a + k_r b$$

$$\frac{db}{dt} = +k_f a - k_r b - k_B b$$

Assuming $b_0 = 0$, the solution is

$$a = \frac{k_f a_0}{S_1 - S_2}\left[\left(1 - \frac{k_B}{S_1}\right)e^{-S_1 t} - \left(1 - \frac{k_B}{S_2}\right)e^{-S_2 t}\right]$$

$$b = \frac{k_f a_0}{S_1 - S_2}\left(e^{-S_2 t} - e^{-S_1 t}\right)$$

(2.24)

where

$$S_1, S_2 = (1/2)\left[k_f + k_r + k_B \pm \sqrt{(k_f + k_r + k_B)^2 - 4 k_f k_B}\right]$$

Suppose that B is highly reactive. When formed, it rapidly reverts back to A or transforms into C. This implies $k_r \gg k_f$ and $k_B \gg k_f$. The *quasi-steady hypothesis* assumes that B is consumed as fast as it is formed so that its time rate of change is zero. More specifically, we assume that the concentration of B rises quickly and achieves a dynamic equilibrium with A, which is consumed at a much slower rate. To apply the quasi-steady hypothesis to component B, we set $db/dt = 0$. The ODE for B then gives

$$b = \frac{k_f a}{k_r + k_B} \tag{2.25}$$

Substituting this into the ODE for A gives

$$a = a_0 \exp\left(\frac{-k_f k_B t}{k_f + k_B}\right) \tag{2.26}$$

After an initial startup period, Equations (2.25) and (2.26) become reasonable approximations of the true solutions. See Figure 2.3 for the case of $k_r = k_B = 10 k_f$. The approximation becomes better when there is a larger difference between k_f and the other two rate constants.

The quasi-steady hypothesis is used when short-lived intermediates are formed as part of a relatively slow overall reaction. The short-lived molecules are hypothesized to achieve an approximate steady state in which they are created at nearly the same rate that they are consumed. Their concentration in this quasi-steady state is necessarily small. A typical use of the quasi-steady

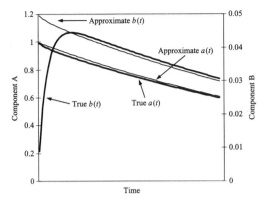

FIGURE 2.3 True solution versus approximation using the quasi-steady hypothesis.

hypothesis is in *chain reactions* propagated by free radicals. Free radicals are molecules or atoms that have an unpaired electron. Many common organic reactions, such as thermal cracking and vinyl polymerization, occur by free-radical processes. The following mechanism has been postulated for the gas-phase decomposition of acetaldehyde.

Initiation

$$CH_3CHO \xrightarrow{k_I} CH_3 \cdot + \cdot CHO$$

This spontaneous or thermal initiation generates two free radicals by breaking a covalent bond. The aldehyde radical is long-lived and does not markedly influence the subsequent chemistry. The methane radical is highly reactive; but rather than disappearing, most reactions regenerate it.

Propagation

$$CH_3 \cdot + CH_3CHO \xrightarrow{k_{II}} CH_4 + CH_3 \cdot CO$$

$$CH_3 \cdot CO \xrightarrow{k_{III}} CH_3 \cdot + CO$$

These propagation reactions are circular. They consume a methane radical but also generate one. There is no net consumption of free radicals, so a single initiation reaction can cause an indefinite number of propagation reactions, each one of which does consume an acetaldehyde molecule. Ignoring any accumulation of methane radicals, the overall stoichiometry is given by the net sum of the propagation steps:

$$CH_3CHO \rightarrow CH_4 + CO$$

The methane radicals do not accumulate because of termination reactions. The concentration of radicals adjusts itself so that the initiation and termination rates are equal. The major termination reaction postulated for the acetaldehyde decomposition is termination by combination.

Termination

$$2CH_3\cdot \xrightarrow{k_{IV}} C_2H_6$$

The assumption of a quasi-steady state is applied to the $CH_3\cdot$ and $CH_3\cdot CO$ radicals by setting their time derivatives to zero:

$$\frac{d[CH_3\cdot]}{dt} = k_I[CH_3CHO] - k_{II}[CH_3CHO][CH_3\cdot]$$
$$+ k_{III}[CH_3\cdot CO] - 2k_{IV}[CH_3\cdot]^2 = 0$$

and

$$\frac{d[CH_3\cdot CO]}{dt} = k_{II}[CH_3CHO][CH_3\cdot] - k_{III}[CH_3\cdot CO] = 0$$

Note that the quasi-steady hypothesis is applied to each free-radical species. This will generate as many algebraic equations as there are types of free radicals. The resulting set of equations is solved to express the free-radical concentrations in terms of the (presumably measurable) concentrations of the long-lived species. For the current example, the solutions for the free radicals are

$$[CH_3\cdot] = \sqrt{\frac{k_I[CH_3CHO]}{2k_{IV}}}$$

and

$$[CH_3\cdot CO] = (k_{II}/k_{III})\sqrt{\frac{k_I[CH_3CHO]^3}{2k_{IV}}}$$

The free-radical concentrations will be small—and the quasi-steady state hypothesis will be justified—whenever the initiation reaction is slow compared with the termination reaction, $k_I \ll k_{IV}[CH_3CHO]$.

Acetaldehyde is consumed by the initiation and propagation reactions.

$$\frac{-d[CH_3CHO]}{dt} = k_I[CH_3CHO] + k_{II}[CH_3CHO][CH_3\cdot]$$

The quasi-steady hypothesis allows the difficult-to-measure free-radical concentrations to be replaced by the more easily measured concentrations of the long-lived species. The result is

$$\frac{-d[CH_3CHO]}{dt} = k_I[CH_3CHO] + \left(\frac{k_{II}^2 k_I}{2k_{IV}}\right)^{1/2} [CH_3CHO]^{3/2}$$

MULTIPLE REACTIONS IN BATCH REACTORS

The first term in this result is due to consumption by the initiation reaction and is presumed to be small compared with consumption by the propagation reactions. Thus, the second term dominates, and the overall reaction has the form

$$A \rightarrow \text{Products} \qquad \mathcal{R} = ka^{3/2}$$

This agrees with experimental findings[1] on the decomposition of acetaldehyde. The appearance of the three-halves power is a wondrous result of the quasi-steady hypothesis. Half-integer kinetics are typical of free-radical systems. Example 2.6 describes a free-radical reaction with an apparent order of one-half, one, or three-halves depending on the termination mechanism.

Example 2.6: Apply the quasi-steady hypothesis to the monochlorination of a hydrocarbon. The initiation step is

$$Cl_2 \xrightarrow{k_I} 2Cl\cdot$$

The propagation reactions are

$$Cl\cdot + RH \xrightarrow{k_{II}} R\cdot + HCl$$

$$R\cdot + Cl_2 \xrightarrow{k_{III}} RCl + Cl\cdot$$

There are three possibilities for termination:

(a) $2Cl\cdot \xrightarrow{k_{IV}} Cl_2$

(b) $Cl\cdot + R\cdot \xrightarrow{k_{IV}} RCl$

(c) $2R\cdot \xrightarrow{k_{IV}} R_2$

Solution: The procedure is the same as in the acetaldehyde example. ODEs are written for each of the free-radical species, and their time derivatives are set to zero. The resulting algebraic equations are then solved for the free-radical concentrations. These values are substituted into the ODE governing RCl production. Depending on which termination mechanism is assumed, the solutions are

(a) $\mathcal{R} = k[Cl_2]^{1/2}[RH]$

(b) $\mathcal{R} = k[Cl_2][RH]^{1/2}$

(c) $\mathcal{R} = k[Cl_2]^{3/2}$

If two or three termination reactions are simultaneously important, an analytical solution for $\mathscr{R}$ is possible but complex. Laboratory results in such situations could probably be approximated as

$$\mathscr{R} = k[\text{Cl}_2]^m[\text{RH}]^n$$

where $1/2 < m < 3/2$ and $0 < n < 1$.

Example 2.7: Apply the quasi-steady hypothesis to the consecutive reactions in Equation (2.20), assuming $k_A \ll k_B$ and $k_A \ll k_C$.

Solution: The assumption of a near steady state is applied to components B and C. The ODEs become

$$\frac{da}{dt} = -k_A a$$
$$\frac{db}{dt} = -k_B b + k_A a = 0$$
$$\frac{dc}{dt} = -k_C c + k_B b = 0$$

The solutions are

$$a = a_0 e^{-k_A t}$$
$$b = \frac{k_A a}{k_B}$$
$$c = \frac{k_B b}{k_C}$$

This scheme can obviously be extended to larger sets of consecutive reactions provided that all the intermediate species are short-lived compared with the parent species, A. See Problem 2.9

Our treatment of chain reactions has been confined to relatively simple situations where the number of participating species and their possible reactions have been sharply bounded. Most free-radical reactions of industrial importance involve many more species. The set of possible reactions is unbounded in polymerizations, and it is perhaps bounded but very large in processes such as naptha cracking and combustion. Perhaps the elementary reactions can be postulated, but the rate constants are generally unknown. The quasi-steady hypothesis provides a functional form for the rate equations that can be used to fit experimental data.

2.5.4 Autocatalytic Reactions

As suggested by the name, the products of an autocatalytic reaction accelerate the rate of the reaction. For example, an acid-catalyzed reaction may produce acid. The rate of most reactions has an initial maximum and then decreases as reaction proceeds. Autocatalytic reactions have an initially increasing rate, although the rate must eventually decline as the reaction goes to completion. A model reaction frequently used to represent autocatalytic behavior is

$$A \rightarrow B + C$$

with an assumed mechanism of

$$A + B \xrightarrow{k} 2B + C \tag{2.27}$$

For a batch system,

$$\frac{da}{dt} = -kab = -ka(b_0 + a_0 - a) \tag{2.28}$$

This ODE has the solution

$$\frac{a}{a_0} = \frac{[1 + (b_0/a_0)]\exp\{-[1 + (b_0/a_0)]a_0kt\}}{(b_0/a_0) + \exp\{-[1 + (b_0/a_0)]a_0kt\}} \tag{2.29}$$

Figure 2.4 illustrates the course of the reaction for various values of b_0/a_0. Inflection points and S-shaped curves are characteristic of autocatalytic behavior. The reaction rate is initially low because the concentration of the catalyst, B, is low. Indeed, no reaction ever occurs if $b_0 = 0$. As B is formed, the rate accelerates and continues to increase so long as the term ab in Equation (2.28) is growing. Eventually, however, this term must decrease as component A is depleted, even though the concentration of B continues to increase. The inflection point is caused by depletion of component A.

Autocatalytic reactions often show higher conversions in a stirred tank than in either a batch flow reactor or a piston flow reactor with the same holding time, $t_{batch} = \bar{t}$. Since $\hat{a} = a_{out}$ in a CSTR, the catalyst, B, is present at the

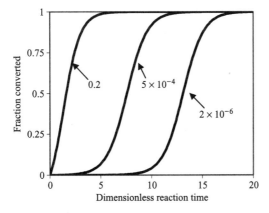

FIGURE 2.4 Conversion versus dimensionless time, a_0kt, for an autocatalytic batch reaction. The parameter is b_0/a_0.

same, high concentrations everywhere within the working volume of the reactor. In contrast, B may be quite low in concentration at early times in a batch reactor and only achieves its highest concentrations at the end of the reaction. Equivalently, the concentration of B is low near the inlet of a piston flow reactor and only achieves high values near the outlet. Thus, the average reaction rate in the CSTR can be higher.

The qualitative behavior shown in Figure 2.4 is characteristic of many systems, particularly biological ones, even though the reaction mechanism may not agree with Equation (2.27). An inflection point is observed in most batch fermentations. Polymerizations of vinyl monomers such as methyl methacrylate and styrene also show autocatalytic behavior when the undiluted monomers react by free-radical mechanisms. A polymerization exotherm for a methyl methacrylate casting system is shown in Figure 2.5. The reaction is approximately adiabatic so that the reaction exotherm provides a good measure of the extent of polymerization. The autocatalytic behavior is caused partially by concentration effects (the "gel effect" is discussed in Chapter 13) and partially by the exothermic nature of the reaction (temperature effects are discussed in Chapter 5). Indeed, heat can be considered a reaction product that accelerates the reaction, and adiabatic reactors frequently exhibit inflection points with respect to both temperature and composition. Autoacceleration also occurs in branching chain reactions where a single chain-propagating species can generate more than one new propagating species. Such reactions are obviously important in nuclear fission. They also occur in combustion processes. For example, the elementary reactions

$$H\cdot + O_2 \rightarrow HO\cdot + O\cdot$$

$$H_2 + O\cdot \rightarrow HO\cdot + H\cdot$$

are believed important in the burning of hydrogen.

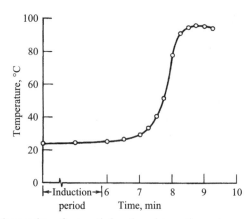

FIGURE 2.5 Reaction exotherm for a methyl methacrylate casting system.

Autocatalysis can cause sustained oscillations in batch systems. This idea originally met with skepticism. Some chemists believed that sustained oscillations would violate the second law of thermodynamics, but this is not true. Oscillating batch systems certainly exist, although they must have some external energy source or else the oscillations will eventually subside. An important example of an oscillating system is the circadian rhythm in animals. A simple model of a *chemical oscillator*, called the Lotka-Volterra reaction, has the assumed mechanism:

$$R + G \xrightarrow{k_I} 2R$$

$$L + R \xrightarrow{k_{II}} 2L$$

$$L \xrightarrow{k_{III}} D$$

Rabbits (R) eat grass (G) to form more rabbits. Lynx (L) eat rabbits to form more lynx. Also, lynx die of old age to form dead lynx (D). The grass is assumed to be in large excess and provides the energy needed to drive the oscillation. The corresponding set of ODEs is

$$\frac{dr}{dt} = k_I g r - k_{II} l r$$

$$\frac{dl}{dt} = k_{II} l r - k_{III} l$$

These equations are nonlinear and cannot be solved analytically. They are included in this section because they are autocatalytic and because this chapter discusses the numerical tools needed for their solution. Figure 2.6 illustrates one possible solution for the initial condition of 100 rabbits and 10 lynx. This model should not be taken too seriously since it represents no known chemistry or

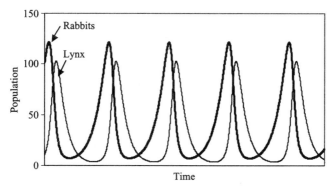

FIGURE 2.6 Population dynamics predicted by the Lotka-Volterra model for an initial population of 100 rabbits and 10 lynx.

ecology. It does show that a relatively simple set of first-order ODEs can lead to oscillations. These oscillations are strictly periodic if the grass supply is not depleted. If the grass is consumed, albeit slowly, both the amplitude and the frequency of the oscillations will decline toward an eventual steady state of no grass and no lynx.

A conceptually similar reaction, known as the Prigogine-Lefver or Brusselator reaction, consists of the following steps:

$$A \xrightarrow{k_I} X$$

$$B + X \xrightarrow{k_{II}} Y + D$$

$$2X + Y \xrightarrow{k_{III}} 3X$$

$$X \xrightarrow{k_{IV}} E$$

This reaction can oscillate in a well-mixed system. In a quiescent system, diffusion-limited spatial patterns can develop, but these violate the assumption of perfect mixing that is made in this chapter. A well-known chemical oscillator that also develops complex spatial patterns is the Belousov-Zhabotinsky or BZ reaction. Flame fronts and detonations are other batch reactions that violate the assumption of perfect mixing. Their analysis requires treatment of mass or thermal diffusion or the propagation of shock waves. Such reactions are briefly touched upon in Chapter 11 but, by and large, are beyond the scope of this book.

2.6 VARIABLE VOLUME BATCH REACTORS

2.6.1 Systems with Constant Mass

The feed is charged all at once to a batch reactor, and the products are removed together, with the mass in the system being held constant during the reaction step. Such reactors usually operate at nearly constant volume. The reason for this is that most batch reactors are liquid-phase reactors, and liquid densities tend to be insensitive to composition. The ideal batch reactor considered so far is perfectly mixed, isothermal, and operates at constant density. We now relax the assumption of constant density but retain the other simplifying assumptions of perfect mixing and isothermal operation.

The component balance for a variable-volume but otherwise ideal batch reactor can be written using moles rather than concentrations:

$$\frac{d(Va)}{dt} = \frac{dN_A}{dt} = V\mathscr{R}_A \qquad (2.30)$$

MULTIPLE REACTIONS IN BATCH REACTORS

where N_A is the number of moles of component A in the reactor. The initial condition associated with Equation (2.30) is that $N_A = (N_A)_0$ at $t = 0$. The case of a first-order reaction is especially simple:

$$\frac{dN_A}{dt} = -Vka = -kN_A$$

so that the solution is

$$N_A = (N_A)_0 e^{-kt} \qquad (2.31)$$

This is a more general version of Equation (1.24). For a first-order reaction, the number of molecules of the reactive component decreases exponentially with time. This is true whether or not the density is constant. If the density happens to be constant, the concentration of the reactive component also decreases exponentially as in Equation (1.24).

Example 2.8: Most polymers have densities appreciably higher than their monomers. Consider a polymer having a density of 1040 kg/m^3 that is formed from a monomer having a density of 900 kg/m^3. Suppose isothermal batch experiments require 2 h to reduce the monomer content to 20% by weight. What is the pseudo-first-order rate constant and what monomer content is predicted after 4 h?

Right Solution: Use a reactor charge of 900 kg as a basis and apply Equation (2.31) to obtain

$$Y_A = \frac{N_A}{(N_A)_0} = \frac{0.2(900)/M_A}{(900)/M_A} = 0.2 = \exp(-2k)$$

This gives $k = 0.8047 \text{ h}^{-1}$. The molecular weight of the monomer, M_A, is not actually used in the calculation. Extrapolation of the first-order kinetics to a 4-h batch predicts that there will be $900 \exp(-3.22) = 36 \text{ kg}$ or 4% by weight of monomer left unreacted. Note that the *fraction unreacted*, Y_A, must be defined as a ratio of moles rather than concentrations because the density varies during the reaction.

Wrong Solution: Assume that the concentration declines exponentially according to Equation (1.24). To calculate the concentration, we need the density. Assume it varies linearly with the weight fraction of monomer. Then $\rho = 1012 \text{ kg/m}^3$ at the end of the reaction. To calculate the monomer concentrations, use a basis of 1 m^3 of reacting mass. This gives

$$\frac{a}{a_0} = \frac{0.2(1012)/M_A}{900/M_A} = 0.225 = \exp(-2k) \quad \text{or} \quad k = 0.746$$

This concentration ratio does not follow the simple exponential decay of first-order kinetics and should not be used in fitting the rate constant. If it were used erroneously, the predicted concentration would be $45.6/M_A$

(kg·mol)/m^3 at the end of the 4 h reaction. The predicted monomer content after 4 h is 4.4% rather than 4.0% as more properly calculated. The difference is small but could be significant for the design of the monomer recovery and recycling system.

For reactions of order other than first, things are not so simple. For a second-order reaction,

$$\frac{d(Va)}{dt} = \frac{dN_A}{dt} = -Vka^2 = -\frac{kN_A^2}{V} = -\frac{kN_A^2 \rho}{\rho_0 V_0} \qquad (2.32)$$

Clearly, we must determine V or ρ as a function of composition. The integration will be easier if N_A is treated as the composition variable rather than a since this avoids expansion of the derivative as a product: $d(Va) = Vda + adV$. The numerical methods in subsequent chapters treat such products as composite variables to avoid expansion into individual derivatives. Here in Chapter 2, the composite variable, $N_A = Va$, has a natural interpretation as the number of moles in the batch system. To integrate Equation (2.32), V or ρ must be determined as a function of N_A. Both liquid- and gas-phase reactors are considered in the next few examples.

Example 2.9: Repeat Example 2.8 assuming that the polymerization is second order in monomer concentration. This assumption is appropriate for a binary polycondensation with good initial stoichiometry, while the pseudo-first-order assumption of Example 2.8 is typical of an addition polymerization.

Solution: Equation (2.32) applies, and ρ must be found as a function of N_A. A simple relationship is

$$\rho = 1040 - 140 N_A/(N_A)_0$$

The reader may confirm that this is identical to the linear relationship based on weight fractions used in Example 2.8. Now set $Y = N_A/(N_A)_0$. Equation (2.32) becomes

$$\frac{dY}{dt} = -k'Y^2 \left[\frac{1040 - 140Y}{900}\right]$$

where $k' = k(N_A)_0/V_0 = ka_0$. The initial condition is $Y=1$ at $t=0$. An analytical solution to this ODE is possible but messy. A numerical solution integrates the ODE for various values of k' until one is found that gives $Y = 0.2$ at $t = 2$. The result is $k' = 1.83$.

Example 2.10: Suppose $2A \xrightarrow{k/2} B$ in the liquid phase and that the density changes from ρ_0 to $\rho_\infty = \rho_0 + \Delta\rho$ upon complete conversion. Find an analytical solution to the batch design equation and compare the results with a hypothetical batch reactor in which the density is constant.

Solution: For a constant mass system,

$$\rho V = \rho_0 V_0 = \text{constant}$$

Assume, for lack of anything better, that the mass density varies linearly with the number of moles of A. Specifically, assume

$$\rho = \rho_\infty - \Delta\rho \left[\frac{N_A}{(N_A)_0} \right]$$

Substitution in Equation (2.32) gives

$$\frac{dN_A}{dt} = -kN_A^2 \left[\frac{\rho_\infty - \Delta\rho N_A/(N_A)_0}{\rho_0 V_0} \right]$$

This messy result apparently requires knowledge of five parameters: k, V_0, $(N_A)_0$, ρ_∞, and ρ_0. However, conversion to dimensionless variables usually reduces the number of parameters. In this case, set $Y = N_A/(N_A)_0$ (the fraction unreacted) and $\tau = t/t_{batch}$ (fractional batch time). Then algebra gives

$$\frac{dY}{d\tau} = \frac{-K^* Y^2 \rho_\infty - \Delta\rho Y}{\rho_0}$$

This contains the dimensionless rate constant, $K^* = a_0 k t_{batch}$, plus the initial and final densities. The comparable equation for reaction at constant density is

$$\frac{dY'}{d\tau} = -K^* Y'^2$$

where Y' would be the fraction unreacted if no density change occurred. Combining these results gives

$$\frac{dY'}{Y'^2} = -K^* d\tau = \frac{\rho_0 dY}{\rho_\infty - \Delta\rho Y Y^2}$$

or

$$\frac{dY'}{dY} = \frac{\rho_0 Y'^2}{\rho_\infty - \Delta\rho Y Y^2}$$

and even K^* drops out. There is a unique relationship between Y and Y' that depends only on ρ_∞ and ρ_0. The boundary condition associated with this ODE is $Y=1$ at $Y'=1$. An analytical solution is possible, but numerical integration of the ODE is easier. Euler's method works, but note that the indepen-dent variable Y' starts at 1.0 and is decreased in small steps until the desired final value is reached. A few results for the case of $\rho_\infty = 1000$ and $\rho_0 = 900$ are

Y	Y'
1.000	1.000
0.500	0.526
0.200	0.217
0.100	0.110
0.050	0.055
0.020	0.022
0.010	0.011

The density change in this example increases the reaction rate since the volume goes down and the concentration of the remaining A is higher than it would be if there were no density change. The effect is not large and would be negligible for many applications. When the real, variable-density reactor has a conversion of 50%, the hypothetical, constant-density reactor would have a conversion of 47.4% ($Y' = 0.526$).

Example 2.11: Suppose initially pure A dimerizes, $2A \xrightarrow{k/2} B$, isothermally in the gas phase at a constant pressure of 1 atm. Find a solution to the batch design equation and compare the results with a hypothetical batch reactor in which the reaction is $2A \to B + C$ so that there is no volume change upon reaction.

Solution: Equation (2.32) is the starting point, as in the previous example, but the ideal gas law is now used to determine V as a function of N_A:

$$V = [N_A + N_B]R_g T/P = \left[N_A + \frac{(N_A)_0 - N_A}{2}\right]R_g T/P$$

$$= \left[\frac{Y+1}{2}\right](N_A)_0 R_g T/P$$

$$= \left[\frac{Y+1}{2}\right]V_0$$

where Y is the fraction unreacted. Substitution into Equation (2.32) gives

$$\frac{dN_A}{dt} = (N_A)_0 \frac{dY}{dt} = \frac{-2kN_A^2}{V_0[Y+1]} = \frac{-2a_0 k Y^2 (N_A)_0}{[1+Y]}$$

Defining τ, K^*, and Y' as in Example 2.10 gives

$$\frac{dY'}{Y'^2} = -K^* d\tau = \frac{[Y+1]dY}{2Y^2}$$

An analytical solution is again possible but messy. A few results are

Y	Y'
1.000	1.000
0.500	0.542
0.200	0.263
0.100	0.150
0.050	0.083
0.020	0.036
0.010	0.019

The effect of the density change is larger than in the previous example, but is still not major. Note that most gaseous systems will have substantial amounts of inerts (e.g. nitrogen) that will mitigate volume changes at constant pressure.

The general conclusion is that density changes are of minor importance in liquid systems and of only moderate importance in gaseous systems at constant pressure. When they are important, the necessary calculations for a batch reactor are easier if compositions are expressed in terms of total moles rather than molar concentrations.

We have considered volume changes resulting from density changes in liquid and gaseous systems. These volume changes were *thermodynamically determined* using an equation of state for the fluid that specifies volume or density as a function of composition, pressure, temperature, and any other state variable that may be important. This is the usual case in chemical engineering problems. In Example 2.10, the density depended only on the composition. In Example 2.11, the density depended on composition and pressure, but the pressure was specified.

Volume changes also can be *mechanically determined*, as in the combustion cycle of a piston engine. If $V = V(t)$ is an explicit function of time, Equations like (2.32) are then variable-separable and are relatively easy to integrate, either alone or simultaneously with other component balances. Note, however, that reaction rates can become dependent on pressure under extreme conditions. See Problem 5.4. Also, the results will not really apply to car engines since mixing of air and fuel is relatively slow, flame propagation is important, and the spatial distribution of the reaction must be considered. The cylinder head is not perfectly mixed.

It is possible that the volume is determined by a combination of thermodynamics and mechanics. An example is reaction in an elastic balloon. See Problem 2.20.

The examples in this section have treated a single, second-order reaction, although the approach can be generalized to multiple reactions with arbitrary kinetics. Equation (2.30) can be written for each component:

$$\frac{d(Va)}{dt} = \frac{dN_A}{dt} = V\mathscr{R}_A(a,b,\ldots) = V\mathscr{R}_A(N_A, N_B, \ldots, V)$$
$$\frac{d(Vb)}{dt} = \frac{dN_B}{dt} = V\mathscr{R}_B(a,b,\ldots) = V\mathscr{R}_B(N_A, N_B, \ldots, V)$$
(2.33)

and so on for components C, D, ... An auxiliary equation is used to determine V. The auxiliary equation is normally an algebraic equation rather than an ODE. In chemical engineering problems, it will usually be an equation of state, such as the ideal gas law. In any case, the set of ODEs can be integrated numerically starting with known initial conditions, and V can be calculated and updated as necessary. Using Euler's method, V is determined at each time step using the "old" values for $N_A, N_B, \ldots$. This method of integrating sets of ODEs with various auxiliary equations is discussed more fully in Chapter 3.

2.6.2 Fed-Batch Reactors

Many industrial reactors operate in the *fed-batch* mode. It is also called the *semibatch* mode. In this mode of operation, reactants are charged to the system at various times, and products are removed at various times. Occasionally, a *heel* of material from a previous batch is retained to start the new batch.

There are a variety of reasons for operating in a semibatch mode. Some typical ones are as follows:

1. A starting material is subjected to several different reactions, one after the other. Each reaction is essentially independent, but it is convenient to use the same vessel.
2. Reaction starts as soon as the reactants come into contact during the charging process. The initial reaction environment differs depending on whether the reactants are charged sequentially or simultaneously.
3. One reactant is charged to the reactor in small increments to control the composition distribution of the product. Vinyl copolymerizations discussed in Chapter 13 are typical examples. Incremental addition may also be used to control the reaction exotherm.
4. A by-product comes out of solution or is intentionally removed to avoid an equilibrium limitation.
5. One reactant is sparingly soluble in the reaction phase and would be depleted were it not added continuously. Oxygen used in an aerobic fermentation is a typical example.
6. The heel contains a biocatalyst (e.g., yeast cells) needed for the next batch.

All but the first of these has chemical reaction occurring simultaneously with mixing or mass transfer. A general treatment requires the combination of transport equations with the chemical kinetics, and it becomes necessary to solve sets of partial differential equations rather than ordinary differential equations. Although this approach is becoming common in continuous flow systems, it remains difficult in batch systems. The central difficulty is in developing good equations for the mixing and mass transfer steps.

The difficulty disappears when the mixing and mass transfer steps are fast compared with the reaction steps. The contents of the reactor remain perfectly mixed even while new ingredients are being added. Compositions and reaction rates will be spatially uniform, and a flow term is simply added to the mass balance. Instead of Equation (2.30), we write

$$\frac{dN_A}{dt} = (Qa)_{in} + V\mathcal{R}_A(N_A, N_B, \ldots, V) \qquad (2.34)$$

where the term $(Qa)_{in}$ represents the molar flow rate of A into the reactor. A fed-batch reactor is an example of the unsteady, variable-volume CSTRs treated in Chapter 14, and solutions to Equation (2.34) are considered there. However, fed-batch reactors are amenable to the methods of this chapter if the charging and discharging steps are fast compared with reaction times. In this special case, the fed-batch reactor becomes a sequence of ideal batch reactors that are reinitialized after each charging or discharging step.

Many semibatch reactions involve more than one phase and are thus classified as *heterogeneous*. Examples are aerobic fermentations, where oxygen is supplied continuously to a liquid substrate, and chemical vapor deposition reactors, where gaseous reactants are supplied continuously to a solid substrate. Typically, the overall reaction rate will be limited by the rate of interphase mass transfer. Such systems are treated using the methods of Chapters 10 and 11. Occasionally, the reaction will be kinetically limited so that the transferred component saturates the reaction phase. The system can then be treated as a batch reaction, with the concentration of the transferred component being dictated by its solubility. The early stages of a batch fermentation will behave in this fashion, but will shift to a mass transfer limitation as the cell mass and thus the oxygen demand increase.

2.7 SCALEUP OF BATCH REACTIONS

Section 1.5 described one basic problem of scaling batch reactors; namely, it is impossible to maintain a constant mixing time if the scaleup ratio is large. However, this is a problem for fed-batch reactors and does not pose a limitation if the reactants are premixed. A single-phase, isothermal (or adiabatic) reaction in batch can be scaled indefinitely if the reactants are premixed and preheated before being charged. The restriction to single-phase systems avoids mass transfer limitations; the restriction to isothermal or, more realistically, adiabatic, systems avoids heat transfer limitations; and the requirement for premixing eliminates concerns about mixing time. All the reactants are mixed initially, the reaction treats all molecules equally, and the agitator may as well be turned off. Thus, within the literal constraints of this chapter, scaleup is not a problem. It is usually possible to preheat and premix the feed streams as they enter the reactor and, indeed, to fill the reactor in a time substantially less than the reaction half-life. Unfortunately, as we shall see in other chapters, real systems can be more complicated. Heat and mass transfer limitations are common. If there is an agitator, it probably has a purpose.

One purpose of the agitator may be to premix the contents after they are charged rather than on the way in. When does this approach, which violates the strict assumptions of an ideal batch reactor, lead to practical scaleup problems? The simple answer is whenever the mixing time, as described in Section 1.5, becomes commensurate with the reaction half-life. If the mixing time threatens to become limiting upon scaleup, try moving the mixing step to the transfer piping.

Section 5.3 discusses a variety of techniques for avoiding scaleup problems. The above paragraphs describe the simplest of these techniques. Mixing, mass transfer, and heat transfer all become more difficult as size increases. To avoid limitations, avoid these steps. Use premixed feed with enough inerts so that the reaction stays single phase and the reactor can be operated adiabatically. This simplistic approach is occasionally possible and even economical.

2.8 STOICHIOMETRY AND REACTION COORDINATES

The numerical methods in this book can be applied to all components in the system, even inerts. When the reaction rates are formulated using Equation (2.8), the solutions automatically account for the stoichiometry of the reaction. We have not always followed this approach. For example, several of the examples have ignored product concentrations when they do not affect reaction rates and when they are easily found from the amount of reactants consumed. Also, some of the analytical solutions have used stoichiometry directly to ease the algebra. This section formalizes the use of stoichiometric constraints.

2.8.1 Stoichiometry of Single Reactions

The general stoichiometric relationships for a single reaction in a batch reactor are

$$\frac{N_A - (N_A)_0}{\nu_A} = \frac{N_B - (N_B)_0}{\nu_B} = \cdots \qquad (2.35)$$

where N_A is the number of moles present in the system at any time. Divide Equation (2.35) by the volume to obtain

$$\frac{\hat{a} - a_0}{\nu_A} = \frac{\hat{b} - b_0}{\nu_B} = \cdots \qquad (2.36)$$

The circumflex over a and b allows for spatial variations. It can be ignored when the contents are perfectly mixed. Equation (2.36) is the form normally used for batch reactors where $\hat{a} = a(t)$. It can be applied to piston flow reactors by setting $a_0 = a_{in}$ and $\hat{a} = a(z)$, and to CSTRs by setting $a_0 = a_{in}$ and $\hat{a} = a_{out}$.

There are two uses for Equation (2.36). The first is to calculate the concentration of components at the end of a batch reaction cycle or at the outlet of a flow reactor. These equations are used for components that do not affect the reaction rate. They are valid for batch and flow systems of arbitrary complexity if the circumflexes in Equation (2.36) are retained. Whether or not there are spatial variations within the reactor makes no difference when $\hat{a}$ and $\hat{b}$ are averages over the entire reactor or over the exiting flow stream. All reactors satisfy *global stoichiometry*.

The second use of Equations (2.36) is to eliminate some of the composition variables from rate expressions. For example, $\mathcal{R}_A(a,b)$ can be converted to $\mathcal{R}_A(a)$ if Equation (2.36) can be applied to each and every point in the reactor. Reactors for which this is possible are said to *preserve local stoichiometry*. This does not apply to real reactors if there are internal mixing or separation processes, such as molecular diffusion, that distinguish between types of molecules. Neither does it apply to multiple reactions, although this restriction can be relaxed through use of the reaction coordinate method described in the next section.

2.8.2 Stoichiometry of Multiple Reactions

Consider a system with N chemical components undergoing a set of M reactions. Obviously, $N > M$. Define the $N \times M$ matrix of stoichiometric coefficients as

$$\mathbf{v} = \begin{pmatrix} \nu_{A,I} & \nu_{A,II} & \cdots \\ \nu_{B,I} & \nu_{B,II} & \\ \vdots & & \ddots \end{pmatrix} \qquad (2.37)$$

Note that the matrix of stoichiometric coefficients devotes a row to each of the N components and a column to each of the M reactions. We require the reactions to be *independent*. A set of reactions is independent if no member of the set can be obtained by adding or subtracting multiples of the other members. A set will be independent if every reaction contains one species not present in the other reactions. The student of linear algebra will understand that the *rank* of $\mathbf{v}$ must equal M.

Using **v**, we can write the design equations for a batch reactor in very compact form:

$$\frac{d(\mathbf{a}V)}{dt} = \mathbf{v}\mathfrak{R}V \qquad (2.38)$$

where **a** is the vector ($N \times 1$ matrix) of component concentrations and $\mathfrak{R}$ is the vector ($M \times 1$ matrix) of reaction rates.

Example 2.12: Consider a constant-volume batch reaction with the following set of reactions:

$$\begin{aligned} A + 2B &\to C & \mathfrak{R}_I &= k_I a \\ A + C &\to D & \mathfrak{R}_{II} &= k_{II} ac \\ B + C &\to E & \mathfrak{R}_{III} &= k_{III} c \end{aligned}$$

These reaction rates would be plausible if B were present in great excess, say as water in an aqueous reaction. Equation (2.38) can be written out as

$$\frac{d}{dt}\begin{pmatrix} a \\ b \\ c \\ d \\ e \end{pmatrix} = \begin{pmatrix} -1 & -1 & 0 \\ -2 & 0 & -1 \\ +1 & -1 & -1 \\ 0 & +1 & 0 \\ 0 & 0 & +1 \end{pmatrix} \begin{pmatrix} k_I a \\ k_{II} ac \\ k_{III} c \end{pmatrix}$$

Expanding this result gives the following set of ODEs:

$$\begin{aligned} \frac{da}{dt} &= -k_I a - k_{II} ac \\ \frac{db}{dt} &= -2k_I a \qquad\qquad - k_{III} c \\ \frac{dc}{dt} &= +k_I a - k_{II} ac - k_{III} c \\ \frac{dd}{dt} &= \qquad\qquad k_{II} ac \\ \frac{de}{dt} &= \qquad\qquad\qquad +k_{III} c \end{aligned}$$

There are five equations in five unknown concentrations. The set is easily solved by numerical methods, and the stoichiometry has already been incorporated. However, it is not the smallest set of ODEs that can be solved to determine the five concentrations. The first three equations contain only a, b, and c as unknowns and can thus be solved independently of the other two equations. The effective dimensionality of the set is only 3.

Example 2.12 illustrates a general result. If local stoichiometry is preserved, no more than M reactor design equations need to be solved to determine all

N concentrations. Years ago, this fact was useful since numerical solutions to ODEs required substantial computer time. They can now be solved in literally the blink of an eye, and there is little incentive to reduce dimensionality in sets of ODEs. However, the theory used to reduce dimensionality also gives global stoichiometric equations that can be useful. We will therefore present it briefly.

The *extent of reaction* or *reaction coordinate*, ε is defined by

$$\hat{\mathbf{N}} - \hat{\mathbf{N}}_0 = \mathbf{v}\boldsymbol{\varepsilon} \tag{2.39}$$

where $\hat{\mathbf{N}}$ and $\hat{\mathbf{N}}_0$ are column vectors ($N \times 1$ matrices) giving the final and initial numbers of moles of each component and $\boldsymbol{\varepsilon}$ is the reaction coordinate vector ($M \times 1$ matrix). In more explicit form,

$$\begin{pmatrix} \hat{N}_A \\ \hat{N}_B \\ \vdots \end{pmatrix} - \begin{pmatrix} \hat{N}_A \\ \hat{N}_B \\ \vdots \end{pmatrix}_0 = \begin{pmatrix} v_{A,I} & v_{A,II} & \cdots \\ v_{B,I} & v_{B,II} & \\ \vdots & & \ddots \end{pmatrix} \begin{pmatrix} \varepsilon_I \\ \varepsilon_{II} \\ \vdots \end{pmatrix} \tag{2.40}$$

Equation (2.39) is a generalization to M reactions of the stoichiometric constraints of Equation (2.35). If the vector $\boldsymbol{\varepsilon}$ is known, the amounts of all N components that are consumed or formed by the reaction can be calculated.

What is needed now is some means for calculating $\boldsymbol{\varepsilon}$. To do this, it is useful to consider some component, H, which is formed only by Reaction I, which does not appear in the feed, and which has a stoichiometric coefficient of $v_{H,I} = 1$ for Reaction I and stoichiometric coefficients of zero for all other reactions. It is always possible to write the chemical equation for Reaction I so that a real product has a stoichiometric coefficient of $+1$. For example, the decomposition of ozone, $2O_3 \to 3O_2$, can be rewritten as $2/3 O_3 \to O_2$. However, you may prefer to maintain integer coefficients. Also, it is necessary that H not occur in the feed, that there is a unique H for each reaction, and that H participates only in the reaction that forms it. Think of H as a kind of chemical neutrino formed by the particular reaction. Since H participates only in Reaction I and does not occur in the feed, Equation (2.40) gives

$$N_H = \varepsilon_I$$

The batch reactor equation gives

$$\frac{d(Vh)}{dt} = \frac{d(N_H)}{dt} = \frac{d\varepsilon_I}{dt} = V \mathscr{R}_I(N_A, N_B, \ldots, V) = V \mathscr{R}_I(\varepsilon_I, \varepsilon_{II}, \ldots, V) \tag{2.41}$$

The conversion from $\mathscr{R}_I(N_A, N_B, \ldots, V)$ to $\mathscr{R}_I(\varepsilon_I, \varepsilon_{II}, \ldots, V)$ is carried out using the algebraic equations obtained from Equation (2.40). The initial condition associated with Equation (2.41) is each $\varepsilon_I = 0$ at $t = 0$. We now consider a different H for each of the M reactions, giving

$$\frac{d\boldsymbol{\varepsilon}}{dt} = V\boldsymbol{\Re} \quad \text{where} \quad \boldsymbol{\varepsilon} = \mathbf{0} \quad \text{at} \quad t = 0 \tag{2.42}$$

Equation (2.42) represents a set of M ODEs in M independent variables, $\varepsilon_I, \varepsilon_{II}, \ldots$ It, like the redundant set of ODEs in Equation (2.38), will normally require numerical solution. Once solved, the values for the ε can be used to calculate the N composition variables using Equation (2.40).

Example 2.13: Apply the reaction coordinate method to the reactions in Example 2.12.

Solution: Equation (2.42) for this set is

$$\frac{d}{dt}\begin{pmatrix} \varepsilon_I \\ \varepsilon_{II} \\ \varepsilon_{III} \end{pmatrix} = V\begin{pmatrix} k_I a \\ k_{II} ac \\ k_{III} c \end{pmatrix} = \begin{pmatrix} k_I N_A \\ k_I N_A N_C / V \\ k_{III} N_C \end{pmatrix} \tag{2.43}$$

Equation (2.40) can be written out for this reaction set to give

$$\begin{aligned}
N_A - (N_A)_0 &= -\varepsilon_I - \varepsilon_{II} \\
N_B - (N_B)_0 &= -2\varepsilon_I \qquad\quad - \varepsilon_{III} \\
N_C - (N_C)_0 &= +\varepsilon_I - \varepsilon_{II} \quad - \varepsilon_{III} \\
N_D - (N_D)_0 &= \qquad\quad +\varepsilon_{II} \\
N_E - (N_E)_0 &= \qquad\qquad\qquad\quad +\varepsilon_{III}
\end{aligned} \tag{2.44}$$

The first three of these equations are used to eliminate N_A, N_B, and N_C from Equation (2.43). The result is

$$\begin{aligned}
\frac{d\varepsilon_I}{dt} &= k_I[(N_A)_0 - \varepsilon_I - \varepsilon_{II}] \\
\frac{d\varepsilon_{II}}{dt} &= \frac{k_{II}}{V}[(N_A)_0 - \varepsilon_I - \varepsilon_{II}][(N_C)_0 + \varepsilon_I - \varepsilon_{II} - \varepsilon_{III}] \\
\frac{d\varepsilon_{III}}{dt} &= k_{III}[(N_C)_0 + \varepsilon_I - \varepsilon_{II} - \varepsilon_{III}]
\end{aligned}$$

Integrate these out to time t_{batch} and then use Equations (2.44) to evaluate $N_A, \ldots, N_E$. The corresponding concentrations can be found by dividing by $V(t_{batch})$.

In a formal sense, Equation (2.38) applies to all batch reactor problems. So does Equation (2.42) combined with Equation (2.40). These equations are perfectly general when the reactor volume is well mixed and the various components are quickly charged. They do not require the assumption of constant reactor volume. If the volume does vary, ancillary, algebraic equations are needed as discussed in Section 2.6.1. The usual case is a thermodynamically imposed volume change. Then, an equation of state is needed to calculate the density.

PROBLEMS

2.1. The following reactions are occurring in a constant-volume, isothermal batch reactor:

$$A + B \xrightarrow{k_I} C$$

$$B + C \xrightarrow{k_{II}} D$$

Parameters for the reactions are $a_0 = b_0 = 10 \, \text{mol/m}^3$, $c_0 = d_0 = 0$, $k_I = k_{II} = 0.01 \, \text{m}^3/(\text{mol} \cdot \text{h})$, $t_{batch} = 4 \, \text{h}$.
(a) Find the concentration of C at the end of the batch cycle.
(b) Find a general relationship between the concentrations of A and C when that of C is at a maximum.

2.2. The following kinetic scheme is postulated for a batch reaction:

$$A + B \rightarrow C \qquad \mathscr{R}_I = k_I a^{1/2} b$$
$$B + C \rightarrow D \qquad \mathscr{R}_{II} = k_{II} c^{1/2} b$$

Determine a, b, c and d as functions of time. Continue your calculations until the limiting reagent is 90% consumed given $a_0 = 10 \, \text{mol/m}^3$, $b_0 = 2 \, \text{mol/m}^3$, $c_0 = d_0 = 0$, $k_I = k_{II} = 0.02 \, \text{m}^{3/2}/(\text{mol}^{1/2} \cdot \text{s})$.

2.3. Refer to Example 2.5. Prepare the plot referred to in the last sentence of that example. Assume $k_{II}/k_I = 0.1$.

2.4. Dimethyl ether thermally decomposes at temperatures above 450°C. The predominant reaction is

$$CH_3OCH_3 \rightarrow CH_4 + H_2 + CO$$

Suppose a homogeneous, gas-phase reaction occurs in a constant-volume batch reactor. Assume ideal gas behavior and suppose pure A is charged to the reactor.

(a) Show how the reaction rate can be determined from pressure measurements. Specifically, relate $\mathscr{R}$ to dP/dt.

(b) Determine $P(t)$, assuming that the decomposition is first order.

2.5. The first step in manufacturing polyethylene terephthalate is to react terephthalic acid with a large excess of ethylene glycol to form diglycol terephthalate:

$$HOOC-\phi-COOH + 2HOCH_2CH_2OH \rightarrow$$
$$HOCH_2CH_2OOC-\phi-COOCH_2CH_2OH + 2H_2O$$

Derive a plausible kinetic model for this reaction. Be sure your model reflects the need for the large excess of glycol. This need is inherent in the chemistry if you wish to avoid by-products.

2.6. Consider the liquid-phase reaction of a diacid with a diol, the first reaction step being

$$HO-R-OH + HOOC-R'-COOH \rightarrow HO-ROOCR'-COOH + H_2O$$

Suppose the desired product is the single-step mixed acidol as shown above. A large excess of the diol is used, and batch reactions are conducted to determine experimentally the reaction time, t_{max}, which maximizes the yield of acidol. Devise a kinetic model for the system and explain how the parameters in this model can be fit to the experimental data.

2.7. The exponential function can be defined as a limit:

$$\lim_{m \to \infty} \left(1 + \frac{z}{m}\right)^m = e^z$$

Use this fact to show that Equation (2.17) becomes Equation (2.16) in the limit as $n \to \infty$.

2.8. Determine the maximum batch reactor yield of B for a reversible, first-order reaction:

$$A \underset{k_r}{\overset{k_f}{\rightleftharpoons}} B$$

Do not assume $b_0 = 0$. Instead, your answer will depend on the amount of B initially present.

2.9. Start with 1 mol of ^{238}U and let it age for 10 billion years or so. Refer to Table 2.1. What is the maximum number of atoms of ^{214}Po that will ever exist? Warning! This problem is monstrously difficult to solve by brute force methods. A long but straightforward analytical solution is possible. See also Section 2.5.3 for a shortcut method.

2.10. Consider the consecutive reactions

$$A \overset{k}{\rightarrow} B \overset{k}{\rightarrow} C$$

where the two reactions have equal rates. Find $b(t)$.

2.11. Find the batch reaction time that maximizes the concentration of component B in Problem 2.10. You may begin with the solution of Problem 2.10 or with Equation (2.23).

2.12. Find $c(t)$ for the consecutive, first-order reactions of Equation (2.20) given that $k_B = k_C$.

2.13. Determine the batch reaction time that maximizes the concentration of component C in Equation (2.20) given that $k_A = 1\,\mathrm{h}^{-1}$, $k_B = 0.5\,\mathrm{h}^{-1}$, $k_C = 0.25\,\mathrm{h}^{-1}$, $k_D = 0.125\,\mathrm{h}^{-1}$.

2.14. Consider the sequential reactions of Equation (2.20) and suppose $b_0 = c_0 = d_0 = 0$, $k_I = 3\,\mathrm{h}^{-1}$, $k_{II} = 2\,\mathrm{h}^{-1}$, $k_{III} = 4\,\mathrm{h}^{-1}$. Determine the ratios a/a_0, b/a_0, c/a_0, and d/a_0, when the batch reaction time is chosen such that

(a) The final concentration of A is maximized.
(b) The final concentration of B is maximized.
(c) The final concentration of C is maximized.
(d) The final concentration of D is maximized.

2.15. Find the value of the dimensionless batch reaction time, $k_f t_{batch}$, that maximizes the concentration of B for the following reactions:

$$A \underset{k_r}{\overset{k_f}{\rightleftarrows}} B \overset{k_B}{\longrightarrow} C$$

Compare this maximum value for b with the value for b obtained using the quasi-steady hypothesis. Try several cases: (a) $k_r = k_B = 10k_f$, (b) $k_r = k_B = 20k_f$, (c) $k_r = 2k_B = 10k_f$.

2.16. The bromine–hydrogen reaction

$$Br_2 + H_2 \rightarrow 2HBr$$

is believed to proceed by the following elementary reactions:

$$Br_2 + M \underset{k_{-I}}{\overset{k_I}{\rightleftarrows}} 2Br\cdot + M \quad \text{(I)}$$

$$Br\cdot + H_2 \underset{k_{-II}}{\overset{k_{II}}{\rightleftarrows}} HBr + H\cdot \quad \text{(II)}$$

$$H\cdot + Br_2 \overset{k_{III}}{\longrightarrow} HBr + Br\cdot \quad \text{(III)}$$

The initiation step, Reaction (I), represents the thermal dissociation of bromine, which is brought about by collision with any other molecule, denoted by M.

(a) The only termination reaction is the reverse of the initiation step and is third order. Apply the quasi-steady hypothesis to [Br•] and [H•] to obtain

$$\mathscr{R} = \frac{k[H_2][Br_2]^{3/2}}{[Br_2] + k_A[HBr]}$$

(b) What is the result if the reverse reaction (I) does not exist and termination is second order, $2Br• \rightarrow Br_2$?

2.17. A proposed mechanism for the thermal cracking of ethane is

$$C_2H_6 + M \xrightarrow{k_I} 2CH_3• + M$$

$$CH_3• + C_2H_6 \xrightarrow{k_{II}} CH_4 + C_2H_5•$$

$$C_2H_5• \xrightarrow{k_{III}} C_2H_4 + H•$$

$$H• + C_2H_6 \xrightarrow{k_{IV}} H_2 + C_2H_5•$$

$$2C_2H_5• \xrightarrow{k_V} C_4H_{10}$$

The overall reaction has variable stoichiometry:

$$C_2H_6 \rightarrow v_B C_2H_4 + v_C C_4H_{10} + (2 - 2v_B - 4v_C)CH_4 \\ + (-1 + 2v_B + 3v_C)H_2$$

The free-radical concentrations are small and are ignored in this equation for the overall reaction.

(a) Apply the quasi-steady hypothesis to obtain an expression for the disappearance of ethane.
(b) What does the quasi-steady hypothesis predict for v_B and v_C?
(c) Ethylene is the desired product. Which is better for this gas-phase reaction, high or low pressure?

2.18. The Lotka-Volterra reaction described in Section 2.5.4 has three initial conditions—one each for grass, rabbits, and lynx—all of which must be positive. There are three rate constants assuming the supply of grass is not depleted. Use dimensionless variables to reduce the number of independent parameters to four. Pick values for these that lead to a sustained oscillation. Then, vary the parameter governing the grass supply and determine how this affects the period and amplitude of the solution.

2.19. It is proposed to study the hydrogenation of ethylene

$$C_2H_4 + H_2 \rightarrow C_2H_6$$

in a constant-pressure, gas-phase batch reactor. Derive an expression for the reactor volume as a function of time, assuming second-order kinetics, ideal gas behavior, perfect stoichiometry, and 50% inerts by volume at $t=0$.

2.20. Suppose a rubber balloon is filled with a gas mixture and that one of the following reactions occurs:

$$2A \xrightarrow{k} B$$

$$A + B \xrightarrow{k} C$$

$$A \xrightarrow{k} B + C$$

Determine $V(t)$.

Hint 1: The pressure difference between the inside and the outside of the balloon must be balanced by the stress in the fabric of the balloon so that $\pi R^2 \, \Delta P = 2\pi R h \sigma$ where h is the thickness of the fabric and σ is the stress.

Hint 2: Assume that the density of the fabric is constant so that $4\pi R^2 h = 4\pi R_0^2 h_0$.

Hint 3: Assume that the fabric is perfectly elastic so that stress is proportional to strain (Hooke's law).

Hint 4: The ideal gas law applies.

2.21. A numerical integration scheme has produced the following results:

Δz	Integral
1.0	0.23749
0.5	0.20108
0.25	0.19298
0.125	0.19104
0.0625	0.19056

(a) What is the apparent order of convergence?
(b) Extrapolate the results to $\Delta z = 0$. (*Note*: Such extrapolation should not be done unless the integration scheme has a theoretical order of convergence that agrees with the apparent order. Assume that it does.)
(c) What value for the integral would you expect at $\Delta z = 1/32$?

2.22. See Example 2.14 in Appendix 2.
(a) Write chemical equations that will give the ODEs of that example.
(b) Rumor has it that there is an error in the Runge-Kutta calculations for the case of $\Delta t = 0.5$. Write or acquire the necessary computer code and confirm or deny the rumor.

2.23. The usual method of testing for convergence and of extrapolating to zero step size assumes that the step size is halved in successive calculations. Example 2.4 quarters the step size. Develop an extrapolation technique for this procedure. Test it using the data in Example 2.15 in Appendix 2.

REFERENCE

1. Boyer, A., Niclause, M., and Letort, M., "Cinetique de pyrolyse des aldehydes aliphatiques en phase gazeuse," *J. Chim. Phys.*, **49**, 345–353 (1952).

SUGGESTIONS FOR FURTHER READING

Most undergraduate texts on physical chemistry give a survey of chemical kinetics and reaction mechanisms. A comprehensive treatment is provided in

Benson, S. W., *Foundations of Chemical Kinetics*, McGraw-Hill, New York, 1960.

A briefer and more recent description is found in

Espenson, J. H., Ed., *Chemical Kinetics and Reaction Mechanisms*, McGraw-Hill, New York, 1995.

A recent, comprehensive treatment of chemical oscillators and assorted esoterica is given in

Epstein, I. R. and Pojman, J. A., Eds., *An Introduction to Nonlinear Chemical Dynamics: Oscillations, Waves, Patterns, and Chaos*, Oxford University Press, New York, 1998.

A classic, mathematically oriented work has been reprinted as a paperback:

Aris, R., *Elementary Chemical Reactor Analysis*, Dover, Mineola, NY, 2000.

An account of the reaction coordinate method as applied to chemical equilibrium is given in Chapter 14 of

Smith, J. M., Van Ness, H. C., and Abbott, M. M., *Introduction to Chemical Engineering Thermodynamics*, 6th ed., McGraw-Hill, New York, 2001.

The Internet has become a wonderful source of (sometimes free) software for numerical analysis. Browse through it, and you will soon see that Fortran remains the programming language for serious numerical computation. One excellent book that is currently available without charge is

Press, W. H., Teukolsky, S. A., Vetterling, W. T., and Flannery, B. P., *Numerical Recipes in Fortran 77: The Art of Scientific Computing*, Vol. 1, 2nd ed., Cambridge University Press, New York, 1992.

This book describes and gives Fortran subroutines for a wide variety of ODE solvers. More to the point, it gives numerical recipes for practically anything you will ever need to compute. Volume 2 is also available online. It discusses Fortran 90 in the context of parallel computing. C, Pascal, and Basic versions of Volume 1 can be purchased.

APPENDIX 2: NUMERICAL SOLUTION OF ORDINARY DIFFERENTIAL EQUATIONS

In this chapter we described Euler's method for solving sets of ordinary differential equations. The method is extremely simple from a conceptual and programming viewpoint. It is computationally inefficient in the sense that a great many arithmetic operations are necessary to produce accurate solutions. More efficient techniques should be used when the same set of equations is to be solved many times, as in optimization studies. One such technique, *fourth-order Runge-Kutta*, has proved very popular and can be generally recommended for all but very stiff sets of first-order ordinary differential equations. The set of equations to be solved is

$$\frac{da}{dt} = \mathscr{R}_A(a, b, \ldots, t)$$
$$\frac{db}{dt} = \mathscr{R}_B(a, b, \ldots, t) \tag{2.45}$$
$$\vdots \quad \vdots$$

A value of Δt is selected, and values for Δa, Δb, ... are estimated by evaluating the functions $\mathscr{R}_A$, $\mathscr{R}_B$ .... In Euler's method, this evaluation is done at the initial point $(a_0, b_0, \ldots, t_0)$ so that the estimate for Δa is just $\Delta t \mathscr{R}_A(a_0, b_0, \ldots, t_0) = \Delta t (\mathscr{R}_A)_0$. In fourth-order Runge-Kutta, the evaluation is done at four points and the estimates for Δa, Δb, ... are based on weighted averages of the $\mathscr{R}_A$, $\mathscr{R}_B$, ... at these four points:

$$\Delta a = \Delta t \frac{(\mathscr{R}_A)_0 + 2(\mathscr{R}_A)_1 + 2(\mathscr{R}_A)_2 + (\mathscr{R}_A)_3}{6}$$

$$\Delta b = \Delta t \frac{(\mathscr{R}_B)_0 + 2(\mathscr{R}_B)_1 + 2(\mathscr{R}_B)_2 + (\mathscr{R}_B)_3}{6} \tag{2.46}$$

$$\vdots \quad \vdots$$

where the various $\mathscr{R}$s are evaluated at the points

$$a_1 = a_0 + \Delta t (\mathscr{R}_A)_0/2$$
$$a_2 = a_0 + \Delta t (\mathscr{R}_A)_1/2 \tag{2.47}$$
$$a_3 = a_0 + \Delta t (\mathscr{R}_A)_2$$

with similar equations for b_1, b_2, b_3, and so on. Time rarely appears explicitly in the $\mathscr{R}$, but, should it appear,

$$t_1 = t_0 + \Delta t/2$$
$$t_2 = t_1 \tag{2.48}$$
$$t_3 = t_0 + \Delta t$$

Example 2.14: Use fourth-order Runge-Kutta integration to solve the following set of ODEs:

$$\frac{da}{dt} = -k_1 a^2$$

$$\frac{db}{dt} = +k_1 a^2 - k_2 bc$$

$$\frac{dc}{dt} = -k_2 bc$$

Use $a_0 = c_0 = 30$, $b_0 = 0$, $k_1 = 0.01$, $k_2 = 0.02$. Find a, b, and c for $t = 1$.

Solution: The coding is left to the reader, but if you really need a worked example of the Runge-Kutta integration, check out Example 6.4. The following are detailed results for $\Delta t = 1.0$, which means that only one step was taken to reach the answer.

j	a_i	b_i	c_i	$(\mathcal{R}_A)_j$	$(\mathcal{R}_B)_j$	$(\mathcal{R}_C)_j$
0	30.000	0	30.000	−18.000	9.000	0
1	21.000	4.500	30.000	−8.820	1.710	−2.700
2	25.590	0.855	28.650	−13.097	6.059	−0.490
3	16.903	6.059	29.510	−5.714	−0.719	−3.576

Final: 18.742   3.970   28.341

For $\Delta t = 0.5$, the results for a, b, and c are

Final: 18.750   4.069   28.445

Results accurate to three places after the decimal are obtained with $\Delta t = 0.25$:

Final: 18.750   4.072   28.448

The fourth Runge-Kutta method converges $O(\Delta t^5)$. Thus, halving the step size decreases the error by a factor of 32. By comparison, Euler's method converges $O(\Delta t)$ so that halving the step size decreases the error by a factor of only 2. These remarks apply only in the limit as $\Delta t \rightarrow 0$, and either method can give anomalous behavior if Δt is large. If you can confirm that the data are converging according to the theoretical order of convergence, the convergence order can be used to extrapolate calculations to the limit as $\Delta t \rightarrow 0$.

Example 2.15: Develop an extrapolation technique suitable for the first-order convergence of Euler integration. Test it for the set of ODEs in Example 2.3.

Solution: Repeat the calculations in Example 2.3 but now reduce Δt by a factor of 2 for each successive calculation rather than by the factor of 4 used in the examples. Calculate the corresponding changes in $a(t_{max})$ and denote these changes by Δ. Then Δ should decrease by a factor of 2 for each calculation of $a(t_{max})$. (The reader interested in rigor will note that the error is halved and will do some algebra to prove that the Δ are halved as well.) If Δ was the change that just occurred, then we would expect the next change to be $\Delta/2$, the one after that to be $\Delta/4$, and so on. The total change yet to come is thus $\Delta/2 + \Delta/4 + \Delta/8 + \cdots$. This is a geometric series that converges to Δ. Using Euler's method, the cumulative change yet to come is equal to the single change that just occurred. Thus, the extrapolated value for $a(t_{max})$ is the value just calculated plus the Δ just calculated. The extrapolation scheme is illustrated for the ODEs in Example 2.3 in the following table:

Number of steps	$a(t_{max})$	Δ	Extrapolated $a(t_{max})$
2	−16.3200		
4	1.5392	17.8591	19.3938
8	2.8245	1.2854	4.1099
16	3.2436	0.4191	3.6626
32	3.4367	0.1931	3.6298
64	3.5304	0.0937	3.6241
128	3.5766	0.0462	3.6228
256	3.5995	0.0229	3.6225
512	3.6110	0.0114	3.6224
1024	3.6167	0.0057	3.6224
2048	3.6195	0.0029	3.6224
4096	3.6210	0.0014	3.6224
8192	3.6217	0.0007	3.6224
16384	3.6220	0.0004	3.6224
32768	3.6222	0.0002	3.6224
65536	3.6223	0.0001	3.6224
131072	3.6223	0.0000	3.6224

Extrapolation can reduce computational effort by a large factor, but computation is cheap. The value of the computational reduction will be trivial for most problems. Convergence acceleration can be useful for complex problems or for the inside loops in optimization studies. For such cases, you should also consider more sophisticated integration schemes such as Runge-Kutta. It too can be extrapolated, although the extrapolation rule is different. The extrapolated factor for Runge-Kutta integration is based on the series

$$1/32 + 1/32^2 + 1/32^3 + \cdots = 0.03226$$

Thus, the total change yet to come is about 3% of the change that just occurred. As a practical matter, your calculations will probably achieve the required accuracy by the time you confirm that successive changes in the integral really are decreasing by a factor of 32 each time. With modern computers and Runge-Kutta integration, extrapolation is seldom needed for the solution of ODEs. It may still be useful in the solution of the second-order, partial differential equations treated in Chapters 8 and 9. Ordinary differential equation solvers are often used as part of the solution technique for PDEs. Extrapolation is used in some highly efficient ODE solvers. A variety of sophisticated integration techniques are available both as freeware and as commercial packages. Their use may be justified for design and optimization studies where the same set of equations must be solved repetitively or when the equations are exceptionally stiff. The casual user need go no further than Runge-Kutta, possibly with adaptive step sizes where Δt is varied from step to step in the calculations based on error estimates. See *Numerical Recipes* by Press et al., as cited in the "Suggestions for Further Reading" section for this chapter, for a usable example of Runge-Kutta integration with adaptive step sizes.

CHAPTER 3
ISOTHERMAL PISTON FLOW REACTORS

Chapter 2 developed a methodology for treating multiple and complex reactions in batch reactors. The methodology is now applied to piston flow reactors. Chapter 3 also generalizes the design equations for piston flow beyond the simple case of constant density and constant velocity. The key assumption of piston flow remains intact: there must be complete mixing in the direction perpendicular to flow and no mixing in the direction of flow. The fluid density and reactor cross section are allowed to vary. The pressure drop in the reactor is calculated. Transpiration is briefly considered. Scaleup and scaledown techniques for tubular reactors are developed in some detail.

Chapter 1 treated the simplest type of piston flow reactor, one with constant density and constant reactor cross section. The reactor design equations for this type of piston flow reactor are directly analogous to the design equations for a constant-density batch reactor. What happens in time in the batch reactor happens in space in the piston flow reactor, and the transformation $t = z/\bar{u}$ converts one design equation to the other. For component A,

$$\bar{u}\frac{da}{dz} = \mathcal{R}_A \qquad \text{where} \qquad a = a_{in} \quad \text{at} \quad z = 0 \qquad (3.1)$$

All the results obtained for isothermal, constant-density batch reactors apply to isothermal, constant-density (and constant cross-section) piston flow reactors. Just replace t with $z/\bar{u}$, and evaluate the outlet concentration at $z = L$. Equivalently, leave the result in the time domain and evaluate the outlet composition $\bar{t} = L/\bar{u}$. For example, the solution for component B in the competitive reaction sequence of

$$A \xrightarrow{k_A} B \xrightarrow{k_B} C \xrightarrow{k_C} D \xrightarrow{k_D} \cdots$$

is given by Equation (2.22) for a batch reactor:

$$b_{batch}(t) = \left[b_0 - \frac{a_0 k_A}{k_B - k_A}\right]e^{-k_B t} + \left[\frac{a_0 k_A}{k_B - k_A}\right]e^{-k_A t}$$

The solution for the same reaction sequence run in a PFR is

$$b_{PFR}(z) = \left[b_{in} - \frac{a_{in}k_A}{k_B - k_A}\right]e^{-k_B z/\bar{u}} + \left[\frac{a_{in}k_A}{k_B - k_A}\right]e^{-k_A z/\bar{u}}$$

The extension to multiple reactions is done by writing Equation (3.1) (or the more complicated versions of Equation (3.1) that will soon be developed) for each of the N components. The component reaction rates are found from Equation (2.7) in exactly the same ways as in a batch reactor. The result is an *initial value problem* consisting of N simultaneous, first-order ODEs that can be solved using your favorite ODE solver. The same kind of problem was solved in Chapter 2, but the independent variable is now z rather than t.

The emphasis in this chapter is on the generalization of piston flow to situations other than constant velocity down the tube. Real reactors can closely approximate piston flow reactors, yet they show many complications compared with the constant-density and constant-cross-section case considered in Chapter 1. Gas-phase tubular reactors may have appreciable density differences between the inlet and outlet. The mass density and thus the velocity down the tube can vary at constant pressure if there is a change in the number of moles upon reaction, but the pressure drop due to skin friction usually causes a larger change in the density and velocity of the gas. Reactors are sometimes designed to have variable cross sections, and this too will change the density and velocity. Despite these complications, piston flow reactors remain closely akin to batch reactors. There is a one-to-one correspondence between time in a batch and position in a tube, but the relationship is no longer as simple as $z = \bar{u}t$.

3.1 PISTON FLOW WITH CONSTANT MASS FLOW

Most of this chapter assumes that the mass flow rate down the tube is constant; i.e., the tube wall is impermeable. The reactor cross-sectional area A_c is allowed to vary as a function of axial position, $A_c = A_c(z)$. Figure 3.1 shows the system and indicates the nomenclature. An overall mass balance gives

$$Q\rho = Q_{in}\rho_{in} = A_c \bar{u} \rho = (A_c)_{in} \bar{u}_{in} \rho_{in} = \text{constant} \quad (3.2)$$

where ρ is the mass density that is assumed to be uniform in the cross section of the reactor but that may change as a function of z. The counterpart for Equation (3.2) in a batch system is just that ρV be constant.

The component balance will be based on the molar flow rate:

$$\dot{N}_A = Qa \quad (3.3)$$

ISOTHERMAL PISTON FLOW REACTORS

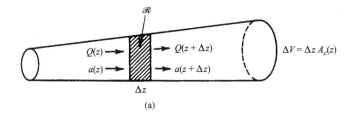

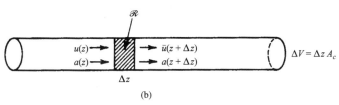

FIGURE 3.1 Differential volume elements in piston flow reactors: (a) variable cross section; (b) constant cross section.

Unlike $Q\rho$, $\dot{N}_A$ is not a conserved quantity and varies down the length of the tube. Consider a differential element of length Δz and volume $\Delta z A_c$. The molar flow entering the element is $\dot{N}_A(z)$ and that leaving the element is $\dot{N}_A(z + \Delta z)$, the difference being due to reaction within the volume element. A balance on component A gives

$$\dot{N}_A(z) + A_c \Delta z \, \mathscr{R}_A = \dot{N}_A(z + \Delta z)$$

or

$$\mathscr{R}_A = \frac{\dot{N}_A(z + \Delta z) - \dot{N}_A(z)}{A_c \Delta z}$$

Taking the limit as $\Delta z \to 0$ gives

$$\frac{1}{A_c}\frac{d(\dot{N}_A)}{dz} = \frac{1}{A_c}\frac{d(Qa)}{dz} = \frac{1}{A_c}\frac{d(A_c \bar{u} a)}{dz} = \mathscr{R}_A \tag{3.4}$$

This is the piston flow analog of the variable-volume batch reactor, Equation (2.30).

The derivative in Equation (3.4) can be expanded into three separate terms:

$$\frac{1}{A_c}\frac{d(A_c \bar{u} a)}{dz} = \bar{u}\frac{da}{dz} + a\frac{d\bar{u}}{dz} + \frac{\bar{u} a}{A_c}\frac{dA_c}{dz} = \mathscr{R}_A \tag{3.5}$$

The first term must always be retained since A is a reactive component and thus varies in the z-direction. The second term must be retained if either the mass density or the reactor cross-sectional area varies with z. The last term is needed for reactors with variable cross sections. Figure 3.2 illustrates an annular flow reactor that is an industrially relevant reason for including this term.

Practical problems involving variable-density PFRs require numerical solutions, and for these it is better to avoid expanding Equation (3.4) into separate derivatives for a and $\bar{u}$. We could continue to use the molar flow rate, $\hat{N}_A$, as the dependent variable, but prefer to use the molar flux,

$$\Phi_A = \bar{u}a \qquad (3.6)$$

The units on Φ_A are mol/(m^2 · s). This is the *convective flux*. The student of mass transfer will recognize that a diffusion term like $-\mathscr{D}_A da/dz$ is usually included in the flux. This term is the *diffusive flux* and is zero for piston flow. The design equation for the variable-density, variable-cross-section PFR can be written as

$$\frac{d\Phi_A}{dz} = \mathscr{R}_A - \frac{\Phi_A}{A_c}\frac{dA_c}{dz} \qquad \text{where} \qquad \Phi_A = (\Phi_A)_{in} \qquad \text{at} \qquad z=0 \qquad (3.7)$$

The dA_c/dz term is usually zero since tubular reactors with constant diameter are by far the most important application of Equation (3.7). For the exceptional case, we suppose that $A_c(z)$ is known, say from the design drawings of the reactor. It must be a smooth (meaning differentiable) and slowly varying function of z or else the assumption of piston flow will run into hydrodynamic as well as mathematical difficulties. Abrupt changes in A_c will create secondary flows that invalidate the assumptions of piston flow.

We can define a new rate expression $\mathscr{R}'_A$ that includes the dA_c/dz term within it. The design equation then becomes

$$\frac{d\Phi_A}{dz} = \mathscr{R}_A - \frac{\Phi_A}{A_c}\frac{dA_c}{dz} = \mathscr{R}'_A = \mathscr{R}'_A(a,b,\ldots,z) \qquad (3.8)$$

where $\mathscr{R}'_A$ has an explicit dependence on z when the cross section is variable and where $\mathscr{R}'_A = \mathscr{R}_A$ for the usual case of constant cross section. The explicit dependence on z causes no problem in numerical integration. Equation (2.48) shows how an explicit dependence on the independent variable is treated in Runge-Kutta integration.

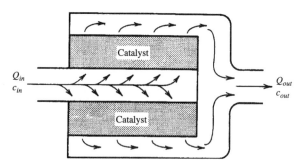

FIGURE 3.2 Annular packed-bed reactor used for adiabatic reactions favored by low pressure.

If there are M reactions involving N components,

$$\frac{d\mathbf{\Phi}}{dz} = \mathbf{v}\mathcal{R}' \quad \text{where} \quad \mathbf{\Phi} = \mathbf{\Phi}_{in} \quad \text{at} \quad z = 0 \tag{3.9}$$

where $\mathbf{\Phi}$ and $\mathbf{\Phi}_{in}$ are $N \times 1$ column vectors of the component fluxes, $\mathbf{v}$ is an $N \times M$ matrix of stoichiometric coefficients, and $\mathcal{R}'$ is an $M \times 1$ column vector of reaction rates that includes the effects of varying the reactor cross section. Equation (3.9) represents a set of first-order ODEs and is the flow analog of Equation (2.38). The dimensionality of the set can be reduced to $M < N$ by the reaction coordinate method, but there is little purpose in doing so. The reduction provides no significant help in a numerical solution, and even the case of one reactant going to one product is difficult to solve analytically when the density or cross section varies. A reason for this difficulty is illustrated in Example 3.1.

Example 3.1: Find the fraction unreacted for a first-order reaction in a variable density, variable-cross-section PFR.

Solution: It is easy to begin the solution. In piston flow, molecules that enter together leave together and have the same residence time in the reactor, $\bar{t}$. When the kinetics are first order, the probability that a molecule reacts depends only on its residence time. The probability that a particular molecule will leave the system without reacting is $\exp(-k\bar{t})$. For the entire collection of molecules, the probability converts into a deterministic fraction. The fraction unreacted for a variable density flow system is

$$Y_A = \frac{(\dot{N}_A)_{out}}{(\dot{N}_A)_{in}} = \frac{(Qa)_{out}}{(Qa)_{in}} = \frac{(A_c \bar{u} a)_{out}}{(A_c \bar{u} a)_{in}} = e^{-k\bar{t}} \tag{3.10}$$

The solution for Y_A is simple, even elegant, but what is the value of $\bar{t}$? It is equal to the mass holdup divided by the mass throughput, Equation (1.41), but there is no simple formula for the holdup when the density is variable. The same gas-phase reactor will give different conversions for A when the reactions are A → 2B and A → B, even though it is operated at the same temperature and pressure and the first-order rate constants are identical.

Fortunately, it is possible to develop a general-purpose technique for the numerical solution of Equation (3.9), even when the density varies down the tube. It is first necessary to convert the component reaction rates from their normal dependence on concentration to a dependence on the molar fluxes. This is done simply by replacing a by $\Phi_A/\bar{u}$, and so on for the various components. This introduces $\bar{u}$ as a variable in the reaction rate:

$$\frac{d\Phi_A}{dz} = \mathcal{R}'_A = \mathcal{R}'_A(a, b, \ldots, z) = \mathcal{R}'_A(\Phi_A, \Phi_B, \ldots, \bar{u}, z) \tag{3.11}$$

To find $\bar{u}$, it is necessary to use some ancillary equations. As usual in solving initial value problems, we assume that all variables are known at the reactor inlet so that $(A_c)_{in}\bar{u}_{in}\rho_{in}$ will be known. Equation (3.2) can be used to calculate $\bar{u}$ at a downstream location if ρ is known. An equation of state will give ρ but requires knowledge of state variables such as composition, pressure, and temperature. To find these, we will need still more equations, but a closed set can eventually be achieved, and the calculations can proceed in a stepwise fashion down the tube.

3.1.1 Gas-Phase Reactions

For gas-phase reactions, the molar density is more useful than the mass density. Determining the equation of state for a nonideal gas mixture can be a difficult problem in thermodynamics. For illustrative purposes and for a great many industrial problems, the ideal gas law is sufficient. Here it is given in a form suitable for flow reactors:

$$\frac{P}{R_g T} = a + b + c + \cdots + i \tag{3.12}$$

where i represents the concentration (molar density) of inerts. Note that Equation (3.9) should include inerts as one of the components when the reaction is gas phase. The stoichiometric coefficient is zero for an inert so that $\mathcal{R}_I = 0$, but if A_c varies with z, then $\mathcal{R}'_I \neq 0$.

Multiply Equation (3.12) by $\bar{u}$ to obtain

$$\frac{P\bar{u}}{R_g T} = \bar{u}a + \bar{u}b + \bar{u}c + \cdots + \bar{u}i = \Phi_A + \Phi_B + \Phi_C + \cdots + \Phi_I \tag{3.13}$$

If the reactor operates isothermally and if the pressure drop is sufficiently low, we have achieved closure. Equations (3.11) and (3.13) together allow a marching-ahead solution. The more common case requires additional equations to calculate pressure and temperature. An ODE is added to calculate pressure $P(z)$, and Chapter 5 adds an ODE to calculate temperature $T(z)$.

For *laminar flow* in a circular tube of radius R, the pressure gradient is given by a differential form of the Poiseuille equation:

$$\frac{dP}{dz} = -\frac{8\mu\bar{u}}{R^2} \tag{3.14}$$

where μ is the viscosity. In the general case, $\bar{u}$, μ, and R will all vary as a function of z and Equation (3.14) must be integrated numerically. The reader may wonder if piston flow is a reasonable assumption for a laminar flow system since laminar flow has a pronounced velocity profile. The answer is not really, but there are exceptions. See Chapter 8 for more suitable design methods and to understand the exceptional—and generally unscalable case—where piston flow is a reasonable approximation to laminar flow.

For *turbulent flow*, the pressure drop is calculated from

$$\frac{dP}{dz} = -\frac{\mathbf{Fa}\rho \bar{u}^2}{R} \tag{3.15}$$

where the Fanning friction factor **Fa** can be approximated as

$$\mathbf{Fa} = \frac{0.079}{\mathbf{Re}^{1/4}} \tag{3.16}$$

More accurate correlations, which take factors like wall roughness into account, are readily available, but the form used here is adequate for most purposes. It has a simple, analytical form that lends itself to conceptual thinking and scaleup calculations, but see Problem 3.14 for an alternative.

For *packed beds* in either turbulent or laminar flow, the Ergun equation is often satisfactory:

$$\begin{aligned}\frac{dP}{dz} &= -\frac{\rho \bar{u}_s^2}{d_p}\frac{(1-\varepsilon)}{\varepsilon^3}\left[\frac{150(1-\varepsilon)\mu}{d_p\rho\bar{u}_s} + 1.75\right] \\ &= -\frac{\rho \bar{u}_s^2}{d_p}\frac{(1-\varepsilon)}{\varepsilon^3}\left[\frac{150(1-\varepsilon)}{(\mathbf{Re})_p} + 1.75\right]\end{aligned} \tag{3.17}$$

where ε is the void fraction of the bed, $(\mathbf{Re})_p$ is the particle Reynolds number, and d_p is the diameter of the packing. For nonspherical packing, use six times the ratio of volume to surface area of the packing as an effective d_p. Note that $\bar{u}_s$ is the superficial velocity, this being the velocity the fluid would have if the tube were empty.

The formulation is now complete. Including the inerts among the N components, there are N ODEs that have the Φ as dependent variables. The general case has two additional ODEs, one for pressure and one for temperature. There are thus $N+2$ first-order ODEs in the general case. There is also an equation of state such as Equation (3.13) and this relates P, T, and the Φ. The marching-ahead technique assumes that all variables are known at the reactor inlet. Pressure may be an exception since the discharge pressure is usually specified and the inlet pressure has whatever value is needed to achieve the requisite flow rate. This is handled by assuming a value for P_{in} and adjusting it until the desired value for P_{out} is obtained.

An analytical solution to a variable-density problem is rarely possible. The following example is an exception that illustrates the solution technique first in analytical form and then in numerical form. It is followed by a description of the general algorithm for solving Equation (3.11) numerically.

Example 3.2: Consider the reaction $2A \xrightarrow{k} B$. Derive an analytical expression for the fraction unreacted in a gas-phase, isothermal, piston flow reactor of length L. The pressure drop in the reactor is negligible.

The reactor cross section is constant. There are no inerts. The feed is pure A and the gases are ideal. Test your mathematics with a numerical solution.

Solution: The design equations for the two components are

$$\frac{d(\bar{u}a)}{dz} = \frac{d\Phi_A}{dz} = -2ka^2 = -2k\frac{\Phi_A^2}{\bar{u}^2}$$

$$\frac{d(\bar{u}b)}{dz} = \frac{d\Phi_B}{dz} = ka^2 = k\frac{\Phi_A^2}{\bar{u}^2}$$

Applying the ideal gas law

$$\frac{P}{R_g T} = \rho_{molar} = a + b$$

Multiplying by $\bar{u}$ gives

$$\frac{P\bar{u}}{R_g T} = \bar{u}\rho_{molar} = \bar{u}(a + b) = \Phi_A + \Phi_B$$

Since the pressure drop is small, $P = P_{in}$, and

$$\bar{u} = \frac{\Phi_A + \Phi_B}{(a + b)} = \frac{\Phi_A + \Phi_B}{a_{in}}$$

The ODEs governing the system are

$$\frac{d\Phi_A}{dz} = -2k\frac{\Phi_A^2}{\bar{u}^2} = -2k\frac{a_{in}^2 \Phi_A^2}{(\Phi_A + \Phi_B)^2}$$

$$\frac{d\Phi_B}{dz} = k\frac{\Phi_A^2}{\bar{u}^2} = k\frac{a_{in}^2 \Phi_A^2}{(\Phi_A + \Phi_B)^2}$$

These equations are the starting point for both the analytical and the numerical solutions.

Analytical Solution: A stoichiometric relationship can be used to eliminate Φ_B. Combine the two ODEs to obtain

$$\frac{-d\Phi_A}{2} = d\Phi_B$$

The initial condition is that $\Phi_A = \Phi_{in}$ when $\Phi_B = 0$. Thus,

$$\Phi_B = \frac{\Phi_{in} - \Phi_A}{2}$$

ISOTHERMAL PISTON FLOW REACTORS

Substituting this into the equation for $\bar{u}$ gives a single ODE:

$$\frac{d\Phi_A}{dz} = \frac{-8ka_{in}^2\Phi_A^2}{(\Phi_A + \Phi_{in})^2}$$

that is variable-separable. Thus,

$$\int_{\Phi_{in}}^{\Phi_A} \frac{(\Phi_A + \Phi_{in})^2}{\Phi_A^2} d\Phi_A = -\int_0^z 8ka_{in}^2\, dz$$

A table of integrals (and a variable substitution, $s = \Phi_A + \Phi_{in}$) gives

$$\frac{\Phi_A}{\Phi_{in}} - \frac{\Phi_{in}}{\Phi_A} - 2\ln\frac{\Phi_{in}}{\Phi_A} = \frac{-8ka_{in}^2 z}{\Phi_{in}} = \frac{-8ka_{in}z}{\bar{u}_{in}}$$

The solution to the constant-density case is

$$\frac{\Phi_A}{\Phi_{in}} = \frac{a}{a_{in}} = \frac{1}{1 + 2ka_{in}z/\bar{u}_{in}}$$

The fraction unreacted is Φ_A/Φ_{in}. Set $z = L$ to obtain it at the reactor outlet. Suppose $\Phi_{in} = 1$ and that $ka_{in}/\bar{u}_{in} = 1$ in some system of units. Then the variable-density case gives $z = 0.3608$ at $\Phi_A = 0.5$. The velocity at this point is $0.75\bar{u}_{in}$. The constant density case gives $z = 0.5$ at $\Phi_A = 0.5$ and the velocity at the outlet is unchanged from $\bar{u}_{in}$. The constant-density case fails to account for the reduction in $\bar{u}$ as the reaction proceeds and thus underestimates the residence time.

Numerical Solution: The following program gives $z = 0.3608$ at $\Phi_A = 0.5$.

```
a=1
b=0
u=1
k=1
dz=.0001
z=0
PAold=u * a
PBold=0
DO
   PAnew=PAold - 2 * k * PAold ^ 2 / u ^ 2 * dz
   PBnew=PBold+k * PAold ^ 2 / u ^ 2 * dz
   u=PAnew+PBnew
   PAold=PAnew
   PBold=PBnew
   z=z+dz
```

```
LOOP WHILE PAold > .5
PRINT USING "###.####"; z
```

Computational Scheme for Gas-Phase PFRs. A general procedure for solving the reactor design equations for a piston flow reactor using the marching-ahead technique (Euler's method) has seven steps:

1. Pick a step size Δz.
2. Calculate initial values for all variables including a guess for P_{in}. Initial values are needed for $a, b, c, \ldots, i, \bar{u}, \Phi_A, \Phi_B, \Phi_C, \ldots, \Phi_I, P$, and T plus physical properties such as μ that are used in the ancillary equations.
3. Take one step, calculating new values for $\Phi_A, \Phi_B, \Phi_C, \ldots, \Phi_I, P$, and T at the new axial location, $z + \Delta z$. The marching-ahead equations for the molar fluxes have the form

$$(\Phi_A)_{new} = (\Phi_A)_{old} + \Delta z \mathcal{R}'_A[(\Phi_A)_{old}, (\Phi_B)_{old}, \ldots, (\Phi_I)_{old}, z] \quad (3.18)$$

The right-hand sides of these equations are evaluated using the old values that correspond to position z. A similar Euler-type solution is used for one of Equations (3.14), (3.15), or (3.17) to calculate P_{new} and an ODE from Chapter 5 is solved in the same way to calculate T_{new}.

4. Update $\bar{u}$ using

$$\bar{u}_{new} = R_g T_{new} (\Phi_A + \Phi_B + \Phi_C + \cdots + \Phi_I)_{new} / P_{new} \quad (3.19)$$

Note that this step uses the ideal gas law. Other equations of state could be substituted.

5. Update all physical property values to the new conditions. The component concentrations are updated using

$$a_{new} = (\Phi_A)_{new} / \bar{u}_{new}, \quad b_{new} = (\Phi_B)_{new} / \bar{u}_{new}, \ldots \quad (3.20)$$

6. If $z < L$, go to Step 3. If $z \geq L$, is P_{out} correct? If not, go to Step 2 and guess another value for P_{in}.
7. Decrease Δz by a factor of 2 and go to Step 1. Repeat until the results converge to three or four significant figures.

The next example applies this general procedure to a packed-bed reactor.

Example 3.3: Fixed-bed reactors are used for the catalytic dehydrogenation of ethylbenzene to form styrene:

$$C_8H_{10} \rightleftarrows C_8H_8 + H_2 \quad (A \rightleftarrows B + C)$$

The reaction is endothermic, but large amounts of steam are used to minimize the temperature drop and, by way of the water–gas shift reaction, to prevent accumulation of coke on the catalyst. Ignore the reverse and competitive reactions and suppose a proprietary catalyst in the form of 3-mm spheres gives a first-order rate constant of $15\,\text{s}^{-1}$ at $625°\text{C}$.

The molar ratio of steam to ethylbenzene at the inlet is 9:1. The bed is 1 m in length and the void fraction is 0.5. The inlet pressure is set at 1 atm and the outlet pressure is adjusted to give a superficial velocity of 9 m/s at the tube inlet. (The real design problem would specify the downstream pressure and the mass flow rate.) The particle Reynolds number is 100 based on the inlet conditions ($\mu \approx 4 \times 10^{-5}\,\text{Pa}\cdot\text{s}$). Find the conversion, pressure, and velocity at the tube outlet, assuming isothermal operation.

Solution: This is a variable-velocity problem with $\bar{u}$ changing because of the reaction stoichiometry and the pressure drop. The flux marching equations for the various components are

$$\Phi_{A_{j+1}} = \Phi_{A_j} - ka\,\Delta z = \Phi_{A_j} - k\frac{\Phi_{A_j}}{\bar{u}}\,\Delta z$$

$$\Phi_{B_{j+1}} = \Phi_{B_j} + ka\,\Delta z = \Phi_{B_j} + k\frac{\Phi_{A_j}}{\bar{u}}\,\Delta z$$

$$\Phi_{C_{j+1}} = \Phi_{C_j} + ka\,\Delta z = \Phi_{C_j} + k\frac{\Phi_{A_j}}{\bar{u}}\,\Delta z$$

$$\Phi_{D_{j+1}} = \Phi_{D_j}$$

where D represents the inerts. There is one equation for each component. It is perfectly feasible to retain each of these equations and to solve them simultaneously. Indeed, this is necessary if there is a complex reaction network or if molecular diffusion destroys local stoichiometry. For the current example, the stoichiometry is so simple it may as well be used. At any step j,

$$\Phi_C = \Phi_B = (\Phi_A)_{in} - \Phi_A$$

Thus, we need retain only the flux marching equation for component A.

The pressure is also given by an ODE. The Ergun equation, Equation (3.17), applies to a packed bed:

$$P_{j+1} = P_j - \frac{\rho \bar{u}_s^2}{d_p}\frac{(1-\varepsilon)}{\varepsilon^3}\left[\frac{150(1-\varepsilon)}{\mathbf{Re}_p} + 1.75\right]\Delta z$$

where $\mathbf{Re}_p = d_p \rho \bar{u}_s/\mu$ is the particle Reynolds number. The viscosity is approximately constant since μ is a function of temperature alone for low-density gases. Also, $\rho \bar{u}_s$ is constant because the mass flow is constant in a tube of constant cross section. These facts justify the assumption that $\mathbf{Re}_p$ is constant. Also, the $\rho \bar{u}_s^2$ term in the Ergun equation is equal to $(\rho \bar{u}_s)_{in}\bar{u}_s$.

The marching equations for flux and pressure contain the superficial velocity $\bar{u}_s$. The ideal gas law in the form of Equation (3.13) is used to relate it to the flux:

$$(\bar{u}_s)_j = \frac{R_g T}{P_j}(\Phi_A + \Phi_B + \Phi_C + \Phi_D) = \frac{R_g T}{P_j}[2(\Phi_A)_{in} - \Phi_A + \Phi_D]$$

The computational scheme marches flux and pressure ahead one step and then updates $\bar{u}_s$.

The various inlet conditions are calculated using the ideal gas law. They are $a_{in} = 1.36$ mol/m^3, $b_{in} = c_{in} = 0$, $d_{in} = 12.2$ mol/m^3, $(\rho\bar{u}_s)_{in} = 1.23$ kg/(m^2·s), $(\Phi_A)_{in} = 12.2$ mol/(m^2·s), and $\Phi_D = 110$ mol/(m^2·s). Substituting known values and being careful with the units gives

$$(\Phi_A)_{j+1} = (\Phi_A)_j \left[1 - \frac{15\,\Delta z}{(\bar{u}_s)_j}\right]$$

$$P_{j+1} = P_j - 0.041(\bar{u}_s)_j\,\Delta z$$

$$(\bar{u}_s)_{j+1} = \frac{0.08}{P_{j+1}}[134 - \Phi_A]_{j+1}$$

These equations are solved, starting with the known initial conditions and proceeding step-by-step down the reactor until the outlet is reached. The solution is

$$X = 1 - \frac{(\Phi_A)_{out}}{(\Phi_A)_{in}} = 0.67 \quad (67\% \text{ conversion})$$

with $P_{out} = 0.4$ atm and $(\bar{u})_{out} = 26$ m/s.

The selectivity is 100% in this simple example, but do not believe it. Many things happen at 625°C, and the actual effluent contains substantial amounts of carbon dioxide, benzene, toluene, methane, and ethylene in addition to styrene, ethylbenzene, and hydrogen. It contains small but troublesome amounts of diethyl benzene, divinyl benzene, and phenyl acetylene. The actual selectivity is about 90%. A good kinetic model would account for all the important by-products and would even reflect the age of the catalyst. A good reactor model would, at a minimum, include the temperature change due to reaction.

The Mean Residence Time in a Gas-Phase Tubular Reactor. Examples such as 3.3 show that numerical solutions to the design equations are conceptually straightforward if a bit cumbersome. The problem with numerical solutions is that they are difficult to generalize. Analytical solutions can provide much greater insight. The next example addresses a very general problem. What is the pressure profile and mean residence time, $\bar{t}$, in a gas-phase tubular reactor? If $\bar{t}$ is known, even approximately, Equations like (3.10) suddenly become useful. The results derived in Example 3.4 apply to any tubular reactor, whether it approximates piston flow or not, provided that the change in moles upon reaction is negligible. This assumption is valid when the reaction stoichiometry gives no change in volume, when inerts are present in large quantities, or when the change in density due to the pressure drop is large compared with the change caused by the reaction. Many gas-phase reactors satisfy at least one of these conditions.

ISOTHERMAL PISTON FLOW REACTORS

Example 3.4: Find the mean residence time in an isothermal, gas-phase tubular reactor. Assume that the reactor has a circular cross section of constant radius. Assume ideal gas behavior and ignore any change in the number of moles upon reaction.

Solution: Begin with laminar flow and Equation (3.14):

$$\frac{dP}{dz} = -\frac{8\mu \bar{u}}{R^2}$$

To integrate this, $\bar{u}$ is needed. When there is no change in the number of moles upon reaction, Equation (3.2) applies to the total molar density as well as to the mass density. Thus, for constant A_c,

$$\bar{u}\rho_{molar} = \bar{u}(a + b + \cdots) = \text{constant} = \bar{u}_{in}(\rho_{molar})_{in}$$

and

$$\frac{\bar{u}(z)}{\bar{u}_{in}} = \frac{\rho_{in}}{\rho(z)} = \frac{(\rho_{molar})_{in}}{\rho_{molar}} = \frac{P_{in}}{P(z)}$$

These relationships result from assuming ideal gas behavior and no change in the number of moles upon reaction. Substituting $\bar{u}$ into the ODE for pressure gives

$$\frac{dP}{dz} = \frac{-\beta}{2P} \tag{3.21}$$

where β is a constant. The same result, but with a different value for β, is obtained for turbulent flow when Equation (3.15) is used instead of Equation (3.14). The values for β are

$$\beta = \frac{16\mu P_{in}\bar{u}_{in}}{R^2} = \frac{16\mu P_{out}\bar{u}_{out}}{R^2} \quad \text{(laminar flow)} \tag{3.22}$$

and

$$\beta = \frac{0.13\mu^{.25} P_{in}(\rho_{in}\bar{u}_{in})^{1.75}}{\rho_{in} R^{1.25}} = \frac{0.13\mu^{.25} P_{out}(\rho_{out}\bar{u}_{out})^{1.75}}{\rho_{out} R^{1.25}} \quad \text{(turbulent flow)} \tag{3.23}$$

Integrating Equation (3.21) and applying the inlet boundary condition gives

$$P^2 - P_{out}^2 = \beta(L - z)$$

Observe that

$$\beta L = P_{in}^2 - P_{out}^2 \tag{3.24}$$

is true for both laminar and turbulent flow.

We are now ready to calculate the mean residence time. According to Equation (1.41), $\bar{t}$ is the ratio of mass inventory to mass throughput. When the number of moles does not change, $\bar{t}$ is also the ratio of molar inventory to molar throughput. Denote the molar inventory (i.e., the total number of moles in the tube) as N_{actual}. Then

$$N_{actual} = \int_0^L A_c \rho_{molar} \, dz = \frac{A_c(\rho_{molar})_{in}}{P_{in}} \int_0^L P \, dz$$

$$= \frac{A_c(\rho_{molar})_{in}}{P_{in}} \int_0^L [P_{out}^2 + \beta(L-z)]^{1/2} \, dz$$

Integration gives

$$\frac{N_{actual}}{N_{inlet}} = \frac{2[P_{in}^3 - P_{out}^3]}{3\beta L P_{in}} = \frac{2[P_{in}^3 - P_{out}^3]}{3(P_{in}^2 - P_{out}^2) P_{in}} \quad (3.25)$$

where $N_{inlet} = A_c(\rho_{molar})_{in} L$ is the number of moles that the tube would contain if its entire length were at pressure P_{in}. When the pressure drop is low, $P_{in} \to P_{out}$ and $\beta \to 0$, and the inventory approaches N_{inlet}. When the pressure drop is high, $P_{in} \to \infty$ and $\beta \to \infty$, and the inventory is two-thirds of N_{inlet}.

The mean residence time is

$$\bar{t} = \frac{N_{actual}}{A_c(\rho_{molar})_{in} \bar{u}_{in}} = \frac{2[P_{in}^3 - P_{out}^3]}{3\beta L P_{in}} L/\bar{u}_{in} = \frac{2[P_{in}^3 - P_{out}^3]}{3(P_{in}^2 - P_{out}^2) P_{in}} L/\bar{u}_{in} \quad (3.26)$$

The term $[L/\bar{u}_{in}]$ is what the residence time would be if the entire reactor were at the inlet pressure. The factor multiplying it ranges from 2/3 to 1 as the pressure drop ranges from large to small and as β ranges from infinity to zero.

The terms *space time* and *space velocity* are antiques of petroleum refining, but have some utility in this example. The space time is defined as V/Q_{in}, which is what $\bar{t}$ would be if the fluid remained at its inlet density. The space time in a tubular reactor with constant cross section is $[L/\bar{u}_{in}]$. The space velocity is the inverse of the space time. The mean residence time, $\bar{t}$, is $V\hat{\rho}/(Q\rho)$ where $\hat{\rho}$ is the average density and ρQ is a constant (because the mass flow is constant) that can be evaluated at any point in the reactor. The mean residence time ranges from the space time to two-thirds the space time in a gas-phase tubular reactor when the gas obeys the ideal gas law.

Equation (3.26) evaluated the mean residence time in terms of the inlet velocity of the gas. The outlet velocity can also be used:

$$\bar{t} = \frac{N_{actual}}{A_c(\rho_{molar})_{out} \bar{u}_{out}} = \frac{2[P_{in}^3 - P_{out}^3]}{3\beta L P_{out}} L/\bar{u}_{out} = \frac{2[P_{in}^3 - P_{out}^3]}{3(P_{in}^2 - P_{out}^2) P_{out}} L/\bar{u}_{out} \quad (3.27)$$

The actual residence time for an ideal gas will always be higher than $[L/\bar{u}_{out}]$ and it will always be lower than $[L/\bar{u}_{in}]$.

Example 3.5: A 1-in i.d coiled tube, 57 m long, is being used as a tubular reactor. The operating temperature is 973 K. The inlet pressure is 1.068 atm; the outlet pressure is 1 atm. The outlet velocity has been measured to be 9.96 m/s. The fluid is mainly steam, but it contains small amounts of an organic compound that decomposes according to first-order kinetics with a half-life of 2.1 s at 973 K. Determine the mean residence time and the fractional conversion of the organic.

Solution: The first-order rate constant is $0.693/2.1 = 0.33$ s^{-1} so that the fractional conversion for a first-order reaction will be $1 - \exp(-0.22\bar{t})$ where $\bar{t}$ is in seconds. The inlet and outlet pressures are known so Equation (3.27) can be used to find $\bar{t}$ given that $[L/\bar{u}_{out}] = 57/9.96 = 5.72$ s. The result is $\bar{t} = 5.91$ s, which is 3.4% higher than what would be expected if the entire reaction was at P_{out}. The conversion of the organic compound is 86 percent.

The ideal gas law can be used to find $[L/\bar{u}_{in}]$ given $[L/\bar{u}_{out}]$. The result is $[L/\bar{u}_{in}] = 6.11$ s. The pressure factor in Equation (3.26) is 0.967, again giving $\bar{t} = 5.91$ s.

Note that the answers do not depend on the tube diameter, the temperature, or the properties of the fluid other than that it is an ideal gas.

Although Example 3.5 shows only a modest effect, density changes can be important for gas-phase reactions. Kinetic measurements made on a flow reactor are likely to be confounded by the density change. In principle, a kinetic model can still be fit to the data, but this is more difficult than when the measurements are made on a batch system where the reaction times are directly measured. When kinetics measurements are made using a flow reactor, $\bar{t}$ will not be known a priori if the density change upon reaction is appreciable. It can be calculated as part of the data fitting process. The equation of state must be known along with the inlet and outlet pressures. The calculations follow the general scheme for gas-phase PFRs given above. Chapter 7 discusses methods for determining kinetic constants using data from a reactor with complications such as variable density. As stated there, it is better to avoid confounding effects. Batch or CSTR experiments are far easier to analyze.

3.1.2 Liquid-Phase Reactions

Solution of the design equations for liquid-phase piston flow reactors is usually easier than for gas-phase reactors because pressure typically has no effect on the fluid density or the reaction kinetics. Extreme pressures are an exception that theoretically can be handled by the same methods used for gas-phase systems. The difficulty will be finding an equation of state. For ordinary pressures, the mass density can usually be estimated as a simple function of composition. This leads to easy and direct use of Equation (3.2).

Computational Scheme for Liquid-Phase PFRs. The following is a procedure for solving the reactor design equations for a moderate-pressure, liquid-phase, piston flow reactor using the marching-ahead technique (Euler's method):

1. Pick a step size Δz.
2. Calculate initial values for all variables. Initial values are needed for $a, b, c, \ldots, i, \rho, \bar{u}, \Phi_A, \Phi_B, \Phi_C, \ldots, \Phi_I$, and T. The pressure can be included if desired but it does not affect the reaction calculations. Also, P_{in} can be set arbitrarily.
3. Take one step, calculating new values for $\Phi_A, \Phi_B, \Phi_C, \ldots, \Phi_I$ at the new axial location, $z + \Delta z$. The current chapter considers only isothermal reactors, but the general case includes an ODE for temperature. The marching-ahead equations have the form

$$(\Phi_A)_{new} = (\Phi_A)_{old} + \Delta z \, \mathscr{R}'_A[(\Phi_A)_{old}, (\Phi_B)_{old}, \ldots, (\Phi_I)_{old}, z] \quad (3.28)$$

The right-hand sides of these equations are evaluated using the old values, which correspond to position z.

4. Update the component concentrations using

$$a_{new} = (\Phi_A)_{new}/\bar{u}_{old}, \; b_{new} = (\Phi_B)_{new}/\bar{u}_{old}, \; \ldots \quad (3.29)$$

5. Use these new concentrations to update the physical properties that appear in ancillary equations. One property that must be updated is ρ.
6. Use the new value for ρ to update $\bar{u}$:

$$\bar{u}_{new} = \frac{\bar{u}_{in}\rho_{in}}{\rho_{new}} \quad (3.30)$$

7. If $z < L$, go to Step 3. If $z \geq L$, decrease Δz by a factor of 2 and go to Step 1. Repeat until the results converge to three or four significant figures.

Note that Step 4 in this procedure uses the old value for $\bar{u}$ since the new value is not yet known. The new value could be used in Equation (3.29) if $\bar{u}_{new}$ is found by simultaneous solution with Equation (3.30). However, complications of this sort are not necessary. Taking the numerical limit as $\Delta z \to 0$ removes the errors. As a general rule, the exact sequence of calculations is unimportant in marching schemes. What is necessary is that each variable be updated once during each Δz step.

Example 3.6: The isothermal batch polymerization in Example 2.8 converted 80% of the monomer in 2 h. You want to do the same thing in

ISOTHERMAL PISTON FLOW REACTORS

a micro-pilot plant using a capillary tube. (If the tube diameter is small enough, assumptions of piston flow and isothermal operation will be reasonable even for laminar flow. Criteria are given in Chapters 8 and 9.) The tube has an i.d. of 0.0015 m and it is 1 m long. The monomer density is 900 kg/m³ and the polymer density is 1040 kg/m³. The pseudo-first-order rate constant is 0.8047 h⁻¹ and the residence time needed to achieve 80% conversion is $\bar{t} = 2$ h. What flow rate should be used?

Solution: The required flow rate is the mass inventory in the system divided by the mean residence time:

$$Q\rho = \frac{\pi R^2 L \hat{\rho}}{\bar{t}}$$

where the composite quantity $Q\rho$ is the mass flow rate and is constant. It is what we want to find. Its value is easily bounded since $\hat{\rho}$ must lie somewhere between the inlet and outlet densities. Using the inlet density,

$$Q\rho = \frac{\pi(0.0015)^2(1)(900)}{2} = 0.00318 \text{ kg/h}$$

The outlet density is calculated assuming the mass density varies linearly with conversion to polymer as in Example 2.8: $\rho_{out} = 1012$ kg/m³. The estimate for $Q\rho$ based on the outlet density is

$$Q\rho = \frac{\pi(0.0015)^2(1)(1012)}{2} = 0.00358 \text{ kg/h}$$

Thus, we can make a reasonably accurate initial guess for $Q\rho$. This guess is used to calculate the conversion in a tubular reactor of the given dimensions. When the right guess is made, the mean residence time will be 2 h and the fraction unreacted will be 20%. The following code follows the general procedure for liquid-phase PFRs. The fraction unreacted is calculated as the ratio of $\Phi_A/(\Phi_A)_{in}$, which is denoted as Phi/PhiIn in the program. A trial-and-error-search gives $Q\rho = 0.003426$ kg/h for the specified residence time of 2 h and a fraction unreacted of 80%. The calculated outlet density is 1012 kg/m³.

```
dz = .00001
1 INPUT Qp  'Replace as necessary depending on the
            'computing platform
R = .0015
L = 1
Pi = 3.14159
k = .8047

rhoin = 900
Qin = Qp / rhoin
```

```
uin = Qin / (Pi * R ^ 2)
ain = 1
PhiIn = uin * ain

a = ain
u = uin
Phi = PhiIn

t = 0
z = 0
DO
   Phinew = Phi - k * Phi / u * dz
   anew = Phinew / u
   rho = 1040 - 140 * Phinew / PhiIn
   unew = uin * rhoin / rho
   z = z + dz
   t = t + dz / unew
   Phi = Phinew
   u = unew
LOOP WHILE z < L
PRINT USING "######.#####"; t, Phi/PhiIn, rho
'Replace as necessary
GOTO 1 'Efficient code even if frowned upon by
       'programming purists
```

Density changes tend to be of secondary importance for liquid-phase reactions and are frequently ignored. They can be confounded in the kinetic measurements (e.g., by using the space time rather than the mean residence time when fitting the data to a kinetic model). If kinetic constants are fit to data from a flow reactor, the density profile in the reactor should be calculated as part of the data-fitting process. The equation of state must be known (i.e., density as a function of composition and temperature). The calculations follow the general scheme for liquid-phase PFRs given above. Chapter 7 discusses methods for fitting data that are confounded by effects such as density changes. It is easier to use a batch reactor or a CSTR for the kinetic measurements even though the final design will be a tubular reactor.

This chapter is restricted to homogeneous, single-phase reactions, but the restriction can sometimes be relaxed. The formation of a second phase as a consequence of an irreversible reaction will not affect the kinetics, except for a possible density change. If the second phase is solid or liquid, the density change will be moderate. If the new phase is a gas, its formation can have a major effect. Specialized models are needed. Two-phase flows of air–water and steam–water have been extensively studied, but few data are available for chemically reactive systems.

3.2 SCALEUP OF TUBULAR REACTORS

There are three conceptually different ways of increasing the capacity of a tubular reactor:

1. Add identical reactors in parallel. The shell-and-tube design used for heat exchangers is a common and inexpensive way of increasing capacity.
2. Make the tube longer. Adding tube length is not a common means of increasing capacity, but it is used. Single-tube reactors exist that are several miles long.
3. Increase the tube diameter, either to maintain a constant pressure drop or to scale with geometric similarity. Geometric similarity for a tube means keeping the same length-to-diameter ratio L/d_t upon scaleup. Scaling with a constant pressure drop will lower the length-to-diameter ratio if the flow is turbulent.

The first two of these methods are preferred when heat transfer is important. The third method is cheaper for adiabatic reactors.

The primary goal of scaleup is to maintain acceptable product quality. Ideally, this will mean making exactly the same product in the large unit as was made in the pilot unit. To this end, it may be necessary to alter the operating conditions in the pilot plant so that product made there can be duplicated upon scaleup. If the pilot plant closely approaches isothermal piston flow, the challenge of maintaining these ideal conditions upon scaleup may be too difficult. The alternative is to make the pilot plant less ideal but more scaleable.

This chapter assumes isothermal operation. The scaleup methods presented here treat relatively simple issues such as pressure drop and in-process inventory. The methods of this chapter are usually adequate if the heat of reaction is negligible or if the pilot unit operates adiabatically. Although included in the examples that follow, laminar flow, even isothermal laminar flow, presents special scaleup problems that are treated in more detail in Chapter 8. The problem of controlling a reaction exotherm upon scaleup is discussed in Chapter 5.

If the pilot reactor is turbulent and closely approximates piston flow, the larger unit will as well. In isothermal piston flow, reactor performance is determined by the feed composition, feed temperature, and the mean residence time in the reactor. Even when piston flow is a poor approximation, these parameters are rarely, if ever, varied in the scaleup of a tubular reactor. The scaleup factor for throughput is S. To keep $\bar{t}$ constant, the inventory of mass in the system must also scale as S. When the fluid is incompressible, the volume scales with S. The general case allows the number of tubes, the tube radius, and the tube length to be changed upon scaleup:

$$S = \frac{V_2}{V_1} = \frac{(N_{tubes})_2 R_2^2 L_2}{(N_{tubes})_1 R_1^2 L_1} = S_{tubes} S_R^2 S_L \qquad \text{(incompressible)} \qquad (3.31)$$

where $S_{tubes} = (N_{tubes})_2/(N_{tubes})_1$ is the scaleup factor for the number of tubes, $S_R = R_2/R_1$ is the scaleup factor for radius, and $S_L = L_2/L_1$ is the scaleup factor for length. For an ideal gas with a negligible change in the number of moles due to reaction, constancy of $\bar{t}$ requires that the molar inventory scale with S. The inventory calculations in Example 3.4 can be used to determine

$$S = S_{tubes} S_R^2 \frac{[P_{in}^3 - P_{out}^3]_2 \beta_1}{[P_{in}^3 - P_{out}^3]_1 \beta_2} \quad \text{(ideal gas)} \quad (3.32)$$

The scaleup strategies that follow have been devised to satisfy Equation (3.31) for liquid systems and Equation (3.32) for gas systems.

3.2.1 Tubes in Parallel

Scaling in parallel gives an exact duplication of reaction conditions. The number of tubes increases in direct proportion to the throughput:

$$S = S_{tubes} = \frac{(N_{tubes})_2}{(N_{tubes})_1} \quad (3.33)$$

Equation (3.31) is satisfied with $S_R = S_L = 1$. Equation (3.32) is satisfied the same way, but with the added provision that the inlet and outlet pressures are the same in the large and small units. Scaling in parallel automatically keeps the same value for $\bar{t}$. The scaleup should be an exact duplication of the pilot plant results but at S times the flow rate.

There are three, somewhat similar, concerns about scaling in parallel. The first concern applies mainly to viscous fluids in unpacked tubes. The second applies mainly to packed tubes.

1. Will the feed distribute itself evenly between the tubes? This is a concern when there is a large change in viscosity due to reaction. The resulting stability problem is discussed in Chapter 13. Feed distribution can also be a concern with very large tube bundles when the pressure drop down the tube is small.

2. Will a single tube in a pilot plant adequately represent the range of behaviors in a multitubular design? This question arises in heterogeneous reactors using large-diameter catalyst particles in small-diameter tubes. The concern is that random variations in the void fraction will cause significant tube-to-tube variations. One suggested solution is to pilot with a minimum of three tubes in parallel. Replicate runs, repacking the tubes between runs, could also be used.

3. Will the distribution of flow on the shell side be uniform enough to give the same heat transfer coefficient for all the tubes?

Subject to resolution of these concerns, scaling in parallel has no obvious limit. Multitubular reactors with 10,000 tubes have been built, e.g., for phthalic anhydride oxidation.

A usual goal of scaleup is to maintain a *single-train* process. This means that the process will consist of a single line of equipment, and will have a single control system and a single operating crew. Single-train processes give the greatest economies of scale and are generally preferred for high-volume chemicals. Shell-and-tube designs are not single-train in a strict sense, but they are cheap to fabricate and operate if all the tubes are fed from a single source and discharge into a common receiver. Thus, shell-and-tube designs are allowed in the usual definition of a single-train process.

Heat transfer limits the maximum tube diameter. If large amounts of heat must be removed, it is normal practice to run the pilot reactor with the same diameter tube as intended for the full-scale reactor. The extreme choices are to scale in complete parallel with $S_{tubes} = S$ or to scale in complete series using a single tube. Occasionally, the scaleup may be a compromise between parallel and series, e.g., double the tube length and set $S_{tubes} = S/2$. Increases in tube diameter are possible if the heat transfer requirements are low to moderate. When adiabatic operation is acceptable, single-tube designs are preferred. The treatment that follows will consider only a single tube, but the results can be applied to multiple tubes just by reducing S so that it becomes the scaleup factor for a single tube. Choose a value for S_{tubes} and use the modified scaleup factor, $S' = S/S_{tubes}$, in the calculations that follow.

3.2.2 Tubes in Series

Scaling in series—meaning keeping the same tube diameter and increasing the tube length—is somewhat unusual but is actually a conservative way of scaling when the fluid is incompressible. It obviously maintains a single-train process. If the length is doubled, the flow rate can be doubled while keeping the same residence time. As will be quantified in subsequent chapters, a liquid-phase tubular reactor that works well in the pilot plant will probably work even better in a production unit that is 100 times longer and has 100 times the output. This statement is true even if the reaction is nonisothermal. The rub, of course, is the pressure drop. Also, even a liquid will show some compressibility if the pressure is high enough. However, single tubes that are several miles long do exist, and a 25% capacity increase at a high-pressure polyethylene plant was achieved by adding an extra mile to the length of the reactor!

The Reynolds number is constant when scaling in parallel, but it increases for the other forms of scaleup. When the large and small reactors both consist of a single tube,

$$\frac{\mathbf{Re}_2}{\mathbf{Re}_1} = \frac{R_2 \bar{u}_2}{R_1 \bar{u}_1} = \left[\frac{R_2}{R_1}\right]^{-1} \left[\frac{Q_2}{Q_1}\right] = S_R^{-1} S$$

For a series scaleup, $S_R = 1$, so that **Re** increases as S. This result ignores possible changes in physical properties. The factor ρ/μ will usually increase with pressure, so **Re** will increase even faster than S.

Series Scaleup of Turbulent Liquid Flows. For series scaleup of an incompressible fluid, the tube length is increased in proportion to the desired increase in throughput. Equation (3.31) is satisfied with $S_R = S_{tubes} = 1$ and $S_L = S$.

To determine the pressure drop, substitute Equation (3.16) into Equation (3.15) to obtain

$$\Delta P = 0.066 L \rho^{0.75} \mu^{0.25} \bar{u}^{1.75} R^{-1.25}$$

This integrated version of Equation (3.15) requires viscosity to be constant as well as density, but this assumption is not strictly necessary. See Problem 3.15. Write separate equations for the pressure drop in the large and small reactors and take their ratio. The physical properties cancel to give the following, general relationship:

$$\frac{\Delta P_2}{\Delta P_1} = \left[\frac{\bar{u}_2}{\bar{u}_1}\right]^{1.75} \left[\frac{L_2}{L_1}\right] \left[\frac{R_2}{R_1}\right]^{-1.25} = \left[\frac{Q_2}{Q_1}\right]^{1.75} \left[\frac{L_2}{L_1}\right] \left[\frac{R_2}{R_1}\right]^{-4.75} = S^{1.75} S_L S_R^{-4.75} \quad (3.34)$$

This section is concerned with the case of $S_R = 1$ and $S_L = S$ so that

$$\frac{\Delta P_2}{\Delta P_1} = S^{2.75} \quad (3.35)$$

A factor of 2 scaleup at constant $\bar{t}$ increases both $\bar{u}$ and L by a factor of 2, but the pressure drop increases by a factor of $2^{2.75} = 6.73$. A factor of 100 scaleup increases the pressure drop by a factor of 316,000! The external area of the reactor, $2\pi RL$, increases as S, apace with the heat generated by the reaction. The Reynolds number also increases as S and the inside heat transfer coefficient increases by $S^{0.8}$ (see Chapter 5). There should be no problem with heat transfer if you can tolerate the pressure drop.

The power input to the fluid by the pump, $Q \Delta P$, increases dramatically upon scaleup, as $S^{3.75}$. The power per unit volume of fluid increases by a factor of $S^{2.75}$.

In turbulent flow, part of this extra energy buys something. It increases turbulence and improves heat transfer and mixing.

Series Scaleup of Laminar Liquid Flows. The pressure drop is given by Equation (3.14). Taking ratios gives

$$\frac{\Delta P_2}{\Delta P_1} = \left[\frac{\bar{u}_2}{\bar{u}_1}\right] \left[\frac{L_2}{L_1}\right] \left[\frac{R_2}{R_1}\right]^{-2} = \left[\frac{Q_2}{Q_1}\right] \left[\frac{L_2}{L_1}\right] \left[\frac{R_2}{R_1}\right]^{-4} = S S_L S_R^{-4} \quad (3.36)$$

Equation (3.36) is the laminar flow counterpart of Equation (3.34). For the current case of $S_R = 1$,

$$\frac{\Delta P_2}{\Delta P_1} = S^2 \quad (3.37)$$

The increase in pumping energy is smaller than for turbulent flow but is still large. The power input to a unit volume of fluid increases by a factor of S^2.

With viscous fluids, pumping energy on the small scale may already be important and will increase upon scaleup. This form of energy input to a fluid is known as *viscous dissipation*. Alas, the increase in energy only buys an increase in fluid velocity unless the Reynolds number—which scales as S—increases enough to cause turbulence. If the flow remains laminar, heat transfer and mixing will remain similar to that observed in the pilot unit. Scaleup should give satisfactory results if the pressure drop and consequent viscous heating can be tolerated.

Series Scaleup of Turbulent Gas Flows. The compressibility causes complications. The form of scaleup continues to set $S_R = S_{tubes} = 1$, but now $S_L < S$. If the reactor length is increased and the exhaust pressure is held constant, the holdup within the reactor will increase more than proportionately because the increased length will force a higher inlet pressure and thus higher densities. When scaling with constant residence time, the throughput increases much faster than length. The scaled-up reactors are remarkably short. They will be highly turbulent since the small reactor is assumed to be turbulent, and the Reynolds number increases by a factor of S upon scaleup.

The discharge pressure for the large reactor, $(P_{out})_2$, may be set arbitrarily. Normal practice is to use the same discharge pressure as for the small reactor, but this is not an absolute requirement. The length of the large reactor, L_2, is chosen to satisfy the inventory constraint of Equation (3.32), and the inlet pressure of the large reactor becomes a dependent variable. The computation procedure actually calculates it first. Substitute Equation (3.23) for β (for turbulent flow) into Equation (3.32) to give

$$\frac{(P_{in}^3)_2 - (P_{out}^3)_2}{(P_{in}^3)_1 - (P_{out}^3)_1} = S^{2.75} S_R^{-6.75} \tag{3.38}$$

Everything is known in this equation but $(P_{out})_2$. Now substitute Equation (3.23) (this uses the turbulent value for β) into Equation (3.24) to obtain

$$\frac{(P_{in}^2)_2 - (P_{out}^2)_2}{(P_{in}^2)_1 - (P_{out}^2)_1} = S^{1.75} S_R^{-4.75} S_L \tag{3.39}$$

Everything is known in this equation but S_L. Note that Equations (3.38) and (3.39) contain S_R as a parameter. When scaling in series, $S_R = 1$, but the same equations can be applied to other scaleup strategies.

Example 3.7: Determine the upstream pressure and the scaling factor for length for gas-phase scaleups that are accomplished by increasing the reactor length at constant diameter. Assume that the pilot reactor is fully turbulent. Assume ideal gas behavior and ignore any change in the number of moles due to reaction. Both the pilot-scale and large-scale reactors will operate with a discharge pressure of 1 (arbitrary units). Consider a variety of throughput scaling factors and observed inlet pressures for the pilot unit.

TABLE 3.1 Series Scaleup of Gas-Phase Reactors in Turbulent Flow

S	$(P_{in}/P_{out})_1$	$(P_{in}/P_{out})_2$	L_2/L_1	$\Delta P_2/\Delta P_1$
2	100	189	1.06	1.90
2	10	18.9	1.07	1.99
2	2	3.6	1.21	2.64
2	1.1	1.48	1.68	4.78
100	100	8813	1.47	68.8
100	10	681	1.48	75.6
100	2	130	1.79	129
100	1.1	47.1	3.34	461

Solution: For this scaleup, $S_R = 1$. Substitute this, the desired value for S, $(P_{out})_1 = (P_{out})_2 = 1$, and the experimental observation for $(P_{in})_1$ into Equation (3.38). Solve for $(P_{in})_2$ and substitute into Equation (3.39) to calculate S_L. Some results are shown in Table 3.1.

At first glance, these results seem fantastic. Look at the case where $S = 100$. When the pressure drop across the pilot reactor is large, a mere 47% increase in length gives a 100-fold increase in inventory! The pressure and the density increase by a factor of about 69. Multiply the pressure increase by the length increase and the factor of 100 in inventory has been found. The reactor volume increases by a factor of only 1.47. The inventory and the throughput scale as S. The scaling factor for volume is much lower, 1.47 instead of 100 in this example.

Table 3.1 suggests that scaling in series could make sense for an adiabatic, gas-phase reaction with no change in the number of moles upon reaction. It would also make sense when the number of moles decreases upon reaction, since the high pressures caused by this form of scaleup will favor the forward reaction. Chapter 5 gives the design equations for nonisothermal reactions and discusses the thermal aspects of scaleup.

Series Scaleup of Laminar Gas Flows. The scaling equations are similar to those used for turbulent gas systems but the exponents are different. The different exponents come from the use of Equation (3.22) for β rather than Equation (3.23). General results, valid for any form of scaleup that uses a single tube, are

$$\frac{(P_{in}^3)_2 - (P_{out}^3)_2}{(P_{in}^3)_1 - (P_{out}^3)_1} = S^2 S_R^{-6} \tag{3.40}$$

$$\frac{(P_{in}^2)_2 - (P_{out}^2)_2}{(P_{in}^2)_1 - (P_{out}^2)_1} = S S_R^{-4} S_L \tag{3.41}$$

Example 3.8: Repeat Example 3.7, now assuming that both the small and large reactors are in laminar flow.

Solution: The approach is similar to that in Example 3.7. The unknowns are S_L and $(P_{in})_2$. Set $(P_{out})_2 = (P_{out})_1$. Equation (3.40) is used to calculate $(P_{in})_2$ and Equation (3.41) is used to calculate S_L. Results are given in Table 3.2. The results are qualitatively similar to those for the turbulent flow of a gas, but the scaled reactors are longer and the pressure drops are lower. In both cases, the reader should recall that the ideal gas law was assumed. This may become unrealistic for higher pressures. In Table 3.2 we make the additional assumption of laminar flow in both the large and small reactors. This assumption will be violated if the scaleup factor is large.

Series Scaleup of Packed Beds. According to the Ergun equation, Equation (3.17), the pressure drop in a packed bed depends on the packing diameter, but is independent of the tube diameter. This is reasonable with small packing. Here, we shall assume that the same packing is used in both large and small reactors and that it is small compared with the tube diameter. Chapter 9 treats the case where the packing is large compared with the tube diameter. This situation is mainly encountered in heterogeneous catalysis with large reaction exotherms. Such reactors are almost always scaled in parallel.

The pressure drop in a packed bed depends on the particle Reynolds number. When $(\mathbf{Re})_p$ is small, Equation (3.17) becomes

$$\frac{dP}{dz} = -\frac{150\mu\bar{u}_s}{d_p^2}\frac{(1-\varepsilon)^2}{\varepsilon}$$

This equation has the same functional dependence on ρ (namely none) and $\bar{u}$ as the Poiseuille equation that governs laminar flow in an empty tube. Thus, laminar flow packed beds scale in series exactly like laminar flow in empty tubes. See the previous sections on series scaleup of liquids and gases in laminar flow.

If $\mathbf{Re}_p$ is large, Equation (3.17) becomes

$$\frac{dP}{dz} = -\frac{1.75\rho\bar{u}_s^2}{d_p}\frac{(1-\varepsilon)}{\varepsilon}$$

TABLE 3.2 Series Scaleup of Gas-Phase Reactors in Laminar Flow

S	$(P_{in}/P_{out})_1$	$(P_{in}/P_{out})_2$	L_2/L_1	$\Delta P_2/\Delta P_1$
2	100	159	1.26	1.90
2	10	15.9	1.27	1.99
2	2	3.1	1.41	2.64
2	1.1	1.3	1.80	4.78
100	100	2154	4.64	21.8
100	10	215	4.69	23.6
100	2	41.2	5.66	40.2
100	1.1	14.9	10.5	139

which has a similar functional dependence on ρ and $\bar{u}$ as Equation (3.15). The dependence on Reynolds number via the friction factor **Fa** is missing, but this quarter-power dependence is weak. To a first approximation, a turbulent packed bed will scale like turbulent flow in an empty tube. See the previous sections on series scaleup of liquids and gases in turbulent flow. To a second approximation, the pressure drop will increase somewhat faster upon scaleup. At high values of $(\mathbf{Re})_p$, the pressure drop shows a scaling exponent of 3 rather than 2.75:

$$\frac{\Delta P_2}{\Delta P_1} \to S^3 \quad \text{as} \quad (\mathbf{Re})_p \to \infty$$

At the other limiting value, Equation (3.17) becomes

$$\frac{\Delta P_2}{\Delta P_1} \to S^2 \quad \text{as} \quad (\mathbf{Re})_p \to 0$$

Once a scaleup strategy has been determined, Equation (3.17)—rather than the limiting cases for laminar and turbulent flow—should be used for the final calculations.

3.2.3 Scaling with Geometric Similarity

Scaling in parallel keeps a constant ΔP upon scaleup, but multitubular designs are not always the best choice. Scaling in series uses a single tube but increases the total pressure drop to what can be excessive levels. One approach to keeping a single-train process is to install booster pumps at intermediate points. This approach is used in some polymer processes. We now consider a single-tube design where the tube diameter is increased in order to limit the pressure in the full-scale plant. This section considers a common but not necessarily good means of scaleup wherein the large and small reactors are geometrically similar. Geometric similarity means that $S_R = S_L$, so the large and small tubes have the same aspect ratio. For incompressible fluids, the volume scales with S, so that $S_R = S_L = S^{1/3}$. The Reynolds number scales as

$$\frac{\mathbf{Re}_2}{\mathbf{Re}_1} = \frac{R_2 \bar{u}_2}{R_1 \bar{u}_1} = \left[\frac{R_2}{R_1}\right]^{-1} \left[\frac{Q_2}{Q_1}\right] = S_R^{-1} S = S^{2/3}$$

The case of a compressible fluid is more complicated since it is the inventory and not the volume that scales with S. The case of laminar flow is the simplest and is one where scaling with geometric similarity can make sense.

Geometrically Similar Scaleups for Laminar Flows in Tubes. The pressure drop for this method of scaleup is found using the integrated form of the Poiseuille equation:

$$\Delta P = \frac{8 \mu \bar{u} L}{R^2}$$

Taking ratios,

$$\frac{\Delta P_2}{\Delta P_1} = \frac{\bar{u}_2 L_2 R_1^2}{\bar{u}_1 L_1 R_2^2} = \left[\frac{R_2^2 \bar{u}_2}{R_1^2 \bar{u}_1}\right]\left[\frac{L_2}{L_1}\right]\left[\frac{R_1^4}{R_2^4}\right] = SS_L S_R^{-4} \qquad (3.42)$$

Substituting $S_R = S_L = S^{1/3}$ gives

$$\frac{\Delta P_2}{\Delta P_1} = S^0 = 1$$

so that the pressure drop remains constant upon scaleup.

The same result is obtained when the fluid is compressible, as may be seen by substituting $S_R = S_L = S^{1/3}$ into Equations (3.40) and (3.41). Thus, using geometric similarity to scale isothermal, laminar flows gives constant pressure drop provided the flow remains laminar upon scaleup. The large and small reactors will have the same inlet pressure if they are operated at the same outlet pressure. The inventory and volume both scale as S.

The external area scales as $S^{2/3}$, so that this design has the usual problem of surface area rising more slowly than heat generation. There is another problem associated with laminar flow in tubes. Although piston flow may be a reasonable approximation for a small-diameter pilot reactor, it will cease to be a reasonable assumption upon scaleup. As described in Chapter 8, radial diffusion of mass and heat gives beneficial effects in small equipment that will decline upon scaleup. Geometrically similar scaleups of laminar flow in tubes cannot be recommended unless radial diffusion was negligible in the pilot-scale reactor. However, if it was negligible at that scale, the reactor cannot be analyzed using the assumptions of piston flow. Instead, there will be pronounced radial gradients in composition and temperature that are analyzed using the methods of Chapter 8.

Geometrically Similar Scaleups for Turbulent Flows in Tubes. Integrating Equation (3.15) for the case of constant density and viscosity gives

$$\Delta P = \frac{0.066 \mu^{0.25} \rho^{0.75} \bar{u}^{1.75} L}{R^{1.25}}$$

and

$$\frac{\Delta P_2}{\Delta P_1} = \frac{S^{1.75} S_L}{S_R^{4.75}} \qquad (3.43)$$

Setting $S_L = S_R = S^{1/3}$ gives a surprisingly simple result:

$$\frac{\Delta P_2}{\Delta P_1} = S^{1/2} \qquad (3.44)$$

In laminar flow, the pressure drop is constant when scaleup is carried out by geometric similarity. In turbulent flow, it increases as the square root of throughput. There is extra pumping energy per unit volume of throughput, which gives somewhat better mixing and heat transfer. The surface area and Reynolds number both scale as $S^{2/3}$. We shall see in Chapter 5 that the increase in heat transfer coefficient is insufficient to overcome the relative loss in surface area. The reaction will become adiabatic if the scaleup factor is large.

Turning to the case where the working fluid is an ideal gas, substituting $S_R = S_L = S^{1/3}$ into Equations (3.38) and (3.39) gives $S^{1/2}$ as the scaling factor for both pressure ratios. This looks neat, but there is no solution to the scaling equations if both reactors have the same discharge pressure. What happens is that the larger reactor has too much inventory to satisfy the condition of constant $\bar{t}$. Scaleup using $S_R = S_L = S^{1/3}$ requires that the discharge pressure be lower in the large unit than in the small one. Even so, scaleup may not be possible because the discharge pressure of the large unit cannot be reduced below zero. Geometrically similar scaleups of turbulent gas flows are possible, but not with $S_R = S_L = S^{1/3}_{Inventory}$.

Geometrically Similar Scaleups for Packed Beds. As was the case for scaling packed beds in series, the way they scale with geometric similarity depends on the particle Reynolds number. The results are somewhat different than those for empty tubes because the bed radius does not appear in the Ergun equation. The asymptotic behavior for the incompressible case is

$$\frac{\Delta P_2}{\Delta P_1} \to S^2 S_L S_R^{-4} = S \quad \text{as} \quad (\text{Re})_p \to \infty$$

Note that S_R appears here even though it is missing from the Ergun equation. It arises because throughput is proportional to $R^2 \bar{u}$.

The other limiting value is

$$\frac{\Delta P_2}{\Delta P_1} \to S S_L S_R^{-2} = S^{2/3} \quad \text{as} \quad (\text{Re})_p \to 0$$

These asymptotic forms may be useful for conceptual studies, but the real design calculations must be based on the full Ergun equation. Turning to the case of compressible fluids, scaleup using geometric similarity with $S_R = S_L = S^{1/3}$ is generally infeasible. Simply stated, the reactors are just too long and have too much inventory.

3.2.4 Scaling with Constant Pressure Drop

This section considers how single tubes can be scaled up to achieve higher capacity at the same residence time and pressure drop. In marked contrast to the previous section, these scaleups are usually feasible, particularly for gas-phase reactions, although they have the common failing of losing heat transfer area relative to throughput.

Constant-Pressure Scaleups for Laminar Flows in Tubes. As shown in the previous section, scaling with geometric similarity, $S_R = S_L = S^{1/3}$, gives constant-pressure drop when the flow is laminar and remains laminar upon scaleup. This is true for both liquids and gases. The Reynolds number and the external area increase as $S^{2/3}$. Piston flow is a poor assumption for laminar flow in anything but small tubes. Thus, the conversion and selectivity of the reaction is likely to worsen upon scaleup. Ways to avoid unpleasant surprises are discussed in Chapter 8.

Constant-Pressure Scaleups for Turbulent Flows in Tubes. Equation (3.34) gives the pressure drop ratio for large and small reactors when density is constant. Set $\Delta P_2 = \Delta P_1$ to obtain $1 = S^{1.75} S_L S_R^{-4.75}$. Equation (3.31) gives the inventory relationship when density is constant. Set $S_{tubes} = 1$ to obtain $S = S_L S_R^2$. Simultaneous solution gives

$$S_R = S^{11/27} \quad \text{and} \quad S_L = S^{5/27} \quad (3.45)$$

The same results are obtained from Equations (3.38) and (3.39), which apply to the turbulent flow of ideal gases. Thus, tube radius and length scale in the same way for turbulent liquids and gases when the pressure drop is constant. For the gas case, it is further supposed that the large and small reactors have the same discharge pressure.

The reactor volume scales as S, and the aspect ratio of the tube decreases upon scaleup. The external surface area scales as $S_R S_L = S^{16/27}$ compared with $S^{2/3}$ for the case with geometric similarity. The Reynolds number also scales as $S^{16/27}$. It increases upon scaleup in both cases, but less rapidly when the pressure drop is held constant than for geometric similarity.

Constant-Pressure Scaleups for Packed Beds. A scaleup with constant pressure drop can be achieved in a packed bed just by increasing the diameter to keep a constant gas velocity $\bar{u}_s$. This gives

$$S_R = S^{1/2} \quad \text{and} \quad S_L = 1$$

Obviously, the ability to transfer heat through the walls drops dramatically when scaling in this fashion, but it is certainly a straightforward and normal method for scaling adiabatic reactions in packed beds. A potential limit arises when the bed diameter becomes so large that even distribution of the entering fluid becomes a problem. Large packed beds are the preferred reactor for heterogeneous catalysis if the reaction (and the catalyst) can tolerate the adiabatic temperature rise. Packed beds are also commonly used for multiphase reactions.

3.2.5 Scaling Down

Small versions of production facilities are sometimes used for product development, particularly in the polymer industries. Single-train plants producing

20–50 t/h are becoming common for the major-volume plastics such as polyethylene, polypropylene, and polystyrene. These plastics are made in many grades, and the optimization of product properties is a means of finding competitive advantage in what would otherwise be a strictly commodity market. Important property changes can result from subtle changes in raw materials, catalysts, and operating conditions.

Multiply the production rate by the selling price and you will understand management's reluctance to conduct product development experiments in the plant. Pilot plants, built and operated after the fact of the production line, are fairly common. Some process licensors include the design of a pilot plant in their technology package for a full-scale plant. The purpose of these pilot plants is to duplicate the performance of the full-scale line at a fraction of the rate. The scaledown factor between the two plants will typically be in the range 100–1000. This would be considered highly ambitious for a scaleup. There is less risk when scaling down, but it may be necessary to adjust the heating and mixing characteristics of the pilot plant to make them as bad as they are in the full-scale facility.

A very different reason for scaling down arises in fields such as biotechnology, microelectronics, and nanotechnology. We are interested in building, altering, or just plain understanding very small reactors, but find it difficult or impossible to do the necessary experiments on the small scale. Measurements made on the "pilot plant" will ultimately be scaled down to the "production plant." One generalization is that the small unit will probably be in laminar flow and, if biological, will be isothermal.

The scaling methods in this chapter work about as well or as poorly when $S < 1$ as when $S > 1$. Scaling down in parallel works until there is only a single tube. Other forms of scaledown cause a decrease in Reynolds number that may cause a transition to laminar flow. Scaling down in series may lead to infeasible L/d_t ratios. Scaling by geometric similarity tends to work better going down than going up. The surface area and Reynolds number both decrease, but they decrease only by $S^{2/3}$ while throughput decreases by S. Thus, heat and mass transfer tend to be better on the small scale. The inventory in a gas system will tend to be too low when scaling down by geometric similarity, but a backpressure valve on the small reactor can be used to adjust it. Scaling at constant pressure drop increases the length-to-diameter ratio in the smaller unit. Packed beds can be scaled down as long as the ratio of bed diameter to packing diameter is reasonable, although values as low as 3 are sometimes used. Scaling down will improve radial mixing and heat transfer. The correlations in Section 9.1 include the effects of packing diameter, although the range of the experimental data on which these correlations are based is small.

As a general rule, scaled-down reactors will more closely approach isothermal operation but will less closely approach ideal piston flow when the large reactor is turbulent. Large scaledowns will lead to laminar flow. If the large system is laminar, the scaled-down version will be laminar as well and will more closely approach piston flow due to greater radial diffusion.

3.3 TRANSPIRED-WALL REACTORS

Tubular reactors sometimes have side entrance points for downstream injection. Like the case of fed-batch reactors, this raises the question of how quickly the new ingredients are mixed. Mixing in the radial direction is the dominant concern. If radial mixing is fast, the assumption of piston flow may be reasonable and the addition of new ingredients merely reinitializes the problem. The equivalent phenomenon was discussed in Section 2.6.2 for fed-batch reactors.

This section considers the case where the tube has a porous wall so that reactants or inerts can be fed gradually. *Transpiration* is used to cool the walls in high-temperature combustions. In this application, there is usually a change of phase, from liquid to gas, so that the cooling benefits from the heat of vaporization. However, we use the term transpiration to include transfer through a porous wall without a phase change. It can provide chemical protection of the wall in extremely reactive systems such as direct fluorinations. There may be selectivity advantages for complex reactions. This possibility is suggested by Example 3.9.

Assume that the entering material is rapidly mixed so that the composition is always uniform in the radial direction. The transpiration rate per unit length of tube is $q = q(z)$ with units of m²/s. Component A has concentration $a_{trans} = a_{trans}(z)$ in the transpired stream. The component balance, Equation (3.4), now becomes

$$\frac{1}{A_c}\frac{d(\dot{N}_A)}{dz} = \frac{1}{A_c}\frac{d(Qa)}{dz} = \frac{1}{A_c}\frac{d(A_c\bar{u}a)}{dz} = \frac{a_{trans}q}{A_c} + \mathcal{R}_A \quad (3.46)$$

We also need a total mass balance. The general form is

$$Q\rho = Q_{in}\rho_{in} + \int_0^z q\rho_{trans}\,dz \quad (3.47)$$

Analytical solutions are possible in special cases. It is apparent that transpiration will lower the conversion of the injected component. It is less apparent, but true, that transpired wall reactors can be made to approach the performance of a CSTR with respect to a transpired component while providing an environment similar to piston flow for components that are present only in the initial feed.

Example 3.9: Solve Equation (3.46) for the case of a first-order reaction where ρ, q and a_{trans} are constant. Then take limits as $Q_{in} \to 0$ and see what happens. Also take the limit as $q \to 0$.

Solution: With constant density, Equation (3.47) becomes

$$Q = Q_{in} + qz$$

Substitute this into the Qa version of Equation (3.46) to obtain a variable-separable ODE. Integrate it subject to the initial condition that $a = a_{in}$ at $z = 0$. The result is

$$a(z) = \frac{qa_{trans}}{A_c k + q} - \frac{\left[\dfrac{qa_{trans}}{A_c k + q} - a_{in}\right]}{\left[1 + \dfrac{qz}{Q_{in}}\right]^{(A_c k + q)/q}} \qquad (3.48)$$

Taking the limit as $Q_{in} \to 0$ gives

$$a = \frac{qa_{trans}}{A_c k + q} = \frac{a_{trans}}{\dfrac{A_c L k}{Q_{out}} + 1}$$

The z dependence has disappeared! The reactor is well mixed and behaves like a CSTR with respect to component A. Noting that $Q_{out} = qL$ gives

$$a_{out} = \frac{a_{trans}}{1 + \dfrac{Vk}{Q_{out}}} = \frac{a_{trans}}{1 + k\bar{t}}$$

which is exactly the behavior of a CSTR. When a transpired-wall reactor has no initial feed, it behaves like a stirred tank. When $Q_{in} > 0$ but $a_{in} = 0$, it will still have a fairly uniform concentration of A inside the reactor while behaving much like a piston flow reactor for component B, which has $b_{in} > 0$ but $b_{trans} = 0$. For this component B,

$$b(z) = \frac{b_{in}}{\left[1 + \dfrac{qz}{Q_{in}}\right]^{(A_c k + q)/q}}$$

Physical insight should tell you what this becomes in the limit as $q \to 0$. Problem 2.7 shows the mathematics of the limit.

This example shows an interesting possibility of achieving otherwise unobtainable products through the use of transpired-wall reactors. They have been proposed for the manufacture of a catalyst used in ammonia synthesis.[1] Transpiration might be useful in maintaining a required stoichiometry in copolymerizations where the two monomers polymerize at different rates, but a uniform product is desired. For the specific case of an anionic polymerization, transpiration of the more reactive monomer could give a chemically uniform copolymer while maintaining a narrow molecular weight distribution. See Section 13.4 for the background to this statement.

Membrane reactors, whether batch or continuous, offer the possibility of selective transpiration. They can be operated in the reverse mode so that some products are selectively removed from the reaction mix in order to avoid an equilibrium limitation. Membrane reactors can be used to separate cell mass from fermentation products. See Section 12.2.2.

PROBLEMS

3.1. The first-order sequence A $\xrightarrow{k_I}$ B $\xrightarrow{k_{II}}$ C is occurring in a constant-density piston flow reactor. The residence time is $\bar{t}$.
   (a) Determine b_{out} and c_{out} given that $b_{in} = c_{in} = 0$ and that $k_I = k_{II}$.
   (b) Find a real chemical example, not radioactive decay, where the assumption that $k_I = k_{II}$ is plausible. As a last resort, you may consider reactions that are only pseudo-first-order.

3.2. Suppose

$$\mathbf{v} = \begin{bmatrix} -1 & 0 \\ -1 & -1 \\ 1 & -1 \\ 0 & 0 \end{bmatrix}$$

gives the stoichiometric coefficients for a set of elementary reactions.

   (a) Determine the elementary reactions and the vector of reaction rates that corresponds to $\mathbf{v}$.
   (b) Write the component balances applicable to these reactions in a PFR with an exponentially increasing reactor cross section, $A_c = A_{inlet} \exp(Bz)$.

3.3. Equation (3.10) can be applied to an incompressible fluid just by setting $\bar{t} = V/Q$. Show that you get the same result by integrating Equation (3.8) for a first-order reaction with arbitrary $A_c = A_c(z)$.

3.4. Consider the reaction B $\xrightarrow{k}$ 2A in the gas phase. Use a numerical solution to determine the length of an isothermal, piston flow reactor that achieves 50% conversion of B. The pressure drop in the reactor is negligible. The reactor cross section is constant. There are no inerts. The feed is pure B and the gases are ideal. Assume $b_{in} = 1$, and $a_{in} = 0$, $\bar{u}_{in} = 1$, and $k = 1$ in some system of units.

3.5. Solve Problem 3.4 analytically rather than numerically.

3.6. Repeat the numerical solution in Example 3.2 for a reactor with variable cross section, $A_c = A_{inlet} \exp(Bz)$. Using the numerical values in that example, plot the length needed to obtain 50% conversion versus B for $-1 < B < 1$ (e.g. $z = 0.3608$ for $B = 0$). Also plot the reactor volume V versus B assuming $A_{inlet} = 1$.

3.7. Rework Example 3.3, now considering reversibility of the reaction.

Assume

$$K_{equil} = \frac{P_{STY}P_{H_2}}{P_{EB}} = 0.61 \text{ atm at } 700°C$$

3.8. Annular flow reactors, such as that illustrated in Figure 3.2, are sometimes used for reversible, adiabatic, solid-catalyzed reactions where pressure near the end of the reactor must be minimized to achieve a favorable equilibrium. Ethylbenzene dehydrogenation fits this situation. Repeat Problem 3.7 but substitute an annular reactor for the tube. The inside (inlet) radius of the annulus is 0.1 m and the outside (outlet) radius is 1.1 m.

3.9. Consider the gas-phase decomposition A → B + C in an isothermal tubular reactor. The tube i.d. is 1 in. There is no packing. The pressure drop is 1 psi with the outlet at atmospheric pressure. The gas flow rate is 0.05 SCF/s. The molecular weights of B and C are 48 and 52, respectively. The entering gas contains 50% A and 50% inerts by volume. The operating temperature is 700°C. The cracking reaction is first order with a rate constant of 0.93 s^{-1}. How long is the tube and what is the conversion? Use $\mu = 5 \times 10^{-5}$ Pa·s. *Answers*: 57 m and 98%.

3.10. Suppose B $\xrightarrow{k}$ 2A in the liquid phase and that the density changes from 1000 kg/m^3 to 900 kg/m^3 upon complete conversion. Find a solution to the batch design equation and compare the results with a hypothetical batch reactor in which the density is constant.

3.11. A pilot-scale, liquid-phase esterification with near-zero heat of reaction is being conduced in a small tubular reactor. The chemist thinks the reaction should be reversible, but the by-product water is sparingly soluble in the reaction mixture and you are not removing it. The conversion is 85%. Your job is to design a 100 × scaleup. The pilot reactor is a 31.8 mm i.d. tube, 4 m long, constructed from 12 BWG (2.769 mm) 316 stainless steel. The feed is preheated to 80°C and the reactor is jacketed with tempered water at 80°C. The material begins to discolor if higher temperatures are used. The flow rate is 50 kg/h and the upstream gauge pressure is 1.2 psi. The density of the mixture is around 860 kg/m^3. The viscosity of the material has not been measured under reaction conditions but is believed to be substantially independent of conversion. The pilot plant discharges at atmospheric pressure.

(a) Propose alternative designs based on scaling in parallel, in series, by geometric similarity, and by constant pressure drop. Estimate the Reynolds number and pressure drop for each case.

(b) Estimate the total weight of metal needed for the reactor in each of the designs. Do not include the metal needed for the water jacket in your weight estimates. Is the 12 BWG tube strong enough for all the designs?

(c) Suppose the full-scale reactor is to discharge directly into a finishing reactor that operates at 100 torr. Could this affect your design? What precautions might you take?

(d) Suppose you learn that the viscosity of the fluid in the pilot reactor is far from constant. The starting raw material has a viscosity of 0.0009 Pa·s at 80°C. You still have no measurements of the viscosity after reaction, but the fluid is obviously quite viscous. What influence will this have on the various forms of scaleup?

3.12. A pilot-scale, turbulent, gas-phase reactor performs well when operated with a inlet pressure of 1.02 bar and an outlet pressure of 0.99 bar. Is it possible to do a geometrically similar scaleup by a factor of 10 in throughput while maintaining the same mean residence time? Assume ideal gas behavior and ignore any change in the number of moles due to reaction. If necessary, the discharge pressure on the large reactor can be something other than 0.99 bar.

3.13. Refer to the results in Example 3.7 for a scaling factor of 100. Suppose that the pilot and large reactors are suddenly capped and the vessels come to equilibrium. Determine the equilibrium pressure and the ratio of equilibrium pressures in these vessels assuming
(a) $P_{in}/P_{out1} = 100$
(b) $P_{in}/P_{out1} = 10$
(c) $P_{in}/P_{out1} = 2$
(d) $P_{in}/P_{out1} = 1.1$

3.14. An alternative to Equation (3.16) is $\mathbf{Fa} = 0.04\mathbf{Re}^{-0.16}$. It is more conservative in the sense that it predicts higher pressure drops at the same Reynolds number. Use it to recalculate the scaling exponents in Section 3.2 for pressure drop. Specifically, determine the exponents for ΔP when scaling in series and with geometric similarity for an incompressible fluid in turbulent flow. Also, use it to calculate the scaling factors for S_R and S_L when scaling at constant pressure.

3.15. An integral form of Equation (3.15) was used to derive the pressure ratio for scaleup in series of a turbulent liquid-phase reactor, Equation (3.34). The integration apparently requires μ to be constant. Consider the case where μ varies down the length of the reactor. Define an average viscosity as

$$\hat{\mu} = \frac{1}{L}\int_0^L \mu(z)\,dz$$

Show that the Equation (3.34) is valid if the large and small reactors have the same value for $\hat{\mu}$ and that this will be true for an isothermal or adiabatic PFR being scaled up in series.

3.16. Suppose an inert material is transpired into a tubular reactor in an attempt to achieve isothermal operation. Suppose the transpiration rate q is independent of z and that $qL = Q_{trans}$. Assume all fluid densities to be constant and equal. Find the fraction unreacted for a first-order reaction. Express your final answer as a function of the two dimensionless parameters, Q_{trans}/Q_{in} and kV/Q_{in} where k is the rate constant and

Q_{in} is the volumetric flow rate at $z=0$ (i.e., $Q_{out} = Q_{in} + Q_{trans}$). *Hint*: the correct formula gives $a_{out}/a_{in} = 0.25$ when $Q_{trans}/Q_{in} = 1$ and $kV/Q_{in} = 1$.

3.17. Repeat Problem (3.16) for a second-order reaction of the $2A \xrightarrow{k/2} B$ type. The dimensionless parameters are now Q_{trans}/Q_{in} and $ka_{in}V/Q_{in}$.

REFERENCE

1. Gens, T. A., "Ammonia synthesis catalysts and process of making and using them," U.S. Patent 4,235,749, 11/25/1980.

SUGGESTIONS FOR FURTHER READING

Realistic examples of variable-property piston flow models, usually nonisothermal, are given in

Froment, G. F. and Bischoff, K. B., *Chemical Reactor Analysis and Design*, 2nd Ed., Wiley, New York, 1990.

Scaleup techniques are discussed in

Bisio, A. and Kabel, R. L., Eds., *Scaleup of Chemical Processes*, Wiley, New York, 1985.

CHAPTER 4
STIRRED TANKS AND REACTOR COMBINATIONS

Chapter 2 treated multiple and complex reactions in an ideal batch reactor. The reactor was ideal in the sense that mixing was assumed to be instantaneous and complete throughout the vessel. Real batch reactors will approximate ideal behavior when the characteristic time for mixing is short compared with the reaction half-life. Industrial batch reactors have inlet and outlet ports and an agitation system. The same hardware can be converted to continuous operation. To do this, just feed and discharge continuously. If the reactor is well mixed in the batch mode, it is likely to remain so in the continuous mode, as least for the same reaction. The assumption of instantaneous and perfect mixing remains a reasonable approximation, but the batch reactor has become a continuous-flow stirred tank.

This chapter develops the techniques needed to analyze multiple and complex reactions in stirred tank reactors. Physical properties may be variable. Also treated is the common industrial practice of using reactor combinations, such as a stirred tank in series with a tubular reactor, to accomplish the overall reaction.

4.1 CONTINUOUS-FLOW STIRRED TANK REACTORS

Perfectly mixed stirred tank reactors have no spatial variations in composition or physical properties within the reactor or in the exit from it. Everything inside the system is uniform except at the very entrance. Molecules experience a step change in environment immediately upon entering. A perfectly mixed CSTR has only two environments: one at the inlet and one inside the reactor and at the outlet. These environments are specified by a set of compositions and operating conditions that have only two values: either $a_{in}, b_{in}, \ldots, P_{in}, T_{in}$ or $a_{out}, b_{out}, \ldots, P_{out}, T_{out}$. When the reactor is at a steady state, the inlet and outlet properties are related by algebraic equations. The piston flow reactors and real flow reactors show a more gradual change from inlet to outlet, and the inlet and outlet properties are related by differential equations.

The component material balances for an ideal CSTR are the following set of algebraic equations:

$$Q_{in}a_{in} + \mathcal{R}_A(a_{out}, b_{out}, \ldots, P_{out}, T_{out})V = Q_{out}a_{out}$$
$$Q_{in}b_{in} + \mathcal{R}_B(a_{out}, b_{out}, \ldots, P_{out}, T_{out})V = Q_{out}b_{out} \quad (4.1)$$
$$\vdots \qquad \vdots \qquad\qquad\qquad \vdots$$

The reaction terms are evaluated at the outlet conditions since the entire reactor inventory is at these conditions. The set of component balances can be summarized as

$$Q_{in}\mathbf{a}_{in} + \mathbf{v}\mathcal{R}V = Q_{out}\mathbf{a}_{out} \quad (4.2)$$

where $\mathbf{v}$ is the $N \times M$ matrix of stoichiometric coefficients (see Equation (2.37)) and $\mathbf{a}_{in}$ and $\mathbf{a}_{out}$ are column vectors of the component concentrations.

For now, we assume that all operating conditions are known. This specifically includes P_{out} and T_{out}, which correspond to conditions within the vessel. There may be a backpressure valve at the reactor exit, but it is ignored for the purposes of the design equations. Suppose also that the inlet concentrations $a_{in}, b_{in}, \ldots$, volumetric flow rate Q_{in}, and working volume V are all known. Then Equations (4.1) or (4.2) are a set of N simultaneous equations in $N+1$ unknowns, the unknowns being the N outlet concentrations $a_{out}, b_{out}, \ldots$, and the one volumetric flow rate Q_{out}. Note that Q_{out} is evaluated at the conditions within the reactor. If the mass density of the fluid is constant, as is approximately true for liquid systems, then $Q_{out} = Q_{in}$. This allows Equations (4.1) to be solved for the outlet compositions. If Q_{out} is unknown, then the component balances must be supplemented by an equation of state for the system. Perhaps surprisingly, the algebraic equations governing the steady-state performance of a CSTR are usually more difficult to solve than the sets of simultaneous, first-order ODEs encountered in Chapters 2 and 3. We start with an example that is easy but important.

Example 4.1: Suppose a liquid-phase CSTR is used for consecutive, first-order reactions:

$$A \xrightarrow{k_A} B \xrightarrow{k_B} C \xrightarrow{k_C} D$$

Determine all outlet concentrations, assuming constant density.

Solution: When density is constant, $Q_{out} = Q_{in} = Q$ and $\bar{t} = V/Q$. Equations (4.1) become

$$a_{in} - k_A \bar{t} a_{out} = a_{out}$$

$$b_{in} + k_A \bar{t} a_{out} - k_B \bar{t} b_{out} = b_{out}$$

$$c_{in} + k_B \bar{t} b_{out} - k_C \bar{t} c_{out} = c_{out}$$

$$d_{in} + k_C \bar{t} c_{out} = c_{out}$$

These equations can be solved sequentially to give

$$a_{out} = \frac{a_{in}}{1 + k_A \bar{t}}$$

$$b_{out} = \frac{b_{in}}{(1 + k_B \bar{t})} + \frac{k_A \bar{t} a_{in}}{(1 + k_A \bar{t})(1 + k_B \bar{t})}$$

$$c_{out} = \frac{c_{in}}{a + k_C \bar{t}} + \frac{k_B \bar{t} b_{in}}{(1 + k_B \bar{t})(1 + k_C \bar{t})} + \frac{k_A k_B \bar{t}^2 a_{in}}{(1 + k_A \bar{t})(1 + k_B \bar{t})(1 + k_C \bar{t})} \quad (4.3)$$

$$d_{out} = d_{in} + (a_{in} - a_{out}) + (b_{in} - b_{out}) + (c_{in} - c_{out})$$

Compare these results with those of Equation (2.22) for the same reactions in a batch reactor. The CSTR solutions do not require special forms when some of the rate constants are equal. A plot of outlet concentrations versus $\bar{t}$ is qualitatively similar to the behavior shown in Figure 2.2, and $\bar{t}$ can be chosen to maximize b_{out} or c_{out}. However, the best values for $\bar{t}$ are different in a CSTR than in a PFR. For the normal case of $b_{in} = 0$, the $\bar{t}$ that maximizes b_{out} is a root-mean, $\bar{t}_{max} = 1/\sqrt{k_A k_B}$, rather than the log-mean of Equation (2.23). When operating at $\bar{t}_{max}$, the CSTR gives a lower yield of B and a lower selectivity than a PFR operating at its $\bar{t}_{max}$.

Competitive first-order reactions and a few other simple cases can be solved analytically, but any reasonably complex kinetic scheme will require a numerical solution. Mathematics programs such as Mathematica, Mathcad, and MatLab offer nearly automatic solvers for sets of algebraic equations. They usually work. Those readers who wish to understand the inner workings of a solution are referred to Appendix 4, where a multidimensional version of Newton's method is described. It converges quickly provided your initial guesses for the unknowns are good, but it will take you to never-never land when your initial guesses are poor. A more robust method of solving the design equations for multiple reactions in a CSTR is given in the next section.

4.2 THE METHOD OF FALSE TRANSIENTS

The method of false transients converts a steady-state problem into a time-dependent problem. Equations (4.1) govern the steady-state performance of a CSTR. How does a reactor reach the steady state? There must be a startup transient that eventually evolves into the steady state, and a simulation of that transient will also evolve to the steady state. The simulation need not be physically exact. Any startup trajectory that is mathematically convenient can be used, even if it does not duplicate the actual startup. It is in this sense that the transient can be false. Suppose at time $t = 0$ the reactor is instantaneously filled to volume V with fluid of initial concentrations $a_0, b_0, \ldots$. The initial concentrations are usually set equal to the inlet concentrations, $a_{in}, b_{in}, \ldots$, but other values can be used. The simulation begins with Q_{in} set to its steady-state value. For constant-density cases, Q_{out} is set to the same value. The variable-density case is treated in Section 4.3.

The ODEs governing the unsteady CSTR are obtained by adding accumulation terms to Equations (4.1). The simulation holds the volume constant, and

$$V \frac{d(a_{out})}{dt} = Q_{in} a_{in} + \mathscr{R}_A(a_{out}, b_{out}, \ldots, P_{out}, T_{out}) V - Q_{out} a_{out}$$
$$V \frac{d(b_{out})}{dt} = Q_{in} b_{in} + \mathscr{R}_B(a_{out}, b_{out}, \ldots, P_{out}, T_{out}) V - Q_{out} b_{out} \quad (4.4)$$
$$\vdots \qquad \vdots \qquad \vdots \qquad \qquad \vdots$$

This set of first-order ODEs is easier to solve than the algebraic equations where all the time derivatives are zero. The initial conditions are that $a_{out} = a_0$, $b_{out} = b_0, \ldots$ at $t = 0$. The long-time solution to these ODEs will satisfy Equations (4.1) provided that a steady-state solution exists and is accessible from the assumed initial conditions. There may be no steady state. Recall the chemical oscillators of Chapter 2. Stirred tank reactors can also exhibit oscillations or more complex behavior known as chaos. It is also possible that the reactor has multiple steady states, some of which are unstable. Multiple steady states are fairly common in stirred tank reactors when the reaction exotherm is large. The method of false transients will go to a steady state that is stable but may not be desirable. Stirred tank reactors sometimes have one steady state where there is no reaction and another steady state where the reaction runs away. Think of the reaction A → B → C. The stable steady states may give all A or all C, and a control system is needed to stabilize operation at a middle steady state that gives reasonable amounts of B. This situation arises mainly in nonisothermal systems and is discussed in Chapter 5.

Example 4.2: Suppose the competing, elementary reactions

$$A + B \xrightarrow{k_I} C$$

$$A \xrightarrow{k_{II}} D$$

occur in a CSTR. Assume density is constant and use the method of false transients to determine the steady-state outlet composition. Suppose $k_I a_{in} \bar{t} = 4$, $k_{II} \bar{t} = 1$, $b_{in} = 1.5 a_{in}$, $c_{in} = 0.1 a_{in}$, and $d_{in} = 0.1 a_{in}$.

Solution: Write a version of Equation (4.4) for each component. Divide through by $Q_{in} = Q_{out}$ and substitute the appropriate reaction rates to obtain

$$\bar{t}\frac{da_{out}}{dt} = a_{in} - a_{out} - k_I \bar{t} a_{out} b_{out} - k_{II} \bar{t} a_{out}$$

$$\bar{t}\frac{db_{out}}{dt} = b_{in} - b_{out} - k_I \bar{t} a_{out} b_{out}$$

$$\bar{t}\frac{dc_{out}}{dt} = c_{in} - c_{out} + k_I \bar{t} a_{out} b_{out}$$

$$\bar{t}\frac{dd_{out}}{dt} = d_{in} - d_{out} + k_{II} \bar{t} a_{out}$$

Then use a first-order difference approximation for the time derivatives, e.g.,

$$\frac{da}{dt} \approx \frac{a_{new} - a_{old}}{\Delta t}$$

The results are

$$\left(\frac{a}{a_{in}}\right)_{new} = \left(\frac{a}{a_{in}}\right)_{old} + \left[1 - (1 + k_{II}\bar{t})\left(\frac{a}{a_{in}}\right)_{old} - k_I a_{in} \bar{t} \left(\frac{a}{a_{in}}\right)_{old}\left(\frac{b}{a_{in}}\right)_{old}\right]\Delta\tau$$

$$\left(\frac{b}{a_{in}}\right)_{new} = \left(\frac{b}{a_{in}}\right)_{old} + \left[\left(\frac{b_{in}}{a_{in}}\right) - \left(\frac{b}{a_{in}}\right)_{old} - k_I a_{in} \bar{t} \left(\frac{a}{a_{in}}\right)_{old}\left(\frac{b}{a_{in}}\right)_{old}\right]\Delta\tau$$

$$\left(\frac{c}{a_{in}}\right)_{new} = \left(\frac{c}{a_{in}}\right)_{old} + \left[\left(\frac{c_{in}}{a_{in}}\right) - \left(\frac{c}{a_{in}}\right)_{old} + k_I a_{in} \bar{t} \left(\frac{a}{a_{in}}\right)_{old}\left(\frac{b}{a_{in}}\right)_{old}\right]\Delta\tau$$

$$\left(\frac{d}{a_{in}}\right)_{new} = \left(\frac{d}{a_{in}}\right)_{old} + \left[\left(\frac{d_{in}}{a_{in}}\right) - \left(\frac{d}{a_{in}}\right)_{old} + k_{II}\bar{t}\left(\frac{a}{a_{in}}\right)_{old}\right]\Delta\tau$$

where $\tau = t/\bar{t}$ is dimensionless time. These equations are directly suitable for solution by Euler's method, although they can be written more compactly as

$$a^*_{new} = a^*_{old} + [1 - 2a^*_{old} - 4a^*_{old}b^*_{old}]\Delta\tau$$

$$b^*_{new} = b^*_{old} + [1.5 - b^*_{old} - 4a^*_{old}b^*_{old}]\Delta\tau$$

$$c^*_{new} = c^*_{old} + [0.1 - c^*_{old} + 4a^*_{old}b^*_{old}]\Delta\tau$$

$$d^*_{new} = d^*_{old} + [0.1 - d^*_{old} + a^*_{old}]\Delta\tau$$

where the various concentrations have been normalized by a_{in} and where numerical values have been substituted. Suitable initial conditions are $a_0^* = 1$, $b_0^* = 1.5$, $c_0^* = 0.1$, and $d_0^* = 0.1$. Figure 4.1 shows the transient approach to steady state. Numerical values for the long-time, asymptotic solutions are also shown in Figure 4.1. They require simulations out to about $\tau = 10$. They could have been found by solving the algebraic equations

$$0 = 1 - 2a_{out}^* - 4a_{out}^* b_{out}^*$$

$$0 = 1.5 - b_{out}^* - 4a_{out}^* b_{out}^*$$

$$0 = 0.1 - c_{out}^* + 4a_{out}^* b_{out}^*$$

$$0 = 0.1 - d_{out}^* + a_{out}^*$$

These equations are obtained by setting the accumulation terms to zero.

Analytical solutions are desirable because they explicitly show the functional dependence of the solution on the operating variables. Unfortunately, they are difficult or impossible for complex kinetic schemes and for the nonisothermal reactors considered in Chapter 5. All numerical solutions have the disadvantage of being case-specific, although this disadvantage can be alleviated through the judicious use of dimensionless variables. Direct algebraic solutions to Equations (4.1) will, in principle, give all the steady states. On the other hand, when a solution is obtained using the method of false transients, the steady state is known to be stable and achievable from the assumed initial conditions.

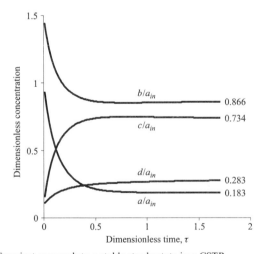

FIGURE 4.1 Transient approach to a stable steady state in a CSTR.

Example 4.2 used the method of false transients to solve a steady-state reactor design problem. The method can also be used to find the equilibrium concentrations resulting from a set of batch chemical reactions. To do this, formulate the ODEs for a batch reactor and integrate until the concentrations stop changing. This is illustrated in Problem 4.6(b). Section 11.1.1 shows how the method of false transients can be used to determine physical or chemical equilibria in multiphase systems.

4.3 CSTRs with Variable Density

The design equations for a CSTR do not require that the reacting mixture has constant physical properties or that operating conditions such as temperature and pressure be the same for the inlet and outlet environments. It is required, however, that these variables be known. Pressure in a CSTR is usually determined or controlled independently of the extent of reaction. Temperatures can also be set arbitrarily in small, laboratory equipment because of excellent heat transfer at the small scale. It is sometimes possible to predetermine the temperature in industrial-scale reactors; for example, if the heat of reaction is small or if the contents are boiling. This chapter considers the case where both P_{out} and T_{out} are known. Density and Q_{out} will not be known if they depend on composition. A steady-state material balance gives

$$\rho_{in} Q_{in} = \rho_{out} Q_{out} \tag{4.5}$$

An equation of state is needed to determine the mass density at the reactor outlet, ρ_{out}. Then, Q_{out} can be calculated.

4.3.1 Liquid-Phase CSTRs

There is no essential difference between the treatment of liquid and gas phase except for the equation of state. Density changes in liquid systems tend to be small, and the density is usually assumed to be a linear function of concentration. This chapter treats single-phase reactors, although some simple multiphase situations are allowed. A solid by-product of an irreversible, liquid-phase reaction will change the density but not otherwise affect the extent of reaction. Gaseous by-products are more of a problem since they cause foaming. The foam density will be affected by the pressure due to liquid head. Also, the gas may partially disengage. Accurate, a priori estimates of foam density are difficult. This is also true in boiling reactors.

A more general treatment of multiphase reactors is given in Chapter 11.

Example 4.3: Suppose a pure monomer polymerizes in a CSTR with pseudo-first-order kinetics. The monomer and polymer have different densities. Assume a linear relationship based on the monomer concentration. Then determine the exit concentration of monomer, assuming that the reaction is first order.

Solution: The reaction is

$$M \longrightarrow P \qquad \mathcal{R} = km$$

The reactor design equation for monomer is

$$0 = m_{in}Q_{in} - Vkm_{out} - m_{out}Q_{out} \tag{4.6}$$

where the unknowns are m_{out} and Q_{out}. A relationship between density and composition is needed. One that serves the purpose is

$$\rho = \rho_{polymer} - \Delta\rho\left(\frac{m}{m_{in}}\right) \tag{4.7}$$

where $\Delta\rho = \rho_{polymer} - \rho_{monomer}$. The procedure from this point is straightforward if algebraically messy. Set $m = m_{out}$ in Equation (4.7) to obtain ρ_{out}. Substitute into Equation (4.5) to obtain Q_{out} and then into Equation (4.6) so that m_{out} becomes the only unknown. The solution for m_{out} is

$$\frac{m_{out}}{m_{in}} = \frac{1 - \sqrt{1 - 4(1-\kappa)Y_0(1-Y_0)}}{2(1-\kappa)(1-Y_0)} \tag{4.8}$$

where

$$\kappa = \frac{\rho_{monomer}}{\rho_{polymer}}$$

and Y' is the fraction unreacted that would be calculated if the density change were ignored. That is,

$$Y' = \frac{Q_{in}}{Q_{in} + kV}$$

This result can be simplified by dividing through by Q_{in} to create the dimensionless group kV/Q_{in}. The quantity V/Q_{in} is the space time, not the mean residence time. See Example 3.4. The mean residence time is

$$\bar{t} = \frac{\hat{\rho}V}{\rho_{out}Q_{out}} = \frac{V}{Q_{out}} \tag{4.9}$$

The first of the relations in Equation (4.9) is valid for any flow system. The second applies specifically to a CSTR since $\hat{\rho} = \rho_{out}$. It is not true for a piston flow reactor. Recall Example 3.6 where determination of $\bar{t}$ in a gas-phase tubular reactor required integrating the local density down the length of the tube.

As a numerical example, suppose $Y' = 0.5$ and $\kappa = 0.9$. Then Equation (4.8) gives $m_{out}/m_{in} = 0.513$. This result may seem strange at first. The density increases upon polymerization so that the reactor has a greater mass inventory when filled with the polymerizing mass than when filled with monomer. More material means a higher residence time, yet m_{out}/m_{in} is higher, suggesting less reaction. The answer, of course, is that m_{out}/m_{in} is not the fraction unreacted when there is a density change. Instead,

$$\text{Fraction unreacted} = Y_M = \frac{Q_{out} m_{out}}{Q_{in} m_{in}} \qquad (4.10)$$

Equation (4.10) uses the general definition of fraction unreacted in a flow system. It is moles out divided by moles in. The corresponding, general definition of conversion is

$$X_M = 1 - \frac{Q_{out} m_{out}}{Q_{in} m_{in}} \qquad (4.11)$$

For the problem at hand,

$$\frac{Q_{out}}{Q_{in}} = \frac{\rho_{in}}{\rho_{out}} = \frac{\kappa}{1 + (1 - \kappa) m_{out}/m_{in}}$$

For the numerical example, $Q_{out}/Q_{in} = 0.949$ and the fraction unreacted is 0.487 compared with 0.5 if there were no change. Thus, the density change causes a modest increase in conversion.

4.3.2 Computational Scheme for Variable-Density CSTRs

Example 4.3 represents the simplest possible example of a variable-density CSTR. The reaction is isothermal, first-order, irreversible, and the density is a linear function of reactant concentration. This simplest system is about the most complicated one for which an analytical solution is possible. Realistic variable-density problems, whether in liquid or gas systems, require numerical solutions. These numerical solutions use the method of false transients and involve sets of first-order ODEs with various auxiliary functions. The solution methodology is similar to but simpler than that used for piston flow reactors in Chapter 3. Temperature is known and constant in the reactors described in this chapter. An ODE for temperature will be added in Chapter 5. Its addition does not change the basic methodology.

The method of false transients begins with the inlet stream set to its steady-state values of $Q_{in}, T_{in}, \rho_{in}, a_{in}, b_{in}, \ldots$. The reactor is full of material having concentrations $a_0, b_0, \ldots$ and temperature T_0.

0. Set the initial values $a_0, b_0, \ldots, T_0$. Use the equation of state to calculate ρ_0 and ρ_{in}. Calculate $Q_0 = \rho_{in} Q_{in}/\rho_0$. Calculate $V\rho_0$.

1. Pick a step size, Δt.
2. Set the initial values for $a_{out}, b_{out}, \ldots, T_{out}$, and Q_{out}.
3. Take one step, calculating new values for $a_{out}, b_{out}, \ldots$, and T_{out} at the new time, $t + \Delta t$. The marching-ahead equations have the form

$$(a_{out})_{new} = (a_{out})_{old} + [Q_{in}a_{in} + \mathscr{R}_A(a_{out}, b_{out}, \ldots, P_{out}, T_{out})V - Q_{out}a_{out}]\, \Delta t / V \quad (4.12)$$

4. Use the equation of state to calculate ρ_{out}.
5. Use Equation (4.5) to calculate $Q_{out} = \rho_{in}Q_{in}/\rho_{out}$.
6. Check if $(a_{out})_{new} \cong (a_{out})_{old}$. If not, go to Step 3.
7. Decrease Δt by a factor of 2 and go to Step 1. Repeat until the results converge to four or five significant figures.
8. Calculate the steady-state value for the reactor volume from $V\rho_0/\rho_{out}$. If this is significantly different than the desired working volume in the reactor, go back to Step 0, but now start the simulation with the tank at the concentrations and temperature just calculated.

Note that an accurate solution is not required for the early portions of the trajectory, and Euler's method is the perfect integration routine. A large step size can be used provided the solution remains stable. As steady state is approached, the quantity in square brackets in Equation (4.12) goes to zero. This allows an accurate solution at the end of the trajectory, even though the step size is large. Convergence is achieved very easily, and Step 7 is included mainly as a matter of good computing practice. Step 8 is needed if there is a significant density change upon reaction and if the initial concentrations were far from the steady-state values. The computational algorithm keeps constant mass in the reactor, not constant volume, so you may wind up simulating a reactor of somewhat different volume than you intended. This problem can be remedied just by rerunning the program. An actual startup transient—as opposed to a false transient used to get steady-state values—can be computed using the methodology of Chapter 14.

Example 4.4: Solve Example 4.3 numerically.

Solution: In a real problem, the individual values for k, V, and Q_{in} would be known. Their values are combined into the dimensionless group, kV/Q_{in}. This group determines the performance of a constant-density reactor and is one of the two parameters needed for the variable-density case. The other parameter is the density ratio, $r = \rho_{monomer}/\rho_{polymer}$. Setting $kV/Q_{in} = 1$ gives $Y' = 0.5$ as the fraction unreacted for the constant-density case. The individual values for k, V, Q_{in}, $\rho_{monomer}$, and $\rho_{polymer}$ can be assigned as convenient, provided the composite values are retained. The following program gives the same results as found in Example 4.3 but with less work:

```
DEFDBL A-Z
dt = .1
Qin = 1
k = 1
V = 1
min = 1
rhom = .9
rhop = 1
rhoin = rhom
Qout = Qin
mold = min
DO
    mnew = mold + (Qin*min − k * V * mold − Qout * mold) * dt/V
    rhoout = rhop − (rhop − rhom) * mnew/min
    Qout = Qin * rhoin / rhoout
    mold = mnew
    PRINT USING ''###.####''; mnew, Qout, Qout * mnew
    t = t + dt
LOOP WHILE t < 10
```

The long-time results to three decimal places are mnew $= 0.513 = m_{out}$, Qout $= 0.949 = Q_{out}/Q_{in}$, and Qout * mnew $= 0.467 = Y_M$.

4.3.3 Gas-Phase CSTRs

Strictly gas-phase CSTRs are rare. Two-phase, gas–liquid CSTRs are common and are treated in Chapter 11. Two-phase, gas–solid CSTRs are fairly common. When the solid is a catalyst, the use of *pseudohomogeneous kinetics* allows these two-phase systems to be treated as though only the fluid phase were present. All concentration measurements are made in the gas phase, and the rate expression is fitted to the gas-phase concentrations. This section outlines the method for fitting pseudo-homogeneous kinetics using measurements made in a CSTR. A more general treatment is given in Chapter 10.

A recycle loop reactor is often used for laboratory studies with gas-phase reactants and a solid, heterogeneous catalyst. See Figure 4.2. Suppose the reactor is a small bed of packed catalyst through which the gas is circulated at a high rate. The high flow rate gives good heat transfer and eliminates gas-phase resistance to mass transfer. The net throughput is relatively small since most of the gas exiting from the catalyst bed is recycled. The per-pass conversion is low, but the overall conversion is high enough that a chemical analysis can be reasonably accurate. Recycle loops behave as CSTRs when the recycle ratio is high. This

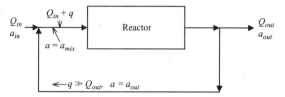

FIGURE 4.2 Reactor in a recycle loop.

fact is intuitively reasonable since the external pump causes circulation similar to that caused by the agitator in a conventional stirred tank reactor. A variant of the loop reactor puts the catalyst in a basket and then rotates the basket at high speed within the gas mixture. This more closely resembles the tank-plus-agitator design of a conventional stirred tank, but the kinetic result is the same. Section 4.5.3 shows the mathematical justification for treating a loop reactor as a CSTR.

A gas-phase CSTR with prescribed values for P_{out} and T_{out} is particularly simple when ideal gas behavior can be assumed. The molar density in the reactor will be known and independent of composition.

Example 4.5: Suppose the recycle reactor in Figure 4.2 is used to evaluate a catalyst for the manufacture of sulfuric acid. The catalytic step is the gas-phase oxidation of sulfur dioxide:

$$SO_2 + \tfrac{1}{2} O_2 \rightarrow SO_3$$

Studies on similar catalysts have suggested a rate expression of the form

$$\mathscr{R} = \frac{k[SO_2][O_2]}{1 + k_C[SO_3]} = \frac{kab}{1 + k_C c}$$

where $a = [SO_2]$, $b = [O_2]$, and $c = [SO_3]$. The object is to determine k and k_C for this catalyst. At least two runs are needed. The following compositions have been measured:

	Concentrations in mole percent			
	Inlet		Outlet	
	Run 1	Run 2	Run 1	Run 2
SO_2	10	5	4.1	2.0
O_2	10	10	7.1	8.6
SO_3	0	5	6.3	8.1
Inerts	80	80	82.5	81.3

The operating conditions for these runs were $Q_{in} = 0.000268 \text{ m}^3/\text{s}$, $P_{in} = 2.04 \text{ atm}$, $P_{out} = 1.0 \text{ atm}$, $T_{in} = 40°C$, $T_{out} = 300°C$, and $V = 0.0005 \text{ m}^3$.

Solution: The analysis could be carried out using mole fractions as the composition variable, but this would restrict applicability to the specific conditions of the experiment. Greater generality is possible by converting to concentration units. The results will then apply to somewhat different pressures. The "somewhat" recognizes the fact that the reaction mechanism and even the equation of state may change at extreme pressures. The results will not apply at different temperatures since k and k_C will be functions of temperature. The temperature dependence of rate constants is considered in Chapter 5.

Converting to standard concentration units, mol/m³, gives the following:

	Molar concentrations			
	Inlet		Outlet	
	Run 1	Run 2	Run 1	Run 2
SO_2	7.94	3.97	0.87	0.43
O_2	7.94	7.94	1.51	1.83
SO_3	0	3.97	1.34	1.72
Inerts	63.51	63.51	17.54	17.28
ρ_{molar}	79.38	79.39	21.26	21.26

The outlet flow rate Q_{out} is required. The easiest way to obtain this is by a molar balance on the inerts:

$$Q_{in}d_{in} = Q_{out}d_{out}$$

which gives $Q_{out} = [(0.000268)(63.51)]/(17.54) = 0.000970 \, m^3/s$ for Run 1 and 0.000985 for Run 2. These results allow the molar flow rates to be calculated:

	Molar flow rates			
	Inlet		Outlet	
	Run 1	Run 2	Run 1	Run 2
SO_2	0.00213	0.00106	0.00085	0.00042
O_2	0.00213	0.00213	0.00146	0.00180
SO_3	0	0.00106	0.00130	0.00169
Inerts	0.01702	0.01704	0.01702	0.01702
Total moles	0.02128	0.02128	0.02063	0.02093

The reader may wish to check these results against the reaction stoichiometry for internal consistency. The results are certainly as good as warranted by the two-place precision of the analytical results.

The reactor design equation for SO_3 is

$$0 = c_{in}Q_{in} + \frac{Vka_{out}b_{out}}{1+k_C c_{out}} - c_{out}Q_{out}$$

Everything in this equation is known but the two rate constants. Substituting the known quantities for each run gives a pair of simultaneous equations:

$$0.00130 + 0.00174k_C = 0.000658k$$

$$0.00063 + 0.00109k_C = 0.000389k$$

Solution gives $k = 8.0\,\text{mol}/(\text{m}^3\cdot\text{s})$ and $k_C = 2.3\,\text{m}^3\,\text{mol}^{-1}$. Be warned that this problem is *ill-conditioned*. Small differences in the input data or rounding errors can lead to major differences in the calculated values for k and k_C. The numerical values in this problem were calculated using greater precision than indicated in the above tables. Also, the values for k and k_C will depend on which component was picked for the component balance. The example used component C, but A or B could have been chosen. Despite this numerical sensitivity, predictions of performance using the fitted values for the rate constants will closely agree within the range of the experimental results. The estimates for k and k_C are *correlated* so that a high value for one will lead to a compensating high value for the other.

Example 4.6: Use the kinetic model of Example 4.5 to determine the outlet concentration for the loop reactor if the operating conditions are the same as in Run 1.

Solution: Example 4.5 was a *reverse problem*, where measured reactor performance was used to determine constants in the rate equation. We now treat the *forward problem*, where the kinetics are known and the reactor performance is desired. Obviously, the results of Run 1 should be closely duplicated. The solution uses the method of false transients for a variable-density system. The ideal gas law is used as the equation of state. The ODEs are

$$\frac{da_{out}}{dt} = \frac{a_{in}Q_{in}}{V} - \frac{ka_{out}b_{out}}{1+k_C c_{out}} - \frac{a_{out}Q_{out}}{V}$$

$$\frac{db_{out}}{dt} = \frac{b_{in}Q_{in}}{V} - \frac{ka_{out}b_{out}}{2(1+k_C c_{out})} - \frac{b_{out}Q_{out}}{V}$$

$$\frac{dc_{out}}{dt} = \frac{c_{in}Q_{in}}{V} + \frac{ka_{out}b_{out}}{1+k_C c_{out}} - \frac{c_{out}Q_{out}}{V}$$

$$\frac{dd_{out}}{dt} = \frac{d_{in}Q_{in}}{V} - \frac{d_{out}Q_{out}}{V}$$

Then add all these together, noting that the sum of the component concentrations is the molar density:

$$\frac{d(\rho_{molar})_{out}}{dt} = \frac{(\rho_{molar})_{in}Q_{in}}{V} - \frac{ka_{out}b_{out}}{2(1+k_C c_{out})} - \frac{(\rho_{molar})_{out}Q_{out}}{V}$$

The ideal gas law says that the molar density is determined by pressure and temperature and is thus known and constant in the reactor. Setting the time derivative of molar density to zero gives an expression for Q_{out} at steady state. The result is

$$Q_{out} = \frac{(\rho_{molar})_{in}Q_{in}}{(\rho_{molar})_{out}} - (1/2)\frac{Vka_{out}b_{out}}{(\rho_{molar})_{out}(1+k_C c_{out})}$$

For the numerical solution, the ODEs for the three reactive components are solved in the usual manner and Q_{out} is updated after each time step. If desired, d_{out} is found from

$$d_{out} = \rho_{molar} - a_{out} - b_{out} - c_{out}$$

The results for the conditions of Run 1 are $a_{out}=0.87$, $b_{out}=1.55$, $c_{out}=1.37$, and $d_{out}=17.47$. The agreement with Example 4.5 is less than perfect because the values for k and k_C were rounded to two places. Better accuracy cannot be expected.

4.4 SCALEUP OF ISOTHERMAL CSTRs

The word "isothermal" in the title of this section eliminates most of the difficulty. The most common problem in scaling up a CSTR is maintaining the desired operating temperature. This is discussed in Chapter 5, along with energy balances in general. The current chapter ignores the energy balance, and there is little to discuss here beyond the mixing time concepts of Section 1.5. Reference is made to that section and to Example 1.7.

A real continuous-flow stirred tank will approximate a perfectly mixed CSTR provided that $t_{mix} \ll t_{1/2}$ and $t_{mix} \ll \bar{t}$. Mixing time correlations are developed using batch vessels, but they can be applied to flow vessels provided the ratio of throughput to circulatory flow is small. This idea is explored in Section 4.5.3 where a recycle loop reactor is used as a model of an internally agitated vessel.

The standard approach to scaling a conventionally agitated stirred tank is to maintain geometric similarity. This means that all linear dimensions—e.g., the impeller diameter, the distance that the impeller is off the bottom, the height of the liquid, and the width of the baffles—scale as the tank diameter; that is, as $S^{1/3}$. As suggested in Section 1.5, the scaleup relations are simple when scaling with geometric similarity and when the small-scale vessel is fully turbulent. The Reynolds number scales as $N_I D_I^2$ and will normally be higher in the large vessel. The mixing time scales as N_I^{-1}, the pumping capacity of the impeller scales as $N_I D_I^3$, and the power to the impeller scales as $N_I^3 D_I^5$. As shown in Example 1.7, it is impractical to maintain a constant mixing time upon scaleup since the power requirements increase too dramatically.

Although experts in agitator design are loath to admit to using such a simplistic rule, most scaleups of conventionally agitated vessels are done at or near constant power per unit volume. The consequences of scaling in this fashion are explored in Example 4.7

Example 4.7: A fully turbulent, baffled vessel is to be scaled up by a factor of 512 in volume while maintaining constant power per unit volume. Determine the effects of the scaleup on the impeller speed, the mixing time, and the internal circulation rate.

Solution: If power scales as $N_I^3 D_I^5$, then power per unit volume scales as $N_I^3 D_I^2$. To maintain constant power per unit volume, N_I must decrease upon scaleup. Specifically, N_I must scale as $D_I^{-2/3}$. When impeller speed is scaled in this manner, the mixing time scales as $D_I^{2/3}$ and the impeller pumping rate scales as $D_I^{7/3}$. To maintain a constant value for $\bar{t}$, the throughput Q scales as $D_I^3 = S$. Results for these and other design and operating variables are shown in Table 4.1.

A volumetric scaleup by a factor of 512 is quite large, and the question arises as to whether the large vessel will behave as a CSTR. The concern is due to the factor of 4 increase in mixing time. Does it remain true that $t_{mix} \ll t_{1/2}$ and $t_{mix} \ll \bar{t}$? If so, the assumption that the large vessel will behave as a CSTR is probably justified. The ratio of internal circulation to net throughput—which is the internal recycle ratio—scales as the inverse of the mixing time and will thus decrease by a factor of 4. The decrease may appear worrisome, but if the increase in mixing time can be tolerated, then it is likely that the decrease in internal recycle ratio is also acceptable.

The above analysis is restricted to high Reynolds numbers, although the definition of high is different in a stirred tank than in a circular pipe. The Reynolds number for a conventionally agitated vessel is defined as

$$(\mathbf{Re})_{impeller} = \frac{\rho N_I D_I^2}{\mu} \tag{4.13}$$

TABLE 4.1 Scaleup Factors for Geometrically Similar Stirred Tanks

	General scaling factor	Scaling factor for constant power per unit volume	Numerical scaling factor for $S=512$
Vessel diameter	$S^{1/3}$	$S^{1/3}$	8
Impeller diameter	$S^{1/3}$	$S^{1/3}$	8
Vessel volume	S	S	512
Throughput	S	S	512
Residence time	1	1	1
Reynolds number	$N_I S^{2/3}$	$S^{4/9}$	8
Froude number	$N_I^2 S^{1/3}$	$S^{-1/9}$	0.5
Agitator speed	N_I	$S^{-2/9}$	0.25
Power	$N_I^3 S^{5/3}$	S	512
Power per volume	$N_I^3 S^{2/3}$	1.0	1
Mixing time	N_I^{-1}	$S^{2/9}$	4
Circulation rate	$N_I S$	$S^{7/9}$	128
Circulation rate/throughput	N_I	$S^{-2/9}$	0.25
Heat transfer area, A_{ext}	$S^{2/3}$	$S^{2/3}$	64
Inside coefficient, h	$N_I^{2/3} S^{1/9}$	$S^{-1/27}$	0.79
Coefficient times area, hA_{ext}	$N_I^{2/3} S^{7/9}$	$S^{17/27}$	50.8
Driving force, ΔT	$N_I^{-2/3} S^{2/9}$	$S^{10/27}$	10.1

where D_I is the diameter of the impeller, not of the tank. The velocity term in the Reynolds number is the tip velocity of the impeller, $N_I D_I$. The transition from laminar to transitional flow occurs when the impeller Reynolds number is less than 100, and the vessel is highly turbulent by $(\text{Re})_{impeller} = 1000$. These statements are true for commercial examples of turbine and paddle agitators. Most industrial stirred tanks operate in the fully turbulent regime. The exceptions are usually polymerization reactors, which often use special types of agitators.

Table 4.1 includes the Froude number, $N_I^2 D_I / g$ where g is the acceleration due to gravity. This dimensionless group governs the extent of swirling and vortexing in an unbaffled stirred tank. Turbulent stirred tanks are normally baffled so that the power from the agitator causes turbulence rather than mere circular motion. Intentional vortexing is occasionally used as a means for rapidly engulfing a feed stream. Table 4.1 shows that the extent of vortexing will decrease for scaleups at constant power per unit volume. Unbaffled tanks will draw somewhat less power than baffled tanks.

Table 4.1 includes scaleup factors for heat transfer. They are discussed in Chapter 5.

4.5 COMBINATIONS OF REACTORS

We have considered two types of ideal flow reactor: the piston flow reactor and the perfectly mixed CSTR. These two ideal types can be connected together in a variety of series and parallel arrangements to give composite reactors that are generally intermediate in performance when compared with the ideal reactors. Sometimes the composite reactor is only conceptual and it is used to model a real reactor. Sometimes the composite reactor is actually built. There are many good reasons for building reactor combinations. Temperature control is a major motivation. The use of standard designs is sometimes a factor, as is the ability to continue operating a plant while adding capacity. Series and parallel scaleups of tubular reactors were considered in Chapter 3. Parallel scaleups of CSTRs are uncommon, but they are sometimes used to gain capacity. Series installations are more common. The series combinations of a stirred tank followed by a tube are also common. This section begins the analysis of composite reactors while retaining the assumption of isothermal operation, at least within a single reactor.

Different reactors in the composite system may operate at different temperatures and thus may have different rate constants.

4.5.1 Series and Parallel Connections

When reactors are connected in series, the output from one serves as the input for the other. For *reactors in series*,

$$(a_{in})_2 = (a_{out})_1 \tag{4.14}$$

The design equations for reactor 1 are solved and used as the input to reactor 2.

Example 4.8: Find the yield for a first-order reaction in a composite reactor that consists of a CSTR followed by a piston flow reactor. Assume that the mean residence time is $\bar{t}_1$ in the CSTR and $\bar{t}_2$ in the piston flow reactor.

Solution: The exit concentration from the perfect mixer is

$$(a_{out})_1 = \frac{a_{in}}{1 + k\bar{t}_1}$$

and that for the piston flow reactor is

$$a_{out} = (a_{in})_2 \exp(-k\bar{t}_2)$$

Using Equation (4.14) to combine these results gives

$$a_{out} = \frac{a_{in} \exp(-k\bar{t}_2)}{1 + k\bar{t}_1}$$

Compare this result with that for a single, ideal reactor having the same input concentration, throughput, and total volume. Specifically, compare the outlet concentration of the composite reactor with that from a single CSTR having a mean residence time of

$$\bar{t} = \frac{V}{Q} = \frac{V_1 + V_2}{Q} = \bar{t}_1 + \bar{t}_2$$

and with that of a piston flow reactor having this same $\bar{t}$. The following inequality is true for physically realistic (meaning positive) values of k, $\bar{t}_1$, and $\bar{t}_2$:

$$\frac{1}{1 + k(\bar{t}_1 + \bar{t}_2)} \geq \frac{\exp(-k\bar{t}_2)}{1 + k\bar{t}_1} \geq \exp[-k(\bar{t}_1 + \bar{t}_2)]$$

Thus, the combination reactor gives intermediate performance. The fraction unreacted from the composite reactor will be lower than that from a single CSTR with $\bar{t} = \bar{t}_1 + \bar{t}_2$ but higher than that from a single PFR with $\bar{t} = \bar{t}_1 + \bar{t}_2$.

For two *reactors in parallel*, the output streams are averaged based on the flow rate:

$$a_{out} = \frac{Q_1(a_{out})_1 + Q_2(a_{out})_2}{Q_1 + Q_2} \tag{4.15}$$

Example 4.9: Find the conversion for a first-order reaction in a composite system that consists of a perfect mixer and a piston flow reactor in parallel.

Solution: Using Equation (4.15),

$$a_{out} = \frac{a_{in}}{Q_1 + Q_2} \left(\frac{Q_1}{1 + k\bar{t}_1} + Q_2 \exp(-k\bar{t}_2) \right)$$

A parallel reactor system has an extra degree of freedom compared with a series system. The total volume and flow rate can be arbitrarily divided between the parallel elements. For reactors in series, only the volume can be divided since the two reactors must operate at the same flow rate. Despite this extra variable, there are no performance advantages compared with a single reactor that has the same total V and Q, provided the parallel reactors are at the same temperature. When significant amounts of heat must be transferred to or from the reactants, identical small reactors in parallel may be preferred because the desired operating temperature is easier to achieve.

The general rule is that combinations of isothermal reactors provide intermediate levels of performance compared with single reactors that have the same total volume and flow rate. The second general rule is that a single, piston flow reactor will give higher conversion and better selectivity than a CSTR. Autocatalytic reactions provide the exception to both these statements.

Example 4.10: Consider a reactor train consisting of a CSTR followed by a piston flow reactor. The total volume and flow rate are fixed. Can series combination offer a performance advantage compared with a single reactor if the reaction is autocatalytic? The reaction is

$$A + B \xrightarrow{k} 2B$$

Treat the semipathological case where $b_{in} = 0$.

Solution: With $b_{in} = 0$, a reaction will never start in a PFR, but a steady-state reaction is possible in a CSTR if the reactor is initially spiked with component B. An analytical solution can be found for this problem and is requested in Problem 4.12, but a numerical solution is easier. The design equations in a form suitable for the method of false transients are

$$\frac{d(a_{out})_1}{dt} = (a_{in})_1 - k\bar{t}_1(a_{out})_1(b_{out})_1 - (a_{out})_1$$

$$\frac{d(b_{out})_1}{dt} = (b_{in})_1 + k\bar{t}_1(a_{out})_1(b_{out})_1 - (b_{out})_1$$

The long-time solution to these ODEs gives $(a_{out})_1$ and $(b_{out})_1$, which are the inlet concentrations for the piston flow portion of the system. The design equations for the PFR are

$$\frac{da_2}{dt} = -ka_2b_2$$

$$\frac{db_2}{dt} = ka_2b_2$$

A simple numerical example sets $a_{in} = 1$, $b_{in} = 0$, and $k = 5$. Suitable initial conditions for the method of false transients are $a_0 = 0$ and $b_0 = 1$. Suppose the residence time for the composite system is $\bar{t}_1 + \bar{t}_2 = 1$. The question is how this total time should be divided. The following results were obtained:

$\bar{t}_1$	$\bar{t}_2$	$(a_{out})_1$	$(b_{out})_1$	$(a_{out})_2$	$(b_{out})_2$
1.0	0	0.2000	0.8000	0.2000	0.8000
0.9	0.1	0.2222	0.7778	0.1477	0.8523
0.8	0.2	0.2500	0.7500	0.1092	0.8908
0.7	0.3	0.2857	0.7143	0.0819	0.9181
0.6	0.4	0.3333	0.6667	0.0634	0.9366
0.5	0.5	0.4000	0.6000	0.0519	0.9481
0.4	0.6	0.5000	0.5000	0.0474	0.9526
0.3	0.7	0.6667	0.3333	0.0570	0.9430
0.2	0.8	1	0	1	0
0.1	0.9	1	0	1	0
0.0	1.0	1	0	1	0

There is an interior optimum. For this particular numerical example, it occurs when 40% of the reactor volume is in the initial CSTR and 60% is in the downstream PFR. The model reaction is chemically unrealistic but illustrates behavior that can arise with real reactions. An excellent process for the bulk polymerization of styrene consists of a CSTR followed by a tubular post-reactor. The model reaction also demonstrates a phenomenon known as *washout* which is important in continuous cell culture. If $k\bar{t}_1$ is too small, a steady-state reaction cannot be sustained even with initial spiking of component B. A continuous fermentation process will have a maximum flow rate beyond which the initial inoculum of cells will be washed out of the system. At lower flow rates, the cells reproduce fast enough to achieve and hold a steady state.

4.5.2 Tanks in Series

For the great majority of reaction schemes, piston flow is optimal. Thus, the reactor designer normally wants to build a tubular reactor and to operate it at high Reynolds numbers so that piston flow is closely approximated. This may not be possible. There are many situations where a tubular reactor is infeasible and where continuous-flow stirred tank reactors must be used instead. Typical examples are reactions involving suspended solids and autorefrigerated reactors where the reaction mass is held at its boiling point. There will usually be a yield advantage, but a cost disadvantage, from using several CSTRs in series. Problems 4.19 and 4.20 show how the cost disadvantage can be estimated.

Example 4.11: Determine the fraction unreacted for a second-order reaction, $2A \xrightarrow{k} B$, in a composite reactor consisting of two equal-volume CSTRs in series. The rate constant is the same for each reactor and $k\bar{t}_1 a_{in} = 0.5$ where $\bar{t}_1 = V_1/Q$ is the mean residence time in a single vessel. Compare your result with the fraction unreacted in a single CSTR that has the same volume as the series combination, $V = 2V_1$. Assume constant mass density.

Solution: Begin by considering the first CSTR. The rate of formation of A is $\mathscr{R}_A = -2ka^2$. For constant ρ, $Q_{in} = Q_{out} = Q$, and the design equation for component A is

$$0 = a_{in} - 2k\bar{t}_1(a_{out}^2)_1 - (a_{out})_1$$

The solution is

$$\frac{(a_{out})_1}{a_{in}} = \frac{-1 + \sqrt{1 + 8k\bar{t}_1 a_{in}}}{4k\bar{t}_1 a_{in}} \qquad (4.16)$$

Set $a_{in} = 1$ for convenience. When $k\bar{t}_1 a_{in} = 0.5$, Equation (4.16) gives

$$(a_{out})_1 = (a_{in})_2 = 0.618 a_{in}$$

The second CSTR has the same rate constant and residence time, but the dimensionless rate constant is now based on $(a_{in})_2 = 0.618 a_{in}$ rather than on a_{in}. Inserting $k\bar{t}_2(a_{in})_2 = k\bar{t}_2 a_{in}(a_{in})_2 = (0.5)(0.618) = 0.309$ into Equation (4.16) gives

$$a_{out} = (a_{out})_2 = (0.698)(a_{in})_2 = 0.432 a_{in}$$

Thus, $a_{out}/a_{in} = 0.432$ for the series combination. A single CSTR with twice the volume has $k\bar{t}_1 a_{in} = 1$. Equation (4.16) gives $a_{out}/a_{in} = 0.5$ so that the composite reactor with two tanks in series gives the higher conversion.

Numerical calculations are the easiest way to determine the performance of CSTRs in series. Simply analyze them one at a time, beginning at the inlet. However, there is a neat analytical solution for the special case of first-order reactions. The outlet concentration from the nth reactor in the series of CSTRs is

$$(a_{out})_n = \frac{(a_{in})_n}{1 + k_n \bar{t}_n} \qquad (4.17)$$

where k_n is the rate constant and $\bar{t}_n$ is the mean residence time ($n = 1, 2, \ldots, N$). Applying Equation (4.14) repeatedly gives the outlet concentration for the entire train of reactors:

$$a_{out} = \frac{a_{in}}{(1 + k_1 \bar{t}_1)(1 + k_2 \bar{t}_2) \cdots (1 + k_N \bar{t}_N)} = a_{in} \prod_{n=1}^{N} (1 + k_n \bar{t}_n)^{-1} \qquad (4.18)$$

When all the k_n are equal (i.e., the reactors are at the same temperature) and all the t_n are equal (i.e., the reactors are the same size),

$$a_{out} = \frac{a_{in}}{(1 + k\bar{t}/N)^N} \qquad (4.19)$$

where $\bar{t}$ is the mean residence time for the entire system. In the limit of many tanks in series,

$$\lim_{N \to \infty} \frac{a_{out}}{a_{in}} = e^{-k\bar{t}} \qquad (4.20)$$

Thus, the limit gives the same result as a piston flow reactor with mean residence time $\bar{t}$. Putting tanks in series is one way to combine the advantages of CSTRs with the better yield of a PFR. In practice, good improvements in yield are possible for fairly small N.

Example 4.12: Suppose the concentration of a toxic substance must be reduced by a factor of 1000. Assuming the substance decomposes with first-order kinetics, compare the total volume requirements when several stirred tanks are placed in series with the volume needed in a PFR to achieve the same factor of 1000 reduction.

Solution: The comparisons will be made at the same k and same throughput (i.e., the same Q). Rearrange Equation (4.19) and take the Nth root to obtain

$$k\bar{t} = N - 1 + \sqrt[N]{a_{in}/a_{out}} = N - 1 + \sqrt[N]{1000}$$

where $k\bar{t}$ is proportional to the volume of the system. Some results are shown below:

Number of tanks in series, N	Value of $k\bar{t}$ to achieve a 1000-fold reduction in concentration	Volume of the composite reactor relative to a PFR
1	999	144.6
2	61.2	8.8
3	27	3.9
4	18.5	2.7
⋮	⋮	⋮
∞	6.9	1

Thus, a single CSTR requires 144.6 times the volume of a single PFR, and the inefficiency of using a CSTR to achieve high conversions is dramatically illustrated. The volume disadvantage drops fairly quickly when CSTRs are put in series, but the economic disadvantage remains great. Cost consequences are explored in Problems 4.19 and 4.20.

4.5.3 Recycle Loops

Recycling of partially reacted feed streams is usually carried out after the product is separated and recovered. Unreacted feedstock can be separated and recycled to (ultimate) extinction. Figure 4.2 shows a different situation. It is a loop reactor where some of the reaction mass is returned to the inlet without separation. Internal recycle exists in every stirred tank reactor. An external recycle loop as shown in Figure 4.2 is less common, but is used, particularly in large plants where a conventional stirred tank would have heat transfer limitations. The net throughput for the system is $Q = Q_{in}$, but an amount q is recycled back to the reactor inlet so that the flow through the reactor is $Q_{in} + q$. Performance of this loop reactor system depends on the *recycle ratio* q/Q_{in} and on the type of reactor that is in the loop. *Fast external recycle* has no effect on the performance of a CSTR but will affect the performance of other reactors. By fast recycle, we mean that no appreciable reaction occurs in the recycle line. The CSTR already has enough internal recycle to justify the assumption of perfect mixing so that fast external recycle does nothing more. If the reactor in the loop is a PFR, the external recycle has a dramatic effect. At high q/Q_{in}, the loop reactor will approach the performance of a CSTR.

A material balance about the mixing point gives

$$a_{mix} = \frac{Q_{in}a_{in} + qa_{out}}{Q_{in} + q} \quad (4.21)$$

The feed to the reactor element within the loop is a_{mix}. The flow rate entering the reactor element is $Q_{in} + q$ and the exit concentration is a_{out}. The relationship between a_{mix} and a_{out} can be calculated without direct consideration of the external recycle. In the general case, this single-pass solution must be obtained numerically. Then the overall solution is iterative. One guesses a_{mix} and solves numerically for a_{out}. Equation (4.21) is then used to calculate a_{mix} for comparison with the original guess. Any good root finder will work. The function to be zeroed is

$$a_{mix} - \frac{Q_{in}a_{in} + qa_{out}}{Q_{in} + q} = 0$$

where a_{out} denotes the solution of the single-pass problem. When a_{out} is known analytically, an analytical solution to the recycle reactor problem is usually possible.

Example 4.13: Determine the outlet concentration from a loop reactor as a function of Q_{in} and q for the case where the reactor element is a PFR and the reaction is first order. Assume constant density and isothermal operation.

Solution: The single-pass solution is

$$a_{out} = a_{mix} \exp\left(\frac{-kV}{Q_{in} + q}\right)$$

Note that $V/(Q_{in} + q)$ is the per-pass residence time and is far different from the mean residence time for the system, $\bar{t} = V/Q_{in}$. Equation (4.21) gives

$$a_{mix} = \frac{a_{in}Q_{in}}{Q_{in} + q - q\,\exp[-kV/(Q_{in} + q)]}$$

and the solution for a_{out} is

$$a_{out} = \frac{a_{in}Q_{in}}{(Q_{in} + q)\,\exp[kV/(Q_{in} + q)] - q} \quad (4.22)$$

Figures 4.3 and 4.4 show how a loop reactor approaches the performance of a CSTR as the recycle rate is increased. Two things happen as $q \to \infty$: $a_{out} \to a_{in} Q_{in}/(Q_{in} + kV)$ and $a_{mix} \to a_{out}$. The specific results in Figures 4.3 and 4.4 apply to a first-order reaction with a piston flow reactor in the recycle loop, but the general concept applies to almost any type of reaction and reactor. High recycle rates mean that perfect mixing will be closely approached. There are two provisos: the mixing point must do a good job of mixing the recycle with the incoming feed and all the volume in the reactor must be accessible to the increased throughput. A rule of thumb is that $q/Q > 8$ will give performance equivalent to a conventionally agitated vessel. This may seem to be belied by the figures since there is still appreciable difference between the loop performance at $q/Q = 8$ and a CSTR. However, the difference will be smaller when a real reactor is put in a recycle loop since, unlike the idealization of piston flow, the real reactor will already have some internal mixing.

The loop reactor is sometimes used to model conventionally agitated stirred tanks. The ratio of internal circulation to net throughput in a large, internally agitated vessel can be as low as 8. The mixing inside the vessel is far from perfect, but assuming that the vessel behaves as a CSTR it may be still be adequate for design purposes. Alternatively, the conventionally agitated vessel could be modeled as a PFR or a composite reactor installed in a recycle loop in order to explore the sensitivity of the system to the details of mixing.

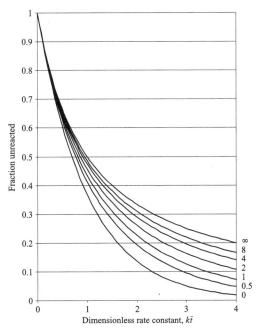

FIGURE 4.3 Effect of recycle rate on the performance of a loop reactor. The dimensionless rate constant is based on the system residence time, $\bar{t} = V/Q$. The parameter is q/Q.

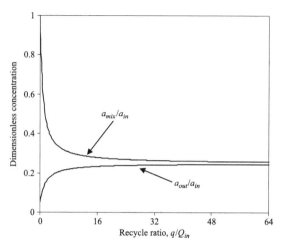

FIGURE 4.4 Extreme concentrations, a_{mix} and a_{out} within a loop reactor. The case shown is for $k\bar{t} = 3$.

PROBLEMS

4.1. Observed kinetics for the reaction

$$A + B \longrightarrow 2C$$

are $\mathcal{R} = 0.43ab^{0.8}\,\text{mol}/(\text{m}^3 \cdot \text{h})$. Suppose the reactor is run in a constant-density CSTR with $a_{in} = 15\,\text{mol/m}^3$, $b_{in} = 20\,\text{mol/m}^3$, $V = 3.5\,\text{m}^3$, and $Q = 125\,\text{m}^3/\text{h}$. Determine the exit concentration of C.

4.2. Find the analytical solution to the steady-state problem in Example 4.2.

4.3. Use Newton's method to solve the algebraic equations in Example 4.2. Note that the first two equations can be solved independently of the second two, so that only a two-dimensional version of Newton's method is required.

4.4. Repeat the false transient solution in Example 4.2 using a variety of initial conditions. Specifically include the case where the initial concentrations are all zero and the cases where the reactor is initially full of pure A, pure B, and so on. What do you conclude from these results?

4.5. Suppose the following reaction network is occurring in a constant-density CSTR:

$$A \rightleftarrows B \qquad \mathcal{R}_I = k_I a^{1/2} - k_{-I} b$$
$$B \longrightarrow C \qquad \mathcal{R}_{II} = k_{II} b^2$$
$$B + D \longrightarrow E \qquad \mathcal{R}_{III} = k_{III} bd$$

The rate constants are $k_I = 3.0 \times 10^{-2}\,\text{mol}^{1/2}/(\text{m}^{3/2} \cdot \text{h})$, $k_{-I} = 0.4\,\text{h}^{-1}$, $k_{II} = 5.0 \times 10^{-4}\,\text{mol}/(\text{m}^3 \cdot \text{h})$, $k_{III} = 3.0 \times 10^{-4}\,\text{mol}/(\text{m}^3 \cdot \text{h})$.

(a) Formulate a solution via the method of false transients. Use dimensionless time, $\tau = t/\bar{t}$, and dimensionless rate constants, e.g., $K_{III}^* = k_{III} a_{in} \bar{t}$.

(b) Solve the set of ODEs for sufficiently long times to closely approximate steady state. Use $a_0 = 3 \, \text{mol/m}^3$, $d_0 = 3 \, \text{mol/m}^3$, $b_0 = c_0 = e_0 = 0$, $\bar{t} = 1 \, \text{h}$. Do vary $\Delta \tau$ to confirm that your solution has converged.

4.6. A more complicated version of Problem 4.5 treats all the reactions as being reversible:

$$A \rightleftarrows B \qquad \mathcal{R}_I = k_I a^{1/2} - k_{-I} b$$
$$B \rightleftarrows C \qquad \mathcal{R}_{II} = k_{II} b^2 - k_{-II} c$$
$$B + D \rightleftarrows E \qquad \mathcal{R}_{III} = k_{III} bd - k_{-III} e$$

Suppose $k_{-II} = 0.08 \, \text{h}^{-1}$ and $k_{-III} = 0.05 \, \text{h}^{-1}$.

(a) Work Problem 4.5(b) for this revised reaction network.

(b) Suppose the reactor is filled but the feed and discharge pumps are never turned on. The reaction proceeds in batch and eventually reaches an equilibrium composition. Simulate the batch reaction to determine the equilibrium concentrations.

4.7. Equation (4.8) appears to be the solution to a quadratic equation. Why was the negative root chosen?

4.8. Are the kinetic constants determined in Example 4.5 accurate? Address this question by doing the following:

(a) Repeat Example 4.5 choosing component A (sulfur dioxide) as the key component rather than component C (sulfur trioxide).

(b) Use these new values for k and k_C to solve the forward problem in Example 4.6.

(c) Suppose a revised compositional analysis for Run I gave $(y_C)_{out} = 0.062$ rather than the original value of 0.063. The inerts change to 0.826. Repeat the example calculation of k and k_C using these new values.

(d) Suppose a repeat of Run 2 gave the following analysis at the outlet:

SO_2	2.2%
O_2	8.7%
SO_3	7.9%
Inerts	81.2%

Find k and k_C.

4.9. The ODE for the inerts was used to calculate Q_{out} in Example 4.6. How would you work the problem if there were no inerts? Use your method to predict reactor performance for the case where the feed contains 67% SO_2 and 33% O_2 by volume.

4.10. The low-temperature oxidation of hydrogen as in the cap of a lead-acid storage battery is an example of heterogeneous catalysis. It is proposed to model this reaction as if it were homogeneous:

$$H_2 + \tfrac{1}{2}O_2 \rightarrow H_2O \qquad \mathcal{R} = k[H_2][O_2] \text{ (nonelementary)}$$

and to treat the cap as if it were a perfect mixer. The following data have been generated on a test rig:

$$T_{in} = 22°C$$
$$T_{out} = 25°C$$
$$P_{in} = P_{out} = 1 \text{ atm}$$
$$H_2 \text{ in} = 2 \text{ g/h}$$
$$O_2 \text{ in} = 32 \text{ g/h } (2/1 \text{ excess})$$
$$N_2 \text{ in} = 160 \text{ g/h}$$
$$H_2O \text{ out} = 16 \text{ g/h}$$

(a) Determine k given $V = 25 \text{ cm}^3$.
(b) Calculate the adiabatic temperature rise for the observed extent of reaction. Is the measured rise reasonable? The test rig is exposed to natural convection. The room air is at 22°C.

4.11. A 100-gal pilot-plant reactor is agitated with a six-blade pitched turbine of 6 in diameter that consumes 0.35 kW at 300 rpm. Experiments with acid–base titrations showed that the mixing time in the vessel is 2 min. Scaleup to a 1000-gal vessel with the same mixing time is desired.
(a) Estimate the impeller size, motor size, and rpm for the larger reactor.
(b) What would be the mixing time if the scaleup were done at constant power per unit volume rather than constant mixing time?

4.12. Solve Example 4.10 algebraically and confirm the numerical example. For $b_{in} = 0$ you should find that the system has two steady states: one with $a_{out} = a_{in}$ that is always possible and one with

$$\frac{a_{out}}{a_{in}} = \frac{1}{1 + (kt_1 a_{in} - 1) \exp(kt_2 a_{in})}$$

that is possible only when $kt_1 > 1$. You should also conclude that the interior optimum occurs when $t_1 = 2/ka_{in}$.

4.13. Generalize the algebraic solution in Problem 4.12 to allow for $b_{in} > 0$.

4.14. Example 4.10 used the initial condition that $a_0 = 0$ and $b_0 = 1$. Will smaller values for b_0 work? How much smaller?

4.15. Suppose you have two identical CSTRs and you want to use these to make as much product as possible. The reaction is pseudo-first-order and the product recovery system requires a minimum conversion of

93.75%. Do you install the reactors in series or parallel? Would it affect your decision if the minimum conversion could be lowered?

4.16. Suppose you have two identical PFRs and you want to use them to make as much product as possible. The reaction is pseudo-first-order and the product recovery system requires a minimum conversion of 93.75%. Assume constant density. Do you install the reactors in series or parallel? Would it affect your decision if the minimum conversion could be lowered?

4.17. Example 4.12 used N stirred tanks in series to achieve a 1000-fold reduction in the concentration of a reactant that decomposes by first-order kinetics. Show how much worse the CSTRs would be if the 1000-fold reduction had to be achieved by dimerization; i.e., by a second order of the single reactant type. The reaction is irreversible and density is constant.

4.18. Suppose you have two CSTRs, one with a volume of $2\,m^2$ and one with a volume of $4\,m^3$. You have decided to install them in series and to operate them at the same temperature. Which goes first if you want to maximize production subject to a minimum conversion constraint? Consider the following cases:
   (a) The reaction is first order.
   (b) The reaction is second order of the form $2A \rightarrow P$.
   (c) The reaction is half-order.

4.19. Equipment costs are sometimes estimated using a scaling rule:

$$\frac{\text{Cost of large unit}}{\text{Cost of small unit}} = S^C$$

where C is the scaling exponent. If $C=1$, twice the size (volume or throughput) means twice the cost and there is no economy of scale. The installed cost of chemical process equipment typically scales as $C=0.6$ to 0.75. Suppose the installed cost of stirred tank reactors varies as $V^{0.75}$. Determine the optimum number of tanks in series for a first-order reaction going to 99.9 % completion.

4.20. Repeat Problem 4.19 for $C=0.6$ and 1.0. Note that more reactors will affect more than just the capital costs. Additional equipment will lower system reliability and increase operating costs. Which value of C is the more conservative? Is this value of C also the more conservative when estimating the installed cost of an entire plant based on the cost of a smaller plant?

4.21. Example 4.13 treated the case of a piston flow reactor inside a recycle loop. Replace the PFR with two equal-volume stirred tanks in series. The reaction remains first order, irreversible, and at constant density.
   (a) Derive algebraic equations for a_{mix} and a_{out} for the composite system.
   (b) Reproduce Figures 4.3 and 4.4 for this case.

4.22. Work Example 4.13 for the case where the reaction is second order of the single reactant type. It is irreversible and density is constant. The reactor element inside the loop is a PFR.

4.23. Find the limit of Equations (4.21) and (4.22) if $q \to \infty$ with Q_{in} fixed. Why would you expect this result?

4.24. The material balance around the mixing point of a loop reactor is given by Equation (4.21) for the case of constant fluid density. How would you work a recycle problem with variable density? Specifically, write the variable-density counterpart of Equation (4.21) and explain how you would use it.

SUGGESTIONS FOR FURTHER READING

Reactor models consisting of series and parallel combinations of ideal reactors are discussed at length in

Levenspiel, O., *Chemical Reaction Engineering*, 3rd ed., Wiley, New York, 1998.

The reaction coordinate, ε, is also call the *molar extent* or *degree of advancement*. It is applied to CSTRs in

Aris, R., *Elementary Chemical Reactor Analysis*, Dover, Mineola, NY, 2000.

APPENDIX 4: SOLUTION OF SIMULTANEOUS ALGEBRAIC EQUATIONS

Consider a set of N algebraic equations of the form

$$F(a, b, \ldots) = 0$$
$$G(a, b, \ldots) = 0$$
$$\vdots \quad \vdots$$

where $a, b, \ldots$ represent the N unknowns. We suppose that none of these equations is easily solvable for any of the unknowns. If an original equation were solvable for an unknown, then that unknown could be eliminated and the dimensionality of the set reduced by 1. Such eliminations are usually worth the algebra when they are possible.

A.4.1 Binary Searches

A binary search is a robust and easily implemented method for finding a root of a single equation, $F(a) = 0$. It is necessary to know bounds, $a_{min} \le a \le a_{max}$, within which the root exists. If $F(a_{min})$ and $F(a_{max})$ differ in sign, there will be an odd number of roots within the bounds and a binary search will find one of them to a specified level of accuracy. It does so by calculating F at the midpoint of the interval; that is, at $a = (a_{min} + a_{max})/2$. The sign of F will be the same as at one of the endpoints. Discard that endpoint and replace it with the midpoint. The sign of F at the two new endpoints will differ, so that the range in which the solution must lie has been halved. This procedure can obviously be repeated J times to reduce the range in which a solution must lie to 2^{-J} of the original range. The accuracy is set in advance by choosing J:

$$J = \ln\left[\frac{a_{max} - a_{min}}{\varepsilon}\right] \bigg/ \ln 2$$

where ε is the uncertainty in the answer. The following code works for any arbitrary function that is specified by the subroutine Func(a, f).

```
DEFDBL A-H, P-Z
DEFLNG I-O

amax = 4 'User supplied value
amin = 1 'User supplied value

er = .0000005# 'User supplied value
X = LOG((amax-amin)/er)/LOG(2)
J = X + 0.5 'Rounds up

CALL Func(amax, Fmax) 'User supplied subroutine
CALL Func(amin, Fmin)

IF Fmax * Fmin >= 0 THEN STOP 'Error condition
FOR jj = 1 TO J
   amid = (amax+amin)/2
   CALL Func(amid, F)
   IF F * Fmin > 0 THEN
      Fmin = F
      amin = amid
   ELSE
      Fmax = F
      amax = amid
   END IF
NEXT jj

PRINT amid
```

A.4.2 Multidimensional Newton's Method

Consider some point $(a_0, b_0, \ldots)$ within the region of definition of the functions $F, G, \ldots$ and suppose that the functions can be represented by an multidimensional Taylor series about this point. Truncating the series after the first-order derivatives gives

$$F(a, b, \ldots) = F(a_0, b_0, \ldots) + \left[\frac{\partial F}{\partial a}\right]_0 (a - a_0) + \left[\frac{\partial F}{\partial b}\right]_0 (b - b_0) + \cdots$$

$$G(a, b, \ldots) = G(a_0, b_0, \ldots) + \left[\frac{\partial G}{\partial a}\right]_0 (a - a_0) + \left[\frac{\partial G}{\partial b}\right]_0 (b - b_0) + \cdots$$

$$\vdots \qquad \vdots$$

where there are as many equations as there are unknowns. In matrix form,

$$\begin{bmatrix} [\partial F/\partial a]_0 & [\partial F/\partial b]_0 & \cdots \\ [\partial G/\partial a]_0 & [\partial G/\partial b]_0 & \cdots \\ \vdots & & \end{bmatrix} \begin{bmatrix} a - a_0 \\ b - b_0 \\ \vdots \end{bmatrix} = \begin{bmatrix} F - F_0 \\ G - G_0 \\ \vdots \end{bmatrix}$$

We seek values for $a, b, \ldots$ which give $F = G = \cdots = 0$. Setting $F = G = \cdots = 0$ and solving for $a, b, \ldots$ gives

$$\begin{bmatrix} a \\ b \\ \vdots \end{bmatrix} = \begin{bmatrix} a_0 \\ b_0 \\ \vdots \end{bmatrix} - \begin{bmatrix} [\partial F/\partial a]_0 & [\partial F/\partial b]_0 & \cdots \\ [\partial G/\partial a]_0 & [\partial G/\partial b]_0 & \cdots \\ \vdots & & \end{bmatrix}^{-1} \begin{bmatrix} F_0 \\ G_0 \\ \vdots \end{bmatrix}$$

For the special case of one unknown,

$$a = a_0 - \frac{F_0}{[dF/da]_0}$$

which is Newton's method for finding the roots of a single equation. For two unknowns,

$$a = a_0 - \frac{F_0 [\partial G/\partial a]_0 - G_0 [\partial F/\partial a]_0}{[\partial F/\partial a]_0 [\partial G/\partial b]_0 - [\partial F/\partial b]_0 [\partial G/\partial a]_0}$$

$$b = b_0 - \frac{-F_0 [\partial G/\partial b]_0 + G_0 [\partial F/\partial b]_0}{[\partial F/\partial a]_0 [\partial G/\partial b]_0 - [\partial F/\partial b]_0 [\partial G/\partial a]_0}$$

which is a two-dimensional generalization of Newton's method.

The above technique can be used to solve large sets of algebraic equations; but, like the ordinary one-dimensional form of Newton's method, the algorithm may diverge unless the initial guess $(a_0, b_0, \ldots)$ is quite close to the final solution. Thus, it might be considered as a method for rapidly improving a good initial guess, with other techniques being necessary to obtain the initial guess.

For the one-dimensional case, dF/da can usually be estimated using values of F determined at previous guesses. Thus,

$$a = a_0 - \frac{F_0}{[(F_0 - F_{-1})/(a_0 - a_{-1})]}$$

where $F_0 = F(a_0)$ is the value of F obtained one iteration ago when the guess was a_0, and $F_{-1} = F(a_{-1})$ is the value obtained two iterations ago when the guess was a_{-1}.

For two- and higher-dimensional solutions, it is probably best to estimate the first partial derivatives by a formula such as

$$\left[\frac{\partial F}{\partial a}\right]_0 \approx \frac{F(a_0, b_0, \ldots) - F(\gamma a_0, b_0, \ldots)}{a_0 - \gamma a_0}$$

where γ is a constant close to 1.0.

CHAPTER 5
THERMAL EFFECTS AND ENERGY BALANCES

This chapter treats the effects of temperature on the three types of ideal reactors: batch, piston flow, and continuous-flow stirred tank. Three major questions in reactor design are addressed. What is the optimal temperature for a reaction? How can this temperature be achieved or at least approximated in practice? How can results from the laboratory or pilot plant be scaled up?

5.1 TEMPERATURE DEPENDENCE OF REACTION RATES

Most reaction rates are sensitive to temperature, and most laboratory studies regard temperature as an important means of improving reaction yield or selectivity. Our treatment has so far ignored this point. The reactors have been isothermal, and the operating temperature, as reflected by the rate constant, has been arbitrarily assigned. In reality, temperature effects should be considered, even for isothermal reactors, since the operating temperature must be specified as part of the design. For nonisothermal reactors, where the temperature varies from point to point within the reactor, the temperature dependence directly enters the design calculations.

5.1.1 Arrhenius Temperature Dependence

The rate constant for elementary reactions is almost always expressed as

$$k = k_0 T^m \exp\left(\frac{-E}{R_g T}\right) = k_0 T^m \exp\left(\frac{-T_{act}}{T}\right) \tag{5.1}$$

where $m = 0$, $1/2$, or 1 depending on the specific theoretical model being used. The quantity E is *activation energy*, although the specific theories interpret this energy term in different ways. The quantity $T_{act} = E/R_g$ has units of temperature (invariably K) and is called the *activation temperature*. The activation temperature should not be interpreted as an actual temperature. It is just a convenient way of expressing the composite quantity E/R_g.

The case of $m = 0$ corresponds to classical Arrhenius theory; $m = 1/2$ is derived from the collision theory of bimolecular gas-phase reactions; and $m = 1$ corresponds to activated complex or transition state theory. None of these theories is sufficiently well developed to predict reaction rates from first principles, and it is practically impossible to choose between them based on experimental measurements. The relatively small variation in rate constant due to the pre-exponential temperature dependence T^m is overwhelmed by the exponential dependence $\exp(-T_{act}/T)$. For many reactions, a plot of $\ln(k)$ versus T^{-1} will be approximately linear, and the slope of this line can be used to calculate E. Plots of $\ln(k/T^m)$ versus T^{-1} for the same reactions will also be approximately linear as well, which shows the futility of determining m by this approach.

Example 5.1: The bimolecular reaction

$$NO + ClNO_2 \rightarrow NO_2 + ClNO$$

is thought to be elementary. The following rate data are available:[1]

T, K	300	311	323	334	344
k, m^3/(mol·s)	0.79	1.25	1.64	2.56	3.40

Fit Equation (5.1) to these data for $m = 0$, 0.5, and 1.

Solution: The classic way of fitting these data is to plot $\ln(k/T^m)$ versus T^{-1} and to extract k_0 and T_{act} from the slope and intercept of the resulting (nearly) straight line. Special graph paper with a logarithmic y-axis and a $1/T$ x-axis was made for this purpose. The currently preferred method is to use nonlinear regression to fit the data. The object is to find values for k_0 and T_{act} that minimize the sum-of-squares:

$$S^2 = \sum_{Data} [\text{Experiment} - \text{model}]^2$$

$$= \sum_{j=1}^{J} \left[k_j - k_0 T_j^m \exp\left(-T_{act}/T_j\right) \right]^2 \quad (5.2)$$

where $J = 5$ for the data at hand. The general topic of nonlinear regression is discussed in Chapter 7 and methods for performing the minimization are described in Appendix 6. However, with only two unknowns, even a manual search will produce the answers in reasonable time. The results of this fitting procedure are:

		k fitted		
T	k experimental	$m=0$	$m=0.5$	$m=1$
300	0.79	0.80	0.80	0.80
311	1.25	1.19	1.19	1.19
323	1.64	1.78	1.78	1.77
334	2.56	2.52	2.52	2.52
344	3.44	3.38	3.37	3.37
	Standard Deviation	0.0808	0.0809	0.0807
	k_0, m³/(mol·s)	64400	2120	71.5
	T_{act}, K	3390	3220	3060

The model predictions are essentially identical. The minimization procedure automatically adjusts the values for k_0 and T_{act} to account for the different values of m. The predictions are imperfect for any value of m, but this is presumably due to experimental scatter. For simplicity and to conform to general practice, we will use $m=0$ from this point on.

Figure 5.1 shows an Arrhenius plot for the reaction $O + N_2 \rightarrow NO + N$; the plot is linear over an experimental temperature range of 1500 K. Note that the rate constant is expressed per molecule rather than per mole. This method for expressing k is favored by some chemical kineticists. It differs by a factor of Avogadro's number from the more usual k.

Few reactions have been studied over the enormous range indicated in Figure 5.1. Even so, they will often show curvature in an Arrhenius plot of $\ln(k)$ versus T^{-1}. The usual reason for curvature is that the reaction is complex with several elementary steps and with different values of E for each step. The overall temperature behavior may be quite different from the simple Arrhenius behavior expected for an elementary reaction. However, a linear Arrhenius plot is neither necessary nor sufficient to prove that a reaction is elementary. It is not sufficient because complex reactions may have one dominant activation energy or several steps with similar activation energies that lead to an overall temperature dependence of the Arrhenius sort. It is not necessary since some low-pressure, gas-phase, bimolecular reactions exhibit distinctly non-Arrhenius behavior, even though the reactions are believed to be elementary. Any experimental study should consider the possibility that k_0 and T_{act} are functions of temperature. A strong dependence on temperature usually signals a change in reaction mechanism, for example, a shift from a kinetic limitation to a mass transfer limitation.

You may recall the rule-of-thumb that reaction rates double for each 10°C increase in temperature. Doubling when going from 20°C to 30°C means

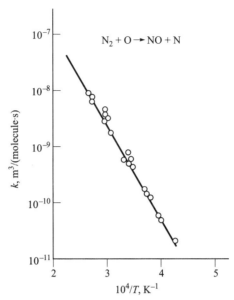

FIGURE 5.1 Arrhenius behavior over a large temperature range. (Data from Monat, J. P., Hanson, R. K., and Kruger, C. H., "Shock tube determination of the rate coefficient for the reaction $N_2 + O \rightarrow NO + N$," *Seventeenth Symposium* (*International*) *on Combustion*, Gerard Faeth, Ed., The Combustion Institute, Pittsburgh, 1979, pp. 543–552.)

$E = 51.2 \, \text{kJ/mol}$ or $T_{act} = 6160 \, \text{K}$. Doubling when going from 100 °C to 110°C means $E = 82.4 \, \text{kJ/mol}$ or $T_{act} = 9910 \, \text{K}$. Activation temperatures in the range 5000–15,000 K are typical of homogeneous reactions, but activation temperatures above 40,000 K are known. The higher the activation energy, the more the reaction rate is sensitive to temperature. Biological systems typically have high activation energies. An activation temperature below about 2000 K usually indicates that the reaction is limited by a mass transfer step (diffusion) rather than chemical reaction. Such limitations are common in heterogeneous systems.

5.1.2 Optimal Temperatures for Isothermal Reactors

Reaction rates almost always increase with temperature. Thus, the best temperature for a single, irreversible reaction, whether elementary or complex, is the highest possible temperature. Practical reactor designs must consider limitations of materials of construction and economic tradeoffs between heating costs and yield, but there is no optimal temperature from a strictly kinetic viewpoint. Of course, at sufficiently high temperatures, a competitive reaction or reversibility will emerge.

Multiple reactions, and reversible reactions, since these are a special form of multiple reactions, usually exhibit an optimal temperature with respect to the yield of a desired product. The reaction energetics are not trivial, even if the reactor is approximately isothermal. One must specify the isotherm at which to operate. Consider the elementary, reversible reaction

$$A \underset{k_r}{\overset{k_f}{\rightleftarrows}} B \qquad (5.3)$$

Suppose this reaction is occurring in a CSTR of fixed volume and throughput. It is desired to find the reaction temperature that maximizes the yield of product B. Suppose $E_f > E_r$, as is normally the case when the forward reaction is endothermic. Then the forward reaction is favored by increasing temperature. The equilibrium shifts in the desirable direction, and the reaction rate increases. The best temperature is the highest possible temperature and there is no interior optimum.

For $E_f < E_r$, increasing the temperature shifts the equilibrium in the wrong direction, but the forward reaction rate still increases with increasing temperature. There is an optimum temperature for this case. A very low reaction temperature gives a low yield of B because the forward rate is low. A very high reaction temperature also gives a low yield of B because the equilibrium is shifted toward the left.

The outlet concentration from the stirred tank, assuming constant physical properties and $b_{in} = 0$, is given by

$$b_{out} = \frac{k_f a_{in} \bar{t}}{1 + k_f \bar{t} + k_r \bar{t}} \qquad (5.4)$$

We assume the forward and reverse reactions have Arrhenius temperature dependences with $E_f < E_r$. Setting $db_{out}/dT = 0$ gives

$$T_{optimal} = \frac{E_r}{R_g \ln[(E_r - E_f)(k_0)_r \bar{t}/E_f]} \qquad (5.5)$$

as the kinetically determined optimum temperature.

The reader who duplicates the algebra needed for this analytical solution will soon appreciate that a CSTR is the most complicated reactor and Equation (5.3) is the most complicated reaction for which an analytical solution for $T_{optimal}$ is likely. The same reaction occurring in a PFR with $b_{in} = 0$ leads to

$$b_{out} = \frac{a_{in} k_f (1 - \exp[-(k_f + k_r)\bar{t}])}{k_f + k_r} \qquad (5.6)$$

Differentiation and setting $db_{out}/dT = 0$ gives a transcendental equation in $T_{optimal}$ that cannot be solved in closed form. The optimal temperature must be found numerically.

Example 5.2: Suppose $k_f = 10^8 \exp(-5000/T)$ and $k_r = 10^{15} \exp(-10000/T), h^{-1}$. Find the temperature that maximizes the concentration of B for the reaction of Equation (5.3). Consider two cases: One where the reaction is carried out in an ideal CSTR with $\bar{t} = 2$ h and one where the reaction is carried out in an ideal PFR with the same 2-h residence time. Assume constant density and a feed of pure A. Calculate the equilibrium concentration at both values for $T_{optimal}$.

Solution: Equation (5.5) can be applied directly to the CSTR case. The result is $T_{optimal} = 283.8\,\text{K}$ for which $b_{out}/a_{in} = 0.691$. The equilibrium concentration is found from

$$K = \frac{k_f}{k_r} = \frac{b_{equil}}{a_{equil}} = \frac{b_{equil}}{a_{in} - b_{equil}} \tag{5.7}$$

which gives $b_{equil}/a_{in} = 0.817$ at 283.8 K.

A PFR reactor gives a better result at the same temperature. Equation (5.6) gives $b_{out}/a_{in} = 0.814$ for the PFR at 283.8 K. However, this is not the optimum. With only one optimization variable, a trial-and-error search is probably the fastest way to determine that $T_{optimal} = 277.5\,\text{K}$ and $b/a_{in} = 0.842$ for the batch case. The equilibrium concentration at 277.5 K is $b_{equil}/a_{in} = 0.870$.

The CSTR operates at a higher temperature in order to compensate for its inherently lower conversion. The higher temperature shifts the equilibrium concentration in an unfavorable direction, but the higher temperature is still worthwhile for the CSTR because equilibrium is not closely approached.

The results of Example 5.2 apply to a reactor with a fixed reaction time, $\bar{t}$ or t_{batch}. Equation (5.5) shows that the optimal temperature in a CSTR decreases as the mean residence time increases. This is also true for a PFR or a batch reactor. There is no interior optimum with respect to reaction time for a single, reversible reaction. When $E_f < E_r$, the best yield is obtained in a large reactor operating at low temperature. Obviously, the kinetic model ceases to apply when the reactants freeze. More realistically, capital and operating costs impose constraints on the design.

Note that maximizing a product concentration such as b_{out} will not maximize the total production rate of component B, $b_{out}Q_{out}$. Total production can normally be increased by increasing the flow rate and thus decreasing the reaction time. The reactor operates nearer to the feed composition so that average reaction rate is higher. More product is made, but it is dilute. This imposes a larger burden on the downstream separation and recovery facilities. Capital and operating costs again impose design constraints. Reactor optimization cannot be achieved without considering the process as a whole. The one-variable-at-a-time optimizations considered here in Chapter 5 are carried out as preludes to the more comprehensive optimizations described in Chapter 6.

Example 5.3: Suppose

$$A \xrightarrow{k_I} B \xrightarrow{k_{II}} C \tag{5.8}$$

with $k_I = 10^8 \exp(-5000/T)$ and $k_{II} = 10^{15} \exp(-10000/T)$, h^{-1}. Find the temperature that maximizes b_{out} for a CSTR with $\bar{t} = 2$ and for a PFR with the same 2-h residence time. Assume constant density with $b_{in} = c_{in} = 0$.

Solution: Use Equation (4.3) with $b_{in} = 0$ for the CSTR to obtain

$$b_{out} = \frac{k_I a_{in} \bar{t}}{(1 + k_I \bar{t})(1 + k_{II} \bar{t})} \quad (5.9)$$

A one-dimensional search gives $T_{optimal} = 271.4$ K and $b_{out} = 0.556 a_{in}$. Convert Equation (2.22) to the PFR form and set $b_{in} = 0$ to obtain

$$b_{out} = \frac{k_I a_{in}[\exp(-k_I \bar{t}) - \exp(-k_{II} \bar{t})]}{k_{II} - k_I} \quad (5.10)$$

Numerical optimization gives $T_{optimal} = 271.7$ and $b = 0.760 a_{in}$.

At a fixed temperature, a single, reversible reaction has no interior optimum with respect to reaction time. If the inlet product concentration is less than the equilibrium concentration, a very large flow reactor or a very long batch reaction is best since it will give a close approach to equilibrium. If the inlet product concentration is above the equilibrium concentration, no reaction is desired so the optimal time is zero. In contrast, there will always be an interior optimum with respect to reaction time at a fixed temperature when an intermediate product in a set of consecutive reactions is desired. (Ignore the trivial exception where the feed concentration of the desired product is already so high that any reaction would lower it.) For the normal case of $b_{in} \ll a_{in}$, a very small reactor forms no B and a very large reactor destroys whatever B is formed. Thus, there will be an interior optimum with respect to reaction time.

Example 5.3 asked the question: If reaction time is fixed, what is the best temperature? Example 5.4 asks a related but different question: If the temperature is fixed, what is the best reaction time? Both examples address maximization of product concentration, not total production rate.

Example 5.4: Determine the optimum reaction time for the consecutive reactions of Example 5.3 for the case where the operating temperature is specified. Consider both a CSTR and a PFR.

Solution: Analytical solutions are possible for this problem. For the CSTR, differentiate Equation (5.9) with respect to $\bar{t}$ and set the result to zero. Solving for $\bar{t}$ gives

$$\bar{t}_{optimal} = \sqrt{\frac{1}{k_I k_{II}}} \quad (5.11)$$

Suppose $T = 271.4$ as for the CSTR case in Example 5.3. Using Equation (5.11) and the same rate constants as in Example 5.3 gives $\bar{t}_{optimal} = 3.17$ h.

The corresponding value for b_{out} is $0.578a_{in}$. Recall that Example 5.3 used $\bar{t} = 2\,h$ and gave $b_{out}/a_{in} = 0.556$. Thus, the temperature that is best for a fixed volume and the volume that is best for a fixed temperature do not correspond.

For a PFR, use Equation (5.10) and set $dp_{out}/d\bar{t} = 0$ to obtain

$$\bar{t}_{optimal} = \frac{\ln(k_I/k_{II})}{k_I - k_{II}} \tag{5.12}$$

Suppose $T = 271.7$ as for the PFR (or batch) case in Example 5.3. Using Equation (5.12) and the same rate constants as in Example 5.3 gives $\bar{t}_{optimal} = 2.50\,h$. The corresponding value for b_{out} is $0.772a_{in}$. Recall that Example 5.3 used $\bar{t} = 2\,h$ and gave $b_{out}/a_{in} = 0.760$. Again, the temperature that is best for a fixed volume does not correspond to the volume that is best for a fixed temperature.

The competitive reactions

$$A \xrightarrow{k_I} B$$
$$A \xrightarrow{k_{II}} C \tag{5.13}$$

will have an intermediate optimum for B only if $E_I < E_{II}$ and will have an intermediate optimum for C only if $E_I > E_{II}$. Otherwise, the yield of the desired product is maximized at high temperatures. If $E_I > E_{II}$, high temperatures maximize the yield of B. If $E_I < E_{II}$, high temperatures maximize the yield of C.

The reader will appreciate that the rules for what maximizes what can be quite complicated to deduce and even to express. The safe way is to write the reactor design equations for the given set of reactions and then to numerically determine the best values for reaction time and temperature. An interior optimum may not exist. When one does exist, it provides a good starting point for the more comprehensive optimization studies discussed in Chapter 6.

5.2 THE ENERGY BALANCE

A reasonably general energy balance for a flow reactor can be written in English as

Enthalpy of input streams − enthalpy of output streams
+ heat generated by reaction − heat transferred out
= accumulation of energy

and in mathematics as

$$Q_{in}\rho_{in}H_{in} - Q_{out}\rho_{out}H_{out} - V\Delta H_R \hat{\mathscr{R}} - \hat{U}A_{ext}(\hat{T} - T_{ext}) = \frac{d(V\hat{\rho}\hat{H})}{dt} \tag{5.14}$$

This is an integral balance written for the whole system. The various terms deserve discussion. The enthalpies are relative to some reference temperature, T_{ref}. Standard tabulations of thermodynamic data (see Chapter 7) make it convenient to choose $T_{ref} = 298$ K, but choices of $T_{ref} = 0$ K or $T_{ref} = 0°$C are also common. The enthalpy terms will normally be replaced by temperature using

$$H = \int_{T_{ref}}^{T} C_P \, dT \tag{5.15}$$

For many purposes, the heat capacity will be approximately constant over the range of temperatures in the system. Then

$$H = C_P(T - T_{ref}) \tag{5.16}$$

where C_P is the average value for the entire reactant mixture, including any inerts. It may be a function of composition as well as temperature. An additional term—e.g, a heat of vaporization—must be added to Equations (5.15) and (5.16) if any of the components undergo a phase change. Also, the equations must be modified if there is a large pressure change during the course of the reaction. See Section 7.2.1.

By thermodynamic convention, $\Delta H_R < 0$ for exothermic reactions, so that a negative sign is attached to the heat-generation term. When there are multiple reactions, the heat-generation term refers to the net effect of all reactions. Thus, the $\Delta H_R \mathscr{R}$ term is an implicit summation over all M reactions that may be occurring:

$$\Delta H_R \mathscr{R} = \sum_{Reactions} (\Delta H_R)_I \mathscr{R}_I = \sum_{I=1}^{M} (\Delta H_R)_I \mathscr{R}_I \tag{5.17}$$

The reaction rates in Equation (5.17) are positive and apply to "the reaction." That is, they are the rates of production of (possibly hypothetical) components having stoichiometric coefficients of +1. Similarly, the heats of reaction are per mole of the same component. Some care is needed in using literature values. See Section 7.2.1.

Chapter 7 provides a review of chemical thermodynamics useful for estimating specific heats, heats of reaction, and reaction equilibria. The examples here in Chapter 5 assume constant physical properties. This allows simpler illustrations of principles and techniques. Example 7.16 gives a detailed treatment of a reversible, gas-phase reaction where there is a change in the number of moles upon reaction and where the equilibrium composition, heat capacities, and reaction rates all vary with temperature. Such rigorous treatments are complicated but should be used for final design calculations. It is better engineering practice to include phenomena than to argue on qualitative grounds that the phenomena are unimportant. Similarly, high numerical precision should be used in the calculations, even though the accuracy of the data may be quite limited. The object is to eliminate sources of error, either physical or numerical, that can be eliminated with reasonable effort. A sensitivity analysis can then be confined to the remaining sources of error that are difficult to eliminate. As a practical matter, few reactor design calculations will have absolute accuracies better than two decimal places. Relative accuracies between similar calculations can be much better and can provide justification for citing values to four or more decimal places, but citing values to full computational precision is a sign of naiveté.

The heat transfer term envisions convection to an external surface, and U is an overall heat transfer coefficient. The heat transfer area could be the reactor jacket, coils inside the reactor, cooled baffles, or an external heat exchanger. Other forms of heat transfer or heat generation can be added to this term; e.g, mechanical power input from an agitator or radiative heat transfer. The reactor is *adiabatic* when $U = 0$.

The accumulation term is zero for steady-state processes. The accumulation term is needed for batch reactors and to solve steady-state problems by the method of false transients.

In practice, the integral formulation of Equation (5.14) is directly useful only when the reactor is a stirred tank with good internal mixing. When there are temperature gradients inside the reactor, as there will be in the axial direction in a nonisothermal PFR, the integral balance remains true but is not especially useful. Instead, a differential energy balance is needed. The situation is exactly analogous to the integral and differential component balances used for the ideal reactors discussed in Chapter 1.

5.2.1 Nonisothermal Batch Reactors

The ideal batch reactor is internally uniform in both composition and temperature. The flow and mixing patterns that are assumed to eliminate concentration gradients will eliminate temperature gradients as well. Homogeneity on a scale approaching molecular dimensions requires diffusion. Both heat and mass diffuse, but thermal diffusivities tend to be orders-of-magnitude higher than molecular diffusivities. Thus, if one is willing to assume compositional uniformity, it is reasonable to assume thermal uniformity as well.

For a perfectly mixed batch reactor, the energy balance is

$$\frac{d(V\rho H)}{dt} = -V \Delta H_R \mathscr{R} - UA_{ext}(T - T_{ext}) \tag{5.18}$$

For constant volume and physical properties,

$$\frac{dT}{dt} = \frac{-\Delta H_R \mathscr{R}}{\rho C_P} - \frac{UA_{ext}(T - T_{ext})}{V \rho C_P} \tag{5.19}$$

THERMAL EFFECTS AND ENERGY BALANCES

Suppose that there is only one reaction and that component A is the limiting reactant. Then the quantity

$$\Delta T_{adiabatic} = \frac{-\Delta H_R a_{in}}{\rho C_P} \tag{5.20}$$

gives the *adiabatic temperature change* for the reaction. This is the temperature that the batch would reach if the physical properties really were constant, if there were no change in the reaction mechanism, and if there were no heat transfer with the environment. Despite all these usually incorrect assumptions, $\Delta T_{adiabatic}$ provides a rough measure of the difficulty in thermal management of a reaction. If $\Delta T_{adiabatic} = 10\,\text{K}$, the reaction is a pussycat. If $\Delta T_{adiabatic} = 1000\,\text{K}$, it is a tiger. When there are multiple reactions, $\Delta H_R \mathscr{R}$ is a sum according to Equation (5.17), and the adiabatic temperature change is most easily found by setting $U=0$ and solving Equation (5.19) simultaneously with the component balance equations. The long-time solution gives $\Delta T_{adiabatic}$.

The N component balances are unchanged from those in Chapter 2, although the reaction rates are now understood to be functions of temperature. In matrix form,

$$\frac{d(\mathbf{a}V)}{dt} = \mathbf{v}\mathscr{R}V \tag{5.21}$$

The design equations for a nonisothermal batch reactor include $N+1$ ODEs, one for each component and one for energy. These ODEs are coupled by the temperature and compositional dependence of $\mathscr{R}$. They may also be weakly coupled through the temperature and compositional dependence of physical properties such as density and heat capacity, but the strong coupling is through the reaction rate.

Example 5.5: Ingredients are quickly charged to a jacketed batch reactor at an initial temperature of $25°\text{C}$. The jacket temperature is $80°\text{C}$. A pseudo-first-order reaction occurs. Determine the reaction temperature and the fraction unreacted as a function of time. The following data are available:

$V = 1\,\text{m}^3 \qquad A_{ext} = 4.68\,\text{m}^2 \qquad U = 1100\,\text{J}/(\text{m}^2 \cdot \text{s} \cdot \text{K}) \qquad \rho = 820\,\text{kg}/\text{m}^3$
$C_p = 3400\,\text{J}/(\text{kg} \cdot \text{K}) \quad k = 3.7 \times 10^8 \exp(-6000/T) \quad \Delta H_R = -108{,}000\,\text{J/mol}$
$a_{in} = 1900.0\,\text{mol}/\text{m}^3$

Physical properties may be assumed to be constant.

Solution: The component balance for A is

$$\frac{da}{dt} = -ka$$

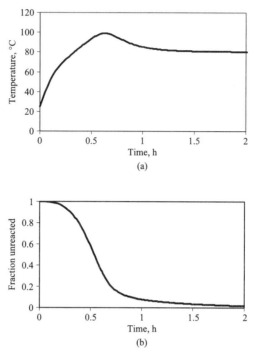

FIGURE 5.2 (a) Temperature and (b) fraction unreacted in a nonisothermal batch reactor with jacket cooling.

and the energy balance is

$$\frac{dT}{dt} = \frac{-\Delta H_R \mathscr{R}}{\rho C_P} - \frac{UA_{ext}(T - T_{ext})}{V\rho C_P} = \Delta T_{adiabatic}\left(\frac{ka}{a_{in}}\right) - \frac{UA_{ext}(T - T_{ext})}{V\rho C_P}$$

where $\Delta T_{adiabatic} = 73.6\,\text{K}$ for the subject reaction. The initial conditions are $a = 1900$ and $T = 298$ at $t = 0$. The Arrhenius temperature dependence prevents an analytical solution. All the dimensioned quantities are in consistent units so they can be substituted directly into the ODEs. A numerical solution gives the results shown in Figure 5.2.

The curves in Figure 5.2 are typical of exothermic reactions in batch or tubular reactors. The temperature overshoots the wall temperature. This phenomenon is called an *exotherm*. The exotherm is moderate in Example 5.2 but becomes larger and perhaps uncontrollable upon scaleup. Ways of managing an exotherm during scaleup are discussed in Section 5.3.

Advice on Debugging and Verifying Computer Programs. The computer programs needed so far have been relatively simple. Most of the problems can

THERMAL EFFECTS AND ENERGY BALANCES

be solved using canned packages for ODEs, although learning how to use the solvers may take more work than writing the code from scratch. Even if you use canned packages, there are many opportunities for error. You have to specify the functional forms for the equations, supply the data, and supply any ancillary functions such as equations of state and physical property relationships. Few programs work correctly the first time. You will need to debug them and confirm that the output is plausible. A key to doing this for physically motivated problems like those in reactor design is simplification. You may wish to write the code all at once, but do not try to debug it all at once. For the nonisothermal problems encountered in this chapter, start by running an isothermal and isobaric case. Set T and P to constant values and see if the reactant concentrations are calculated correctly. If the reaction network is complex, you may need to simplify it, say by dropping some side reactions, until you find a case that you know is giving the right results. When the calculated solution for an isothermal and isobaric reaction makes sense, put an ODE for temperature or pressure back into the program and see what happens. You may wish to test the adiabatic case by setting $U=0$ and to retest the isothermal case by setting U to some large value. Complications like variable physical properties and variable reactor cross sections are best postponed until you have a solid base case that works. If something goes wrong when you add a complication, revert to a simpler case to help pinpoint the source of the problem.

Debugging by simplifying before complicating is even more important for the optimization problems in Chapter 6 and the nonideal reactor design problems in Chapters 8 and 9. When the reactor design problem is embedded as a subroutine inside an optimization routine, be sure that the subroutine will work for any parameter values that the optimization routine is likely to give it. Having trouble with axial dispersion? Throw out the axial dispersion terms for heat and mass and confirm that you get the right results for a nonisothermal (or even isothermal) PFR. Having trouble with the velocity profile in a laminar flow reactor? Get the reactor program to work with a parabolic or even a flat profile. Separately test the subroutine for calculating the axial velocity profile by sending it a known viscosity profile. Put it back into the main program only after it works on its own. Additional complications like radial velocity components are added still later.

Long programs will take hours and even days to write and test. A systematic approach to debugging and verification will reduce this time to a minimum. It will also give you confidence that the numbers are right when they finally are produced.

5.2.2 Nonisothermal Piston Flow

Steady-state temperatures along the length of a piston flow reactor are governed by an ordinary differential equation. Consider the differential reactor element shown in Figure 5.3. The energy balance is the same as Equation (5.14) except

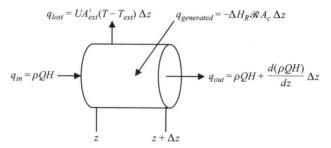

FIGURE 5.3 Differential element in a nonisothermal piston flow reactor.

that differential quantities are used. The ρQH terms cancel and Δz factors out to give:

$$\frac{d(\rho Q H)}{dz} = \rho Q \frac{dH}{dz} = \rho A_c \bar{u} \frac{dH}{dz} = -\Delta H_R \mathscr{R} A_c - UA'_{ext}(T - T_{ext}) \quad (5.22)$$

Unlike a molar flow rate—e.g, aQ—the mass flow rate, ρQ, is constant and can be brought outside the differential. Note that $Q = \bar{u}A_c$ and that A'_{ext} is the external surface area per unit length of tube. Equation (5.22) can be written as

$$\frac{dH}{dz} = \frac{-\Delta H_R \mathscr{R}_c}{\rho \bar{u}} - \frac{UA'_{ext}}{\rho \bar{u} A_c}(T - T_{ext}) \quad (5.23)$$

This equation is coupled to the component balances in Equation (3.9) and with an equation for the pressure; e.g., one of Equations (3.14), (3.15), (3.17). There are $N+2$ equations and some auxiliary algebraic equations to be solved simultaneously. Numerical solution techniques are similar to those used in Section 3.1 for variable-density PFRs. The dependent variables are the component fluxes Φ, the enthalpy H, and the pressure P. A necessary auxiliary equation is the thermodynamic relationship that gives enthalpy as a function of temperature, pressure, and composition. Equation (5.16) with $T_{ref}=0$ is the simplest example of this relationship and is usually adequate for preliminary calculations.

With a constant, circular cross section, $A'_{ext} = 2\pi R$ (although the concept of piston flow is not restricted to circular tubes). If C_P is constant,

$$\frac{dT}{dz} = \frac{-\Delta H_R \mathscr{R}}{\bar{u}\rho C_P} - \frac{2U}{\bar{u}\rho C_P R}(T - T_{ext}) \quad (5.24)$$

This is the form of the energy balance that is usually used for preliminary calculations. Equation (5.24) does not require that $\bar{u}$ be constant. If it is constant, we can set $dz = \bar{u}dt$ and $2/R = A_{ext}/A_c$ to make Equation (5.24) identical to Equation (5.19). A constant-velocity, constant-properties PFR behaves

THERMAL EFFECTS AND ENERGY BALANCES   165

identically to a constant-volume, constant-properties batch reactor. The curves in Figure 5.2 could apply to a piston flow reactor as well as to the batch reactor analyzed in Example 5.5. However, Equation (5.23) is the appropriate version of the energy balance when the reactor cross section or physical properties are variable.

The solution of Equations (5.23) or (5.24) is more straightforward when temperature and the component concentrations can be used directly as the dependent variables rather than enthalpy and the component fluxes. In any case, however, the initial values, $T_{in}, P_{in}, a_{in}, b_{in}, \ldots$ must be known at $z=0$. Reaction rates and physical properties can then be calculated at $z=0$ so that the right-hand side of Equations (5.23) or (5.24) can be evaluated. This gives ΔT, and thus $T(z + \Delta z)$, directly in the case of Equation (5.24) and implicitly via the enthalpy in the case of Equation (5.23). The component equations are evaluated similarly to give $a(z + \Delta z), b(z + \Delta z), \ldots$ either directly or via the concentration fluxes as described in Section 3.1. The pressure equation is evaluated to give $P(z + \Delta z)$. The various auxiliary equations are used as necessary to determine quantities such as $\bar{u}$ and A_c at the new axial location. Thus, $T, a, b, \ldots$ and other necessary variables are determined at the next axial position along the tubular reactor. The axial position variable z can then be incremented and the entire procedure repeated to give temperatures and compositions at yet the next point. Thus, we march down the tube.

Example 5.6: Hydrocarbon cracking reactions are endothermic, and many different techniques are used to supply heat to the system. The maximum inlet temperature is limited by problems of materials of construction or by undesirable side reactions such as coking. Consider an adiabatic reactor with inlet temperature T_{in}. Then $T(z) < T_{in}$ and the temperature will gradually decline as the reaction proceeds. This decrease, with the consequent reduction in reaction rate, can be minimized by using a high proportion of inerts in the feed stream.

Consider a cracking reaction with rate

$$\mathscr{R} = [10^{14} \exp(-24{,}000/T)]a, \text{g}/(\text{m}^3 \cdot \text{s})$$

where a is in g/m^3. Suppose the reaction is conducted in an adiabatic tubular reactor having a mean residence time of 1 s. The crackable component and its products have a heat capacity of 0.4 cal/(g·K), and the inerts have a heat capacity of 0.5 cal/(g·K); the entering concentration of crackable component is 132 g/m^3 and the concentration of inerts is 270 g/m^3; $T_{in} = 525°$C. Calculate the exit concentration of A given $\Delta H_R = 203$ cal/g. Physical properties may be assumed to be constant. Repeat the calculation in the absence of inerts.

Solution: Aside from the temperature calculations, this example illustrates the systematic use of mass rather than molar concentrations for reactor calculations. This is common practice for mixtures having ill-defined molecular weights. The energy balance for the adiabatic reactor gives

$$\frac{dT}{dt} = \frac{-\Delta H_R \mathscr{R}}{\rho C_P} = \Delta T_{adiabatic}\left(\frac{ka}{a_{in}}\right)$$

Note that ρ and C_P are properties of the reaction mixture. Thus, $\rho = 132 + 270 = 402 \, \text{g/m}^3$ and $C_P = [0.4(132) + 0.5(270)]/402 = 0.467 \, \text{cal/(g·K)}$. This gives $\Delta T_{adiabatic} = -142.7 \, \text{K}$. If the inerts are removed, $\rho \, 132 \, \text{g/m}^3$, $C_P = 0.4 \, \text{cal/(g·K)}$, and $\Delta T_{adiabatic} = -507.5 \, \text{K}$.

Figure 5.4 displays the solution. The results are $a_{out} = 57.9 \, \text{g/m}^3$ and $T_{out} = 464.3°\text{C}$ for the case with inerts and $a_{out} = 107.8 \, \text{g/m}^3$ and $T_{out} = 431.9°\text{C}$ for the case without inerts. It is apparent that inerts can have a remarkably beneficial effect on the course of a reaction.

In the general case of a piston flow reactor, one must solve a fairly small set of simultaneous, ordinary differential equations. The minimum set (of one) arises for a single, isothermal reaction. In principle, one extra equation must be added for each additional reaction. In practice, numerical solutions are somewhat easier to implement if a separate equation is written for each reactive component. This ensures that the stoichiometry is correct and keeps the physics and chemistry of the problem rather more transparent than when the reaction coordinate method is used to obtain the smallest possible set of differential

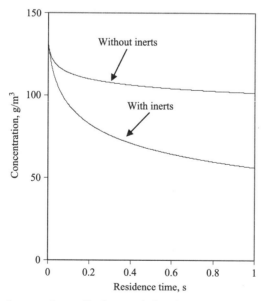

FIGURE 5.4 Concentration profiles for an endothermic reaction in an adiabatic reactor.

equations. Computational speed is rarely important in solving design problems of this type. The work involved in understanding and assembling and data, writing any necessary code, debugging the code, and verifying the results takes much more time than the computation.

5.2.3 Nonisothermal CSTRs

Setting $\hat{T} = T_{out}, \hat{H} = H_{out}$, and so on, specializes the integral energy balance of Equation (5.14) to a perfectly mixed, continuous-flow stirred tank:

$$\frac{d(V\rho_{out}H_{out})}{dt} = Q_{in}\rho_{in}H_{in} - Q_{out}\rho_{out}H_{out} - V\Delta H_R \mathscr{R} - UA_{ext}(T_{out} - T_{ext}) \quad (5.25)$$

where $\Delta H_R \mathscr{R}$ denotes the implied summation of Equation (5.17). The corresponding component balance for component A is

$$\frac{d(Va)}{dt} = Q_{in}a_{in} - Q_{out}a_{out} + V\mathscr{R}_A \quad (5.26)$$

and also has an implied summation

$$\mathscr{R}_A = \nu_{A,I}\mathscr{R}_I + \nu_{A,II}\mathscr{R}_{II} + \cdots \quad (5.27)$$

The simplest, nontrivial version of these equations is obtained when all physical properties and process parameters (e.g., Q_{in}, a_{in}, and T_{in}) are constant. The energy balance for this simplest but still reasonably general case is

$$\bar{t}\frac{dT_{out}}{dt} = T_{in} - T_{out} - \frac{\Delta H_R \mathscr{R}\bar{t}}{\rho C_P} - \frac{UA_{ext}(T_{out} - T_{ext})\bar{t}}{V\rho C_p} \quad (5.28)$$

The time derivative is zero at steady state, but it is included so that the method of false transients can be used. The computational procedure in Section 4.3.2 applies directly when the energy balance is given by Equation (5.28). The same basic procedure can be used for Equation (5.25). The enthalpy rather than the temperature is marched ahead as the dependent variable, and then T_{out} is calculated from H_{out} after each time step.

The examples that follow assume constant physical properties and use Equation (5.28). Their purpose is to explore nonisothermal reaction phenomena rather than to present detailed design calculations.

Example 5.7: A CSTR is commonly used for the bulk polymerization of styrene. Assume a mean residence time of 2 h, cold monomer feed (300 K), adiabatic operation ($UA_{ext} = 0$), and a pseudo-first-order reaction with rate constant

$$k = 10^{10} \exp(-10{,}000/T)$$

where T is in kelvins. Assume constant density and heat capacity. The adiabatic temperature rise for complete conversion of the feed is about 400 K for undiluted styrene.

Solution: The component balance for component A (styrene) for a first-order reaction in a constant-volume, constant-density CSTR is

$$\bar{t}\frac{da_{out}}{dt} = a_{in} - a_{out} - k\bar{t}a_{out}$$

The temperature balance for the adiabatic case is

$$\bar{t}\frac{dT_{out}}{dt} = T_{in} - T_{out} - \frac{\Delta H_R \mathscr{R} \bar{t}}{\rho C_P} = T_{in} - T_{out} + \Delta T_{adiabatic}\left(\frac{k\bar{t}a}{a_{in}}\right)$$

Substituting the given values,

$$\frac{da_{out}}{d\tau} = a_{in} - a_{out} - 2 \times 10^{10} \exp(-10{,}000/T_{out})a_{out} \qquad (5.29)$$

and

$$\frac{dT_{out}}{d\tau} = T_{in} - T_{out} + 8 \times 10^{12} \exp(-10{,}000/T_{out})a_{out}/a_{in} \qquad (5.30)$$

where $\tau = t/\bar{t}$ and $T_{in} = 300$ K. The problem statement did not specify a_{in}. It happens to be about 8700 mol/m^3 for styrene; but, since the reaction is first order, the problem can be worked by setting $a_{in} = 1$ so that a_{out} becomes equal to the fraction unreacted. The initial conditions associated with Equations (5.29) and (5.30) are $a_{out} = a_0$ and $T_{out} = T_0$ at $\tau = 0$. Solutions for $a_0 = 1$ (pure styrene) and various values for T_0 are shown in Figure 5.5.

The behavior shown in Figure 5.5 is typical of systems that have two stable steady states. The realized steady state depends on the initial conditions. For this example with $a_0 = 1$, the upper steady state is reached if T_0 is greater than about 398 K, and the lower steady state is reached if T_0 is less than about 398 K. At the lower steady state, the CSTR acts as a styrene monomer storage vessel with $T_{out} \approx T_{in}$ and there is no significant reaction. The upper steady state is a runaway where the reaction goes to near completion with $T_{out} \approx T_{in} + \Delta T_{adiabatic}$. (In actuality, the styrene polymerization is reversible at very high temperatures, with a ceiling temperature of about 625 K.)

There is a middle steady state, but it is metastable. The reaction will tend toward either the upper or lower steady states, and a control system is needed to maintain operation around the metastable point. For the styrene polymerization, a common industrial practice is to operate at the metastable point, with temperature control through autorefrigeration (cooling by boiling). A combination of feed preheating and jacket heating ensures that the uncontrolled reaction would tend toward the upper, runaway condition. However,

THERMAL EFFECTS AND ENERGY BALANCES

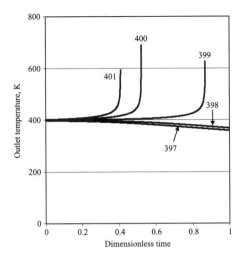

FIGURE 5.5 Method of false transients applied to a system having two stable steady states. The parameter is the initial temperature T_0.

the reactor pressure is set so that the styrene boils when the desired operating temperature is exceeded. The latent heat of vaporization plus the return of subcooled condensate maintains the temperature at the boiling point.

The method of false transients cannot be used to find a metastable steady state. Instead, it is necessary to solve the algebraic equations that result from setting the derivatives equal to zero in Equations (5.29) and (5.30). This is easy in the current example since Equation (5.29) (with $da_{out}/d\tau = 0$) can be solved for a_{out}. The result is substituted into Equation (5.30) (with $dT_{out}/d\tau = 0$) to obtain a single equation in a single unknown. The three solutions are

T_{out}, K	a_{out}/a_{in}
300.03	0.99993
403	0.738
699.97	0.00008

The existence of three steady states, two stable and one metastable, is common for exothermic reactions in stirred tanks. Also common is the existence of only one steady state. For the styrene polymerization example, three steady states exist for a limited range of the process variables. For example, if T_{in} is sufficiently low, no reaction occurs, and only the lower steady state is possible. If T_{in} is sufficiently high, only the upper, runaway condition can be realized. The external heat transfer term, $UA_{ext}(T_{out} - T_{ext})$, in Equation (5.28) can also be used to vary the location and number of steady states.

Example 5.8: Suppose that, to achieve a desired molecular weight, the styrene polymerization must be conducted at 413 K. Use external heat transfer to achieve this temperature as the single steady state in a stirred tank.

Solution: Equation (5.29) is unchanged. The heat transfer term is added to Equation (5.30) to give

$$\frac{dT_{out}}{d\tau} = 300 - T_{out} + 8 \times 10^{12} \exp(-10{,}000/T_{out}) a_{out}/a_{in} - \frac{UA_{ext}}{\rho Q C_P}(T_{out} - T_{ext}) \quad (5.31)$$

We consider T_{ext} to be an *operating variable* that can be manipulated to achieve $T_{out} = 413$ K. The dimensionless heat transfer group $UA_{ext}/\rho Q C_P$ is considered a *design variable*. It must be large enough that a single steady state can be imposed on the system. In small equipment with good heat transfer, one simply sets $T_{ext} \approx T_{out}$ to achieve the desired steady state. In larger vessels, $UA_{ext}/\rho Q C_P$ is finite, and one must find set $T_{ext} < T_{out}$ such that the steady state is 413 K.

Since a stable steady state is sought, the method of false transients could be used for the simultaneous solution of Equations (5.29) and (5.31). However, the ease of solving Equation (5.29) for a_{out} makes the algebraic approach simpler. Whichever method is used, a value for $UA_{ext}/\rho Q C_P$ is assumed and then a value for T_{ext} is found that gives 413 K as the single steady state. Some results are

$UA_{ext}/\rho Q C_P$	T_{ext} that gives $T_{out} = 413$ K
100	412.6
50	412.3
20	411.1
10	409.1
5	405.3
4	No solution

Thus, the minimum value for $UA_{ext}/\rho Q C_P$ is about 5. If the heat transfer group is any smaller than this, stable operation at $T_{out} = 413$ K by manipulation of T_{ext} is no longer possible because the temperature driving force, $\Delta T = T_{out} - T_{ext}$, becomes impossibly large. As will be seen in Section 5.3.2, the quantity $UA_{ext}/\rho Q C_P$ declines on a normal scaleup.

At a steady state, the amount of heat generated by the reaction must exactly equal the amount of heat removed by flow plus heat transfer to the environment: $q_{generated} = q_{removed}$. The heat generated by the reaction is

$$q_{generated} = -V \Delta H_R \mathscr{R} \quad (5.32)$$

THERMAL EFFECTS AND ENERGY BALANCES

This generation term will be an S-shaped curve when plotted against T_{out}. When T_{out} is low, reaction rates are low, and little heat is generated. When T_{out} is high, the reaction goes to completion, the entire exotherm is released, and T_{out} reaches a maximum. A typical curve for the rate of heat generation is plotted in Figure 5.6(a). The shape of the curve can be varied by changing the reaction mechanism and rate constant.

The rate of heat removal is given by

$$q_{removed} = -Q_{in}\rho_{in}H_{in} + Q_{out}\rho_{out}H_{out} + UA_{ext}(T_{out} - T_{ext}) \quad (5.33)$$

As shown in Figure 5.6(b), the rate of heat removal is a linear function of T_{out} when physical properties are constant:

$$q_{removed} = \rho Q C_P(T_{out} - T_{in}) + UA_{ext}(T_{out} - T_{ext})$$
$$= -(\rho Q C_P T_{in} + UA_{ext} T_{ext}) + (\rho Q C_P + UA_{ext})T_{out} = C_0 + C_1 T_{out}$$

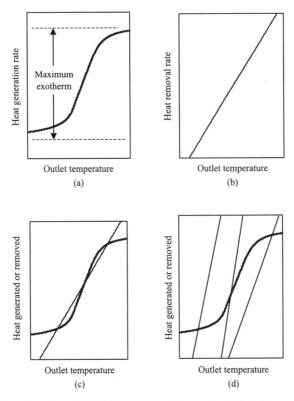

FIGURE 5.6 Heat balance in a CSTR: (a) heat generated by reaction; (b) heat removed by flow and transfer to the environment; (c) superposition of generation and removal curves. The intersection points are steady states. (d) Superposition of alternative heat removal curves that give only one steady state.

where C_0 and C_1 are the slope and intercept of the heat absorption line, respectively. They can be manipulated by changing either the design or the operating variables.

Setting Equation (5.32) equal to Equation (5.33) gives the general heat balance for a steady-state system. Figure 5.6(c) shows the superposition of the heat generation and removal curves. The intersection points are steady states. There are three in the illustrated case, but Figure 5.6(d) illustrates cases that have only one steady state.

More than three steady states are sometimes possible. Consider the reaction sequence

$$A + B \rightarrow C \qquad (I)$$

$$A \rightarrow D \qquad (II)$$

where Reaction (I) occurs at a lower temperature than Reaction (II). It is possible that Reaction (I) will go to near-completion, consuming all the B, while still at temperatures below the point where Reaction (II) becomes significant. This situation can generate up to five steady states as illustrated in Figure 5.7. A practical example is styrene polymerization using component B as an initiator at low temperatures, < 120°C, and with spontaneous (thermal) initiation at higher temperatures. The lower S-shaped portion of the heat-generation curve consumes all the initiator, B; but there is still unreacted styrene, A. The higher S-shaped portion consumes the remaining styrene.

To learn whether a particular steady state is stable, it is necessary to consider small deviations in operating conditions. Do they decline and damp out or do they lead to larger deviations? Return to Figure 5.6(c) and suppose that the reactor has somehow achieved a value for T_{out} that is higher than the upper steady state. In this region, the heat-removal line is above the heat-generation line so that the reactor will tend to cool, approaching the steady state from above. Suppose, on the other hand, that the reactor becomes cooler than the upper steady state but remains hotter than the central, metastable state. In this region, the heat-removal line is below the heat-generation line so that the temperature will increase, heading back to the upper steady state. Thus, the upper steady state is stable when subject to small disturbances, either positive or negative. The same reasoning can be applied to the lower steady state. However, the middle steady state is unstable. A small positive disturbance will send the system toward the upper steady state and a small negative disturbance will send the system toward the lower steady state. Applying this reasoning to the system in Figure 5.7 with five steady states shows that three of them are stable. These are the lower, middle, and upper ones that can be numbered 1, 3, and 5. The two even-numbered steady states, 2 and 4, are metastable.

The dynamic behavior of nonisothermal CSTRs is extremely complex and has received considerable academic study. Systems exist that have only a metastable state and no stable steady states. Included in this class are some chemical oscillators that operate in a reproducible limit cycle about their metastable

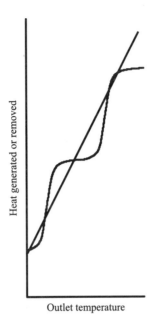

FIGURE 5.7 Consecutive reactions with five steady states.

state. Chaotic systems have discernible long-term patterns and average values but have short-term temperature-composition trajectories that appear essentially random. Occasionally, such dynamic behavior is of practical importance for industrial reactor design. A classic situation of a sustained oscillation occurs in emulsion polymerizations. These are complex reactions involving both kinetic and mass transfer limitations, and a stable-steady-state conversion is difficult or impossible to achieve in a single CSTR. It was reasoned that if enough CSTRs were put in series, results would average out so that effectively stable, high conversions could be achieved. For a synthetic rubber process built during a wartime emergency, "enough" stirred tanks turned out to be 25–40. Full-scale production units were actually built in this configuration! More elegant solutions to continuous emulsion polymerizations are now available.

5.3 SCALEUP OF NONISOTHERMAL REACTORS

Thermal effects can be the key concern in reactor scaleup. The generation of heat is proportional to the volume of the reactor. Note the factor of V in Equation (5.32). For a scaleup that maintains geometric similarity, the surface area increases only as $V^{2/3}$. Sooner or later, temperature can no longer be controlled, and the reactor will approach adiabatic operation. There are relatively few reactions where the full adiabatic temperature change can be tolerated. Endothermic reactions will have poor yields. Exothermic reactions will have thermal runaways giving undesired by-products. It is the reactor designer's job to avoid limitations of scale or at least to understand them so that a desired product will result. There are many options. The best process and the best equipment at the laboratory scale are rarely the best for scaleup. Put another way, a process that is less than perfect at a small scale may be the best for scaleup, precisely because it is scalable.

5.3.1 Avoiding Scaleup Problems

Scaleup problems are sometimes avoidable. A few simple possibilities are:

1. Use enough diluents so that the adiabatic temperature change is acceptable.
2. Scale in parallel; e.g., shell-and-tube designs.
3. Depart from geometric similarity so that V and A_{ext} both increase in direct proportion to the throughput scaling factor S. Scaling a tubular reactor by adding length is a possibility for an incompressible fluid.
4. Use temperature-control techniques that inherently scale as S; e.g., cold feed to a CSTR, or autorefrigeration.
5. Intentionally degrade the performance of the small unit so that the same performance and product quality can be achieved upon scaleup.

Use Diluents. In a gas system, inerts such as nitrogen, carbon dioxide, or steam can be used to mitigate the reaction exotherm. In a liquid system, a solvent can be used. Another possibility is to introduce a second liquid phase that has the function of absorbing and transferring heat; i.e., water in an emulsion or suspension polymerization. Adding an extraneous material will increase cost, but the increase may be acceptable if it allows scaleup. Solvents have a deservedly bad name in open, unconfined applications; but these applications are largely eliminated. In a closed environment, solvent losses are small and the cost of confining the solvent is often borne by the necessary cost of confining the reactants.

Scale in Parallel. This common scaling technique was discussed in Section 3.2.1. Subject to possible tube-to-tube distribution problems, it is an inexpensive way of gaining capacity in what is otherwise a single-train plant.

Depart from Geometric Similarity. Adding length to a tubular reactor while keeping the diameter constant allows both volume and external area to scale as S if the liquid is incompressible. Scaling in this manner gives poor results for gas-phase reactions. The quantitative aspects of such scaleups are discussed in Section 5.3.3. Another possibility is to add stirred tanks, or, indeed, any type of reactor in series. Two reactors in series give twice the volume, have twice the external surface area, and give a closer approach to piston flow than a single, geometrically similar reactor that has twice the volume but only 1.59 times the surface area of the smaller reactor. Designs with several reactors in series are quite common. Multiple pumps are sometimes used to avoid high pressures. The apparent cost disadvantage of using many small reactors rather than one large one can be partially offset by standardizing the design of the small reactors.

If a single, large CSTR is desired, internal heating coils or an external, pump-around loop can be added. This is another way of departing from geometric similarity and is discussed in Section 5.3.2.

Use Scalable Heat Transfer. The feed flow rate scales as S and a cold feed stream removes heat from the reaction in direct proportion to the flow rate. If the energy needed to heat the feed from T_{in} to T_{out} can absorb the reaction exotherm, the heat balance for the reactor can be scaled indefinitely. Cooling costs may be an issue, but there are large-volume industrial processes that have $T_{in} \approx -40°C$ and $T_{out} \approx 200°C$. Obviously, cold feed to a PFR will not work since the reaction will not start at low temperatures. Injection of cold reactants at intermediate points along the reactor is a possibility. In the limiting case of many injections, this will degrade reactor performance toward that of a CSTR. See Section 3.3 on transpired-wall reactors.

Autorefrigeration or boiling is another example of heat transfer that scales as S. The chemist calls it refluxing and routinely uses it as a method of temperature control. Laboratory glassware is usually operated at atmospheric pressure so the temperature is set by the normal boiling point of the reactants. Chemists sometimes choose solvents that have a desired boiling point. Process equipment can operate at a regulated pressure so the boiling point can be adjusted. On the basis of boiling point, toluene at about 0.4 atm can replace benzene. The elevation of boiling point with pressure does impose a scaleup limitation. A tall reactor will have a temperature difference between top and bottom due to the liquid head.

Use Diplomatic Scaleup. This possibility is called diplomatic scaleup because it may require careful negotiations to implement. The idea is that thermal effects are likely to change the distribution of by-products or the product properties upon scaleup. The economics of the scaled process may be perfectly good and the product may be completely satisfactory, but it will be different than what the chemist could achieve in glassware. Setting appropriate and scalable expectations for product properties can avoid surprises and the cost of requalifying the good but somewhat different product that is made in the larger reactor. Diplomacy may be needed to convince the chemist to change the glassware to lower its performance with respect to heat transfer. A recycle loop reactor is one way of doing this in a controlled fashion.

5.3.2 Scaling Up Stirred Tanks

This section is concerned with the $UA_{ext}(T - T_{ext})$ term in the energy balance for a stirred tank. The usual and simplest case is heat transfer from a jacket. Then A_{ext} refers to the inside surface area of the tank that is jacketed on the outside and in contact with the fluid on the inside. The temperature difference, $T - T_{ext}$, is between the bulk fluid in the tank and the heat transfer medium in the jacket. The overall heat transfer coefficient includes the usual contributions from wall resistance and jacket-side coefficient, but the inside coefficient is normally limiting. A correlation applicable to turbine, paddle, and propeller agitators is

$$\mathbf{Nu} = \frac{hD_I}{\lambda} = C_h \left(\frac{D_I^2 N_I \rho}{\mu} \right)^{2/3} \left(\frac{\mu}{\mu_{wall}} \right)^{0.14} \quad (5.34)$$

where **Nu** is the Nusselt number and λ is the thermal conductivity. The value for C_h is needed for detailed design calculations but factors out in a scaling analysis; $C_h \approx 0.5$ for turbines and propellers. For a scaleup that maintains constant fluid properties,

$$\frac{(hD_I)_{large}}{(hD_I)_{small}} = \left[\frac{(D_I^2 N_I)_{large}}{(D_I^2 N_I)_{small}} \right]^{2/3}$$

Assuming geometric similarity and recalling that D_I scales as $S^{1/3}$ gives

$$\frac{h_{large}}{h_{small}} = \left[\frac{(D_I N_I^2)_{large}}{(D_I N_I^2)_{small}} \right]^{1/3} = S^{1/9} N^{2/3}$$

For a scaleup with constant power per unit volume, Example 4.7 showed that N_I must scale as $D_I^{-2/3}$. Thus,

$$\frac{h_{large}}{h_{small}} = \left[\frac{(D_I)_{large}}{(D_I)_{small}} \right]^{-1/9} = S^{-1/27}$$

and h decreases slightly upon scaleup. Assuming h controls the overall coefficient,

$$\frac{(UA_{ext})_{large}}{(UA_{ext})_{small}} = S^{-1/27} D_I^2 = S^{17/27}$$

If we want $UA_{ext}(T - T_{ext})$ to scale as S, the driving force for heat transfer must be increased:

$$\frac{(T - T_{ext})_{large}}{(T - T_{ext})_{small}} = S^{10/27}$$

These results are summarized in the last four rows of Table 4.1. Scaling the volume by a factor of 512 causes a large loss in hA_{ext} per unit volume. An increase in the temperature driving force (e.g., by reducing T_{ext}) by a factor of 10 could compensate, but such a large increase is unlikely to be possible. Also, with cooling at the walls, the viscosity correction term in Equation (5.34) will become important and will decrease hA_{ext} still more.

This analysis has been carried out for a batch reactor, but it applies equally well to a CSTR. The heat transfer coefficient is the same because the agitator dominates the flow inside the vessel, with little contribution from the net throughput. The analysis also applies to heat transfer using internal coils or baffles. The equations for the heat transfer coefficients are similar in form to Equation (5.34). Experimental results for the exponent on the impeller Reynolds number vary from 0.62 to 0.67 and are thus close to the semitheoretical value of 2/3 used in Equation (5.34). The results in Table 4.1 are generally restricted to turbulent flow. The heat transfer coefficient in laminar flow systems scales with impeller Reynolds number to the 0.5 power. This causes an even greater loss in heat transfer capability upon scaleup than in a turbulent system, although a transition to turbulence will occur if S is large enough. Close-clearance impellers such as anchors and helical ribbons are frequently used in laminar systems. So are pitched-blade turbines with large ratios of the impeller to tank diameter. This improves the absolute values for h but has a minor effect on the scaling relationships. Several correlations for **Nu** in laminar flow show a dependence on **Re** to the 0.5 power rather than the 0.67 power.

It is sometimes proposed to increase A_{ext} by adding internal coils or increasing the number of coils upon scaleup. This is a departure from geometric similarity that will alter flow within the vessel and reduce the heat transfer coefficient for the jacket. It can be done within reason; but to be safe, the coil design should be tested on the small scale using dummy coils or by keeping a low value for $T-T_{ext}$. A better approach to maintaining good heat transfer upon scaleup is to use a heat exchanger in an external loop as shown in Figure 5.8. The illustrated case is for a CSTR, but the concept can also be used for a batch reactor. The per-pass residence time in the loop should be small compared to the residence time in the reactor as a whole. A rule-of-thumb for a CSTR is

$$\bar{t}_{loop} = \frac{\text{Volume of loop}}{\text{Flow rate through loop}} < \bar{t}/10 \tag{5.35}$$

Reaction occurs in the loop as well as in the stirred tank, and it is possible to eliminate the stirred tank so that the reactor volume consists of the heat exchanger and piping. This approach is used for very large reactors. In the limiting case where the loop becomes the CSTR without a separate agitated vessel, Equation (5.35) becomes $q/Q > 10$. This is similar to the rule-of-thumb discussed in Section 4.5.3 that a recycle loop reactor approximates a CSTR. The reader may wonder why the rule-of-thumb proposed a minimum recycle ratio of 8 in Chapter 4 but 10 here. Thumbs vary in size. More conservative designers have proposed a minimum recycle ratio of 16, and designs with recycle ratios above 100 are known. The real issue is how much conversion per pass can be tolerated in the more-or-less piston flow environment of the heat exchanger. The same issue arises in the stirred tank reactor itself since the internal pumping rate is finite and intense mixing occurs only in the region of the impeller. In a loop reactor, the recirculation pump acts as the impeller and provides a local zone of intense mixing.

Example 5.9: This is a consultant's war story. A company had a brand-name product for which they purchased a polymer additive. They decided to create their own proprietary additive, and assigned the task to a synthetic chemist who soon created a fine polymer in a 300-ml flask. Scaleup was assigned to engineers who translated the chemistry to a 10-gal steel reactor. The resulting polymer was almost as good as what the chemist had made. Enough polymer was made in the 10-gal reactor for expensive qualification trials. The trials were a success. Management was happy and told the engineers to design a 1000-gal vessel.

Now the story turns bad. The engineers were not rash enough to attempt a direct scaleup with $S=100$, but first went to a 100-gal vessel for a test with $S=10$. There they noted a significant exotherm and found that the polymer had a broader molecular-weight distribution than achieved on the small scale. The product was probably acceptable but was different from what had

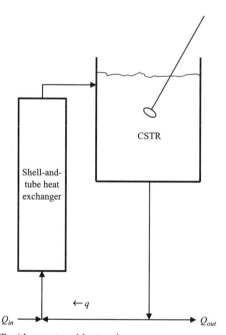

FIGURE 5.8 A CSTR with an external heat exchanger.

THERMAL EFFECTS AND ENERGY BALANCES

been so carefully tested. Looking back at the data from the 10-gal runs, yes there was a small exotherm but it had seemed insignificant. Looking ahead to a 1000-gal reactor and (finally) doing the necessary calculations, the exotherm would clearly become intolerable. A mixing problem had also emerged. One ingredient in the fed-batch recipe was reacting with itself rather than with the target molecule. Still, the engineers had designed a 2000-gal reactor that might have handled the heat load. The reactor volume was 2000 gal rather than 1000 gal to accommodate the great mass of cooling coils. Obviously, these coils would significantly change the flow in the vessel so that the standard correlation for heat transfer to internal coils could not be trusted. What to do?

Solution: There were several possibilities, but the easiest to design and implement with confidence was a shell-and-tube heat exchanger in an external loop. Switching the feed point for the troublesome ingredient to the loop also allowed its rapid and controlled dilution even though the overall mixing time in the vessel was not significantly changed by the loop.

There is one significant difference between batch and continuous-flow stirred tanks. The heat balance for a CSTR depends on the inlet temperature, and T_{in} can be adjusted to achieve a desired steady state. As discussed in Section 5.3.1, this can eliminate scaleup problems.

5.3.3 Scaling Up Tubular Reactors

Convective heat transfer to fluid inside circular tubes depends on three dimensionless groups: the Reynolds number, $\mathbf{Re} = \rho d_t \bar{u}/\mu$, the Prandtl number, $\mathbf{Pr} = C_P \mu / \lambda$ where λ is the thermal conductivity, and the length-to-diameter ratio, L/D. These groups can be combined into the Graetz number, $\mathbf{Gz} = \mathbf{Re}\mathbf{Pr}d_t/L$. The most commonly used correlations for the inside heat transfer coefficient are

$$hd_t/\lambda = 3.66 + \frac{0.085\mathbf{Gz}}{1 + 0.047\mathbf{Gz}^{2/3}} \left(\frac{\mu_{bulk}}{\mu_{wall}}\right)^{0.14} \quad \text{(Deep laminar)} \quad (5.36)$$

for laminar flow and $\mathbf{Gz} < 75$,

$$hd_t/\lambda = 1.86\mathbf{Gz}^{1/3}\left(\frac{\mu_{bulk}}{\mu_{wall}}\right)^{0.14} \quad \text{(Laminar)} \quad (5.37)$$

for laminar flow and $\mathbf{Gz} > 75$ and

$$hd_t/\lambda = 0.023\mathbf{Re}^{0.8}\mathbf{Pr}^{1/3}\left(\frac{\mu_{bulk}}{\mu_{wall}}\right)^{0.14} \quad \text{(Fully turbulent)} \quad (5.38)$$

for $Re > 10{,}000$, $0.7 < Pr < 700$ and $L/d_t > 60$. These equations apply to ordinary fluids (not liquid metals) and ignore radiative transfer. Equation (5.36) is rarely used. It applies to very low Re or very long tubes. No correlation is available for the transition region, but Equation (5.37) should provide a lower limit on Nu in the transition region.

Approximate scaling behavior for incompressible fluids based on Equations (5.36)–(5.38) is given in Table 5.1. Scaling in parallel is not shown since all scaling factors would be 1. Scaleups with constant pressure drop give the same results for gases as for liquids. Scaleups with geometric similarity also give the same results if the flow is laminar. Other forms of gas-phase scaleup are rarely possible if significant amounts of heat must be transferred to or from the reactants. The reader is reminded of the usual caveat: detailed calculations are needed to confirm any design. The scaling exponents are used for

TABLE 5.1 Scaleup Factors for Liquid-Phase Tubular Reactors.

Flow regime	General scaleup factors	Series scaleup	Geometric similarity	Constant pressure scaleup
Deep laminar				
Diameter scaling factor	S_R	1	$S^{1/3}$	$S^{1/3}$
Length scaling factor	S_L	S	$S^{1/3}$	$S^{1/3}$
Length-to-diameter ratio	$S_L S_R^{-1}$	S	1	1
Pressure scaling factor, ΔP	$S S_R^{-4} S_L$	S^2	1	1
Heat transfer area, A_{ext}	$S_R S_L$	S	$S^{2/3}$	$S^{2/3}$
Inside coefficient, h	S_R^{-1}	1	$S^{-1/3}$	$S^{-1/3}$
Coefficient times area, $h A_{ext}$	S_L	S	$S^{1/3}$	$S^{1/3}$
Driving force, ΔT	$S S_L^{-1}$	1	$S^{2/3}$	$S^{2/3}$
Laminar				
Diameter scaling factor	S_R	1	$S^{1/3}$	$S^{1/3}$
Length scaling factor	S_L	S	$S^{1/3}$	$S^{1/3}$
Length-to-diameter ratio	$S_L S_R^{-1}$	S	1	1
Pressure scaling factor, ΔP	$S S_R^{-4} S_L$	S^2	1	1
Heat transfer area, A_{ext}	$S_R S_L$	S	$S^{2/3}$	$S^{2/3}$
Inside coefficient, h	$S^{1/3} S_R^{-1} S_L^{-1/2}$	1	$S^{-1/9}$	$S^{-1/9}$
Coefficient times area, $h A_{ext}$	$S^{1/3} S_L^{2/3}$	S	$S^{5/9}$	$S^{5/9}$
Driving force, ΔT	$S^{2/3} S_L^{-2/3}$	1	$S^{4/9}$	$S^{4/9}$
Fully turbulent				
Diameter scaling factor	S_R	1	$S^{1/3}$	$S^{11/27}$
Length scaling factor	S_L	S	$S^{1/3}$	$S^{5/27}$
Length-to-diameter ratio	$S_L S_R^{-1}$	S	1	$S^{-2/9}$
Pressure scaling factor, ΔP	$S^{1.75} S_R^{-4.75} S_L$	$S^{2.75}$	$S^{1/2}$	1
Heat transfer area, A_{ext}	$S_R S_L$	S	$S^{2/3}$	$S^{0.59}$
Inside coefficient, h	$S^{0.8} S_R^{-1.8}$	$S^{0.8}$	$S^{0.2}$	$S^{0.07}$
Coefficient times area, $h A_{ext}$	$S^{0.8} S_R^{-0.8} S_L$	$S^{1.8}$	$S^{0.87}$	$S^{0.66}$
Driving force, ΔT	$S^{0.2} S_R^{0.8} S_L^{-1}$	$S^{-0.8}$	$S^{0.13}$	$S^{0.34}$

conceptual studies and to focus attention on the most promising options for scaleup. Recall also that these scaleups maintain a constant value for T_{out}. The scaleup factors for the driving force, ΔT, maintain a constant T_{out} and a constant rate of heat transfer per unit volume of fluid.

Example 5.10: A liquid-phase, pilot-plant reactor uses a 12-ft tube with a 1.049-in i.d. The working fluid has a density of 860 kg/m^3, the residence time in the reactor is 10.2 s, and the Reynolds number is 8500. The pressure drop in the pilot plant has not been accurately measured, but is known to be less than 1 psi. The entering feed is preheated and premixed. The inlet temperature is 60°C and the outlet temperature is 64°C. Tempered water at 55°C is used for cooling. Management loves the product and wants you to design a plant that is a factor of 128 scaleup over the pilot plant. Propose scaleup alternatives and explore their thermal consequences.

Solution: Table 5.1 provides the scaling relationships. The desired throughput and volume scaling factor is $S = 128$.

Some alternatives for the large plant are as follows:

Parallel—put 128 identical tubes in parallel using a shell-and-tube design. The total length of tubes will be 1536 ft, but they are compactly packaged. All operating conditions are identical on a per-tube basis to those used in the pilot plant.

Series—build a reactor that is 1536 ft long. Use U-bends or coiling to make a reasonable package. The length-to-diameter ratio increases to $137S = 17{,}600$. The Reynolds number increases to $8500S = 1.1 \times 10^6$, and the pressure drop will be $S^{2.75} = 623{,}000$ times greater than it was in the pilot plant. The temperature driving force changes by a factor of $S^{-0.8} = 0.021$ from 7°C to 0.14°C. The production unit would have to restrict the water flow rate to hold this low a ΔT. Note that we used Equation (5.38) to scale the heat transfer coefficient even though the pilot plant was in the transitional region. Also, the driving force for turbulent flow should be based on the log-mean ΔT. The difference is minor, and approximations can be justified in a scaling study. When a reasonable scaleup is found, more accurate estimates can be made. The current calculations are accurate enough to show that a series scaleup is unreasonable.

Geometric similarity—build a reactor that is nominally $12S^{1/3} = 61$ ft long and $1.049S^{1/3} = 5.3$ inches in diameter. Use U-bends to give a reasonable footprint. Correct to a standard pipe size in the detailed design phase. The length-to-diameter ratio is unchanged in a geometrically similar scaleup. The Reynolds number increases to $8500S^{2/3} = 216{,}000$ and the pressure drop increases by factor of $S^{1/2} = 11.2$. The temperature driving force will increase by a factor of $S^{0.13} = 1.9$ to about 13°C so that the jacket temperature would be about 49°C. This design seems reasonable.

Constant pressure—build a reactor that is nominally $12S^{5/27} = 29$ ft long and $1.049S^{11/27} = 7.6$ in in diameter. The length-to-diameter ratio decreases by a factor of $S^{-2/9}$ to 47. The Reynolds number increases to $8500S^{16/27} = 151{,}000$. The temperature driving force must increase by a factor of $S^{0.34} = 5.2$ to about 36°C so that the jacket temperature would be about 26°C. This design is also reasonable, but the jacket temperature is a bit lower than is normally possible without a chiller.

There is no unique solution to this or most other design problems. Any design using a single tube with an i.d. of about 7.5 in or less and with a volume scaled by S will probably function from a reaction engineering viewpoint.

Example 5.11: The results of Table 5.1 suggest that scaling a tubular reactor with *constant heat transfer* per unit volume is possible, even with the further restriction that the temperature driving force be the same in the large and small units. Find the various scaling factors for this form of scaleup for turbulent liquids and apply them to the pilot reactor in Example 5.10.

Solution: Table 5.1 gives the driving-force scaling factor as $S^{0.2}S_R^{0.8}S_L^{-1}$. This is set to 1. A constant residence time is imposed by setting $S_R^2 S_L = S$. There are two equations and two unknowns, S_R and S_L. The solution is $S_R = S^{0.28}$ and $S_L = S^{0.44}$. The length-to-diameter ratio scales as $S^{0.16}$. Equation (3.43) can be used to determine that the pressure scaling factor is $S^{0.86}$. The Reynolds number scales as $S/S_R = S^{0.72}$.

Applying these factors to the $S = 128$ scaleup in Example 5.10 gives a tube that is nominally $12S^{0.44} = 101$ ft long and $1.049S^{0.28} = 4.1$ inches in diameter. The length-to-diameter ratio increases to 298. The Reynolds number increases to $8500S^{0.72} = 278,000$. The pressure drop would increase by a factor of $S^{0.86} = 65$. The temperature driving force would remain constant at 7°C so that the jacket temperature would remain 55°C.

Example 5.12: Repeat Examples 5.10 and 5.11 for $T_{in} = 160°C$ and $T_{out} = 164°C$. The coolant temperature remains at 55°C.

Solution: Now, $\Delta T = 107°C$. Scaling with geometric similarity would force the temperature driving force to increase by $S^{0.13} = 1.9$, as before, but the scaled-up value is now 201°C. The coolant temperature would drop to $-39°C$, which is technically feasible but undesirable. Scaling with constant pressure forces an even lower coolant temperature. A scaleup with constant heat transfer becomes attractive.

These examples show that the ease of scaling up of tubular reactors depends on the heat load. With moderate heat loads, single-tube scaleups are possible. Multitubular scaleups, $S_{tubes} > 1$, become attractive when the heat load is high, although it may not be necessary to go to full parallel scaling using S tubes. The easiest way to apply the scaling relations in Table 5.1 to multitubular reactors is to divide S by the number of tubes to obtain S'. Then S' is the volumetric and throughput scaling factor per tube.

THERMAL EFFECTS AND ENERGY BALANCES

Example 5.13: An existing shell-and-tube heat exchanger is available for the process in Example 5.10. It has 20 tubes, each 2 in i.d. and 18 ft long. How will it perform?

Solution: The volume of the existing reactor is 7.85 ft^3. The volume of the pilot reactor is 0.072 ft^3. Thus, at constant $\bar{t}$, the scaleup is limited to a factor of 109 rather than the desired 128. The per-tube scaling factor is $S' = 109/20 = 5.45$. $S_R = 1.91$ and $S_L = 1.5$. The general scaling factor for pressure drop in turbulent, incompressible flow is $(S')^{1.75} S_R^{-4.75} S_L = 1.35$, so that the upstream pressure increases modestly. The scaling factor for ΔT is $(S')^{0.2} S_R^{0.8} S_L^{-1} = 1.57$, so $\Delta T = 11°C$ and the coolant temperature will be 51°C. What about the deficiency in capacity? Few marketing estimates are that accurate. When the factor of 109 scaleup becomes inadequate, a second or third shift can be used. If operation on a 24/7 basis is already planned—as is common in the chemical industry—the operators may nudge the temperatures a bit in an attempt to gain capacity. Presumably, the operating temperature was already optimized in the pilot plant, but it is a rare process that cannot be pushed a bit further.

This section has based scaleups on pressure drops and temperature driving forces. Any consideration of mixing, and particularly the closeness of approach to piston flow, has been ignored. Scaleup factors for the extent of mixing in a tubular reactor are discussed in Chapters 8 and 9. If the flow is turbulent and if the Reynolds number increases upon scaleup (as is normal), and if the length-to-diameter ratio does not decrease upon scaleup, then the reactor will approach piston flow more closely upon scaleup. Substantiation for this statement can be found by applying the axial dispersion model discussed in Section 9.3. All the scaleups discussed in Examples 5.10–5.13 should be reasonable from a mixing viewpoint since the scaled-up reactors will approach piston flow more closely.

PROBLEMS

5.1. A reaction takes 1 h to complete at 60°C and 50 min at 65°C. Estimate the activation energy. What assumptions were necessary for your estimate?

5.2. Dilute acetic acid is to be made by the hydrolysis of acetic anhydride at 25°C. Pseudo-first-order rate constants are available at 10°C and 40°C. They are $k = 3.40\,h^{-1}$ and $22.8\,h^{-1}$, respectively. Estimate k at 25°C.

5.3. Calculate b_{out}/a_{in} for the reversible reaction in Example 5.2 in a CSTR at 280 K and 285 K with $\bar{t} = 2\,h$. Suppose these results were actual measurements and that you did not realize the reaction was reversible. Fit a first-order model to the data to find the apparent activation energy. Discuss your results.

5.4. At extreme pressures, liquid-phase reactions exhibit pressure effects. A suggested means for correlation is the *activation volume*, ΔV_{act}. Thus,

$$k = k_0 \exp\left(\frac{-E}{R_g T}\right) \exp\left(\frac{-\Delta V_{act} P}{R_g T}\right)$$

Di-*t*-butyl peroxide is a commonly used free-radical initiator that decomposes according to first-order kinetics. Use the following data[2] to estimate ΔV_{act} for the decomposition in toluene at 120°C:

P, kg/cm^2	k, s^{-1}
1	13.4×10^{-6}
2040	9.5×10^{-6}
2900	8.0×10^{-6}
4480	6.6×10^{-6}
5270	5.7×10^{-6}

5.5. Consider the consecutive reactions, $A \xrightarrow{k_I} B \xrightarrow{k_{II}} C$, with rate constants of $k_I = 10^{15} \exp(-10{,}000/T)$ and $k_{II} = 10^8 \exp(-5000/T)$. Find the temperature that maximizes b_{out} for a CSTR with $\bar{t} = 2$ and for a batch reactor with a reaction time of 2 h. Assume constant density with $b_{in} = c_{in} = 0$.

5.6. Find the temperature that maximizes b_{out} for the competitive reactions of Equation (5.13). Do this for a CSTR with $\bar{t} = 2$ and for a batch reactor with a reaction time of 2 h. Assume constant density with $b_{in} = c_{in} = 0$. The rate constants are $k_I = 10^8 \exp(-5000/T)$ and $k_{II} = 10^{15} \exp(-10000/T)$.

5.7. The reaction $A \xrightarrow{k_I} B \xrightarrow{k_{II}} C$ is occurring in an isothermal, piston flow reactor that has a mean residence time of 2 min. Assume constant cross section and physical properties and

$$k_I = 1.2 \times 10^{15} \exp(-12{,}000/T), \text{min}^{-1}$$
$$k_{II} = 9.4 \times 10^{15} \exp(-14{,}700/T), \text{min}^{-1}$$

(a) Find the operating temperature that maximizes b_{out} given $b_{in} = 0$.
(b) The laboratory data were confused: k_I was interchanged with k_{II}. Revise your answer accordingly.

5.8. Repeat the analysis of hydrocarbon cracking in Example 5.6 with $a_{in} = 100$ g/m^3.

5.9. Repeat the analysis of hydrocarbon cracking in Example 5.6 for the case where there is external heat exchange. Suppose the reaction is conducted in tubes that have an i.d. of 0.012 m and are 3 m long. The inside heat transfer coefficient is 9.5 cal/(K · m^2 · s) and the wall temperature is 525°C. The inerts are present.

THERMAL EFFECTS AND ENERGY BALANCES   **185**

5.10. For the styrene polymerization in Example 5.7, determine that value of T_{in} below which only the lower steady state is possible. Also determine that value of T_{in} above which only the upper steady state is possible.

5.11. For the styrene polymerization in Example 5.7, determine those values of the mean residence time that give one, two, or three steady states.

5.12. The pressure drop was not measured in the pilot plant in Example 5.10, but the viscosity must be known since the Reynolds number is given. Use it to calculate the pressure drop. Does your answer change the feasibility of any of the scaleups in Examples 5.10–5.13?

5.13. Determine the reactor length, diameter, Reynolds number, and scaling factor for pressure drop for the scaleup with constant heat transfer in Example 5.12.

5.14. Your company is developing a highly proprietary new product. The chemistry is complicated, but the last reaction step is a dimerization:

$$2A \xrightarrow{k} B$$

Laboratory kinetic studies gave $a_0 k = 1.7 \times 10^{13} \exp(-14000/T), \text{s}^{-1}$. The reaction was then translated to the pilot plant and reacted in a 10-liter batch reactor according to the following schedule:

Time from Start of Batch (min)	Action
0	Begin charging raw materials
15	Seal vessel; turn on jacket heat (140°C steam)
90	Vessel reaches 100°C and reflux starts
180	Reaction terminated; vessel discharge begins
195	Vessel empty; washdown begins
210	Reactor clean, empty, and cool

Management likes the product and has begun to sell it enthusiastically. The pilot-plant vessel is being operated around the clock and produces two batches per shift for a total of 42 batches per week. It is desired to increase production by a factor of 1000, and the engineer assigned to the job orders a geometrically similar vessel that has a working capacity of 10,000 liters.

(a) What production rate will actually be realized in the larger unit? Assume the heat of reaction is negligible.

(b) You have replaced the original engineer and have been told to achieve the forecast production rate of 1000 times the pilot rate. What might you do to achieve this? (You might think that the original engineer was fired. More likely, he was promoted based on the commercial success of the pilot-plant work, is now your boss, and will expect you to deliver planned capacity from the reactor that he ordered.)

5.15. A liquid-phase, pilot-plant reactor uses a 0.1-m^3 CSTR with cooling at the walls. The working fluid has water-like physical properties. The residence time in the reactor is 3.2 h. The entering feed is preheated and premixed. The inlet temperature is 60°C and the outlet temperature is 64°C. Tempered water at 55°C is used for cooling. The agitator speed is 600 rpm. Management loves the product and wants you to scaleup by a modest factor of 20. However, for reasons obscure to you, they insist that you maintain the same agitator tip speed. Thus, the scaleup will use a geometrically similar vessel with $N_I D$ held constant.

(a) Assuming highly turbulent flow, by what factor will the total power to the agitator increase in the larger, 2-m^3 reactor?

(b) What should be the temperature of the cooling water to keep the same inlet and outlet temperatures for the reactants?

REFERENCES

1. Freiling, E. C., Johnson, H. C., and Ogg, R. A., Jr., "The kinetics of the fast gas-phase reaction between nitryl chloride and nitric oxide," *J. Chem. Phys.*, **20**, 327–329 (1952).
2. Walling, C. and Metzger, G., "Organic reactions under high pressure. V. The decomposition of di-t-butyl peroxide," *J. Am. Chem. Soc.*, **81**, 5365–5369 (1959).

SUGGESTIONS FOR FURTHER READING

The best single source for design equations remains

Perry's Handbook, 7th ed., D. W. Green, Ed., McGraw-Hill, New York, 1997.

Use it or other detailed sources after preliminary scaling calculations have been made.

CHAPTER 6
DESIGN AND OPTIMIZATION STUDIES

The goal of this chapter is to provide semirealistic design and optimization exercises. Design is a creative endeavor that combines elements of art and science. It is hoped that the examples presented here will provide some appreciation of the creative process.

This chapter also introduces several optimization techniques. The emphasis is on robustness and ease of use rather than computational efficiency.

6.1 A CONSECUTIVE REACTION SEQUENCE

The first consideration in any design and optimization problem is to decide the boundaries of "the system." A reactor can rarely be optimized without considering the upstream and downstream processes connected to it. Chapter 6 attempts to integrate the reactor design concepts of Chapters 1–5 with process economics. The goal is an optimized process design that includes the costs of product recovery, in-process recycling, and by-product disposition. The reactions are

$$A \xrightarrow{k_I} B \xrightarrow{k_{II}} C \qquad (6.1)$$

where A is the raw material, B is the desired product, and C is an undesired by-product. The process flow diagram is given in Figure 6.1. For simplicity, the recovery system is assumed to be able to make a clean separation of the three components without material loss.

Note that the production of C is not stoichiometrically determined but that the relative amounts of B and C can be changed by varying the reaction conditions. Had C been stoichiometrically determined, as in the production of by-product HCl when hydrocarbons are directly chlorinated, there is nothing that can be done short of very fundamental changes to the chemistry, e.g., using ClO_2 rather than Cl_2. Philosophically, at least, this is a problem for a chemist rather than a chemical engineer. In the present example, component C is a secondary or side product such as a dichlorinated compound when

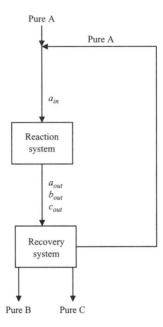

FIGURE 6.1 Simplified process flow diagram for consecutive reaction process.

monochlorination is desired, and the chemical reaction engineer has many options for improving performance without changing the basic chemistry.

Few reactions are completely clean in the sense of giving only the desired product. There are some cases where the side products have commensurate value with the main products, but these cases are becoming increasingly rare, even in the traditional chemical industry, and are essentially nonexistent in fields like pharmaceuticals. Sometimes, C is a hazardous waste and has a large, negative value.

The structure of the reactions in Equation (6.1) is typical of an immense class of industrially important reactions. It makes little difference if the reactions are all second order. Thus, the reaction set

$$A_1 + A_2 \rightarrow B_1 + B_2 \rightarrow C_1 + C_2 \qquad (6.2)$$

has essentially the same structure. The As can be lumped as the raw material, the Bs can be lumped as product, even though only one may be useful, and the Cs can be lumped as undesired. The reaction mechanism and the kinetics are different in detail, but the optimization methodology and economic analysis will be very similar.

Example 6.1: Show by example that it is generally necessary to include the cost of recovering the product and recycling unused reactants in the reactor design optimization.

Solution: Suppose component C in Equation (6.1) is less valuable than A. Then, if the cost of the recovery step is ignored, the optimal design is a high-throughput but low-conversion reactor. Presumably, this will be cheap to build since it produces low concentrations of B and thus can be a simple design such as an adiabatic tube. Since b_{out} is low, c_{out} will be lower yet, and essentially all the incoming A will be converted to B or recycled. Thus, the reaction end of the process will consist of a cheap reactor with nearly 100% raw-material efficiency after recycling. Of course, huge quantities of reactor effluent must be separated, with the unreacted A being recycled, but that is the problem of the separations engineer.

In fairness, processes do exist where the cost of the recovery step has little influence on the reactor design, but these are the exceptions.

The rest of this chapter is a series of examples and problems built around semirealistic scenarios of reaction characteristics, reactor costs, and recovery costs. The object is not to reach general conclusions, but to demonstrate a method of approaching such problems and to provide an introduction to optimization techniques.

The following are some data applicable to a desired plant to manufacture component B of Equation (6.1):

   Required production rate = 50,000 t/yr (metric tons) = 6250 kg/h

   Cost of raw material A = $1.50/kg

   Value of side product C = $0.30/kg

Note that 8000 h is a commonly used standard "year" for continuous processes. The remainder of the time is for scheduled and random maintenance. In a good year when demand is high, production personnel have the opportunity to exceed their plan.

You can expect the cost of A and the value of C to be fairly accurate. The required production rate is a marketing guess. So is the selling price of B, which is not shown above. For now, assume it is high enough to justify the project. Your job is the conceptual design of a reactor to produce the required product at minimum total cost.

The following are capital and operating cost estimates for the process:

Reactor capital costs = $500,000 $V^{0.6}$

Reactor operating costs (excluding raw materials)
   = $0.08 per kg of reactor throughput

Recovery system capital cost = $21,000 $W^{0.6}$

Recovery system operating costs
   = $0.20 per kg of recovery system throughput

where V is the reactor volume in cubic meters and W is the total mass flow rate (virgin+recycle) in t/yr. Options in reactor design can include

CSTRs, shell-and-tube reactors, and single-tube reactors, particularly a single adiabatic tube. Realistically, these different reactors may all scale similarly e.g., as $V^{0.6}$, but the dollar premultipliers will be different, with CSTRs being more expensive than shell-and-tube reactors, which are more expensive than adiabatic single tubes. However, in what follows, the same capital cost will be used for all reactor types in order to emphasize inherent kinetic differences. This will bias the results toward CSTRs and toward shell-and-tube reactors over most single-tube designs.

Why are the CSTRs worth considering at all? They are more expensive per unit volume and less efficient as chemical reactors (except for autocatalysis). In fact, CSTRs are useful for some multiphase reactions, but that is not the situation here. Their potential justification in this example is temperature control. Boiling (autorefrigerated) reactors can be kept precisely at the desired temperature. The shell-and-tube reactors cost less but offer less effective temperature control. Adiabatic reactors have no control at all, except that T_{in} can be set.

As shown in Figure 6.1, the separation step has been assumed to give clean splits, with pure A being recycled back to the reactor. As a practical matter, the B and C streams must be clean enough to sell. Any C in the recycle stream will act as an inert (or it may react to component D). Any B in the recycle stream invites the production of undesired C. A realistic analysis would probably have the recovery system costs vary as a function of purity of the recycle stream, but we will avoid this complication for now.

The operating costs are based on total throughput for the unit. Their main components are utilities and maintenance costs, along with associated overheads. Many costs, like labor, will be more or less independent of throughput in a typical chemical plant. There may be some differences in operating costs for the various reactor types, but we will worry about them, like the difference in capital costs, only if the choice is a close call. The total process may include operations other than reaction and recovery and will usually have some shared equipment such as the control system. These costs are ignored since the task at hand is to design the best reaction and recovery process and not to justify the overall project. That may come later. The dominant uncertainty in justifying most capacity expansions or new-product introductions is marketing. How much can be sold at what price?

Some of the costs are for capital and some are operating costs. How to convert apples to oranges? The proper annualization of capital costs is a difficult subject. Economists, accountants, and corporate managers may have very different viewpoints. Your company may have a cast-in-stone rule. Engineers tend to favor precision and have invented a complicated, time-dependent scheme (*net present value* or *NPV* analysis) that has its place (on the Engineer-in-Training exam among other places), but can impede understanding of cause and effect. We will adopt the simple rule that the annual cost associated with a capital investment is 25% of the investment. This accounts for depreciation plus a return on fixed capital investment. Working capital items (cash, inventory, accounts receivable) will be ignored on the grounds that they will be similar for all the options under consideration.

Assume for now that the reactions in Equation (6.1) are elementary first order with rate constants

$$k_I = 4.5 \times 10^{11} \exp(-10000/T) \text{ h}^{-1}$$
$$k_{II} = 1.8 \times 10^{12} \exp(-12000/T) \text{ h}^{-1} \qquad (6.3)$$

Table 6.1 illustrates the behavior of the rate constants as a function of absolute temperature. Low temperatures favor the desired, primary reaction, but the rate is low. Raise the rate enough to give a reasonable reactor volume and the undesired, secondary reaction becomes significant. There is clearly an interior optimum with respect to temperature.

Both reactions are endothermic:

$$\frac{(\Delta H_R)_I a_{in}}{\rho C_P} = \frac{(\Delta H_R)_{II} a_{in}}{\rho C_P} = 30 \text{ K} \qquad (6.4)$$

All three components, A, B, and C, have a molecular weight of 200 Da.

Example 6.2: Cost-out a process that uses a single CSTR for the reaction.

Solution: The reactor design equations are very simple:

$$a_{out} = \frac{a_{in}}{1 + k_I \bar{t}}$$
$$b_{out} = \frac{b_{in} + k_I \bar{t}(a_{in} + b_{in})}{(1 + k_I \bar{t})(1 + k_{II} \bar{t})} \qquad (6.5)$$
$$c_{out} = c_{in} + a_{in} - a_{out} + b_{in} - b_{out}$$

TABLE 6.1 Effect of Temperature on Rate Constants

T, K	k_I, h^{-1}	k_I/k_{II}
300	0.002	196.4
320	0.012	129.5
340	0.076	89.7
360	0.389	64.7
380	1.677	48.3
400	6.250	37.1
420	20.553	29.2
440	60.657	23.6
460	162.940	19.3
480	403.098	16.1
500	927.519	13.6

The total product demand is fixed. The unknowns are the reactor volume V (by way of $\bar{t}$), and the temperature, $T_{in} = T_{out}$ (by way of k_I and k_{II}). These are the variables that determine the production cost, but calculating the cost is complicated because the output of B is specified and the necessary input of A must be found. Assume that V and T_{in} are known. Then guess a value for the total flow rate W, which is the sum of virgin A plus recycled A. The amount of B is calculated and compared with the required value of 6250 kg/h. The guessed value for W is then adjusted. The following Basic program uses a *binary search* to adjust the guess. See Appendix 6 for a description of the method or reason your way through the following code. The program uses three subroutines: Reactor, Cost, and Cprint. Reactor is shown at the end of the main program, and can be replaced with suitable, albeit more complicated, subroutine to treat CSTRs in series, or PFRs. The subroutine Cost calculates the total cost and Cprint displays the results.

```
DEFDBL A-H, P-Z
DEFLNG I-O
COMMON SHARED MwA, MwB, MwC, rho, ain, hr1, hr2

'Simple evaluation of a single CSTR using a binary
'search
MwA = 200 'kg/kg moles
MwB = 200
MwC = 200
rho = 900 'kg/m^3
ain = rho / MwA 'kg moles/m^3
bin = 0
cin = 0
V = 10
Tin = 400

'Binary search to find WAin
Wmin = 6250 'lower bound, kg/hr
Wmax = 100000 'upper bound
FOR I = 1 TO 24
    WAin = (Wmin+Wmax)/2
    Q = WAin/rho
    tbar = V/Q
    Call Reactor (tbar, Tin, ain, bin, cin, Tout, aout,
+   bout, cout)
    Wbout = bout * Q * MwB
    IF WBout > 6250 THEN
        Wmax = WAin
    ELSE
```

```
        Wmin = WAin
    END IF
NEXT I

CALL Cost(WAin, V, aout, bout, cout, total )
CALL Cprint(WAin, V, aout, bout, cout, Tin, Tout)
END

SUB Reactor (tbar, Tin, ain, bin, cin, Tout, aout,
+              bout, cout)
'Single CSTR version
    xk1 = 450000000000 * EXP(-10000/Tin)
    xk2 = 1800000000000 * EXP(-12000/Tin)
    aout = ain/(1+xk1 * tbar)
    bout = (bin+xk1 * tbar * (ain+bin))/(1+xk1 *
    tbar)/(1+xk2 * tbar)
    cout = cin+ain - aout + bin - bout
END SUB
```

The results for a single CSTR operating at $T_{out} = 400$ K and $V = 10$ m^3 are shown below:

Throughput	8478 kg/h
Product rate	6250 kg/h
Reactor $\bar{t}$	1.06 h
Raw materials cost	88.41 MM\$/yr
By-product credit	2.68 MM\$/yr
Throughput cost	18.99 MM\$/yr
Annualized reactor capital	0.50 MM\$/yr (1.99 MM\$ capital)
Annualized recovery capital	2.62 MM\$/yr (10.50 MM\$ capital)
Total annual cost	107.84 MM\$/yr
Unit cost of product	2.157 \$/kg

Note that MM\$ or \$MM are commonly used shorthand for millions of dollars.

This example found the reactor throughput that would give the required annual capacity. For prescribed values of the design variables T and V, there is only one answer. The program uses a binary search to find that answer, but another *root-finder* could have been used instead. Newton's method (see Appendix 4) will save about a factor of 4 in computation time.

The next phase of the problem is to find those values for T and V that will give the lowest product cost. This is a problem in *optimization* rather than root-finding. Numerical methods for optimization are described in Appendix 6. The present example of consecutive, mildly endothermic reactions provides exercises for these optimization methods, but the example reaction sequence is not especially sensitive to operating conditions. Thus, the minimums tend to be quite shallow.

Example 6.3: Find the values of $T_{in} = T_{out}$ and V that give the lowest production cost for the consecutive reactions of Example 6.2.

Solution: The most straightforward way to optimize a function is by a brute force search. Results from such a search are shown in Table 6.2.

The lowest cost corresponds to $V = 58 \, m^3$ and $T_{out} = 364 \, K$, but the minimum is very flat so that there is essentially no difference in cost over a wide range of reactor volumes and operating temperatures. The good news is that an error in determining the minimum will have little effect on plant economics or the choice of operating conditions. The bad news is that perfectionists will need to use very precise numerical methods to find the true minimum.

The data in Table 6.2 illustrate a problem when optimizing a function by making one-at-a-time guesses. The cost at $V = 50 \, m^3$ and $T_{out} = 366 \, K$ is not the minimum, but is lower than the entries above and below it, on either side of it, or even diagonally above or below it. Great care must be taken to avoid false optimums. This is tedious to do manually, even with only two variables, and quickly becomes unmanageable as the number of variables increases.

More or less automatic ways of finding an optimum are described in Appendix 6. The simplest of these by far is the *random search* method. It can be used for any number of optimization variables. It is extremely inefficient from the viewpoint of the computer but is joyously simple to implement. The following program fragment illustrates the method.

TABLE 6.2 Results of a Comprehensive Search for the Case of a Single CSTR

Volume, m^3	Temperature, K					
	362	363	364	365	366	367
44	2.06531	2.05348	2.04465	2.03840	2.03440	2.03240
46	2.05817	2.04808	2.04074	2.03577	2.03292	**2.03196**
48	2.05232	2.04374	2.03771	2.03390	2.03208	2.03206
50	2.04752	2.04028	2.03542	2.03265	**2.03178**	2.03263
52	2.04361	2.03757	2.03376	2.03194	2.03193	2.03359
54	2.04044	2.03548	2.03263	**2.03168**	2.03247	2.03488
56	2.03790	2.03392	2.03195	2.03180	2.03334	2.03645
58	2.03590	2.03282	**2.03166**	2.03226	2.03450	2.03828
60	2.03437	2.03212	2.03172	2.03302	2.03591	2.04031
62	2.03325	2.03176	2.03206	2.03402	2.03753	2.04254
64	2.03248	**2.03171**	2.03267	2.03525	2.03935	2.04492
66	2.03202	2.03192	2.03350	2.03667	2.04134	2.04745
68	**2.03183**	2.03236	2.03454	2.03826	2.04347	2.05011

Values in bold indicate local minimums for fixed combinations of volume and temperature. They are potentially false optimums.

```
Maxtrials = 10000
BestTotal = 1000000000 'an arbitrary high value
                       'for the total cost
T = 400 'Initial guess
V = 10 'Initial guess

DO
'The reactor design calculations of Example 6.2 go here.
'They produce the total annualized cost, Total, that is the
'objective function for this optimization
  IF Total < BestTotal THEN
    BestTotal = Total
    BestT = Tin
    BestV = V
  END IF
  Tin = BestT+.5 * (.5 − RND)
  V = BestV + .1 * (.5 − RND)
  Ntrials = Ntrials + 1
Loop while Ntrials < Maxtrials
```

Applying the random search technique to the single CSTR case gives $V = 58.1 \, m^3$, $T = 364.1 \, K$, and a unit cost of 2.0316 \$/kg. These results are achieved very quickly because the design equations for the CSTR are simple algebraic equations. More complicated reactions in a CSTR may need the method of false transients, and any reaction in a nonisothermal PFR will require the solution of simultaneous ODEs. Computing times may become annoyingly long if crude numerical methods continue to be used. However, crude methods are probably best when starting a complex program. Get them working, get a feel for the system, and then upgrade them.

The general rule in speeding up a computation is to start by improving the innermost loops. For the example problem, the subroutine Reactor cannot be significantly improved for the case of a single CSTR, but Runge-Kutta integration is far better than Euler integration when solving ODEs. The next level of code is the overall materials balance used to calculate the reactor throughput and residence time. Some form of Newton's method can replace the binary search when you have a feel for the system and know what are reasonable initial guesses. Finally, tackle the outer loop that comprises the optimization routine.

The next example treats isothermal and adiabatic PFRs. Newton's method is used to determine the throughput, and Runge-Kutta integration is used in the Reactor subroutine. (The analytical solution could have been used for the isothermal case as it was for the CSTR.) The optimization technique remains the random one.

The temperature profile down the reactor is the issue. The CSTR is isothermal but selectivity is inherently poor when the desired product is an intermediate in a consecutive reaction scheme. An isothermal PFR is often better for selectivity and can be approximated in a shell-and-tube design by using many small tubes. Before worrying about the details of the shell-and-tube design, calculate the performance of a truly isothermal PFR and compare it with that of a CSTR and an adiabatic reactor. If the isothermal design gives a significant advantage, then tube size and number can be selected as a separate optimization exercise.

Example 6.4: Find the best combination of reaction temperature and volume for the example reaction using isothermal and adiabatic PFRs.

Solution: A program for evaluating the adiabatic reactor is given below. Subroutine Reactor solves the simultaneous ODEs for the concentrations and temperature. The equation for temperature includes contributions from both reactions according to the methods of Section 5.2.

```
DEFDBL A-H, P-Z
DEFLNG I-O
COMMON SHARED MwA, MwB, MwC, rho, Ain, hr1, hr2

'Random optimization of an adiabatic PFR
'using a Newton's search to close the material balance

MwA = 200
MwB = 200
MwC = 200
rho = 900
ain = rho/MwA
hr1 = 30/ain    'This is the adiabatic temperature change
' (a decrease is positive) per unit concentration of
'component A. Refer to Equation 6.4
hr2 = 30/ain    'Same for the second reaction
Maxtrials = 10000
BestTotal = 1000000000
V = 30
Tin = 390

  DO   'Main Loop
       'Newton's method to find WAin
       WA = 6250 'lower bound, kg/hr
       Q = WA/rho
       tbar = V/Q
       CALL Reactor(tbar, Tin, ain, Tout, aout, bout, cout)
       WB = bout * Q * MwB
```

```
    WAin = 2 * 6250 'lower bound, kg/hr
    Q = WAin/rho
    tbar = V/Q
    CALL Reactor(tbar, Tin, ain, Tout, aout, bout, cout)
    WBout = bout * Q * MwB
    DO
      Del = WAin - WA
      IF ABS(WBout-6250)<.001 THEN EXIT DO
        WA = WAin
        WAin = WAin-(WBout-6250)/(WBout-WB) * Del
        WB = WBout
        Q = WAin/rho
        tbar = V/Q
        CALL Reactor (tbar, Tin, ain, bin, cin, Tout,
        aout, bout, cout)
        WBout = bout * Q * MwB
    LOOP 'End of Newton's method
    CALL Cost(WAin, V, aout, bout, cout, total)

    IF total < BestTotal THEN
      BestTotal = total
      BestT = Tin
      BestV = V
    END IF
    Tin = BestT + .5 * (.5 - RND)
    V = BestV + .5 * (.5 - RND)
    Ntrials = Ntrials + 1
LOOP WHILE Ntrials < Maxtrials
'Output results here.

END

SUB Reactor (tbar, Tin, ain, bin, cin, Tout, aout, bout,
cout)

'Adiabatic version of PFR equations solved by Runge-Kutta
integration

N = 128
dtau = tbar/N
a = ain
T = Tin
FOR i = 1 TO N
    xk1 = 450000000000# * EXP(-10000/T)
```

```
    xk2 = 1800000000000# * EXP(-12000#/T)
    RA0 = -xk1 * a
    RB0 = xk1 * a - xk2 * b
    RT0 = -xk1 * a * hr1 - xk2 * b * hr2
    a1 = a + dtau * RA0/2
    b1 = b + dtau * RB0/2
    T1 = T + dtau * RT0/2
    RA1 = -xk1 * a1
    RB1 = xk1 * a1 - xk2 * b1
    RT1 = xk1 * a1 * hr1 - xk2 * b1 * hr2
    a2 = a + dtau * RA1/2
    b2 = b + dtau * RB1/2
    T2 = T + dtau * RT1/2
    RA2 = -xk1 * a2
    RB2 = xk1 * a2 - xk2 * b2
    RT2 = -xk1 * a2 * hr1 - xk2 * b2 * hr2
    a3 = a + dtau * RA2
    b3 = b + dtau * RB2
    T3 = T + dtau * RT2/2
    RA3 = -xk1 * a3
    RB3 = xk1 * a3 - xk2 * b3
    RT3 = -xk1 * a3 * hr1 - xk2 * b3 * hr2
    a = a + dtau * (RA0 + 2 * RA1 + 2 * RA2 + RA3)/6
    b = b + dtau * (RB0 + 2 * RB1 + 2 * RB2 + RB3)/6
    T = T + dtau * (RT0 + 2 * RT1 + 2 * RT2 + RT3)/6
NEXT
aout = a
bout = b
out = ain - aout - bout
Tout = T
END SUB
```

The above computation is quite fast. Results for the three ideal reactor types are shown in Table 6.3. The CSTR is clearly out of the running, but the difference between the isothermal and adiabatic PFR is quite small. Any reasonable shell-and-tube design would work. A few large-diameter tubes in parallel would be fine, and the limiting case of one tube would be the best. The results show that a close approach to adiabatic operation would reduce cost. The cost reduction is probably real since the comparison is nearly "apples-to-apples."

The results in Table 6.3 show that isothermal piston flow is not always the best environment for consecutive reactions. The adiabatic temperature profile gives better results, and there is no reason to suppose that it is the best

TABLE 6.3 Comparison of Ideal Reactors for Consecutive, Endothermic Reactions

	Single CSTR	Isothermal PFR	Adiabatic PFR
T_{in}, K	364	370	392
T_{out}, K	364	370	363
V, m^3	58.1	24.6	24.1
W, kg/h	8621	6975	6974
Unit cost, \$/kg	2.0316	1.9157	1.9150

TABLE 6.4 Optimal Zone Temperatures for Consecutive Reactions

		Zone temperatures, K					
N_{zones}	b_{out}	1	2	3	4	5	6
1	8.3165	376.2					
2	8.3185	378.4	371.7				
3	8.3196	380.0	374.4	373.4			
4	8.3203	381.3	375.12	373.8	373.3		
5	8.3207	382.4	375.8	374.2	373.6	373.2	
6	8.3210	383.3	376.4	374.7	373.9	373.4	373.2

possible profile. Finding the best temperature profile is a problem in *functional optimization*.

Example 6.5: Find the optimal temperature profile, $T(z)$, that maximizes the concentration of component B in the competitive reaction sequence of Equation (6.1) for a piston flow reactor subject to the constraint that $\bar{t} = 3$ h.

Solution: This mouthful of a problem statement envisions a PFR operating at a fixed flow rate. The wall temperature can be adjusted as an arbitrary function of position z, and the heat transfer coefficient is so high that the fluid temperature exactly equals the wall temperature. What temperature profile maximizes b_{out}? The problem is best solved in the time domain $t = z/\bar{u}$, since the results are then independent of tube diameter and flow rate. Divide the reactor into N_{zones} equal-length zones each with residence time $\bar{t}/N_{zones}$. Treat each zone as an isothermal reactor operating at temperature T_n, where $n = 1, 2, \ldots, N_{zones}$. The problem in functional optimization has been converted to a problem in parameter optimization, with the parameters being the various T_n. The computer program of Example 6.4 can be converted to find these parameters. The heart of the program is shown in the following segment. Given $\bar{t}_n = \bar{t}/N_{zones}$, T_n, and the three inlet concentrations to each zone, it calculates the outlet concentrations for that zone, assuming isothermal piston flow within the zone. Table 6.4 shows the results.

```
Maxtrials = 20000
Nzones = 6
BestBout = 0
FOR nz = 1 TO Nzones
   Tin(nz) = 382
   BestT(nz) = Tin(nz)
NEXT nz
tbar = 3/Nzones

DO 'Main Loop
   a = ain
   b = 0
   c = 0
   FOR nz = 1 TO Nzones
      CALL ZoneReactor(tbar, Tin(nz), a, b, c, Tout,
+     aout, bout, cout)
      a = aout
      b = bout
      c = cout
   NEXT nz

   IF bout > BestBout THEN
      BestBout = bout
      FOR nz = 1 TO Nzones
         BestT(nz) = Tin(nz)
      NEXT nz
   END IF
   FOR nz = 1 TO Nzones
      Tin(nz) = BestT(nz) +.01 * (.5 − RND)
   NEXT
   Ntrials = Ntrials + 1
LOOP WHILE Ntrials < Maxtrials

'output goes here

END
```

Figure 6.2 displays the temperature profile for a 10-zone case and for a 99-zone case. The 99-zone case is a tour de force for the optimization routine that took a few hours of computing time. It is not a practical example since such a multizone design would be very expensive to build. More practical designs are suggested by Problems 6.11–6.13.

Example 6.6: Suppose the reactions in Equation (6.1) are exothermic rather than endothermic. Specifically, reverse the sign on the heat of reaction terms

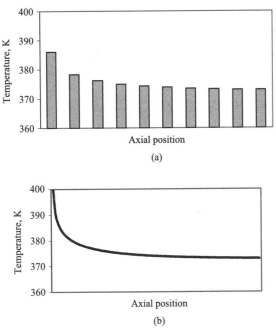

FIGURE 6.2 Piecewise-constant approximations to an optimal temperature profile for consecutive reactions: (a) 10-zone optimization; (b) 99-zone optimization.

so that the adiabatic temperature rise for complete conversion of A to B (but no C) is +30 K rather than −30 K. How does this change the results of Examples 6.2 through 6.5?

Solution: The temperature dependence of the reaction rates is unchanged. When temperatures can be imposed on the system, as for the CSTR and isothermal reactor examples, the results are unchanged from the endothermic case. The optimal profile results in Example 6.5 are identical for the same reason. The only calculation that changes is that for an adiabatic reactor. The program in Example 6.4 can be changed just by setting hr1 and hr2 to −30 rather than +30. The resulting temperature profile is increasing rather than decreasing, and this hurts selectivity. The production cost for an adiabatic reactor would be nearly 2 cents per kilogram higher than that for an isothermal reactor. Thus, a shell-and-tube design that approximates isothermal operation or even one that imposes a decreasing temperature profile is the logical choice for the process. The required volume for this reactor will be on the order of $24\,m^3$ as per Example 2.4. The specific choice of number of tubes, tube length, and tube diameter depends on the fluid properties, the economics of manufacturing heat exchangers, and possibly even the prejudgment of plant management regarding minimum tube diameters.

6.2 A COMPETITIVE REACTION SEQUENCE

Suppose the reactions are elementary, competitive, and of the form

$$A \xrightarrow{k_I} B$$
$$A \xrightarrow{k_{II}} C \qquad (6.6)$$

The rate constants are given by Equation (6.3), and both reactions are endothermic as per Equation (6.4). The flow diagram is identical to that in Figure 6.1, and all cost factors are the same as for the consecutive reaction examples. Table 6.1 also applies, and there is an interior optimum for any of the ideal reactor types.

Example 6.7: Determine optimal reactor volumes and operating temperatures for the three ideal reactors: a single CSTR, an isothermal PFR, and an adiabatic PFR.

Solution: The computer programs used for the consecutive reaction examples can be used. All that is needed is to modify the subroutine Reactor. Results are shown in Table 6.5.

All other things being equal, as they are in this contrived example, the competitive reaction sequence of Equation (6.6) is superior for the manufacture of B than the consecutive sequence of Equation (6.1). The CSTR remains a doubtful choice, but the isothermal PFR is now better than the adiabatic PFR. The reason for this can be understood by repeating Example 6.5 for the competitive reaction sequence.

Example 6.8: Find the optimal temperature profile, $T(t)$, that maximizes the concentration of component B in the competitive reaction sequence of Equation (6.6) for a piston flow reactor subject to the constraint that $\bar{t} = 1.8$ h.

Solution: The computer program used for Example 6.5 will work with minor changes. It is a good idea to start with a small number of zones until you get some feel for the shape of the profile. This allows you to input a

TABLE 6.5 Comparison of Ideal Reactors for Competitive, Endothermic Reactions

	Single CSTR	Isothermal PFR	Adiabatic PFR
T_{in}, K	411	388	412
T_{out}, K	411	388	382
V, m³	20.9	13.0	14.1
W, kg/h	6626	6420	6452
Unit cost, $/kg	1.8944	1.8716	1.8772

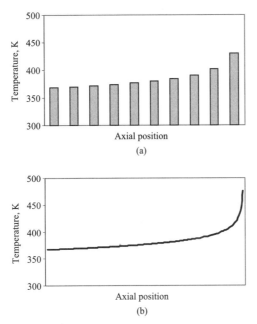

FIGURE 6.3 Piecewise-constant approximations to an optimal temperature profile for competitive reactions: (a) 10-zone optimization; (b) 99-zone optimization.

reasonable starting estimate for the profile and greatly speeds convergence when the number of zones is large. It also ensures that you converge to a local optimum and miss a better, global optimum that, under quite rare circumstances, may be lurking somewhere.

Results are shown in Figure 6.3.

The optimal profile for the competitive reaction pair is an increasing function of t (or z). An adiabatic temperature profile is a decreasing function when the reactions are endothermic, so it is obviously worse than the constant temperature, isothermal case. However, reverse the signs on the heats of reactions, and the adiabatic profile is preferred although still suboptimal.

PROBLEMS

6.1. Repeat Example 6.2 but change all the molecular weights to 100. Explain your results.

6.2. Determine the minimum operating cost for the process of Example 6.2 when the reactor consists of two equal-volume CSTRs in series. The capital cost per reactor is the same as for a single reactor.

6.3. Add a parameter to Problem 6.2 and study the case where the CSTRs can have different volumes.

6.4. The following sets of rate constants give nearly the same values for k_I and k_{II} at 360 K:

k_I	k_{II}
$4.2 \times 10^5 \exp(-5000/T)$	$1.04 \times 10^5 \exp(-6000/T)$
$4.5 \times 10^{11} \exp(-10000/T)$	$1.8 \times 10^{12} \exp(-12000/T)$
$5.2 \times 10^{23} \exp(-20000/T)$	$5.4 \times 10^{26} \exp(-24000/T)$

There are nine possible combinations of rate constants. Pick (or be assigned) a combination other than the base case of Equation (6.3) that was used in the worked examples. For the new combination:

(a) Do a comprehensive search similar to that shown in Table 6.2 for the case of a single CSTR. Find the volume and temperature that minimizes the total cost. Compare the relative flatness or steepness of the minimum to that of the base case.

(b) Repeat the comparison of reactor types as in Example 6.4.

(c) Determine the optimum set of temperatures for a six-zone reactor as in Example 6.4. Discuss the shape of the profile compared with that of the base case. Computer heroes may duplicate the 99-zone case instead.

6.5. Repeat Example 6.5 for the three-parameter problem consisting of two temperature zones, but with a variable zone length, and with $\bar{t}$ fixed at 3 h. Try a relatively short and hot first zone.

6.6. Work the five-parameter problem consisting of three variable-length zones.

6.7. Repeat Example 6.5 using 10 zones of equal length but impose the constraint that no zone temperature can exceed 373 K.

6.8. Determine the best value for T_{in} for an adiabatic reactor for the exothermic case of the competitive reactions in Equation (6.6).

6.9. Compare the (unconstrained) optimal temperature profiles of 10-zone PFRs for the following cases where: (a) the reactions are consecutive as per Equation (6.1) and endothermic; (b) the reactions are consecutive and exothermic; (c) the reactions are competitive as per Equation (6.6) and endothermic; and (d) the reactions are competitive and exothermic.

6.10. Determine the best two-zone PFR strategy for the competitive, endothermic reactions of Equation (6.6).

6.11. Design a shell-and-tube reactor that has a volume of 24 m³ and evaluate its performance as the reactor element in the process of Example 6.2. Use tubes with an i.d. of 0.0254 m and a length of 5 m. Assume components A, B, and C all have a specific heat of 1.9 kJ/(kg·K) and a thermal conductivity of 0.15 W/(m·K). Assume $T_{in} = 70°C$. Run the reaction on the tube side and assume that the shell-side temperature is constant (e.g., use condensing steam). Do the consecutive, endothermic case.

6.12. Extend Problem 6.12 to a two-zone shell-and-tube reactor with different shell-side temperatures in the zones.
6.13. Switch to oil heat in Problem 6.11 in order to better tailor the temperature profile down the tube. Choices include co- or countercurrent flow, the oil flow rate, and the oil inlet temperature.
6.14. Can the calculus of variations be used to find the optimal temperature profile in Example 6.5?

SUGGESTIONS FOR FURTHER READING

A good place to begin a more comprehensive study of chemical engineering optimization is

Edgar, T. F. and Himmelblau, D. M., *Optimization of Chemical Processes*, 2nd ed., McGraw-Hill, New York, 2001.

Two books with a broader engineering focus that have also survived the test of time are

Rao, S. S., *Engineering Optimization: Theory and Practice*, 3rd ed., John Wiley & Sons, New York, 1996.

Fletcher, R., *Practical Methods of Optimization*, 2nd ed., John Wiley & Sons, New York, 2000.

The bible of numerical methods remains

Press, W. H., Teukolsky, S. A., Vetterling, W. T., and Flannery, B. P., *Numerical Recipes in Fortran 77: The Art of Scientific Computing*, Vol. 1, 2nd ed., Cambridge University Press, New York, 1992.

Versions of Volume I exist for C, Basic, and Pascal. Matlab enthusiasts will find some coverage of optimization (and nonlinear regression) techniques in

Constantinides, A. and Mostoufi, N., *Numerical Methods for Chemical Engineers with Matlab Applications*, Prentice Hall, New York, 1999.

Mathematica fans may consult

Bhatti, M. A., *Practical Optimization Methods with Mathematica Applications*, Springer-Verlag, New York, 1999.

APPENDIX 6: NUMERICAL OPTIMIZATION TECHNIQUES

Optimization is a complex and sometimes difficult topic. Many books and countless research papers have been written about it. This appendix section discusses *parameter optimization*. There is a function, $F(p_1, p_2, \ldots)$, called the *objective function* that depends on the parameters $p_1, p_2, \ldots$. The goal is to determine the best values for the parameters, best in the sense that these parameter values will maximize or minimize F. We normally assume that the parameters can assume any values that are physically possible. For the single CSTR of Example 6.2, the two parameters are T and $\bar{t}$ and the objective function is the unit cost of production. The parameters must be positive, but there are no other restrictions, and the optimization is *unconstrained*. Suppose that the reactor has a limit on operating temperature, say 373 K. The problem becomes a *constrained* optimization, but the constraint has no effect on the result. The constraint is not *active*. Lower the temperature limit to 360 K, and it becomes active. It then forces a slightly lower temperature (namely 360 K) and slightly higher volume than found for the unconstrained optimization in Example 6.2. Multidimensional optimization problems usually have some active constraints.

Numerical optimization techniques find local optima. They will find the top of a hill or the bottom of a valley. In constrained optimizations, they may take you to a boundary of the parameter space. The objective function will get worse when moving a small amount in any direction. However, there may be a higher hill or a deeper valley or even a better boundary. There can be no guarantee that the global minimum will be found unless $F(p_1, p_2, \ldots)$ belongs to a restricted class of functions. If $F(p_1, p_2, \ldots)$ is linear in its parameters, there are no interior optima, and no hills or valleys, just slopes. Linear programming techniques will then find the global optimum that will be at an intersection of constraints. However, problems in reactor design can be aggressively nonlinear, and interior optima are fairly common.

A.6.1 Random Searches

The random search technique can be applied to constrained or unconstrained optimization problems involving any number of parameters. The solution starts with an initial set of parameters that satisfies the constraints. A small random change is made in each parameter to create a new set of parameters, and the objective function is calculated. If the new set satisfies all the constraints and gives a better value for the objective function, it is accepted and becomes the starting point for another set of random changes. Otherwise, the old parameter set is retained as the starting point for the next attempt. The key to the method is the step that sets the new, trial values for the parameters:

$$p_{trial} = p_{old} + \Delta_p(0.5 - \text{RND}) \qquad (6.7)$$

where RND is a random number uniformly distributed over the range 0–1. It is called RAND in C and RAN in Fortran. Equation (6.7) generates values of p_{trial} in the range $p_{trial} \pm \Delta_p/2$. Large values of Δ_p are desirable early in the search and small values are desirable toward the end, but the algorithm will eventually converge to a local optimum for any Δ_p. Repeated numerical experiments with different initial values can be used to search for other local optima.

A.6.2 Golden Section Search

The golden section search is the optimization analog of a binary search. It is used for functions of a single variable, $F(a)$. It is faster than a random search, but the difference in computing time will be trivial unless the objective function is extremely hard to evaluate.

To know that a minimum exists, we must find three points $a_{min} < a_{int} < a_{max}$ such that $F(a_{int})$ is less than either $F(a_{min})$ or $F(a_{max})$. Suppose this has been done. Now choose another point $a_{min} < a_{new} < a_{max}$ and evaluate $F(a_{new})$. If $F(a_{new}) < F(a_{int})$, then a_{new} becomes the new interior point. Otherwise a_{new} will become one of the new endpoints. Whichever the outcome, the result is a set of three points with an interior minimum and with a smaller distance between the endpoints than before. This procedure continues until the distance between a_{min} and a_{max} has been narrowed to an acceptable extent. The way of choosing a_{new} is not of critical importance, but the range narrows fastest if a_{new} is chosen to be at 0.38197 of the distance between the interior point and the more distant of the endpoints a_{min} and a_{max}.

A.6.3 Sophisticated Methods of Parameter Optimization

If the objective function is very complex or if the optimization must be repeated a great many times, the random search method should be replaced with something more efficient computationally. For a minimization problem, all the methods search for a way downhill. One group of methods uses nothing but function evaluations to find the way. Another group combines function evaluations with derivative calculations—e.g., $\partial F/\partial a$—to speed the search. All these methods are complicated. The easiest to implement is the simplex method of Nelder and Mead. (It is different than the simplex algorithm used to solve linear programming problems.) A subroutine is given in the book by Press et al.[A1] Other sources and codes for other languages are available on the web and in some versions of commercial packages, e.g., *Matlab*. More efficient but more complicated, gradient-based methods are available from the same sources.

A.6.4 Functional Optimization

A function $f(x)$ starts with a number, x, performs mathematical operations, and produces another number, f. It transforms one number into another. A *functional* starts with a function, performs mathematical operations, and produces a number. It transforms an entire function into a single number. The simplest and most common example of a functional is a definite integral. The goal in Example 6.5 was to maximize the integral

$$b_{out} - b_{in} = \int_0^{\bar{t}} \mathscr{R}_B(a, b, T)\, dt \qquad (6.8)$$

Equation (6.8) is a functional. There are several functions, $a(t)$, $b(t)$, $T(t)$, that contribute to the integral, but $T(t)$ is the one function directly available to the reactor designer as a manipulated variable. Functional optimization is used to determine the best function $T(t)$. Specification of this function requires that $T(t)$ be known at every point within the interval $0 < \bar{t} < L$.

Some problems in functional optimization can be solved analytically. A topic known as the *calculus of variations* is included in most courses in advanced calculus. It provides ground rules for optimizing integral functionals. The ground rules are necessary conditions analogous to the derivative conditions (i.e., $df/dx = 0$) used in the optimization of ordinary functions. In principle, they allow an exact solution; but the solution may only be implicit or not in a useful form. For problems involving Arrhenius temperature dependence, a numerical solution will be needed sooner or later.

Example 6.5 converted the functional optimization problem to a parameter optimization problem. The function $T(t)$ was assumed to be piecewise-constant. There were N pieces, the nth piece was at temperature T_n, and these N temperatures became the optimization parameters. There are other techniques for numerical functional optimization, including some gradient methods; but conversion to parameter optimization is by far the easiest to implement and the most reliable. In the limit as N grows large, the numerical solution will presumably converge to the true solution. In Example 6.5, no constraints were imposed on the temperature, and the parameter optimization appears to be converging to a smooth function with a high-temperature spike at the inlet. In constrained optimizations, the optimal solution may be at one of the constraints and then suddenly shift to the opposite constraint. This is called *bang-bang* control and is studied in courses in advanced process control. The best strategy for a constrained optimization may be to have a small number of different-length zones with the temperature in each zone being at either one constraint or the other. This possibility is easily explored using parameter optimization.

Reference

A1. Press, W. H., Teukolsky, S. A., Vetterling, W. T., and Flannery, B. P., *Numerical Recipes in Fortran 77: The Art of Scientific Computing*, Vol. 1, 2nd ed., Cambridge University Press, New York, 1992.

CHAPTER 7
FITTING RATE DATA AND USING THERMODYNAMICS

Chapter 7 has two goals. The first is to show how reaction rate expressions, $\mathscr{R}(a, b, \ldots, T)$, are obtained from experimental data. The second is to review the thermodynamic underpinnings for calculating reaction equilibria, heats of reactions and heat capacities needed for the rigorous design of chemical reactors.

7.1 ANALYSIS OF RATE DATA

With two adjustable constants, you can fit a straight line. With five, you can fit an elephant. With eight, you can fit a running elephant or a cosmological model of the universe.[1]

Section 5.1 shows how nonlinear regression analysis is used to model the temperature dependence of reaction rate constants. The functional form of the reaction rate was assumed; e.g., $\mathscr{R} = kab$ for an irreversible, second-order reaction. The rate constant k was measured at several temperatures and was fit to an Arrhenius form, $k = k_0 \exp(-T_{act}/T)$. This section expands the use of nonlinear regression to fit the compositional and temperature dependence of reaction rates. The general reaction is

$$\nu_A A + \nu_B B + \cdots \rightleftarrows \nu_R R + \nu_S S + \cdots \qquad (7.1)$$

and the rate expression can take several possible forms.

If the reaction is known to be elementary, then

$$\mathscr{R} = k_f [A]^{-\nu_A}[B]^{-\nu_B} \cdots - k_r[R]^{\nu_R}[S]^{\nu_S} \cdots \qquad (7.2)$$

where the stoichiometric coefficients are known small integers. Experimental data will be used to determine the rate constants k_f and k_r. A more general form for the rate expression is

$$\mathscr{R} = k_f [A]^m [B]^n \cdots - k_r [R]^r [S]^s \cdots \qquad (7.3)$$

where $m, n, \ldots, r, s, \ldots$ are empirical constants that may or may not be integers. These constants, together with k_f and k_r, must be determined from the data. An alternative form that may fit the data reasonably well is

$$\mathscr{R} = k[A]^m[B]^n[R]^r[S]^s \cdots \quad (7.4)$$

where some of the exponents (e.g. $r, s, \ldots$) can be negative. The virtue of this form is that it has one fewer empirical constants than Equation (7.3). Its fault is that it lacks the mechanistic basis of Equation (7.3) and will not perform as well near the equilibrium point of a reversible reaction.

For enzymatic and other heterogeneously catalyzed reactions, there may be competition for active sites. This leads to rate expressions with forms such as

$$\mathscr{R} = \frac{k[A]^m[B]^n[R]^r[S]^s \cdots}{(1 + k_A[A] + k_B[B] + k_R[R] + k_S[S] + \ldots)} \quad (7.5)$$

All the rate constants should be positive so the denominator in this expression will always retard the reaction. The same denominator can be used with Equation (7.3) to model reversible heterogeneous reactions:

$$\mathscr{R} = \frac{k_f[A]^m[B]^n \cdots - k_r[R]^r[S]^s \cdots}{(1 + k_A[A] + k_B[B] + k_R[R] + k_S[S] + \ldots)} \quad (7.6)$$

More complicated rate expressions are possible. For example, the denominator may be squared or square roots can be inserted here and there based on theoretical considerations. The denominator may include a term $k_I[I]$ to account for compounds that are nominally inert and do not appear in Equation (7.1) but that occupy active sites on the catalyst and thus retard the rate. The forward and reverse rate constants will be functions of temperature and are usually modeled using an Arrhenius form. The more complex kinetic models have enough adjustable parameters to fit a stampede of elephants. Careful analysis is needed to avoid being crushed underfoot.

7.1.1 Least-Squares Analysis

The goal is to determine a functional form for $\mathscr{R}(a, b, \ldots, T)$ that can be used to design reactors. The simplest case is to suppose that the reaction rate $\mathscr{R}$ has been measured at various values $a, b, \ldots, T$. A CSTR can be used for these measurements as discussed in Section 7.1.2. Suppose J data points have been measured. The jth point in the data is denoted as $\mathscr{R}_{data}(a_j, b_j, \ldots, T_j)$ where $a_j, b_j, \ldots, T_j$ are experimentally observed values. Corresponding to this measured reaction rate will be a predicted rate, $\mathscr{R}_{model}(a_j, b_j, \ldots, T_j)$. The predicted rate depends on the parameters of the model e.g., on $k, m, n, r, s, \ldots$ in Equation (7.4) and these parameters are chosen to obtain the best fit of the experimental data to the model. Specifically, we seek values for $k, m, n, r, s, \ldots$ that will minimize the sum-of-squares:

$$\begin{aligned}S^2 &= \sum_{j=1}^{J}[\mathscr{R}_{data}(a_j, b_j, \ldots, T_j) - \mathscr{R}_{model}(a_j, b_j, \ldots, T_j)]^2 \\ &= \sum_{j=1}^{J}[(\mathscr{R}_{data})_j - \mathscr{R}_{model}(k, m, n, r, s, \ldots, k_0, T_{act})]^2\end{aligned} \quad (7.7)$$

The first equation shows that the data and model predictions are compared at the same values of the (nominally) independent variables. The second equation explicitly shows that the sum-of-squares depends on the parameters in the model.

Any of Equations (7.2)–(7.6) may be used as the model. The parameters in the model are adjusted to minimize the sum-of-squares using any of the optimization methods discussed in Chapter 6. An analytical solution to the minimization problem is possible when the model has a linear form. The fitting process is then known as *linear regression analysis*. This book emphasizes *nonlinear regression* because it is generally more suitable for fitting rate data. However, rate expressions can often be transformed to a linear form, and there are many canned computer programs for linear regression analysis. These programs can be useful for obtaining preliminary estimates of the model parameters that can subsequently be refined using nonlinear regression. Appendix 7 gives the rudiments of linear regression analysis.

When kinetic measurements are made in batch or piston flow reactors, the reaction rate is not determined directly. Instead, an integral of the rate is measured, and the rate itself must be inferred. The general approach is as follows:

1. Conduct kinetic experiments and *measure* some response of the system, such as a_{out}. Call this "data."
2. Pick a rate expression and assume values for its parameters. Solve the reactor design equations to *predict* the response. Call this "prediction."
3. Adjust the parameters to minimize *the sum-of-squares*:

$$S^2 = \sum_{j=1}^{J}[\text{data} - \text{prediction}]^2 \quad (7.8)$$

The sum of squares as defined by Equation 7.8 is the general form for the objective function in nonlinear regression. Measurements are made. Models are postulated. Optimization techniques are used to adjust the model parameters so that the sum-of-squares is minimized. There is no requirement that the model represent a simple reactor such as a CSTR or isothermal PFR. If necessary, the model could represent a nonisothermal PFR with variable physical properties. It could be one of the distributed parameter models in Chapters 8 or 9. The model parameters can include the kinetic parameters in Equations (7.2)–(7.6) together with unknown transport properties such as a heat transfer coefficient. However, the simpler the better.

To fit the parameters of a model, there must be at least as many data as there are parameters. There should be many more data. The case where the number of data equals the number of points can lead to exact but spurious fits. Even a perfect model cannot be expected to fit all the data because of experimental error. The *residual sum-of-squares* $S^2_{residual}$ is the value of S^2 after the model has been fit to the data. It is used to calculate the *residual standard deviation*:

$$\sigma_{residual} = \sqrt{\frac{S^2_{residual}}{J-1}} \qquad (7.9)$$

where J is the number of data points. When $\sigma_{residual}$ equals what would be expected from experimental error, the model has done all it should do. Values of $\sigma_{residual}$ less than expected experimental error mean that there are too few data or that the model has too many adjustable parameters.

A good model is consistent with physical phenomena (i.e., $\mathscr{R}$ has a physically plausible form) and reduces $\sigma_{residual}$ to experimental error using as few adjustable parameters as possible. There is a philosophical principle known as Occam's razor that is particularly appropriate to statistical data analysis: when two theories can explain the data, the simpler theory is preferred. In complex reactions, particularly heterogeneous reactions, several models may fit the data equally well. As seen in Section 5.1 on the various forms of Arrhenius temperature dependence, it is usually impossible to distinguish between mechanisms based on goodness of fit. The choice of the simplest form of Arrhenius behavior ($m=0$) is based on Occam's razor.

The experimental basis for the model should span a broader range of the independent variables than will be encountered in the use of the model. To develop a comprehensive model, it is often necessary to add components to the feed in amounts that would not normally be present. For A → B, the concentration of B is correlated to that of A: $a_{in} - a = b - b_{in}$. Varying b_{in} will lessen the correlation and will help distinguish between rate expressions such as $\mathscr{R} = ka$ or $\mathscr{R} = k_f a - k_r b$ or $\mathscr{R} = ka/(1 + k_B b)$. Books and courses on the *design of experiments* can provide guidance, although our need for formalized techniques is less than that in the social and biological sciences, where experiments are much more difficult to control and reproduce.

7.1.2 Stirred Tanks and Differential Reactors

A component balance for a steady-state CSTR gives

$$\mathscr{R}_A(a_{out}, b_{out}, \ldots, T_{out}) = \frac{Q_{in}a_{in} - Q_{out}a_{out}}{V} = \frac{Q_{in}a_{in}/Q_{out} - a_{out}}{\bar{t}} \qquad (7.10)$$

where $\bar{t} = V/Q_{out}$. Equation (7.10) does not require constant density, but if it varies significantly, Q_{out} or ρ_{out} will have to be measured or calculated from an equation of state. In a normal experimental design, the inlet conditions Q_{in}, a_{in}, b_{in}, ... are specified, and outlet concentrations a_{out}, b_{out}, ... are measured. The experimental plan will also specify approximate values for T_{out}. The reaction rate for a key component is calculated using Equation (7.10), and the results are regressed against measured values of a_{out}, b_{out}, ..., and T_{out}.

Example 7.1: The following data have been measured in a CSTR for a reaction having the form A → B.

Run number	a_{in}	b_{in}	a_{out}	b_{out}
1	0.200	0	0.088	0.088
2	0.400	0	0.191	0.206
3	0.600	0	0.307	0.291
4	0.800	0	0.390	0.400
5	1.000	0	0.493	0.506

The density is constant and the mean residence time is 2 h, as determined from the known volume of the reactor and the outlet flow rate. The temperature was the same for all runs.

Solution: An overall material balance gives $a_{in} + b_{in} = a_{out} + b_{out}$. The data are obviously imperfect, but they will be accepted as is for this example. The following program fragment uses the random search technique to fit the general form $\mathcal{R} = k a_{out}^m b_{out}^n$.

```
DefDbl A-Z
Dim ain(100), aout(100), bout(100)
'Data
ain(1) = 0.2: aout(1) = 0.088: bout(1) = 0.088
ain(2) = 0.4: aout(2) = 0.191: bout(2) = 0.206
ain(3) = 0.6: aout(3) = 0.307: bout(3) = 0.291
ain(4) = 0.8: aout(4) = 0.390: bout(4) = 0.400
ain(5) = 1.0: aout(5) = 0.493: bout(5) = 0.506

tbar = 2
Jdata = 5

bestsd = 1
Ntrials = 10000
k = 1
m = 0
n = 0

For nr = 1 To Ntrials
```

```
  ss = 0
  For j = 1 To Jdata
    RA = (aout(j) - ain(j))/tbar
    ss = ss + (RA + k * aout(j)^m * bout(j)^n)^2
Next
  sd = Sqr(ss/(Jdata - 1))
  If sd < bestsd Then
    bestk = k
    bestm = m
    bestn = n
    bestsd = sd
End If
  'm = bestm + 0.05 * (0.5 - Rnd) 'adjusts m randomly
  'n = bestn + 0.05 * (0.5 - Rnd) 'adjusts n randomly
  k = bestk + 0.05 * (0.5 - Rnd) 'adjusts k randomly
Next nr
'Output results
```

As given above, the statements that adjust the exponents m and n have been "commented out" and the initial values for these exponents are zero. This means that the program will fit the data to $\mathcal{R} = k$. This is the form for a zero-order reaction, but the real purpose of running this case is to calculate the standard deviation of the experimental rate data. The object of the fitting procedure is to add functionality to the rate expression to reduce the standard deviation in a manner that is consistent with physical insight. Results for the zero-order fit are shown as Case 1 in the following data:

Case	k	m	n	σ
1	0.153	0	0	0.07841
2	0.515	1	0	0.00871
3	0.490	0.947	0	0.00813
4	0.496	1	−0.040	0.00838
5	0.478	−0.086	1.024	0.00468
6	0.507	0	1	0.00602

Results for a first-order fit—corresponding to Equation (7.2) for an irreversible first-order reaction—are shown as Case 2. This case is obtained by setting $m = 1$ as an initial value in the program fragment. Case 2 reduces the standard deviation of data versus model by nearly an order of magnitude using a single, semitheoretical parameter. The residual standard deviation is probably as low as can be expected given the probable errors in the concentration measurements, but the remaining cases explore various embellishments to the model. Case 3 allows m to vary by enabling the statement m = bestm + 0.05 * (0.5 - Rnd). The results show a small reduction in the standard deviation. A statistician could attempt to see if the change from $m=1$ to $m=0.947$ was statistically significant. A chemist or chemical engineer would most likely prefer to keep $m=1$. Case 4 sets $m=1$ but now allows n to vary. The small negative exponent might remind the experimenter that the reaction could be reversible, but the effect is too small to be of much concern. Cases 5 and 6 illustrate a weakness of statistic analysis. Case 5 is obtained by minimizing the sum-of-squares when k, m, and n are all allowed to vary. The reaction rate better correlates with the product concentration than the reactant concentration! Case 6 carries this physical absurdity to an extreme by showing that a first-order dependence on product concentration gives a good fit to the data for an essentially irreversible reaction. The reason for these spurious fits is that a_{out} and b_{out} are strongly correlated.

The conclusion, based on a mixture of physical insight and statistical analysis, is that $\mathcal{R} = 0.515a$ is close to the truth, but further experiments can be run.

Example 7.2: The nagging concern that the reaction of Example 7.1 may somehow depend on the product concentration prompted the following additional runs. These runs add product to the feed in order to destroy the correlation between a_{out} and b_{out}.

Run number	a_{in}	b_{in}	a_{out}	b_{out}
6	0.500	0.200	0.248	0.430
7	0.500	0.400	0.246	0.669
8	0.500	0.600	0.239	0.854
9	0.500	0.800	0.248	1.052
10	0.500	1.000	0.247	1.233

Solution: The new data are combined with the old, and the various cases are rerun. The results are:

Case	k	m	n	σ
1	0.140	0.000	0.000	0.05406
2	0.516	1.000	0.000	0.00636
3	0.496	0.963	0.000	0.00607
4	0.514	1.000	−0.007	0.00636
5	0.403	0.963	−0.007	0.00605
6	0.180	0.000	1.000	0.09138

The retrograde behavior of Case 5 has vanished, and Case 6 has become worse than the zero-order fit of Case 1. The recommended fit for the reaction rate at this point in the analysis, $\mathcal{R} = 0.516a$, is very similar to the original recommendation, but confidence in it has increased.

We turn now to the issue of *material balance closure*. Material balances can be perfect when one of the flow rates and one of the components is unmeasured. The keen experimenter for Examples 7.1 and 7.2 measured the outlet concentration of both reactive components and consequently obtained a less-than-perfect balance. Should the measured concentrations be adjusted to achieve closure and, if so, how should the adjustment be done? The general rule is that a material balance should be closed if it is reasonably possible to do so. It is necessary to know the number of inlet and outlet flow streams and the various components in these streams. The present example has one inlet stream, one outlet stream, and three components. The components are A, B, and I, where I represents all inerts.

Closure normally begins by satisfying the overall mass balance; i.e., by equating the input and outlet mass flow rates for a steady-state system. For the present case, the outlet flow was measured. The inlet flow was unmeasured so it must be assumed to be equal to the outlet flow. We suppose that A and B are the only reactive components. Then, for a constant-density system, it must be that

$$a_{in} + b_{in} = a_{out} + b_{out} \tag{7.11}$$

This balance is not quite satisfied by the experimental data, so an adjustment is needed. Define material balance fudge factors by

$$f_{in} f_{out} = \left[\frac{a_{in} + b_{in}}{a_{out} + b_{out}}\right]_{measured} \tag{7.12}$$

and then adjust the component concentrations using

$$[a_{in}]_{adjusted} = [a_{in}]_{measured}/f_{in}$$

and $$\tag{7.13}$$

$$[a_{out}]_{adjusted} = f_{out}[a_{out}]_{measured}$$

with similar adjustments for component B. When the adjustments are made, Equation (7.11) will be satisfied. The apportionment of the total imbalance between the inlet and outlet streams is based on judgment regarding the relative accuracy of the measurements. If the inlet measurements are very accurate—i.e., when the concentrations are set by well-calibrated proportioning pumps—set $f_{in} = 1$ and let f_{out} absorb the whole error. If the errors are similar, the two factors are equal to the square root of the concentration ratio in Equation (7.12).

Example 7.3: Close the material balance and repeat Example 7.2.

Solution: Suppose $f_{in} = 1$ so that f_{out} is equal to the concentration ratio in Equation (7.12). Equations (7.13) are applied to each experimental run using the value of f_{out} appropriate to that run. The added code is

```
For j = 1 To Jdata
  fudgeout = (ain(j) + bin(j)) / (aout(j) + bout(j))
  aout(j) = fudgeout * aout(j)
```

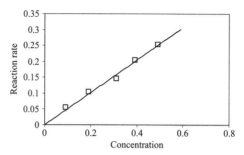

FIGURE 7.1 Final correlation for $\mathcal{R}(a)$.

```
bout(j) = fudgeout* bout(j)
Next j
```

The results show that closing the material balance improves the fit. The recommended fit becomes $\mathcal{R} = 0.509a$. It is shown in Figure 7.1. As a safeguard against elephant stampedes and other hazards of statistical analysis, a *graphical view of a correlation* is always recommended. However, graphical techniques are not recommended for the fitting process.

Case	k	m	n	σ
1	0.139	0.000	0.000	0.03770
2	0.509	1.000	0.000	0.00598
3	0.509	1.000	0.000	0.00598
4	0.515	1.000	0.018	0.00583
5	0.516	1.002	0.018	0.00583
6	0.178	0.000	1.000	0.09053

All these examples have treated kinetic data taken at a single temperature. Most kinetic studies will include a variety of temperatures so that two parameters, k_0 and $E/R_g = T_{act}$, are needed for each rate constant. The question now arises as to whether all the data should be pooled in one glorious minimization, or if you should conduct separate analyses at each temperature and then fit the resulting rate constants to the Arrhenius form. The latter approach was used in Example 5.1 (although the preliminary work needed to find the rate constants was not shown), and it has a major advantage over the combined approach. Suppose Equation (7.4) is being fit to the data. Are the exponents $m, n, \ldots$ the same at each temperature? If not, the reaction mechanism is changing and the possibility of consecutive or competitive reactions should be explored. If the exponents are the same within reasonable fitting accuracy, the data can be pooled or kept separate as desired. Pooling will give the best overall fit, but a better fit in some regions of the experimental space might be desirable for scaleup. Problem 7.3, although for batch data, offers scope to try a variety of fitting strategies.

The CSTRs are wonderful for kinetic experiments since they allow a direct determination of the reaction rate at known concentrations of the reactants.

One other type of reactor allows this in principle. *Differential reactors* are so short that concentrations and temperatures do not change appreciably from their inlet values. However, the small change in concentration makes it very hard to determine an accurate rate. The use of differential reactors is not recommended. If a CSTR cannot be used, a batch or piston flow reactor is preferred over a differential reactor even though the reaction rate is not measured directly but must be inferred from measured outlet concentrations.

7.1.3 Batch and Piston Flow Reactors

Most kinetic experiments are run in batch reactors for the simple reason that they are the easiest reactor to operate on a small, laboratory scale. Piston flow reactors are essentially equivalent and are implicitly included in the present treatment. This treatment is confined to constant-density, isothermal reactions, with nonisothermal and other more complicated cases being treated in Section 7.1.4. The batch equation for component A is

$$\frac{da}{dt} = \mathscr{R}_A(a, b, \ldots, T) \tag{7.14}$$

subject to the initial condition that $a = a_0$ at $t = 0$.

Batch and piston flow reactors are called *integral reactors* because the rate expression must be integrated to determine reactor performance. When an integral reactor is used for a kinetic study, the procedure for determining parameters in the rate expression uses Equation (7.8) for the regression analysis. Do not attempt to differentiate the experimental data to allow the use of Equation (7.7). Instead, assume a functional form for $\mathscr{R}_A$ together with initial guesses for the parameters. Equation (7.14) is integrated to obtain predictions for $a(t)$ at the various experimental values of t. The predictions are compared with the experimental data using Equation (7.8), and the assumed parameters are adjusted until the sum-of-squares is a minimum. The various caveats regarding overfitting of the data apply as usual.

Example 7.4: The following data have been obtained in a constant-volume, isothermal reactor for a reaction with known stoichiometry: A → B + C. The initial concentration of component A was 2200 mol/m^3. No B or C was charged to the reactor.

Sample number j	Time t, min	Fraction unreacted Y_A
1	0.4	0.683
2	0.6	0.590
3	0.8	0.513
4	1.0	0.445
5	1.2	0.381

Solution: A suitable rate expression is $\mathscr{R}_A = -ka^n$. Equation (7.14) can be integrated analytically or numerically. Equation (7.8) takes the following form for $n \neq 1$:

$$S^2 = \sum_{j=1}^{J} \left(Y_A(j) - \left[\frac{1}{1+(n-1)a_0^{n-1}kt_j}\right]^{\frac{1}{n-1}} \right)^2$$

where t_i is the time at which the ith sample was taken. The special form for $n=1$ is

$$S^2 = \sum_{j=1}^{J} Y_A(j) - \exp(-kt_j)^2$$

There are two adjustable parameters, n and k. Results for various kinetic models are shown below and are plotted in Figure 7.2.

Reaction order n	Rate constant $a_0^{n-1}k$	Standard deviation σ
0	0.572	0.06697
1	0.846	0.02024
1.53	1.024	0.00646
2	1.220	0.01561

The fit with $n=1.53$ is quite good. The results for the fits with $n=1$ and $n=2$ show systematic deviations between the data and the fitted model. The reaction order is approximately 1.5, and this value could be used instead of $n=1.53$ with nearly the same goodness of fit, $\sigma = 0.00654$ versus 0.00646. This result should motivate a search for a mechanism that predicts an order of 1.5. Absent such a mechanism, the best-fit value of 1.53 may as well be retained.

The curves in Figure 7.2 plot the natural variable $a(t)/a_0$, versus time. Although this accurately portrays the goodness of fit, there is a classical technique for plotting batch data that is more sensitive to reaction order for irreversible nth-order reactions. The reaction order is assumed and the experimental data are transformed to one of the following forms:

$$\left[\frac{a(t)}{a_0}\right]^{1-n} - 1 \quad \text{for} \quad n \neq 1 \quad \text{and} \quad -\ln\left[\frac{a(t)}{a_0}\right] \quad \text{for} \quad n=1 \quad (7.15)$$

Plot the transformed variable versus time. A straight line is a visually appealing demonstration that the correct value of n has been found. Figure 7.3 shows these plots for the data of Example 7.4. The central line in Figure 7.3 is for $n=1.53$. The upper line shows the curvature in the data that results from assuming an incorrect order of $n=2$, and the lower line is for $n=1$.

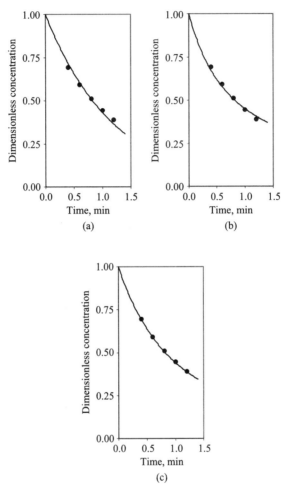

FIGURE 7.2 Experiment versus fitted batch reaction data: (a) first-order fit; (b) second-order fit; (c) 1.53-order fit.

The reaction of Example 7.4 is not elementary and could involve short-lived intermediates, but it was treated as a single reaction. We turn now to the problem of fitting kinetic data to multiple reactions. The multiple reactions listed in Section 2.1 are consecutive, competitive, independent, and reversible. Of these, the consecutive and competitive types, and combinations of them, pose special problems with respect to kinetic studies. These will be discussed in the context of integral reactors, although the concepts are directly applicable to the CSTRs of Section 7.1.2 and to the complex reactors of Section 7.1.4.

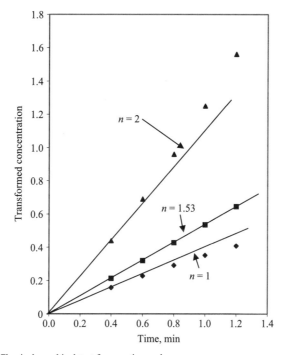

FIGURE 7.3 Classical graphical test for reaction order.

Consecutive Reactions. The prototypical reaction is A → B → C, although reactions like Equation (6.2) can be treated in the same fashion. It may be that the first reaction is independent of the second. This is the normal case when the first reaction is irreversible and homogeneous (so that component B does not occupy an active site). A kinetic study can then measure the starting and final concentrations of component A (or of A_1 and A_2 as per Equation (6.2)), and these data can be used to fit the rate expression. The kinetics of the second reaction can be measured independently by reacting pure B. Thus, it may be possible to perform completely separate kinetic studies of the reactions in a consecutive sequence. The data are fit using two separate versions of Equation (7.8), one for each reaction. The "data" will be the experimental values of a_{out} for one sum-of-squares and b_{out} for another.

If the reactions cannot be separated, it is not immediately clear as to what sum-of-squares should be minimized to fit the data. Define

$$S_A^2 = \sum_{Data} [a_{experiment} - a_{model}]^2 \qquad (7.16)$$

with similar equations for S_B^2 and S_C^2. If only b_{out} has been measured, there is no choice but to use S_B^2 to fit both reactions. If both a_{out} and b_{out} have been measured, S_A^2 can be used to find $\mathcal{R}$ for the first reaction. The fitted rate expression becomes part of the model used to calculate b_{out}. The other part of the model is the assumed rate expression for the second reaction, the parameters of which are found by minimizing S_B^2.

Example 7.5: Suppose the consecutive reactions $2A \rightarrow B \rightarrow C$ are elementary. Determine the rate constants from the following experimental data obtained with an isothermal, constant-volume batch reactor:

Time, min	$a(t)$	$b(t)$
15	1.246	0.305
30	0.905	0.347
45	0.715	0.319
60	0.587	0.268
75	0.499	0.221
90	0.435	0.181

The concentrations shown are dimensionless. Actual concentrations have been divided by $a_0/2$ so that the initial conditions are $a=2$, $b=0$ at $t=0$. The long-time value for $c(t)$ is 1.0.

Solution: The component balances for the batch reaction are

$$\frac{da}{dt} = -2k_I a^2$$

$$\frac{db}{dt} = k_I a^2 - k_{II} b$$

Values for k_I and k_{II} are assumed and the above equations are integrated subject to the initial conditions that $a=2$, $b=0$ at $t=0$. The integration gives the model predictions amodel(j) and bmodel(j). The random search technique is used to determine optimal values for the rate constants based on minimization of S_A^2 and S_B^2. The following program fragment shows the method used to adjust k_I and k_{II} during the random search. The specific version shown is used to adjust k_I based on the minimization of S_A^2, and those instructions concerned with the minimization of S_B^2 appear as comments.

```
ssa = 0
ssb = 0
For j = 1 To Jdata
   ssa = ssa + (adata(j) - amodel(j))^2
   'ssb = ssb + (bdata(j) - bmodel(j))^2
Next
    If ssa < bestssa Then
    'If ssb < bestssb Then
```

```
    bestxk1 = xk1
    'bestxk2 = xk2
    bestssa = ssa
    'bestssb = ssb
End If
xka = bestxka + 0.005 * (0.5 - Rnd)
'xkb = bestxkb + 0.005 * (0.5 - Rnd)
```

The results are

Minimization method	k_I	k_{II}	σ_A	σ_B
Minimize S_B^2	1.028	2.543	0.01234	0.00543
Minimize S_A^2 and then S_B^2	1.016	2.536	0.01116	0.00554

There is little difference between the two methods in the current example since the data are of high quality. However, the sequential approach of first minimizing S_A^2 and then minimizing S_B^2 is somewhat better for this example and is preferred in general. Figure 7.4 shows the correlation. It is theoretically possible to fit both k_I and k_{II} by minimizing S_C^2, but this is prone to great error.

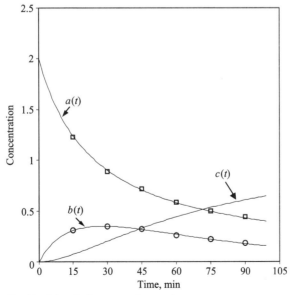

FIGURE 7.4 Combined data fit for consecutive reactions.

Competitive Reactions. The prototypical reactions are A → B and A → C. At least two of the three component concentrations should be measured and the material balance closed. Functional forms for the two reaction rates are assumed, and the parameters contained within these functional forms are estimated by minimizing an objective function of the form $w_A S_A^2 + w_B S_B^2 + w_C S_C^2$ where w_A, w_B, and w_C are positive weights that sum to 1. Weighting the three sums-of-squares equally has given good results when the rates for the two reactions are similar in magnitude.

7.1.4 Confounded Reactors

There are many attempts to extract kinetic information from pilot-plant or plant data. This may sound good to parsimonious management, but it is seldom a good alternative to doing the kinetic measurements under controlled conditions in the laboratory. Laboratory studies can usually approximate isothermal operation of an ideal reactor, while measurements on larger equipment will be confounded by heat transfer and mixing effects. The laboratory studies can cover a broader range of the experimental variables than is possible on the larger scale. An idealized process development sequence has the following steps:

1. Determine physical property and kinetic data from the literature or laboratory studies.
2. Combine these data with estimates of the transport parameters to model the desired full-scale plant.
3. Scale down the model to design a pilot plant that is scalable upward and that will address the most significant uncertainties in the model of the full-scale facility.
4. Operate the pilot plant to determine the uncertain parameters. These will usually involve mixing and heat transfer, not basic kinetics.
5. Revise the model and build the full-scale plant.

Ideally, measurements on a pilot- or full-scale plant can be based on known reaction kinetics. If the kinetics are unknown, experimental limitations will usually prevent their accurate determination. The following section describes how to make the best of a less-than-ideal situation.

A relatively simple example of a confounded reactor is a nonisothermal batch reactor where the assumption of perfect mixing is reasonable but the temperature varies with time or axial position. The experimental data are fit to a model using Equation (7.8), but the model now requires a heat balance to be solved simultaneously with the component balances. For a batch reactor,

$$\frac{d(V\rho H)}{dt} = -V \Delta H_R \mathscr{R} - UA_{ext}(T - T_{ext}) \tag{7.17}$$

FITTING RATE DATA AND USING THERMODYNAMICS

Equation (7.17) introduces a number of new parameters, although physical properties such as ΔH_R should be available. If all the parameters are all known with good accuracy, then the introduction of a heat balance merely requires that the two parameters k_0 and $E/R_g = T_{act}$ be used in place of each rate constant. Unfortunately, parameters such as UA_{ext} have $\pm 20\%$ error when calculated from standard correlations, and such errors are large enough to confound the kinetics experiments. As a practical matter, T_{out} should be measured as an experimental response that is used to help determine UA_{ext}. Even so, fitting the data can be extremely difficult. The sum-of-squares may have such a shallow minimum that essentially identical fits can be achieved over a broad range of parameter values.

Example 7.6: Suppose a liquid–solid, heterogeneously catalyzed reaction is conducted in a jacketed, batch vessel. The reaction is $A \to B$. The reactants are in the liquid phase, and the catalyst is present as a slurry. The adiabatic temperature rise for complete conversion is 50 K. The reactants are charged to the vessel at 298 K. The jacket temperature is held constant at 343 K throughout the reaction. The following data were measured:

t, h	$a(t)$	$T(t)$, K
0.4	0.967	313
0.8	0.887	327
1.0	0.816	333
1.2	0.719	339
1.4	0.581	345
1.6	0.423	352
1.8	0.254	358
2.2	0.059	362

where $a(t) = [A]/[A]_0$. Use these data to fit a rate expression of the form $\mathcal{R}_A = ka/(1 + k_A a)$.

Solution: The equations to be solved are

$$\frac{da}{dt} = -\mathcal{R}_A$$

and

$$\frac{dT}{dt} = 50 \mathcal{R}_A - U'(T - T_{ext})$$

where $\mathcal{R}_A = k_0 \exp(-T_{act}/T) a/(1 + k_A a)$. There are four adjustable constants. A least-squares minimization based on S_A^2 heads toward $k_A < 0$. Stopping the optimizer at $k_A \approx 0$ gives $k_0 = 5.37 \times 10^9 \, \text{h}^{-1}$, $T_{act} = 7618 \, \text{K}$, $k_A = 0.006$, and $U' = 0.818\,\text{h}^{-1}$. The standard deviations are $\sigma_A = 0.0017$ and $\sigma_T = 1.8\,\text{K}$. The results are given below:

Time	Experimental data		Fitted results		Error-free results	
t, h	$a(t)$	$T(t)$, K	$a(t)$	$T(t)$, K	$a(t)$	$T(t)$, K
0.4	0.967	313	0.968	312	0.970	314
0.8	0.887	327	0.887	324	0.889	327
1.0	0.816	333	0.816	330	0.817	333
1.2	0.719	339	0.717	337	0.716	340
1.4	0.581	345	0.585	344	0.584	346
1.6	0.423	352	0.422	351	0.422	353
1.8	0.254	358	0.254	358	0.256	359
2.2	0.059	362	0.060	362	0.061	361

The fit is excellent. The parameters have physically plausible values, and the residual standard deviations are reasonable compared to likely experimental error. If the data were from a real reactor, the fitted values would be perceived as close to the truth, and it would be concluded that the k_A term is negligible. In fact, the data are not from a real reactor but were contrived by adding random noise to a simulated process. The true parameters are $k_0 = 4 \times 10^9\,\text{h}^{-1}$, $T_{act} = 7500\,\text{K}$, $k_A = 0.5$, and $U' = 1\,\text{h}^{-1}$, and the k_A term has a significant effect on the reaction rate. When the error-free results are compared with the "data," the standard deviation is higher than that of the fitted model for concentration, $\sigma_A = 0.0024$, but lower for temperature, $\sigma_T = 0.9\,\text{K}$. A fit closer to the truth can be achieved by using a weighted sum of σ_A and σ_T as the objective function, but it would be hard to anticipate the proper weighting in advance.

Confounded reactors are likely to stay confounded. Data correlations can produce excellent fits and can be useful for predicting the response of the particular system on which the measurements were made to modest changes in operating conditions. They are unlikely to produce any fundamental information regarding the reaction rate, and have very limited utility in scaleup calculations.

7.2 THERMODYNAMICS OF CHEMICAL REACTIONS

Thermodynamics is a fundamental engineering science that has many applications to chemical reactor design. Here we give a summary of two important topics: determination of heat capacities and heats of reaction for inclusion in energy balances, and determination of free energies of reaction to calculate equilibrium compositions and to aid in the determination of reverse reaction rates. The treatment in this book is brief and is intended as a review. Details are available in any standard textbook on chemical engineering thermodynamics, e.g., Smith et al.[2] Tables 7.1 and 7.2 provide selected thermodynamic data for use in the examples and for general use in reaction engineering.

7.2.1 Terms in the Energy Balance

The design equations for a chemical reactor contain several parameters that are functions of temperature. Equation (7.17) applies to a nonisothermal batch reactor and is exemplary of the physical property variations that can be important even for ideal reactors. Note that the word "ideal" has three uses in this chapter. In connection with reactors, ideal refers to the quality of mixing in the vessel. Ideal batch reactors and CSTRs have perfect internal mixing. Ideal PFRs are perfectly mixed in the radial direction and have no mixing in the axial direction. These ideal reactors may be nonisothermal and may have physical properties that vary with temperature, pressure, and composition.

Ideal gases obey the ideal gas law, $PV = N_{total}R_g T$, and have internal energies that are a function of temperature alone. Ideal solutions have no enthalpy change upon mixing and have a special form for the entropy change upon mixing, $\Delta S_{mix} = R_g \Sigma x_A \ln x_A$, where x_A is the mole fraction of component A in the mixture. Ideal gases form ideal solutions. Some liquid mixtures approximate ideal solutions, but this is relatively uncommon.

Enthalpy. Enthalpy is calculated relative to a standard state that is normally chosen as $T_0 = 298.15\,\text{K} = 25°\text{C}$ and $P_0 = 1$ bar pressure. The change in enthalpy with pressure can usually be ignored. For extreme changes in pressure, use

$$\left(\frac{\partial H}{\partial P}\right)_T = V - T\left(\frac{\partial V}{\partial T}\right)_P = V(1 - \beta T) \qquad (7.18)$$

where β is the volumetric coefficient of thermal expansion. β can be evaluated from the equation of state for the material and is zero for an ideal gas. The standard state for gases is actually that for a hypothetical, ideal gas. Real gases are not perfectly ideal at 1 bar. Thus, H for a real gas at 298.15 K and 1 bar will not be exactly zero. The difference is usually negligible.

The change in enthalpy with respect to temperature is not negligible. It can be calculated for a pure component using the specific heat correlations like those in Table 7.1:

$$H = \int_{T_0}^{T} C_P\,dt = R_g\left[AT + \frac{BT^2}{2 \times 10^3} + \frac{CT^3}{3 \times 10^6} - \frac{10^5 D}{T}\right]_{T_0}^{T} \qquad (7.19)$$

TABLE 7.1 Heat Capacities at Low Pressures

		T_{max}	Std.	A	B	C	D
Gaseous alkanes							
Methane	CH_4	1500	4.217	1.702	9.081	−2.164	
Ethane	C_2H_6	1500	6.369	1.131	19.225	−5.561	
Propane	C_3H_8	1500	9.001	1.213	28.785	−8.824	
n-Butane	C_4H_{10}	1500	11.928	1.935	36.915	−11.402	
iso-Butane	C_4H_{10}	1500	11.901	1.677	37.853	−11.945	
n-Pentane	C_5H_{12}	1500	14.731	2.464	45.351	−14.111	
n-Hexane	C_6H_{14}	1500	17.550	3.025	53.722	−16.791	
n-Heptane	C_7H_{16}	1500	20.361	3.570	62.127	−19.486	
n-Octane	C_8H_{18}	1500	23.174	4.108	70.567	−22.208	
Gaseous alkenes							
Ethylene	C_2H_4	1500	5.325	1.424	14.394	−4.392	
Propylene	C_3H_6	1500	7.792	1.637	22.706	−6.915	
1-Butene	C_4H_8	1500	10.520	1.967	31.630	−9.873	
1-Pentene	C_5H_{10}	1500	13.437	2.691	39.753	−12.447	
1-Hexene	C_6H_{12}	1500	16.240	3.220	48.189	−15.157	
1-Heptene	C_7H_{14}	1500	19.053	3.768	56.588	−17.847	
1-Octene	C_8H_{16}	1500	21.868	4.324	64.960	−20.521	
Organic gases							
Acetaldehyde	C_2H_4O	1000	6.506	1.693	17.978	−6.158	
Acetylene	C_2H_2	1500	5.253	6.132	1.952		−1.299
Benzene	C_6H_6	1500	10.259	−0.206	39.064	−13.301	
1,3-Butadiene	C_4H_6	1500	10.720	2.734	26.786	−8.882	
Cyclohexane	C_6H_{12}	1500	13.121	3.876	63.249	−20.928	
Ethanol	C_2H_6O	1500	8.948	3.518	20.001	−6.002	
Ethylbenzene	C_8H_{10}	1500	15.993	1.124	55.380	−18.476	
Ethylene oxide	C_2H_4O	1000	5.784	0.385	23.463	−9.296	
Formaldehyde	CH_2O	1500	4.191	2.264	7.022	−1.877	
Methanol	CH_4O	1500	5.547	2.211	12.216	−3.450	
Styrene	C_8H_8	1500	15.534	2.050	50.192	−16.662	
Toluene	C_7H_8	1500	12.922	0.290	47.052	−15.716	
Inorganic gases							
Air		2000	3.509	3.355	0.575		−0.016
Ammonia	NH_3	1800	4.269	3.578	3.020		−0.186
Bromine	Br_2	3000	4.337	4.493	0.056		−0.154
Carbon monoxide	CO	2500	3.507	3.376	0.557		−0.031
Carbon dioxide	CO_2	2000	4.467	5.457	1.045		−1.157
Carbon disulfide	CS_2	1800	5.532	6.311	0.805		−0.906
Chlorine	Cl_2	3000	4.082	4.442	0.089		−0.344
Hydrogen	H_2	3000	3.468	3.249	0.422		0.083
Hydrogen sulfide	H_2S	2300	4.114	3.931	1.490		−0.232
Hydrogen chloride	HCl	2000	3.512	3.156	0.623		−0.151
Hydrogen cyanide	HCN	2500	4.326	4.736	1.359		−0.725
Nitrogen	N_2	2000	3.502	3.280	0.593		−0.040
Nitrous oxide	N_2O	2000	4.646	5.328	1.214		−0.928
Nitric oxide	NO	2000	3.590	3.387	0.629		−0.014
Nitrogen dioxide	NO_2	2000	4.447	4.982	1.195		−0.792

continued

TABLE 7.1 Continued

		T_{max}	Std.	A	B	C	D
Dinitrogen tetroxide	N_2O_4	2000	9.198	11.660	2.257		−2.787
Oxygen	O_2	2000	3.535	3.639	0.506		−0.227
Sulfur dioxide	SO_2	2000	4.796	5.699	0.801		−1.015
Sulfur trioxide	SO_3	2000	6.094	8.060	1.056		−2.028
Water	H_2O	2000	4.038	3.470	1.450		0.121
Liquids							
Ammonia	NH_3	373	9.718	22.626	−100.75	192.71	
Aniline	C_6H_7N	373	23.070	15.819	29.03	−15.80	
Benzene	C_6H_6	373	16.157	−0.747	67.96	−37.78	
1,3-Butadiene	C_4H_6	373	14.779	22.711	−87.96	205.79	
Carbon tetrachloride	CCl_4	373	15.751	21.155	−48.28	101.14	
Chlorobenzene	C_6H_5Cl	373	18.240	11.278	32.86	−31.90	
Chloroform	$CHCl_3$	373	13.806	19.215	−42.89	83.01	
Cyclohexane	C_6H_{12}	373	18.737	−9.048	141.38	−161.62	
Ethanol	C_2H_6O	373	13.444	33.866	−172.60	349.17	
Ethylene oxide	C_2H_4O	373	10.590	21.039	−86.41	172.28	
Methanol	CH_4O	373	9.798	13.431	−51.28	131.13	
n-Propanol	C_3H_8O	373	16.921	41.653	−210.32	427.20	
Sulfur trioxide	SO_3	373	30.408	−2.930	137.08	−84.73	
Toluene	C_7H_8	373	18.611	15.133	6.79	16.35	
Water	H_2O	373	9.069	8.712	1.25	−0.18	
Solids							
Carbon (graphite)	C	2000	1.026	1.771	0.771		−0.867
Sulfur (rhombic)	S	368	3.748	4.114	−1.728		−0.783

This table provides data for calculating molar heat capacities at low pressures according to the empirical formula

$$\frac{C_P}{R_g} = A + \frac{BT}{10^3} + \frac{CT^2}{10^6} + \frac{10^5 D}{T^2}$$

The column marked "Std." shows the calculated value of C_P/R_g at 298.15 K.
Source: Data selected from Smith, J. M., Van Ness, H. C., and Abbott, M. M., *Introduction to Chemical Engineering Thermodynamics*, 6th ed., McGraw-Hill, New York, 2001.

where the constants are given in Table 7.1. Note that these are molar heat capacities. For reactions involving a change of phase, Equation (7.19) must be modified to include the heat associated with the phase transition (e.g., a heat of vaporization). The enthalpy term in the heat balance applies to the entire reacting mixture, and thus heats of mixing may warrant inclusion. However, they are usually small compared with the heats of reaction and are generally ignored in reaction engineering calculations. The normal assumption is that

$$H = aH_A + bH_B + \cdots + iH_I = \sum_{\text{Species}} aH_A \qquad (7.20)$$

where the summation extends over all reactants and inerts.

TABLE 7.2 Standard Enthalpies and Gibbs Free Energies of Formation (Values are joules per mole of the substance formed)

		ΔH_F°	ΔG_F°
Gaseous alkanes			
Methane	CH_4	−74,520	−50,460
Ethane	C_2H_6	−83,820	−31,855
Propane	C_3H_8	−104,680	−24,290
n-Butane	C_4H_{10}	−125,790	−16,570
n-Pentane	C_5H_{12}	−146,760	−8,650
n-Hexane	C_6H_{14}	−166,920	150
n-Heptane	C_7H_{16}	−187,780	8,260
n-Octane	C_8H_{18}	−208,750	16,260
Gaseous alkenes			
Ethylene	C_2H_4	52,510	68,460
Propylene	C_3H_6	19,710	62,205
1-Butene	C_4H_8	−540	70,340
1-Pentene	C_5H_{10}	−21,820	78,410
1-Hexene	C_6H_{12}	−41,950	86,830
Other organic gases			
Acetaldehyde	C_2H_4O	−166,190	−128,860
Acetylene	C_2H_2	227,480	209,970
Benzene	C_6H_6	82,930	129,665
1,3-Butadiene	C_4H_6	109,240	149,795
Cyclohexane	C_6H_{12}	−123,140	31,920
Ethanol	C_2H_6O	−235,100	−168,490
Ethylbenzene	C_8H_{10}	29,920	130,890
Ethylene oxide	C_2H_4O	−52,630	−13,010
Formaldehyde	CH_2O	−108,570	−102,530
Methanol	CH_4O	−200,660	−161,960
Methylcyclohexane	C_7H_{14}	−154,770	27,480
Styrene	C_8H_8	147,360	213,900
Toluene	C_7H_8	50,170	122,050
Inorganic gases			
Ammonia	NH_3	−46,110	−16,450
Carbon dioxide	CO_2	−393,509	−394,359
Carbon monoxide	CO	−110,525	−137,169
Hydrogen chloride	HCl	−92,307	−95,299
Hydrogen cyanide	HCN	135,100	124,700
Hydrogen sulfide	H_2S	−20,630	−33,560
Nitrous oxide	N_2O	82,050	104,200
Nitric oxide	NO	90,250	86,550
Nitrogen dioxide	NO_2	33,180	51,310
Dinitrogen tetroxide	N_2O_4	9,160	97,540
Sulfur dioxide	SO_2	−296,830	−300,194
Sulfur trioxide	SO_3	−395,720	−371,060
Water	H_2O	−241,818	−228,572

continued

TABLE 7.2 Continued

		ΔH_F°	ΔG_F°
Organic liquids			
Acetic acid	$C_2H_4O_2$	−484,500	−389,900
Benzene	C_6H_6	49,080	124,520
Cyclohexane	C_6H_{12}	−156,230	26,850
Ethanol	C_2H_6O	−277,690	−174,780
Ethylene glycol	$C_2H_6O_2$	−454,800	−323,080
Ethylene oxide	C_2H_4O	−52,630	−13,010
Methanol	CH_4O	−238,660	−166,270
Methylcyclohexane	C_7H_{14}	−190,160	20,560
Toluene	C_7H_8	12,180	113,630
Other liquids			
Nitric acid	HNO_3	−174,100	−80,710
Sulfuric acid	H_2SO_4	−813,989	−690,003
Water	H_2O	−285,830	−237,129

Source: Data selected from Smith, J. M., Van Ness, H. C., and Abbott, M. M., *Introduction to Chemical Engineering Thermodynamics*, 6th ed., McGraw-Hill, New York, 2001.

Heats of Reaction. Chemical reactions absorb or liberate energy, usually in the form of heat. The heat of reaction, ΔH_R, is defined as the amount of energy absorbed or liberated if the reaction goes to completion at a fixed temperature and pressure. When $\Delta H_R > 0$, energy is absorbed and the reaction is said to be *endothermic*. When $\Delta H_R < 0$, energy is liberated and the reaction is said to be *exothermic*. The magnitude of ΔH_R depends on the temperature and pressure of the reaction and on the phases (e.g., gas, liquid, solid) of the various components. It also depends on an arbitrary constant multiplier in the stoichiometric equation.

Example 7.7: The reaction of hydrogen and oxygen is highly exothermic. At 298.15 K and 1 bar,

$$H_2(g) + \tfrac{1}{2}O_2(g) \rightarrow H_2O(g) \qquad \Delta H_R = -241,818\,J \qquad (I)$$

Alternatively,

$$2H_2(g) + O_2(g) \rightarrow 2H_2O(g) \qquad \Delta H_R = -483,636\,J \qquad (II)$$

The reverse reaction, the decomposition of water is highly endothermic:

$$H_2O(g) \rightarrow H_2(g) + \tfrac{1}{2}O_2(g) \qquad \Delta H_R = +241,818\,J \qquad (III)$$

$$H_2O(g) \rightarrow 2H_2(g) + O_2(g) \qquad \Delta H_R = +483,636\,J \qquad (IV)$$

These equations differ by constant factors, but all the heats of reaction become equal when expressed in joules per mole of water formed, −241,818. They are also equal when expressed in joules per mole of oxygen formed, +483,636, or in joules per mole of hydrogen formed, +241,818. Any of these values can be used provided $\mathscr{R}$ is the rate at which a reaction product with a stoichiometric coefficient of +1 is being produced. Thus, $\mathscr{R}_I$ should be the rate at which water is being formed; $\mathscr{R}_{III}$ should be the rate at which hydrogen is being produced; and $\mathscr{R}_{IV}$ should be the rate at which oxygen is being produced. Even $\mathscr{R}_{II}$ can be made to fit the scheme, but it must be the rate at which a hypothetical component is being formed.

Suppose ΔH_R for Reaction (I) was measured in a calorimeter. Hydrogen and oxygen were charged at 298.15 K and 1 bar. The reaction occurred, the system was restored to 298.15 K and 1 bar, but the product water was not condensed. This gives the heat of reaction for Reaction (I). Had the water been condensed, the measured exothermicity would have been larger:

$$H_2(g) + \tfrac{1}{2}O_2(g) \to H_2O(l) \qquad \Delta H_R = -285{,}830 \text{ J} \qquad \text{(V)}$$

Reactions (I) and (V) differ by the heat of vaporization:

$$H_2O(g) \to H_2O(l) \qquad \Delta H_R = +44{,}012 \text{ J} \qquad \text{(VI)}$$

Reactions (V) and (VI) can obviously be summed to give Reaction (I).

The heats of reaction associated with stoichiometric equations are additive just as the equations themselves are additive. Some authors illustrate this fact by treating the evolved heat as a product of the reaction. Thus, they write

$$H_2(g) + \tfrac{1}{2}O_2(g) \to H_2O(g) + 241{,}818 \text{ J}$$

This is beautifully correct in terms of the physics, and is a very useful way to include heats of reaction when summing chemical equations. It is confused by the thermodynamic convention that heat is positive when absorbed by the system. The convention may have been logical for mechanical engineers concerned with heat engines, but chemists and chemical engineers would have chosen the opposite convention. Once a convention is adopted, it is almost impossible to change. Electrical engineers still pretend that current flows from positive to negative.

The additive nature of stoichiometric equations and heats of reactions allows the tabulation of ΔH_R for a relatively few canonical reactions that can be algebraically summed to give ΔH_R for a reaction of interest. The canonical reactions represent the formation of compounds directly from their elements. The participating species in these reactions are the elements as reactants and a single chemical compound as the product. The heats of reactions for these mainly hypothetical reactions are called *heats of formation*. Table 7.2 gives *standard heats of formation* ΔH_F° for a variety of compounds. The reacting elements and the product compound are all assumed to be at standard conditions of $T_0 = 298.15$ K and $P_0 = 1$ bar. In addition to directly tabulated data, heats of formation can be calculated from heats of combustion and can be estimated using group contribution theory.

FITTING RATE DATA AND USING THERMODYNAMICS

Example 7.8: Determine ΔH_R for the dehydrogenation of ethylbenzene to styrene at 298.15 K and 1 bar.

Solution: Table 7.2 gives ΔH_F° for styrene at 298.15 K. The formation reaction is

$$8\text{C(graphite)} + 4\text{H}_2(\text{g}) \rightarrow \text{Styrene(g)} \qquad \Delta H_R = 147{,}360 \text{ J} \qquad (7.21)$$

For ethylbenzene, $\Delta H_F^\circ = 29{,}920$ J, but we write the stoichiometric equation using a multiplier of -1. Thus,

$$-8\text{C(graphite)} - 5\text{H}_2(\text{g}) \rightarrow -\text{Ethylbenzene(g)} \qquad \Delta H_R = -29{,}920 \text{ J} \qquad (7.22)$$

The stoichiometry and heats of reaction in Equations (7.21) and (7.22) are algebraically summed to give

$$\text{Ethylbenzene(g)} \rightarrow \text{Styrene(g)} + \text{H}_2(\text{g}) \qquad \Delta H_R = 117{,}440 \text{ J} \qquad (7.23)$$

so that $\Delta H_R = 117{,}440$ J per mole of styrene produced. Note that the species participating in Equation (7.23) are in their standard states since standard heats of formation were used in Equations (7.21) and (7.22). Thus, we have obtained the *standard heat of reaction*, ΔH_R°, at $T_0 = 298.15$ K and $P_0 = 1$ bar.

It does not matter that there is no known catalyst that can accomplish the reaction in Equation (7.21) directly. Heats of reaction, including heats of formation, depend on conditions before and after the reaction but not on the specific reaction path. Thus, one might imagine a very complicated chemistry that starts at standard conditions, goes through an arbitrary trajectory of temperature and pressure, returns to standard conditions, and has Equation (7.21) as its overall effect. $\Delta H_F^\circ = +147{,}360$ J/mol of styrene formed is the net heat effect associated with this overall reaction.

The reaction in Equation (7.23) is feasible as written but certainly not at temperatures as low as 25°C, and it must be adjusted for more realistic conditions. The adjustment for temperature uses

$$\left(\frac{\partial \Delta H_R}{\partial T}\right)_P = \sum_{\text{Species}} \nu_A \left[\left(\frac{\partial H}{\partial T}\right)_P\right]_A = \sum_{\text{Species}} \nu_A (C_P)_A = \Delta C_P \qquad (7.24)$$

So that the corrected heat of reaction is

$$\Delta H_R = \Delta H_R^\circ + \int_{T_0}^{T} \Delta C_P \, dT = \Delta H_R^\circ + \sum_{\text{Species}} \nu_A H_A \qquad (7.25)$$

The summations in these equations include only those chemical species that directly participate in the reaction, and the weighting is by stoichiometric coefficient. Compare this with Equation (7.20) where the summation includes everything in the reactor and the weighting is by concentration. Equation (7.25) is used to determine the heat generated by the reaction. Equation (7.20) is used to determine how the generated heat affects the entire reacting mass.

A pressure adjustment to the heat of reaction may be needed at high pressures. The adjustment is based on

$$\left(\frac{\partial \Delta H_R}{\partial P}\right)_T = \sum_{Species} \nu_A \left[\left(\frac{\partial H}{\partial P}\right)_T\right]_A = \Delta\left(\frac{\partial H}{\partial P}\right)_T \quad (7.26)$$

See Equation (7.18) to evaluate this expression.

Example 7.9: Determine ΔH_R for the ethylbenzene dehydrogenation reaction at 973 K and 0.5 atm.

Solution: From Example 7.8, $\Delta H_R^\circ = 117{,}440$ J at $T_0 = 298.15$ K. We need to calculate ΔC_P. Using Equation (7.24),

$$\Delta C_P = (C_P)_{styrene} + (C_P)_{hydrogen} - (C_P)_{ethylbenzene}$$

The data of Table 7.1 give

$$\frac{\Delta C_P}{R_g} = 4.175 - \frac{4.766T}{10^3} + \frac{1.814T^2}{10^6} + \frac{8300}{T^2}$$

From this,

$$\Delta H_R = \Delta H_R^\circ + \int_{T_0}^{T} \Delta C_P dT = 117{,}440 + 8.314$$

$$\times \left[4.175T - \frac{4.766T^2}{2 \times 10^3} + \frac{1.814T^3}{3 \times 10^6} - \frac{8300}{T}\right]_{T_0}^{T} \quad (7.27)$$

Setting $T = 973$ K gives $\Delta H_R = 117{,}440 + 11{,}090 = 128{,}530$ J. The temperature is high and the pressure is low relative to critical conditions for all three components. Thus, an ideal gas assumption is reasonable, and the pressure change from 1 bar to 0.5 atm does not affect the heat of reaction.

7.2.2 Reaction Equilibria

Many reactions show appreciable reversibility. This section introduces thermodynamic methods for estimating equilibrium compositions from free energies of reaction, and relates these methods to the kinetic approach where the equilibrium composition is found by equating the forward and reverse reaction rates.

Equilibrium Constants. We begin with the kinetic approach. Refer to Equations (1.14) and (1.15) and rewrite (1.15) as

$$K_{kinetic} = \prod_{Species} [A]^{v_A} = \prod_{Species} a^{v_A} \qquad (7.28)$$

This is the expected form of the *kinetic equilibrium constant* for elementary reactions. $K_{kinetic}$ is a function of the temperature and pressure at which the reaction is conducted.

In principle, Equation (7.28) is determined by equating the rates of the forward and reverse reactions. In practice, the usual method for determining $K_{kinetic}$ is to run batch reactions to completion. If different starting concentrations give the same value for $K_{kinetic}$, the functional form for Equation (7.28) is justified. Values for chemical equilibrium constants are routinely reported in the literature for specific reactions but are seldom compiled because they are hard to generalize.

The reactant mixture may be so nonideal that Equation (7.28) is inadequate. The rigorous thermodynamic approach is to replace the concentrations in Equation (7.28) with chemical activities. This leads to the *thermodynamic equilibrium constant*:

$$K_{thermo} = \prod_{Species} \left[\frac{\hat{f}_A}{f_A^\circ}\right]^{v_A} = \exp\left(\frac{-\Delta G_R^\circ}{R_g T}\right) \qquad (7.29)$$

where $\hat{f}_A$ is the fugacity of component A in the mixture, f_A° is the fugacity of pure component A at the temperature and pressure of the mixture, and ΔG_R° is the *standard free energy of reaction* at the temperature of the mixture. The thermodynamic equilibrium constant is a function of temperature but not of pressure. A form of Equation (7.29) suitable for gases is

$$K_{thermo} = \left[\frac{P}{P_0}\right]^v \prod_{Species} [y_A \hat{\phi}_A]^{v_A} = \exp\left(\frac{-\Delta G_R^\circ}{R_g T}\right) \qquad (7.30)$$

where $v = \Sigma v_A$; y_A is the mole fraction of component A, $\hat{\phi}_A$ is its fugacity coefficient and P_0 is the pressure used to determine the *standard free energy of formation* ΔG_F°. Values for ΔG_F° are given in Table 7.2. They can be algebraically summed, just like heats of formation, to obtain ΔG_R° for reactions of interest.

Example 7.10: Determine ΔG_R° for the dehydrogenation of ethylbenzene to styrene at 298.15 K.

Solution: Table 7.2 gives ΔG_F° for styrene at 298.15 K. The formation reaction is

$$8C(graphite) + 4H_2(g) \rightarrow Styrene(g) \qquad \Delta G_R - 213,900 \text{ J}$$

For ethylbenzene,

$$-8\text{C(graphite)} - 5\text{H}_2(g) \rightarrow -\text{Ethylbenzene(g)} \qquad \Delta G_R = -130{,}890 \text{ J}$$

These equations are summed to give

$$\text{Ethylbenzene(g)} \rightarrow \text{Styrene(g)} + \text{H}_2(g) \qquad \Delta G_R = 83{,}010 \text{ J}$$

so that $\Delta G_R = 83{,}010$ J per mole of styrene produced. Since the species are in their standard states, we have obtained ΔG_R°.

The fugacity coefficients in Equation (7.29) can be calculated from pressure-volume-temperature data for the mixture or from generalized correlations. It is frequently possible to assume ideal gas behavior so that $\hat{\phi}_A = 1$ for each component. Then Equation (7.29) becomes

$$K_{thermo} = \left[\frac{P}{P_0}\right]^\nu \prod_{Species} [y_A]^{\nu_A} = \exp\left(\frac{-\Delta G_R^\circ}{R_g T}\right) \qquad (7.31)$$

For incompressible liquids or solids, the counterpart to Equation (7.30) is

$$K_{thermo} = \exp\left[\frac{P - P_0}{R_g T} \sum_{Species} \nu_A V_A\right] \prod_{Species} [x_A \gamma_A]^{\nu_A} = \exp\left(\frac{-\Delta G_F^\circ}{R_g T}\right) \qquad (7.32)$$

where x_A is the mole fraction of component A, V_A is its molar volume, and γ_A is its activity coefficient in the mixture. Except for high pressures, the exponential term containing $P - P_0$ is near unity. If the mixture is an ideal solution, $\gamma_A = 1$ and

$$K_{thermo} = \prod_{Species} [x_A]^{\nu_A} = \exp\left(\frac{-\Delta G_F^\circ}{R_g T}\right) \qquad (7.33)$$

As previously noted, the equilibrium constant is independent of pressure as is ΔG_R°. Equation (7.33) applies to ideal solutions of incompressible materials and has no pressure dependence. Equation (7.31) applies to ideal gas mixtures and has the explicit pressure dependence of the P/P_0 term when there is a change in the number of moles upon reaction, $\nu \neq 0$. The temperature dependence of the thermodynamic equilibrium constant is given by

$$\frac{d \ln K_{thermo}}{dT} = \frac{\Delta H_R}{R_g T^2} \qquad (7.34)$$

This can be integrated to give

$$K_{thermo} = K_0 K_1 K_2 K_3 =$$

$$\exp\left(\frac{-\Delta G_R^\circ}{R_g T_0}\right) \exp\left[\frac{\Delta H_R^\circ}{R_g T_0}\left(1 - \frac{T_0}{T}\right)\right] \exp\left[-\frac{1}{T}\int_{T_0}^{T} \frac{\Delta C_P}{R_g} dT\right] \exp\left[\int_{T_0}^{T} \frac{\Delta C_P dT}{R_g T}\right]$$

(7.35)

Equation (7.35) is used to find K_{thermo} as a function of reaction temperature T. Only the first two factors are important when $\Delta C_P^\circ \approx 0$, as is frequently the case. Then $\ln(K_{thermo})$ will be a linear function of T^{-1}. This fact justifies Figure 7.5, which plots the equilibrium constant as a linear function of temperature for some gas-phase reactions.

Reconciliation of Equilibrium Constants. The two approaches to determining equilibrium constants are consistent for ideal gases and ideal solutions of incompressible materials. For a reaction involving ideal gases, Equation (7.29) becomes

$$K_{thermo} = \left[\frac{P}{P_0}\right]^\nu \rho_{molar}^{-\nu} \prod_{Species} [A]^{\nu_A} = \left[\frac{P}{P_0}\right]^\nu \rho_{molar}^{-\nu} K_{kinetic} = \left[\frac{R_g T}{P_0}\right]^\nu K_{kinetic} \quad (7.36)$$

and the explicit pressure dependence vanishes. Since K_{thermo} is independent of pressure, so is $K_{kinetic}$ for an ideal gas mixture.

For ideal solutions of incompressible materials,

$$K_{thermo} = \rho_{molar}^{-\nu} \prod_{Species} [A]^{\nu_A} = \rho_{molar}^{-\nu} K_{kinetic} = \exp\left(\frac{-\Delta G_F^\circ}{R_g T}\right) \quad (7.37)$$

which is also independent of pressure.

For nonideal solutions, the thermodynamic equilibrium constant, as given by Equation (7.29), is fundamental and $K_{kinetic}$ should be reconciled to it even though the exponents in Equation (7.28) may be different than the stoichiometric coefficients. As a practical matter, the equilibrium composition of nonideal solutions is usually found by running reactions to completion rather than by thermodynamic calculations, but they can also be predicted using generalized correlations.

Reverse Reaction Rates. Suppose that the kinetic equilibrium constant is known both in terms of its numerical value and the exponents in Equation (7.28). If the solution is ideal and the reaction is elementary, then the exponents in the reaction rate—i.e., the exponents in Equation (1.14)—should be the stoichiometric coefficients for the reaction, and $K_{kinetic}$ should be the ratio of

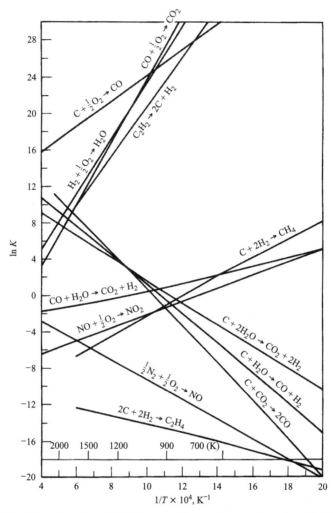

FIGURE 7.5 Thermodynamic equilibrium constant for gas-phase reactions. (From Smith, J. M. and Van Ness, H. C., *Introduction to Chemical Engineering Thermodynamics*, 4th Ed., McGraw-Hill, New York, 1986.)

forward-to-reverse rate constants as in Equation (1.15). If the reaction is complex, the kinetic equilibrium constant may still have the ideal form of Equation (7.28). The appropriateness of Equation (7.28) is based on the ideality of the mixture at equilibrium and not on the kinetic path by which equilibrium was reached. However, the forward and reverse reaction rates must still be equal at equilibrium, and this fact dictates the functional form of the rate expression near the equilibrium point.

Example 7.11: Suppose A ⇔ B + C at high temperatures and low pressures in the gas phase. The reaction rate is assumed to have the form

$$\mathscr{R} = k_f a^n - \mathscr{R}_r$$

where the various constants are to be determined experimentally. The kinetic equilibrium constant as defined by Equation (7.28) is

$$K_{kinetic} = \frac{bc}{a}$$

and has been measured to be $50 \, \text{mol/m}^3$ at 1 atm pressure and 550 K. Find the appropriate functional form for the overall rate equation in the vicinity of the equilibrium point as a function of temperature, pressure, and composition

Solution: Assume the reverse reaction has the form $\mathscr{R}_r = k_r \, a^m b^r c^s$. Setting the overall reaction rate equal to zero at the equilibrium point gives a second expression for $K_{kinetic}$:

$$K_{kinetic} = \frac{k_f}{k_r} = \frac{a^m b^r c^s}{a^n}$$

Equating the two expressions for $K_{kinetic}$ gives $m = n-1$ and $r = s = 1$. Also, $k_r = k_f K_{kinetic}$. Thus,

$$\mathscr{R} = k_f \left[a^n - \frac{a^{n-1} bc}{K_{kinetic}} \right]$$

This is the required form with $K_{kinetic} = 50 \, \text{mol/m}^3$ at 1 atm and 550 K. According to Equation (7.36), $K_{kinetic}$ is a function of temperature but not of pressure. (This does not mean that the equilibrium composition is independent of pressure. See Example 7.12.) To evaluate the temperature dependence, it is useful to replace $K_{kinetic}$ with K_{thermo}. For $\nu = 1$:

$$\mathscr{R} = k_f \left[a^n - \frac{R_g T}{P_o K_{thermo}} a^{n-1} bc \right]. \tag{7.38}$$

Equation (7.35) is used to find K_{thermo} as a function of temperature. Since $K_{kinetic}$ was given, and K_{thermo} can be calculated from it, Equation 7.38 contains only n and k_f as adjustable constants, although k_f can be divided between k_0 and T_{act} if measurements are made at several temperatures.

Example 7.11 showed how reaction rates can be adjusted to account for reversibility. The method uses a single constant, $K_{kinetic}$ or K_{thermo}, and is rigorous for both the forward and reverse rates when the reactions are elementary. For complex reactions with fitted rate equations, the method should produce good results provided the reaction always starts on the same side of equilibrium.

A separate fitting exercise and a separate rate expression are needed for reactions starting on the other side of equilibrium.

Equation (7.28) may not provide a good fit for the equilibrium data if the equilibrium mixture is nonideal. Suppose that the proper form for $K_{kinetic}$ is determined through extensive experimentation or by using thermodynamic correlations. It could be a version of Equation (7.28) with exponents different from the stoichiometric coefficients, or it may be a different functional form. Whatever the form, it is possible to force the reverse rate to be consistent with the equilibrium constant, and this is recommended whenever the reaction shows appreciable reversibility.

Equilibrium Compositions for Single Reactions. We turn now to the problem of calculating the equilibrium composition for a single, homogeneous reaction. The most direct way of estimating equilibrium compositions is by simulating the reaction. Set the desired initial conditions and simulate an isothermal, constant-pressure, batch reaction. If the simulation is accurate, a real reaction could follow the same trajectory of composition versus time to approach equilibrium, but an accurate simulation is unnecessary. The solution can use the method of false transients. The rate equation must have a functional form consistent with the functional form of K_{thermo}; e.g., Equation (7.38). The time scale is unimportant and even the functional forms for the forward and reverse reactions have some latitude, as will be illustrated in the following example.

Example 7.12: Use the method of false transients to determine equilibrium concentrations for the reaction of Example 7.11. Specifically, determine the equilibrium mole fraction of component A at $T = 550\,\text{K}$ as a function of pressure, given that the reaction begins with pure A.

Solution: The obvious way to solve this problem is to choose a pressure, calculate a_0 using the ideal gas law, and then conduct a batch reaction at constant T and P. Equation (7.38) gives the reaction rate. Any reasonable values for n and k_f can be used. Since there is a change in the number of moles upon reaction, a variable-volume reactor is needed. A straightforward but messy approach uses the methodology of Section 2.6 and solves component balances in terms of the number of moles, N_A, N_B, and N_C.

A simpler method arbitrarily picks values for a_0 and reacts this material in a batch reactor at constant V and T. When the reaction is complete, P is calculated from the molar density of the equilibrium mixture. As an example, set $a_0 = 22.2$ ($P = 1\,\text{atm}$) and react to completion. The long-time results from integrating the constant-volume batch equations are $a = 5.53$, $b = c = 16.63$, $\rho_{molar} = 38.79\,\text{mol/m}^3$, and $y_A = 0.143$. The pressure at equilibrium is 1.75 atm.

The curve shown in Figure 7.6 is produced, whichever method is used. The curve is independent of n and k_f in Equation (7.38).

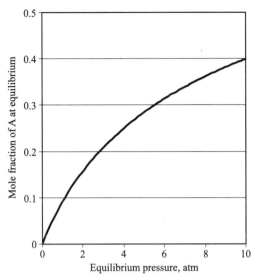

FIGURE 7.6 Equilibrium concentrations calculated by the method of false transients for a non-elementary reaction.

The reaction coordinate defined in Section 2.8 provides an algebraic method for calculating equilibrium concentrations. For a single reaction,

$$N_A = (N_A)_0 + \nu_A \varepsilon \tag{7.39}$$

and mole fractions are given by

$$y_A = \frac{N_A}{N_0 + \nu \varepsilon} = \frac{(N_A)_0 + \nu_A \varepsilon}{N_0 + \nu \varepsilon} \tag{7.40}$$

Suppose the numerical value of the thermodynamic equilibrium constant is known, say from the free energy of formation. Then Equation (7.40) is substituted into Equation (7.31) and the result is solved for ε.

Example 7.13: Use the reaction coordinate method to determine equilibrium concentrations for the reaction of Example 7.11. Specifically, determine the equilibrium mole fraction of component A at $T = 550\,K$ as a function of pressure, given that the reaction begins with pure A.

Solution: The kinetic equilibrium constant is $50\,\text{mol/m}^3$. It is converted to mole fraction form using

$$\prod_{\text{Species}} [y_A]^{\nu_A} = \rho_{molar}^{-\nu} K_{kinetic} = \frac{\prod\limits_{\text{Species}} [a]^{\nu_A}}{\left[\dfrac{P}{R_g T}\right]^{\nu}} \tag{7.41}$$

For the reaction at hand,

$$\frac{y_B y_C}{y_A} = \frac{[(N_B)_0 + \varepsilon][(N_C)_0 + \varepsilon]}{[(N_A)_0 - \varepsilon][N_0 + \varepsilon]} = 50 \times 8.205 \times 10^{-5} \times 550/P = 2.256/P$$

where P is in atmospheres. This equation is a quadratic in ε that has only one root in the physically realistic range of $-1 \leq \varepsilon \leq 1$. The root depends on the pressure and the relative values for N_A, N_B, and N_C. For a feed of pure A, set $N_A = 1$ and $N_B = N_C = 0$. Solution gives

$$\varepsilon = \sqrt{\frac{2.256}{P + 2.256}}$$

Set $P = 1.75$ atm. Then $\varepsilon = 0.750$ and $y_A = 0.143$ in agreement with Example 7.12.

Examples 7.12 and 7.13 treated the case where the kinetic equilibrium constant had been determined experimentally. The next two examples illustrate the case where the thermodynamic equilibrium constant is estimated from tabulated data.

Example 7.14: Estimate the equilibrium composition of the ethylbenzene dehydrogenation reaction at 298.15 K and 0.5 atm. Consider two cases:

1. The initial composition is pure ethylbenzene.
2. The initial composition is 1 mol each of ethylbenzene and styrene and 0.5 mol of hydrogen.

Solution: Example 7.10 found $\Delta G_R = 83{,}010$ J. Equation (7.29) gives $K_{thermo} = 2.8 \times 10^{-15}$ so that equilibrium at 298.15 K overwhelmingly favors ethylbenzene. Suppose the ideal gas assumption is not too bad, even at this low temperature ($T_c = 617$ K for ethylbenzene). The pressure is 0.5066 bar and $v = 1$. The reaction has the form A → B + C so the reaction coordinate formulation is similar to that in Example 7.13. When the feed is pure ethylbenzene, Equation (7.31) becomes

$$2.86 \times 10^{-15} = \left[\frac{0.5066}{1}\right] \frac{y_{H_2} y_{styrene}}{y_{ethylbenzene}} = 0.5066 \frac{\varepsilon^2}{(1-\varepsilon)(1+\varepsilon)}$$

Solution gives $\varepsilon = 7.5 \times 10^{-8}$. The equilibrium mole fractions are $y_{ethylbenzene} \approx 1$ and $y_{styrene} = y_{hydrogen} = 7.5 \times 10^{-8}$.

The solution for Case 1 is obtained from

$$2.8 \times 10^{-15} = \left[\frac{0.5066}{1}\right] \frac{y_{H_2} y_{styrene}}{y_{ethylbenzene}} = 0.5066 \frac{(1+\varepsilon)(0.5+\varepsilon)}{(1-\varepsilon)(2.5+\varepsilon)}$$

FITTING RATE DATA AND USING THERMODYNAMICS **243**

Solution of the quadratic gives $\varepsilon \approx -0.5$ so that $y_{ethylbenzene} \approx 0.75$, $y_{styrene} \approx 0.25$, and $y_{hydrogen} \approx 0$. The equilibrium is shifted so strongly toward ethylbenzene that essentially all the hydrogen is used to hydrogenate styrene.

Example 7.15: Estimate the equilibrium composition from the ethylbenzene dehydrogenation reaction at 973 K and 0.5 atm. The starting composition is pure ethylbenzene.

Solution: This problem illustrates the adjustment of K_{thermo} for temperature. Equation (7.35) expresses it as the product of four factors. The results in Examples 7.10 and 7.11 are used to evaluate these factors.

$$K_0 = \exp\left(\frac{-\Delta G_R^\circ}{R_g T_0}\right) = \exp\left(\frac{-83{,}010}{8.314 T_0}\right) = 2.86 \times 10^{-15}$$

$$K_1 = \exp\left[\frac{\Delta H_R^\circ}{R_g T_0}\left(1 - \frac{T_0}{T}\right)\right] = \exp\left[\frac{117{,}440}{8.314 T_0}\left(1 - \frac{T_0}{T}\right)\right] = 1.87 \times 10^{14}$$

$$K_2 = \exp\left[-\frac{1}{R_g T}\int_{T_0}^{T} \Delta C_P \, dt\right] = \exp\left[\frac{\Delta H_R^\circ - \Delta H_R}{8.314 T}\right] = 0.264$$

$$K_3 = \exp\left[\int_{T_0}^{T}\frac{\Delta C_P dT}{R_g T}\right] = \exp\left[4.175 \ln T - \frac{4.766 T}{10^3} + \frac{1.814 T^2}{2\times 10^6} - \frac{8300}{2T^2}\right]_{T_0}^{T} = 12.7$$

and $K_{thermo} = K_0 K_1 K_2 K_3 = 1.72$. Proceeding as in Example 7.14, Case 1,

$$1.72 = \left[\frac{0.5066}{1}\right]\frac{y_{H_2} y_{styrene}}{y_{ethylbenzene}} = 0.5066 \frac{\varepsilon^2}{(1-\varepsilon)(1+\varepsilon)}$$

Solution gives $\varepsilon = 0.879$. The equilibrium mole fractions are $y_{ethylbenzene} = 0.064$ and $y_{styrene} = y_{hydrogen} = 0.468$.

Example 7.16: Pure ethylbenzene is contacted at 973 K with a 9:1 molar ratio of steam and a small amount of a dehydrogenation catalyst. The reaction rate has the form

$$A \underset{k_r}{\overset{k_f}{\rightleftharpoons}} B + C$$

where $k_f = k_0 \exp(-T_{act}/T) = 160{,}000 \exp(-9000/T)$ s^{-1} and k_r is determined from the equilibrium relationship according to Equation (7.38). The mixture is charged at an initial pressure of 0.1 bar to an adiabatic, constant-volume, batch reactor. The steam is inert and the thermal mass of the catalyst can be neglected. Calculate the reaction trajectory. Do not assume constant physical properties.

Solution: A rigorous treatment of a reversible reaction with variable physical properties is fairly complicated. The present example involves just two ODEs: one for composition and one for enthalpy. Pressure is a dependent variable. If the rate constants are accurate, the solution will give the actual reaction trajectory (temperature, pressure, and composition as a function of time). If k_0 and T_{act} are wrong, the long-time solution will still approach equilibrium. The solution is then an application of the method of false transients.

An Excel macro is given in Appendix 7.2, and some results are shown in Figure 7.7. The macro is specific to the example reaction with $\nu = +1$ but can be generalized to other reactions. Components of the macro illustrate many of the previous examples. Specific heats and enthalpies are calculated analytically using the functional form of Equation (7.19) and the data in Tables 7.1 and 7.2. The main computational loop begins with the estimation of K_{thermo} using the methodology of Example (7.15).

The equilibrium composition corresponding to instantaneous values of T and P is estimated using the methodology of Example 7.13. These calculations are included as a point of interest. They are not needed to find the reaction trajectory. Results are reported as the mole fraction of styrene in the organic mixture of styrene plus ethylbenzene. The initial value, corresponding to $T = 973$ K and $P = 0.1$ bar, is 0.995. This equilibrium value gradually declines, primarily due to the change in temperature. The final value is 0.889, which is closely approximated by the long-time solution to the batch reactor equations.

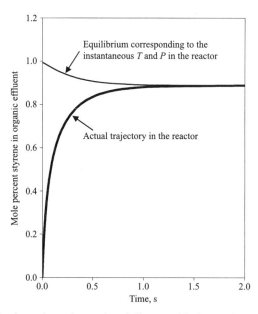

FIGURE 7.7 Batch reaction trajectory for ethylbenzene dehydrogenation.

The kinetic equilibrium constant is estimated from the thermodynamic equilibrium constant using Equation (7.36). The reaction rate is calculated and compositions are marched ahead by one time step. The energy balance is then used to march enthalpy ahead by one step. The energy balance in Chapter 5 used a mass basis for heat capacities and enthalpies. A molar basis is more suitable for the current problem. The molar counterpart of Equation (5.18) is

$$\frac{d(V \rho_{molar} H)}{dt} = -V \Delta H_R \mathscr{R} - UA_{ext}(T - T_{ext}) \qquad (7.42)$$

where $U = 0$ in the current example and H is the enthalpy per mole of the reaction mixture:

$$H = \int_{T_0}^{T} (C_p)_{mix} dT' \qquad (7.43)$$

The quantity $V \rho_{molar}$ is a not constant since there is a change in moles upon reaction, $\nu = 1$. Expanding the derivative gives

$$\frac{d(V\rho_{molar}H)}{dt} = V\rho_{molar}\frac{dH}{dt} + H\frac{d(V\rho_{molar})}{dt} = \left[V\rho_{molar}\frac{dH}{dT} + H\frac{d(V\rho_{molar})}{dT}\right]\frac{dT}{dt}$$

The dH/dT term is evaluated by differentiating Equation (7.43) with respect to the upper limit of the integral. This gives

$$\left[\rho_{molar}(C_P)_{mix} + H\frac{d\rho_{molar}}{dT}\right]\frac{dT}{dt} = -\Delta H_R \mathscr{R} - \frac{UA_{ext}(T - T_{ext})}{V} \qquad (7.44)$$

This result is perfectly general for a constant-volume reactor. It continues to apply when ρ, C_P, and H are expressed in mass units, as is normally the case for liquid systems. The current example has a high level of inerts so that the molar density shows little variation. The approximate heat balance

$$\frac{dT}{dt} = \frac{-\Delta H_R \mathscr{R}}{\rho_{molar}(C_P)_{mix}} - \frac{UA_{ext}(T - T_{ext})}{V\rho_{molar}(C_P)_{mix}} \qquad (7.45)$$

gives a result that is essentially identical to using Equation (7.42) to march the composite variable $V\rho_{molar}H$.

Equilibrium Compositions for Multiple Reactions. When there are two or more independent reactions, Equation (7.29) is written for each reaction:

$$(K_{thermo})_I = \exp\left[\frac{(\Delta G_R^\circ)_I}{R_g T}\right]$$

$$(K_{thermo})_{II} = \exp\left[\frac{(\Delta G_R^\circ)_{II}}{R_g T}\right] \qquad (7.46)$$

$$\vdots \qquad \vdots$$

so that there are M thermodynamic equilibrium constants associated with M reactions involving N chemical components. The various equilibrium constants can be expressed in terms of the component mole fractions, for suitable ideal cases, using Equation (7.31) or Equation (7.33). There will be N such mole fractions, but these can be expressed in terms of M reaction coordinates by using the reaction coordinate method. For multiple reactions, there is a separate reaction coordinate for each reaction, and Equation (7.40) generalizes to

$$y_A = \frac{(N_A)_0 + \sum\limits_{Reactions} v_{A,I}\varepsilon_I}{N_0 + \sum\limits_{Reactions} v_I \varepsilon_I} \qquad (7.47)$$

Example 7.17: At high temperatures, atmospheric nitrogen can be converted to various oxides. Consider only two: NO and NO_2. What is their equilibrium in air at 1500 K and 1 bar pressure?

Solution: Two independent reactions are needed that involve all four components. A systematic way of doing this begins with the formation reactions; but, for the present, fairly simple case, Figure 7.5 includes two reactions that can be used directly:

$$\tfrac{1}{2}N_2 + \tfrac{1}{2}O_2 \to NO \qquad (I)$$

$$NO + \tfrac{1}{2}O_2 \to NO_2 \qquad (II)$$

The plots in Figure 7.5 give $(K_{thermo})_I = 0.0033$ and $(K_{thermo})_{II} = 0.011$. The ideal gas law is an excellent approximation at the reaction conditions so that Equation (7.31) applies. Since $P = P_0$, there is no correction for pressure. Thus,

$$0.0033 = \frac{y_{NO}}{y_{N_2}^{1/2} y_{O_2}^{1/2}}$$

$$0.011 = \frac{y_{NO_2}}{y_{NO} y_{O_2}^{1/2}}$$

A solution using the reaction coordinate method will be illustrated. Equation (2.40) is applied to a starting mixture of 0.21 mol of oxygen and 0.79 mol of nitrogen. Nitrogen is not an inert in these reactions, so the lumping of atmospheric argon with nitrogen is not strictly justified, but the error will be small. Equation (2.40) gives

$$\begin{bmatrix} N_{N_2} \\ N_{O_2} \\ N_{NO} \\ N_{NO_2} \end{bmatrix} = \begin{bmatrix} 0.79 \\ 0.21 \\ 0 \\ 0 \end{bmatrix} + \begin{bmatrix} -0.5 & 0 \\ -0.5 & -0.5 \\ 1 & -1 \\ 0 & 1 \end{bmatrix} \begin{bmatrix} \varepsilon_I \\ \varepsilon_{II} \end{bmatrix}$$

or

$$N_{N_2} = 0.79 - 0.5\varepsilon_I$$
$$N_{O_2} = 0.21 - 0.5\varepsilon_I - 0.5\varepsilon_{II}$$
$$N_{NO} = \varepsilon_I - \varepsilon_{II}$$
$$N_{NO_2} = -\varepsilon_I - \varepsilon_{II}$$
$$N_{total} = 1 - 0.5\varepsilon_{II}$$

where the last row was obtained by summing the other four. The various mole fractions are

$$y_{N_2} = \frac{0.79 - 0.5\varepsilon_I}{1 - 0.5\varepsilon_{II}}$$

$$y_{O_2} = \frac{0.21 - 0.5\varepsilon_I - 0.5\varepsilon_{II}}{1 - 0.5\varepsilon_{II}}$$

$$y_{NO} = \frac{\varepsilon_I - \varepsilon_{II}}{1 - 0.5\varepsilon_{II}}$$

$$y_{NO_2} = \frac{\varepsilon_{II}}{1 - 0.5\varepsilon_{II}}$$

Substitution into the equilibrium conditions gives

$$0.0033 = \frac{\varepsilon_I - \varepsilon_{II}}{(0.79 - 0.5\varepsilon_I)^{1/2}(0.21 - 0.5\varepsilon_I - 0.5\varepsilon_{II})^{1/2}}$$

$$0.011 = \frac{\varepsilon_{II}(1 - 0.5\varepsilon_{II})^{1/2}}{(\varepsilon_I - \varepsilon_{II})(0.21 - 0.5\varepsilon_I - 0.5\varepsilon_{II})^{1/2}}$$

This pair of equations can be solved simultaneously to give $\varepsilon_I = 0.0135$ and $\varepsilon_{II} = 6.7 \times 10^{-6}$. The mole fractions are $y_{N_2} = 0.7893$, $y_{O_2} = 0.2093$, $y_{NO} = 0.00135$, and $y_{NO_2} = 7 \times 10^{-6}$.

Example 7.17 illustrates the utility of the reaction coordinate method for solving equilibrium problems. There are no more equations than there are independent chemical reactions. However, in practical problems such as atmospheric chemistry and combustion, the number of reactions is very large. A relatively complete description of high-temperature equilibria between oxygen and nitrogen might consider the concentrations of N_2, O_2, N_2O, N_2O_4, NO, NO_2, N, O, N_2O_2, N_2O_3, N_2O_5, NO_3, O_3, and possibly others. The various reaction coordinates will differ by many orders of magnitude; and the numerical solution would be quite difficult even assuming that the various equilibrium constants could be found. The method of false transients would ease the numerical solution but would not help with the problem of estimating the equilibrium constants.

Independent Reactions. In this section, we consider the number of independent reactions that are necessary to develop equilibrium relationships between N chemical species. A systematic approach is the following:

1. List all chemical species, both elements and compounds, that are believed to exist at equilibrium. By "element" we mean the predominant species at standard-state conditions, for example, O_2 for oxygen at 1 bar and 298.15 K.
2. Write the formation reactions from the elements for each compound. The term "compound" includes elemental forms other than the standard one; for example, we would consider monatomic oxygen as a compound and write $\frac{1}{2}O_2 \rightarrow O$ as one of the reactions.
3. The stoichiometric equations are combined to eliminate any elements that are not believed to be present in significant amounts at equilibrium.

The result of the above procedures is M equations where $M < N$.

Example 7.18: Find a set of independent reactions to represent the equilibrium of CO, CO_2, H_2, and H_2O.

Solution: Assume that only the stated species are present at equilibrium. Then there are three formation reactions:

$$H_2 + \tfrac{1}{2}O_2 \rightarrow H_2O$$

$$C + \tfrac{1}{2}O_2 \rightarrow CO$$

$$C + O_2 \rightarrow CO_2$$

The third reaction is subtracted from the second to eliminate carbon, giving the following set:

$$H_2 + \tfrac{1}{2}O_2 \rightarrow H_2O$$

$$-\tfrac{1}{2}O_2 \rightarrow CO - CO_2$$

These are now added together to eliminate oxygen. The result can be rearranged to give

$$H_2 + CO_2 \rightarrow H_2O + CO$$

Thus $N = 4$ and $M = 1$. The final reaction is the water–gas shift reaction.

Example 7.19: Find a set of independent reactions to represent the equilibrium products for a reaction between 1 mol of methane and 0.5 mol of oxygen.

Solution: It is difficult to decide a priori what species will be present in significant concentrations. Experimental observations are the best guide to constructing an equilibrium model. Lacking this, exhaustive calculations or chemical insight must be used. Except at very high temperatures, free-radical concentrations will be quite low, but free radicals could provide the reaction mechanisms by which equilibrium is approached. Reactions such as $2CH_3 \cdot \rightarrow C_2H_6$ will yield higher hydrocarbons so that the number of theoretically possible species is unbounded. In a low-temperature oxidation, such reactions may be impossible. However, the impossibility is based on kinetic considerations, not thermodynamics.

Assume that oxygen and hydrogen will not be present as elements but that carbon may be. Nonelemental compounds to be considered are CH_4, CO_2, CO, H_2O, CH_3OH, and CH_2O, each of which has a formation reaction:

$$C + 2H_2 \rightarrow CH_4$$

$$C + O_2 \rightarrow CO_2$$

$$C + \tfrac{1}{2}O_2 \rightarrow CO$$

$$H_2 + \tfrac{1}{2}O_2 \rightarrow H_2O$$

$$C + 2H_2 + \tfrac{1}{2}O_2 \rightarrow CH_3OH$$

$$C + H_2 + \tfrac{1}{2}O_2 \rightarrow CH_2O$$

If carbon, hydrogen, and oxygen were all present as elements, none of the formation reactions could be eliminated. We would then have $N = 9$ and $M = 6$. With elemental hydrogen and oxygen assumed absent, two species and two equations can be eliminated, giving $N = 7$ and $M = 4$. Pick any equation containing oxygen—there are five choices—and use it to eliminate oxygen from the other equations. Discard the equation used for the elimination. This reduces M to 5. Now pick any equation containing hydrogen and use it to eliminate hydrogen from the other equations. Discard the equation used for the elimination. This gives $M = 4$. One of the many possible results is

$$3C + 2H_2O \rightarrow CH_4 + 2CO$$

$$2CO \rightarrow C + CO_2$$

$$2C + 2H_2O \rightarrow CH_3OH + CO$$

$$C + H_2O \rightarrow CH_2O$$

These four equations are perfectly adequate for equilibrium calculations although they are nonsense with respect to mechanism. Table 7.2 has the data needed to calculate the four equilibrium constants at the standard state of 298.15 K and 1 bar. Table 7.1 has the necessary data to correct for temperature. The composition at equilibrium can be found using the reaction coordinate method or the method of false transients. The four chemical equations are not unique since various members of the set can be combined algebraically without reducing the dimensionality, $M = 4$. Various equivalent sets can be derived, but none can even approximate a plausible mechanism since one of the starting materials, oxygen, has been assumed to be absent at equilibrium. Thermodynamics provides the destination but not the route.

We have considered thermodynamic equilibrium in homogeneous systems. When two or more phases exist, it is necessary that the requirements for reaction equilibria (i.e., Equations (7.46)) be satisfied simultaneously with the requirements for phase equilibria (i.e., that the component fugacities be equal in each phase). We leave the treatment of chemical equilibria in multiphase systems to the specialized literature, but note that the method of false transients normally works quite well for multiphase systems. The simulation includes reaction—typically confined to one phase—and mass transfer between the phases. The governing equations are given in Chapter 11.

PROBLEMS

7.1. Suppose the following data on the iodination of ethane have been obtained at 603 K using a recirculating gas-phase reactor that closely approximates a CSTR. The indicated concentrations are partial pressures in atmospheres and the mean residence time is in seconds.

$[I_2]_{in}$	$[C_2H_6]_{in}$	$\bar{t}$	$[I_2]_{out}$	$[C_2H_6]_{out}$	$[HI]_{out}$	$[C_2H_5I]_{out}$
0.1	0.9	260	0.0830	0.884	0.0176	0.0162
0.1	0.9	1300	0.0420	0.841	0.0615	0.0594
0.1	0.9	2300	0.0221	0.824	0.0797	0.0770

Use nonlinear regression to fit these data to a plausible functional form for $\mathscr{R}$. See Example 7.20 for linear regression results that can provide good initial guesses.

7.2. The disproportionation of p-toluenesulfonic acid has the following stoichiometry:

$$3(CH_3C_6H_4SO_2H) \rightarrow CH_3C_6H_4SO_2\ SC_6H_4CH_3 + CH_3C_6H_4SO_3H + H_2O$$

Kice and Bowers[3] obtained the following batch data at 70°C in a reaction medium consisting of acetic acid plus 0.56-molar H_2O plus 1.0-molar H_2SO_4:

Time, h	$[CH_3C_6H_4SO_2H]^{-1}$
0	5
0.5	8
1.0	12
1.5	16
4.0	36
5.0	44
6.0	53

The units on $[CH_3C_6H_4SO_2H]^{-1}$ are inverse molarity. Reciprocal concentrations are often cited in the chemical kinetics literature for second-order reactions. Confirm that second-order kinetics provide a good fit and determine the rate constant.

7.3. The decolorization of crystal violet dye by reaction with sodium hydroxide is a convenient means for studying mixing effects in continuous-flow reactors. The reaction is

$$(C_6H_4N(CH_3)_2)_3CCl + NaOH \rightarrow (C_6H_4N(CH_3)_2)_3COH + NaCl$$

The first step is to obtain a good kinetic model for the reaction. To this end, the following batch experiments were conducted in laboratory glassware:

Run no.:	B1		B2		B3		B4	
$[NaOH]_0$:	0.02 N		0.04 N		0.04 N		0.04 N	
Temp.:	30°C		30°C		38°C		45°C	
	t	[dye]	t	[dye]	t	[dye]	t	[dye]
	0	13.55	0	13.55	0	13.55	0	13.55
	2.0	7.87	3.0	2.62	0.5	9.52	0.5	8.72
	4.0	4.62	3.6	1.85	1.0	6.68	1.0	5.61
	5.0	3.48	4.5	1.08	2.0	3.3	2.0	2.33
	6.0	2.65	6.0	0.46	3.0	1.62	3.0	0.95

The times t are in minutes and the dye concentrations [dye] are in milliliters of stock dye solution per 100 ml of the reactant mixture. The stock dye solution was 7.72×10^{-5} molar. Use these data to fit a rate expression of the form

$$\mathscr{R} = k_0[\exp(-T_{act}/T)][\text{dye}]^n[\text{NaOH}]^m$$

The unknown parameters are k_0, T_{act}, n, and m. There are several ways they could be found. Use at least two methods and compare the results. Note that the NaOH is present in great excess.

7.4. Use stoichiometry to calculate $c(t)$ for the data of Example 7.5. Then fit k_I and k_{II} by minimizing S_C^2.

7.5. The following data were collected in an isothermal, constant-volume batch reactor. The stoichiometry is known and the material balance has been closed. The reactions are A → B and A → C. Assume they are elementary. Determine the rate constants k_I and k_{II}.

Time, h	$a(t)$	$b(t)$	$c(t)$
0.1	0.738	0.173	0.089
0.2	0.549	0.299	0.152
0.3	0.408	0.394	0.198
0.4	0.299	0.462	0.239
0.5	0.222	0.516	0.262
0.6	0.167	0.557	0.276
0.7	0.120	0.582	0.298
0.8	0.088	0.603	0.309
0.9	0.069	0.622	0.309
1.0	0.047	0.633	0.320

7.6. The data on the iodination of ethane given in Problem 7.1 have been supplemented by three additional runs done at total pressures of 2 atm:

$[I_2]_{in}$	$[C_2H_6]_{in}$	$\bar{t}$	$[I_2]_{out}$	$[C_2H_6]_{out}$	$[HI]_{out}$	$[C_2H_5I]_{out}$
0.1	0.9	260	0.0830	0.884	0.0176	0.0162
0.1	0.9	1300	0.0420	0.841	0.0615	0.0594
0.1	0.9	2300	0.0221	0.824	0.0797	0.0770
0.1	1.9	150	0.0783	1.878	0.0222	0.0220
0.1	1.9	650	0.0358	1.839	0.0641	0.0609
0.1	1.9	1150	0.0200	1.821	0.0820	0.0803

Repeat Problem 7.1 using the entire set. First do a preliminary analysis using linear regression and then make a final determination of the model parameters using nonlinear regression.

7.7. The following mechanism has been reported for ethane iodination:

$$I_2 + M \rightleftarrows 2I^\bullet + M$$

$$I^\bullet + C_2H_6 \rightarrow C_2H_5^\bullet + HI$$

$$C_2H_5^\bullet + I_2 \rightarrow C_2H_5I + I^\bullet$$

Apply the pseudo-steady hypothesis to the free-radical concentrations to determine a functional form for the reaction rate. Note that M represents any molecule. Use the combined data in Problem 7.6 to fit this mechanism.

7.8. Hinshelwood and Green[4] studied the homogeneous, gas-phase reaction

$$2NO + 2H_2 \rightarrow N_2 + 2H_2O$$

at 1099 K in a constant-volume batch reactor. The reactor was charged with known partial pressures of NO and H_2, and the course of the reaction was monitored by the total pressure. The following are the data from one of their runs. Pressures are in millimeters of mercury (mm Hg). The initial partial pressures were $(P_{NO})_0 = 406$ mm and $(P_{H_2})_0 = 289$. Suppose $\mathscr{R} = k[NO]^m [H_2]^n$. Determine the constants in the rate expression.

T (s)	$\Delta P = P - P_0$
8	10
13	20
19	30
26	40
33	50
43	60
54	70
69	80
87	90
110	100
140	110
204	120
310	127
∞	144.5

7.9. The kinetic study by Hinshelwood and Green cited in Problem 7.8 also included initial rate measurements over a range of partial pressures.

$(P_{NO})_0$	$(P_{H_2})_0$	$\mathscr{R}_0$, mm/s
359	400	1.50
300	400	1.03
152	400	0.25
400	300	1.74
310	300	0.92
232	300	0.45
400	289	1.60
400	205	1.10
400	147	0.79

Use these initial rate data to estimate the constants in the rate expression $\mathscr{R} = k[NO]^m [H_2]^n$.

7.10. The ordinary burning of sulfur produces SO_2. This is the first step in the manufacture of sulfuric acid. The second step oxidizes SO_2 to SO_3 in a gas–solid catalytic reactor. The catalyst increases the reaction rate but does not change the equilibrium compositions in the gas phase.

(a) Determine the heat of reaction for SO_2 oxidation at 600 K and 1 atm.

(b) Determine the mole fractions at equilibrium of N_2, O_2, SO_2, and SO_3 at 600 K and 1 atm given an initial composition of 79 mol% N_2, 15 mol% O_2, and 6 mol% SO_2. Assume that the nitrogen is inert.

7.11. Critique the enthalpy calculation in the alternative solution of Example 7.16 that is based on Equation (7.45) rather than Equation (7.42).

7.12. Rework Example 7.16 without inerts. Specifically, determine whether this case shows any discernable difference between solutions based on Equation (7.42) and Equation (7.45).

7.13. Determine the equilibrium distribution of the three pentane isomers given the following data on free energies of formation at 600 K. Assume ideal gas behavior.

$$\Delta G_F^\circ = 40{,}000 \text{ J/mol of } n\text{-pentane}$$

$$\Delta G_F^\circ = 34{,}000 \text{ J/mol of isopentane}$$

$$\Delta G_F^\circ = 37{,}000 \text{ J/mol of neopentane}$$

7.14. Example 7.17 treated the high-temperature equilibrium of four chemical species: N_2, O_2, NO, and NO_2. Extend the analysis to include N_2O and N_2O_4.

7.15. The following reaction has been used to eliminate NO_x from the stack gases of stationary power plants:

$$NO_x + NH_3 + 0.5(1.5 - x)O_2 \Longleftrightarrow N_2 + 1.5H_2O$$

A zeolite catalyst operated at 1 atm and 325–500 K is so active that the reaction approaches equilibrium. Suppose that stack gas having the equilibrium composition calculated in Example 7.17 is cooled to 500 K. Ignore any reactions involving CO and CO_2. Assume the power plant burns methane to produce electric power with an overall efficiency of 70%. How much ammonia is required per kilowatt-hour (kWh) in order to reduce NO_x emissions by a factor of 10, and how much will the purchased ammonia add to the cost of electricity. Obtain the cost of tank car quantities of anhydrous ammonia from the *Chemical Market Reporter* or from the web.

REFERENCES

1. Hogan, C. J., "Cosmic discord," *Nature*, **408**, 47–48 (2000).
2. Smith, J. M., Van Ness, H. C., and Abbott, M. M., *Introduction to Chemical Engineering Thermodynamics*, 6th ed., McGraw-Hill, New York, 2001.
3. Kice, J. L. and Bowers, K. W., "The mechanism of the disproportionation of sulfinic acids," *J. Am. Chem. Soc.*, **84**, 605–610 (1962).
4. Hinshelwood, C. N. and Green, T. W., "The interaction of nitric oxide and hydrogen and the molecular statistics of termolecular gaseous reactions," *J. Chem. Soc.*, **1926**, 730–739 (1926).

SUGGESTIONS FOR FURTHER READING

A massive but readable classic on chemical kinetics and the extraction of rate data from batch experiments is

Laidler, K. J., *Reactor Kinetics* (in two volumes), Pergamon, London, 1963.

This book, and many standard texts, emphasizes graphical techniques for fitting data. These methods give valuable qualitative insights that may be missed with too much reliance on least-squares analysis.

The classic text on chemical engineering thermodynamics is now in its sixth edition:

Smith, J. M., Van Ness, H. C., and Abbott, M. M., *Introduction to Chemical Engineering Thermodynamics*, 6th ed., McGraw-Hill, New York, 2001.

Chapters 4 and 13 of that book treat chemical reaction thermodynamics in much greater detail than given here.

The Internet has become the best source for thermodynamic data. Run a search on something like "chemical thermodynamic data" on any serious search engine, and you will find multiple sources, most of which allow free downloads. The data in the standard handbooks, e.g. *Perry's Handbook* (see "Suggestions for Further Reading" section of Chapter 5), are still correct but rather capricious in scope and likely to be expressed in archaic units like those sprinkled here and there in this book.

APPENDIX 7.1: LINEAR REGRESSION ANALYSIS

Determination of the model parameters in Equation (7.7) usually requires numerical minimization of the sum-of-squares, but an analytical solution is possible when the model is a linear function of the independent variables. Take the logarithm of Equation (7.4) to obtain

$$\ln \mathscr{R} = \ln k + m \ln[A] + n \ln[B] + r \ln[R] + s \ln[S] + \cdots \quad (7.48)$$

Define $Y = \ln \mathscr{R}$, $C = \ln k$, $X_1 = \ln[A]$, $X_2 = \ln[B]$, and so on. Then,

$$Y = C + mX_1 + nX_2 + rX_3 \ldots \quad (7.49)$$

Thus, Y is a linear function of the new independent variables, $X_1, X_2, \ldots$. Linear regression analysis is used to fit linear models to experimental data. The case of three independent variables will be used for illustrative purposes, although there can be any number of independent variables provided the model remains linear. The dependent variable Y can be directly measured or it can be a mathematical transformation of a directly measured variable. If transformed variables are used, the fitting procedure minimizes the sum-of-squares for the differences between the transformed data and the transformed model. Nonlinear regression minimizes the sum-of-squares between the data as actually measured and the model in untransformed form. The results may be substantially different. In particular, a logarithmic transformation will weight small numbers more heavily than large numbers.

The various independent variables can be the actual experimental variables or transformations of them. Different transformations can be used for different variables. The "independent" variables need not be actually independent. For example, linear regression analysis can be used to fit a cubic equation by setting X, X^2, and X^3 as the independent variables.

The sum-of-squares to be minimized is

$$S^2 = \sum_{Data} (Y - C - mX_1 - nX_2 - rX_3)^2 \qquad (7.50)$$

We now regard the experimental data as fixed and treat the model parameters as the variables. The goal is to choose C, m, n, and r such that $S^2 > 0$ achieves its minimum possible value. A necessary condition for S^2 to be a minimum is that

$$\frac{\partial S^2}{\partial C} = \frac{\partial S^2}{\partial m} = \frac{\partial S^2}{\partial n} = \frac{\partial S^2}{\partial r} = 0$$

For the assumed linear form of Equation (7.50),

$$\frac{\partial S^2}{\partial C} = 2 \sum_{Data} (Y - C - mX_1 - nX_2 - rX_3)(-1) = 0$$

$$\frac{\partial S^2}{\partial m} = -2 \sum_{Data} (Y - C - mX_1 - nX_2 - rX_3)(-X_1) = 0$$

$$\frac{\partial S^2}{\partial n} = -2 \sum_{Data} (Y - C - mX_1 - nX_2 - rX_3)(-X_2) = 0$$

$$\frac{\partial S^2}{\partial r} = -2 \sum_{Data} (Y - C - mX_1 - nX_2 - rX_3)(-X_3) = 0$$

Rearrangement gives

$$\begin{aligned} JC + m\sum X_1 + n\sum X_2 + r\sum X_3 &= \sum Y \\ C\sum X_1 + m\sum X_1^2 + n\sum X_1 X_2 + r\sum X_1 X_3 &= \sum X_1 Y \\ C\sum X_2 + m\sum X_1 X_2 + n\sum X_2^2 + r\sum X_2 X_3 &= \sum X_2 Y \\ C\sum X_3 + m\sum X_1 X_3 + n\sum X_2 X_3 + r\sum X_3^2 &= \sum X_3 Y \end{aligned} \qquad (7.51)$$

where J is the number of data and the summations extend over the data. The various sums can be calculated from the data, and Equations (7.51) can be solved for C, m, n, and r. Equations (7.51) are linear in the unknown parameters and can be solved by matrix inversion. See any text on linear algebra. No solution will exist if there are fewer observations than model parameters, and the model will fit the data exactly if there are as many parameters as observations.

Example 7.20: Use linear regression analysis to determine k, m, and n for the data taken at 1 atm total pressure for the ethane iodination reaction in Problem 7.1.

Solution: The assumed linear form is

$$\ln \mathcal{R} = \ln k + m \ln[I_2] + n \ln[C_2H_5]$$

The data are:

$\bar{t}$ (s)	$\mathcal{R}$(atm/s)	$Y = \ln \mathcal{R}$	$X_1 = \ln[I_2]$	$X_2 = \ln[C_2H_6]$
240	7.08×10^{-5}	-9.56	-2.49	-0.123
1300	4.60×10^{-5}	-9.99	-3.21	-0.173
2300	3.39×10^{-5}	-10.29	-3.81	-0.194

Suppose we attempt to evaluate all three constants, k, m, and n. Then the first three components of Equations (7.51) are needed. Evaluating the various sums gives

$$3 \ln k - 9.51m - 0.49n = -29.84$$
$$-9.51 \ln k + 31.0203m + 1.60074n = 95.07720$$
$$-0.49 \ln k + 1.60074m + 0.082694n = 4.90041$$

The solution is $\ln k = -8.214$, $m = 0.401$, and $n = 2.82$. This model uses as many parameters as there are observations and thus fits the data exactly, $S^2 = 0$. One can certainly doubt the significance of such a fit. It is clear that the data are not perfect, since the material balance is not perfect. Additional data could cause large changes in the parameter values. Problem 7.6 addresses this issue. Certainly, the value for n seems high and is likely to be an artifact of the limited range over which $[C_2H_6]$ was varied. Suppose we pick $n = 1$ on semitheoretical grounds. Then regression analysis can be used to find best values for the remaining parameters. The dependent variable is now $Y = \ln \mathcal{R} - \ln[C_2H_6]$. There is now only one independent variable, $X_1 = \ln[I_2]$. The data are

$Y = \ln \mathcal{R} - \ln[C_2H_6]$	$X_1 = \ln[I_2]$
-9.44	-2.49
-9.82	-3.21
-10.10	-3.81

Now only the first two components of Equations (7.51) are used. Evaluating the various sums gives

$$3\ln k - 9.51m = -29.36$$
$$-9.51\ln k + 31.0203m = 93.5088$$

Solution gives $\ln k = -9.1988$ and $m = 0.5009$. Since there are now only two fitted parameters, the model does not fit the data exactly, $S^2 > 0$, but the fit is quite good:

$(\ln \mathscr{R})_{observed}$	$(\ln \mathscr{R})_{predicted}$
−9.56	−9.57
−9.99	−9.98
−10.29	−10.30

The predictions with $n = 1$ are essentially as good as those with $n = 2.82$. An excellent fit is also obtained with $n = 2$. Thus, the data do not allow n to be determined with any confidence. However, a kineticist would probably pick $m = 0.5$ and $n = 1$ based on the simple logic that these values replicate the experimental measurements and are physically plausible.

Regression analysis is a powerful tool for fitting models but can obviously be misused. In the above example, physical reasoning avoids a spurious result. Statistical reasoning is also helpful. Confidence intervals and other statistical measures of goodness of fit can be used to judge whether or not a given parameter is *statistically significant* and if it should be retained in the model. Also, statistical analysis can help in the planning of experiments so that the new data will remove a maximum amount of uncertainty in the model. See any standard text on the *statistical design of experiments*.

APPENDIX 7.2: CODE FOR EXAMPLE 7.16

```
DefDbl A-L, P-Z
DefLng M-O
Dim conc(4), yinit(4)
Public A(5), B(5), C(5), D(5), y(4)
Sub Exp7_16()

'Data from Table 7.1
'Ethylbenzene is 1, Styrene is 2, Hydrogen is 3,
'Water is 4.
A(1) =1.124: B(1) =55.38: C(1) =-18.476: D(1) =0
A(2) =2.05: B(2) =50.192: C(2) =-16.662: D(2) =0
A(3) =3.249: B(3) =0.422: C(3) =0: D(3) =0.083
A(4) =3.47: B(4) =1.45: C(4) =0: D(4) =0.121
'Calculate delta Cp for C1 reacting to C2 + C3
```

```
A(5) = A(2) + A(3) - A(1)
B(5) = B(2) + B(3) - B(1)
C(5) = C(2) + C(3) - C(1)
D(5) = D(2) + D(3) - D(1)
For n = 1 To 5
  A(n) = A(n)
  B(n) = B(n)/1000#
  C(n) = C(n)/1000000#
  D(n) = D(n) * 100000#
Next n
Rg = 8.314

'Results from Examples 7.8 and 7.10.
DeltaHR0 = 117440
DeltaGR0 = 83010

'Starting conditions
y(1) = 0.1
y(2) = 0
y(3) = 0
y(4) = 0.9
Tinit = 973
T = Tinit
T0 = 298.15
P0 = 1
P = 0.1
'Calculate molar density using bar as the pressure unit
Rgg = 0.00008314
rhoinit = P / Rgg / T
rho = rhoinit
For n = 1 To 4
  yinit(n) = y(n)
  conc(n) = rho * y(n)
Next
'Initial condition used for enthalpy marching
'For n = 1 To 4
'Enthalpy = Enthalpy + y(n) * rho * Rg * (CpInt(n, T)
'+           - CpInt(n, T0))
'Next

'Time step and output control
dtime = 0.00001
ip = 2
Tp = Tinit
```

```
Do 'Main Loop

'Thermodynamic equilibrium constant calculated as in
'Example 7.15
  K0 = Exp(-DeltaGR0/Rg/T0)
  K1 = Exp(DeltaHR0/Rg/T0 * (1 - T0/T))
  K2 = Exp(-(CpInt(5, T) - CpInt(5, T0))/T)
  K3 = Exp(DCpRTInt(T) - DCpRTInt(T0))
  Kthermo = K0 * K1 * K2 * K3

'Equilibrium mole fractions calculated using method of
'Example 7.13. These results are calculated for
'interest only. They are not needed for the main
'calculation. The code is specific to initial conditions
  G = Kthermo * P0/P
  eps = (-0.9 * G + Sqr(0.81 * G * G + 0.4 * (1 + G) * G))/2/
+       (1 + G)
  eyEB = (0.1 - eps)/(1 + eps)
  eySty = eps/(1 + eps)

'Kinetic equilibrium constant from Equation 7.36
KK = Kthermo * P0/Rgg/T

'Reaction
  kf = 160000 * Exp(-9000/T)
  RRate = kf * (conc(1) - conc(2) * conc(3)/KK)
  DeltaHR = 117440 + (CpInt(5, T) - CpInt(5, T0)) * Rg

'Approximate solution based on marching ahead in
'temperature, Equation 7.45
T = T - DeltaHR * RRate * dtime/rho/CpMix(T)/Rg

'A more rigorous solution based on marching ahead in
'enthalpy according to Equation 7.42 is given in the
'next 16 lines of code. The temperature is found from
'the enthalpy using a binary search. The code is specific
'to the initial conditions of this problem. Results are
'very similar to those for marching temperature
'directly.
'     Enthalpy = Enthalpy- DeltaHR * RRate * dtime
'     Thigh = T
'     Tlow = T - 1
'     Txx = Tlow
'     For m = 1 To 20
'       Tx = (Thigh + Tlow)/2#
```

```
'       DeltaHR=117440# + (CpInt(5, Tx) - CpInt(5, T0)) *
'+              Rg
'       DHR=DeltaHR * (rhoinit * 0.1 - rho * y(1))/rhoinit
'       DHS=Rg * (0.1 * (CpInt(1, Tx) - CpInt(1, Tinit))
'+            +0.9 * (CpInt(4, Tx) - CpInt(4, Tinit)))
'       If DHR + DHS > Enthalpy Then
'          Thigh=Tx
'       Else
'          Tlow=Tx
'       End If
'       Next m
'       T=Tx
conc(1) =conc(1) - RRate * dtime
conc(2) =conc(2) + RRate * dtime
conc(3) =conc(3) + RRate * dtime
rho=conc(1) + conc(2) + conc(3) + conc(4)
y(1) =conc(1)/rho
y(2) =conc(2)/rho
y(3) =conc(3)/rho
y(4) =conc(4)/rho

'Pressure
  P=rho * Rgg * T

'Output trajectory results when temperature has
'decreased by 1 degree
If T <= Tp Then
  GoSub Output
  Tp=Tp - 1
End If
Rtime=Rtime + dtime
Loop While Abs(y(1) - eyEB) > 0.0000001 'End of main loop

GoSub Output 'Output final values

Exit Sub

Output:
    ip=ip + 1
    Range("A"& CStr(ip)).Select
    ActiveCell.FormulaR1C1=Rtime
    Range("B"& CStr(ip)).Select
  ActiveCell.FormulaR1C1=y(2)/(y(1) + y(2))
  Range("C"& CStr(ip)).Select
  ActiveCell.FormulaR1C1=eySty/(eyEB + eySty)
```

```
    Range("D"& CStr(ip)).Select
    ActiveCell.FormulaR1C1 = T
    Range("E"& CStr(ip)).Select
    ActiveCell.FormulaR1C1 = P
    Range("F"& CStr(ip)).Select
    ActiveCell.FormulaR1C1 = y(1)
Return

End Sub
Function Cp(n, T)
    Cp = A(n) + B(n) * T + C(n) * T * T + D(n)/T/T
End Function
Function CpInt(n, T)
    CpInt = A(n) * T + B(n) * T * T/2 + C(n) * T * T * T/3 -
+          D(n)/T
End Function
Function DCpRTInt(T)
    DCpRTInt = A(5) * Log(T) + B(5) * T + C(5) * T * T/2 -
+           D(5)/2/T^2
End Function
Function CpMix(T)
    CpMix = y(1) * Cp(1, T) + y(2) * Cp(2, T) + y(3) *
+          Cp(3, T) + y(4) * Cp(4, T)
End Function
```

CHAPTER 8
REAL TUBULAR REACTORS IN LAMINAR FLOW

Piston flow is a convenient approximation of a real tubular reactor. The design equations for piston flow are relatively simple and are identical in mathematical form to the design equations governing batch reactors. The key to their mathematical simplicity is the assumed absence of any radial or tangential variations within the reactor. The dependent variables $a, b, \ldots, T, P$, change in the axial, down-tube direction but are completely uniform across the tube. This allows the reactor design problem to be formulated as a set of ordinary differential equations in a single independent variable, z. As shown in previous chapters, such problems are readily solvable, given the initial values $a_{in}, b_{in}, \ldots, T_{in}, P_{in}$.

Piston flow is an accurate approximation for some practical situations. It is usually possible to avoid tangential (θ-direction) dependence in practical reactor designs, at least for the case of premixed reactants, which we are considering throughout most of this book. It is harder, but sometimes possible, to avoid radial variations. A long, highly turbulent reactor is a typical case where piston flow will be a good approximation for most purposes. Piston flow will usually be a bad approximation for laminar flow reactors since radial variations in composition and temperature can be large.

Chapters 8 and 9 discuss design techniques for real tubular reactors. By "real," we mean reactors for which the convenient approximation of piston flow is so inaccurate that a more realistic model must be developed. By "tubular," we mean reactors in which there is a predominant direction of flow and a reasonably high aspect ratio, characterized by a length-to-diameter ratio, L/d_t, of 8 or more, or its equivalent, an L/R ratio of 16 or more. Practical designs include straight and coiled tubes, multitubular heat exchangers, and packed-bed reactors. Chapter 8 starts with isothermal laminar flow in tubular reactors that have negligible molecular diffusion. The complications of significant molecular diffusion, nonisothermal reactions with consequent diffusion of heat, and the effects of temperature and composition on the velocity profile are subsequently introduced. Chapter 9 treats turbulent reactors and packed-bed reactors of both the laminar and turbulent varieties. The result of these two chapters is a comprehensive design methodology that is applicable to many design problems in the traditional chemical industry and which forms a conceptual framework for extension to nontraditional industries. The major limitation of the methodology is its restriction to reactors that have a single mobile phase. Reactors with two or three mobile phases, such as gas–liquid reactors, are considered in Chapter 11, but the treatment is necessarily less comprehensive than for the reactors of Chapters 8 and 9 that have only one mobile phase.

8.1 ISOTHERMAL LAMINAR FLOW WITH NEGLIGIBLE DIFFUSION

Consider isothermal laminar flow of a Newtonian fluid in a circular tube of radius R, length L, and average fluid velocity $\bar{u}$. When the viscosity is constant, the axial velocity profile is

$$V_z(r) = 2\bar{u}\left[1 - \frac{r^2}{R^2}\right] \tag{8.1}$$

Most industrial reactors in laminar flow have pronounced temperature and composition variations that change the viscosity and alter the velocity profile from the simple parabolic profile of Equation (8.1). These complications are addressed in Section 8.7. However, even the profile of Equation (8.1) presents a serious complication compared with piston flow. There is a velocity gradient across the tube, with zero velocity at the wall and high velocities near the centerline. Molecules near the center will follow high-velocity streamlines and will undergo relatively little reaction. Those near the tube wall will be on low-velocity streamlines, will remain in the reactor for long times, and will react to near-completion. Thus, a gradient in composition develops across the radius of the tube. Molecular diffusion acts to alleviate this gradient but will not completely eliminate it, particularly in liquid-phase systems with typical diffusivities of 1.0×10^{-9} to 1.0×10^{-10} for small molecules and much lower for polymers.

When diffusion is negligible, the material moving along a streamline is isolated from material moving along other streamlines. The streamline can be treated as if it were a piston flow reactor, and the system as a whole can be regarded as a large number of piston flow reactors in parallel. For the case of straight streamlines and a velocity profile that depends on radial position alone, concentrations along the streamlines at position r are given by

$$V_z(r)\frac{\partial a}{\partial z} = \mathscr{R}_A \tag{8.2}$$

This result is reminiscent of Equation (1.36). We have replaced the average velocity with the velocity corresponding to a particular streamline. Equation (8.2) is written as a partial differential equation to emphasize the fact that the concentration $a = a(r, z)$ is a function of both r and z. However, Equation (8.2) can be integrated as though it were an ordinary differential equation. The inlet boundary condition associated with the streamline at position r is $a(r, 0) = a_{in}(r)$. Usually, a_{in} will be same for all values of r, but it is possible to treat the more general case. The outlet concentration for a particular streamline is found by solving Equation (8.2) and setting $z = L$. The outlet concentrations for the various streamlines are averaged to get the outlet concentration from the reactor as a whole.

8.1.1 A Criterion for Neglecting Diffusion

The importance of diffusion in a tubular reactor is determined by a dimensionless parameter, $\mathscr{D}_A \bar{t}/R^2 = \mathscr{D}_A L/(\bar{u} R^2)$, which is the molecular diffusivity of component A scaled by the tube size and flow rate. If $\mathscr{D}_A \bar{t}/R^2$ is small, then the effects of diffusion will be small, although the definition of small will depend on the specific reaction mechanism. Merrill and Hamrin[1] studied the effects of diffusion on first-order reactions and concluded that molecular diffusion can be ignored in reactor design calculations if

$$\mathscr{D}_A \bar{t}/R^2 < 0.003 \qquad (8.3)$$

Equation (8.3) gives the criterion for neglecting diffusion. It is satisfied in many industrial-scale, laminar flow reactors, but may not be satisfied in laboratory-scale reactors since they operate with the same values for $\mathscr{D}_A$ and $\bar{t}$ but generally use smaller diameter tubes. Molecular diffusion becomes progressively more important as the size of the reactor is decreased. The effects of molecular diffusion are generally beneficial, so that a small reactor will give better results than a large one, a fact that has proved distressing to engineers attempting a scaleup. For the purposes of scaleup, it may be better to avoid diffusion and accept the composition gradients on the small scale so that they do not cause unpleasant surprises on the large scale. One approach to avoiding diffusion in the small reactor is to use a short, fat tube. If diffusion is negligible in the small reactor, it will remain negligible upon scaleup. The other approach is to accept the benefit of diffusion and to scaleup at constant tube diameter, either in parallel or in series as discussed in Chapter 3. This will maintain a constant value for the dimensionless diffusivity, $\mathscr{D}_A \bar{t}/R^2$.

The Merrill and Hamrin criterion was derived for a first-order reaction. It should apply reasonably well to other simple reactions, but reactions exist that are quite sensitive to diffusion. Examples include the decomposition of free-radical initiators where a few initial events can cause a large number of propagation reactions, and coupling or cross-linking reactions where a few events can have a large effect on product properties.

8.1.2 Mixing-Cup Averages

Suppose Equation (8.2) is solved either analytically or numerically to give $a(r, z)$. It remains to find the average outlet concentration when the flows from all the streamlines are combined into a single stream. This average concentration is the *convected-mean* or *mixing-cup average concentration*. It is the average concentration, $a_{mix}(L)$, of material leaving the reactor. This material could be collected in a bucket (a mixing cup) and is what a company is able to sell. It is not the spatial average concentration inside the reactor, even at the reactor outlet. See Problem 8.5 for an explanation of this distinction.

The convected mean at position z is denoted by $a_{mix}(z)$ and is found by multiplying the concentration on a streamline, $a(r, z)$, by the volumetric flow rate associated with that streamline, $dQ(r) = V_z(r)dA_c$, and by summing over all the streamlines. The result is the molar flow rate of component A. Dividing by the total volumetric flow, $Q = \bar{u}A_c$, gives the convected-mean concentration:

$$a_{mix}(z) = \frac{1}{\bar{u}A_c} \iint_{A_c} aV_z \, dA_c = \frac{1}{\bar{u}R^2} \int_0^R a(r, z) V_z(r) 2r \, dr \qquad (8.4)$$

The second integral in Equation (8.4) applies to the usual case of a circular tube with a velocity profile that is a function of r and not of θ. When the velocity profile is parabolic,

$$a_{mix}(z) = \frac{4}{R^2} \int_0^R a(r, z)\left[1 - \frac{r^2}{R^2}\right] r \, dr = 4 \int_0^1 a(\imath, z)\left[1 - \imath^2\right] \imath \, d\imath \qquad (8.5)$$

where $\imath = r/R$ is the dimensionless radius.

The mixing-cup average outlet concentration $a_{mix}(L)$ is usually denoted just as a_{out} and the averaging is implied. The averaging is necessary whenever there is a radial variation in concentration or temperature. Thus, Equation (8.4) and its obvious generalizations to the concentration of other components or to the mixing-cup average temperature is needed throughout this chapter and much of Chapter 9. If in doubt, calculate the mixing-cup averages. However, as the next example suggests, this calculation can seldom be done analytically.

Example 8.1: Find the mixing-cup average outlet concentration for an isothermal, first-order reaction with rate constant k that is occurring in a laminar flow reactor with a parabolic velocity profile as given by Equation (8.1).

Solution: This is the simplest, nontrivial example of a laminar flow reactor. The solution begins by integrating Equation (8.2) for a specific streamline that corresponds to radial position r. The result is

$$a(r, z) = a_{in} \exp\left[\frac{-kz}{V_z(r)}\right] \qquad (8.6)$$

where k is the first-order rate constant. The mixing-cup average outlet concentration is found using Equation (8.5) with $z = L$:

$$a_{out} = a_{mix}(L) = 4a_{in} \int_0^1 \exp\left[\frac{-kL}{2\bar{u}(1-\imath^2)}\right][1-\imath^2]\imath\, d\imath$$

This integral can be solved analytically. Its solution is a good test for symbolic manipulators such as Mathematica or Maple. We illustrate its solution using classical methods. Differentiating Equation (8.1) gives

$$\imath\, d\imath = -\frac{dV_z}{4\bar{u}}$$

This substitution allows the integral to be expressed as a function of V_z:

$$a_{out} = \frac{a_{in}}{2\bar{u}^2} \int_0^{2\bar{u}} \exp[-kL/V_z] V_z\, dV_z$$

A second substitution is now made,

$$t = L/V_z \tag{8.7}$$

to obtain an integral with respect to t. Note that t ranges from $\bar{t}/2$ to ∞ as V_z ranges from $2\bar{u}$ to 0 as $\imath$ ranges from 0 to 1. Some algebra gives the final result:

$$\frac{a_{out}}{a_{in}} = \int_{\bar{t}/2}^{\infty} \exp(-kt)\frac{\bar{t}^2}{2t^3}\, dt \tag{8.8}$$

This integral is a special function related to the incomplete gamma function. The solution can be considered to be analytical even though the function may be unfamiliar. Figure 8.1 illustrates the behavior of Equation (8.8) as compared with CSTRs, PFRs, and laminar flow reactors with diffusion.

Mixing-cup averages are readily calculated for any velocity profile that is axisymmetric—i.e., has no θ-dependence. Simply use the appropriate functional form for V_z in Equation (8.4). However, analytical integration as in Example 8.1 is rarely possible. Numerical integration is usually necessary, and the trapezoidal rule described in Section 8.3.4 is recommended because it converges $O(\Delta r^2)$, as do the other numerical methods used in Chapters 8 and 9. Example 8.3 includes a sample computer code. Use of the rectangular rule (see Figure 2.1) is not recommended because it converges $O(\Delta r)$ and would limit the accuracy of other calculations. Simpson's rule converges $O(\Delta r^3)$ and

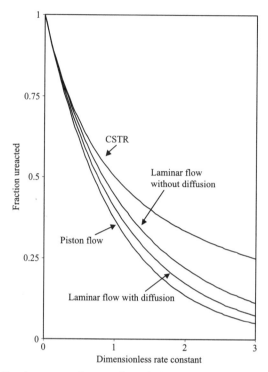

FIGURE 8.1 Fraction unreacted versus dimensionless rate constant for a first-order reaction in various isothermal reactors. The case illustrated with diffusion is for $\mathscr{D}_A \bar{t}/R^2 = 0.1$.

will calculate $\bar{u}$ exactly when the velocity profile is parabolic, but ceases to be exact for the more complex velocity profiles encountered in real laminar flow reactors. The use of Simpson's rule then does no harm but offers no real advantage. The convergence order for a complex calculation is determined by the most slowly converging of the computational components.

The double integral in Equation (8.4) is a fairly general definition of the mixing-cup average. It is applicable to arbitrary velocity profiles and noncircular cross sections but does assume straight streamlines of equal length. Treatment of curved streamlines requires a precise and possibly artificial definition of the system boundaries. See Nauman and Buffham.[2]

8.1.3 A Preview of Residence Time Theory

Example 8.1 derived a specific example of a powerful result of *residence time theory*. The residence time associated with a streamline is $t = L/V_z$. The outlet concentration for this streamline is $a_{batch}(t)$. This is a general result applicable to diffusion-free laminar flow. Example 8.1 treated the case of a first-order reaction where $a_{batch}(t) = \exp(-kt)$. Repeating Example 8.1 for the general case gives

$$a_{out} = \int_{\bar{t}/2}^{\infty} a_{batch}(t) \frac{\bar{t}^2}{2t^3} dt \qquad (8.9)$$

Equation (8.9) can be applied to any reaction, even a complex reaction where $a_{batch}(t)$ must be determined by the simultaneous solution of many ODEs. The restrictions on Equation (8.9) are isothermal laminar flow in a circular tube with a parabolic velocity profile and negligible diffusion.

The condition of negligible diffusion means that the reactor is *completely segregated*. A further generalization of Equation (8.9) applies to any completely segregated reactor:

$$a_{out} = \int_{0}^{\infty} a_{batch}(t) f(t) \, dt \qquad (8.10)$$

where $f(t)$ is the *differential distribution function of residence times*. In principle, $f(t)$ is a characteristic of the reactor, not of the reaction. It can be used to predict conversions for any type of reaction in the same reactor. Chapter 15 discusses ways of measuring $f(t)$. For a parabolic velocity profile in a diffusion-free tube,

$$\begin{aligned} f(t) &= 0 & t \leq \bar{t}/2 \\ f(t) &= \frac{\bar{t}^2}{2t^3} & t > \bar{t}/2 \end{aligned} \qquad (8.11)$$

8.2 CONVECTIVE DIFFUSION OF MASS

Molecules must come into contact for a reaction to occur, and the mechanism for the contact is molecular motion. This is also the mechanism for diffusion. Diffusion is inherently important whenever reactions occur, but there are some reactor design problems where diffusion need not be explicitly considered, e.g., tubular reactors that satisfy the Merrill and Hamrin criterion, Equation (8.3). For other reactors, a detailed accounting for molecular diffusion may be critical to the design.

Diffusion is important in reactors with unmixed feed streams since the initial mixing of reactants must occur inside the reactor under reacting conditions. Diffusion can be a slow process, and the reaction rate will often be limited by diffusion rather than by the intrinsic reaction rate that would prevail if the reactants were premixed. Thus, diffusion can be expected to be important in tubular reactors with unmixed feed streams. Its effects are difficult to calculate, and normal design practice is to use premixed feeds whenever possible.

With premixed reactants, molecular diffusion has already brought the reacting molecules into close proximity. In an initially mixed batch reactor, various portions of the reacting mass will start at the same composition, will react at the same rate, and will thus have the same composition at any time. No concentration gradients develop, and molecular diffusion is unimportant during the reaction step of the process even though it was important during the premixing step. Similarly, mechanical mixing is unnecessary for an initially mixed batch reactor, although mixing must be good enough to eliminate temperature gradients if there is heating or cooling at the wall. Like ideal batch reactors, CSTRs lack internal concentration differences. The agitator in a CSTR brings fluid elements into such close contact that mixing is complete and instantaneous.

Tubular reactors are different. They must have concentration gradients in the axial direction since the average concentration changes from a_{in} to a_{out} along the length of the reactor. The nonisothermal case will have an axial temperature gradient as well. Piston flow reactors are a special case of tubular reactor where radial mixing is assumed to be complete and instantaneous. They continue to have axial gradients.

Laminar flow reactors have concentration and temperature gradients in both the radial and axial directions. The radial gradient normally has a much greater effect on reactor performance. The diffusive flux is a vector that depends on concentration gradients. The flux in the axial direction is

$$J_z = -\mathscr{D}_A \frac{\partial a}{\partial z}$$

As a first approximation, the concentration gradient in the axial direction is

$$\frac{\partial a}{\partial z} \approx \frac{a_{out} - a_{in}}{L}$$

and since L is large, the diffusive flux will be small and can be neglected in most tubular reactors. Note that the piston flow model ignores axial diffusion even though it predicts concentration gradients in the axial direction.

The flux in the radial direction is

$$J_r = -\mathscr{D}_A \frac{\partial a}{\partial r}$$

A first approximation to the radial concentration gradient is

$$\frac{\partial a}{\partial r} \approx \frac{a_{wall} - a_{in}}{R} \approx \frac{-a_{in}}{R}$$

where we have assumed component A to be consumed by the reaction and to have a concentration near zero at the tube wall. The concentration differences in the radial and axial directions are similar in magnitude, but the length scales are very different. It is typical for tubular reactors to have $L/R \gg 1$.

The relatively short distance in the radial direction leads to much higher diffusion rates. In most of what follows, axial diffusion will be ignored.

To account for molecular diffusion, Equation (8.2), which governs reactant concentrations along the streamlines, must be modified to allow diffusion between the streamlines; i.e., in the radial direction. We ignore axial diffusion but add a radial diffusion term to obtain

$$V_z(r)\frac{\partial a}{\partial z} = \mathscr{D}_A\left[\frac{1}{r}\frac{\partial a}{\partial r} + \frac{\partial^2 a}{\partial r^2}\right] + \mathscr{R}_A \tag{8.12}$$

A derivation of this equation is given in Appendix 8.1.

Equation (8.12) is a form of the *convective diffusion equation*. More general forms can be found in any good textbook on transport phenomena, but Equation (8.12) is sufficient for many practical situations. It assumes constant diffusivity and constant density. It is written in cylindrical coordinates since we are primarily concerned with reactors that have circular cross sections, but Section 8.4 gives a rectangular-coordinate version applicable to flow between flat plates.

Equation (8.12) is a partial differential equation that includes a first derivative in the axial direction and first and second derivatives in the radial direction. Three boundary conditions are needed: one axial and two radial. The axial boundary condition is

$$a(r, 0) = a_{in}(r) \tag{8.13}$$

As noted earlier, a_{in} will usually be independent of r, but the numerical solution techniques that follow can easily accommodate the more general case. The radial boundary conditions are

$$\frac{\partial a}{\partial r} = 0 \text{ at the wall}, \qquad r = R \tag{8.14}$$

$$\frac{\partial a}{\partial r} = 0 \text{ at the centerline}, \qquad r = 0 \tag{8.15}$$

The wall boundary condition applies to a solid tube without transpiration. The centerline boundary condition assumes symmetry in the radial direction. It is consistent with the assumption of an axisymmetric velocity profile without concentration or temperature gradients in the θ-direction. This boundary condition is by no means inevitable since gradients in the θ-direction can arise from natural convection. However, it is desirable to avoid θ-dependency since appropriate design methods are generally lacking.

A solution to Equation (8.12) together with its boundary conditions gives $a(r, z)$ at every point in the reactor. An analytical solution is possible for the special case of a first-order reaction, but the resulting infinite series is cumbersome to evaluate. In practice, numerical methods are necessary.

If several reactive components are involved, a version of Equation (8.12) should be written for each component. Thus, for complex reactions involving N components, it is necessary to solve N simultaneous PDEs (partial differential equations). For batch and piston flow reactors, the task is to solve N simultaneous ODEs. Stoichiometric relationships and the reaction coordinate method can be used to eliminate one or more of the ODEs, but this elimination is not generally possible for PDEs. Except for the special case where all the diffusion coefficients are equal, $\mathscr{D}_A = \mathscr{D}_B = \cdots$, stoichiometric relationships should not be used to eliminate any of the PDEs governing reaction with diffusion. When the diffusion coefficients are unequal, the various species may separate due to diffusion. Overall stoichiometry, as measured by $a_{in} - a_{out}$, $b_{in} - b_{out}, \ldots$ is preserved and satisfies Equation (2.39). However, convective diffusion does not preserve local stoichiometry. Thus, the reaction coordinate method does not work locally; and if N components affect reaction rates, then all N simultaneous equations should be solved. Even so, great care must be taken with multicomponent systems when the diffusivities differ significantly in magnitude unless there is some dominant component, the "solvent," that can be assumed to distribute itself to satisfy a material balance constraint such as constant density. The general case of multicomponent diffusion remains an area of research where reliable design methods are lacking.[3]

8.3 NUMERICAL SOLUTION TECHNIQUES

Many techniques have been developed for the numerical solution of partial differential equations. The best method depends on the type of PDE being solved and on the geometry of the system. Partial differential equations having the form of Equation (8.12) are known as parabolic PDEs and are among the easiest to solve. We give here the simplest possible method of solution, one that is directly analogous to the marching-ahead technique used for ordinary differential equations. Other techniques should be considered (but may not be much better) if the computing cost becomes significant. The method we shall use is based on finite difference approximations for the partial derivatives. Finite element methods will occasionally give better performance, although typically not for parabolic PDEs.

The technique used here is a variant of *the method of lines* in which a PDE is converted into a set of simultaneous ODEs. The ODEs have z as the independent variable and are solved by conventional means. We will solve them using Euler's method, which converges $O(\Delta z)$. Higher orders of convergence, e.g., Runge-Kutta, buy little for reasons explained in Section 8.3.3. The ODEs obtained using the method of lines are very stiff, and computational efficiency can be gained by using an ODE-solver designed for stiff equations. However, for a solution done only once, programming ease is usually more important than computational efficiency.

8.3.1 The Method of Lines

Divide the tube length into a number of equally sized increments, $\Delta z = L/J$, where J is an integer. A finite difference approximation for the partial derivative of concentration in the axial direction is

$$\frac{\partial a}{\partial z} \approx \frac{a(r, z + \Delta z) - a(r, z)}{\Delta z} \tag{8.16}$$

This approximation is called a *forward difference* since it involves the forward point, $z + \Delta z$, as well as the central point, z. (See Appendix 8.2 for a discussion of finite difference approximations.) Equation (8.16) is the simplest finite difference approximation for a first derivative.

The tube radius is divided into a number of equally sized increments, $\Delta r = R/I$, where I is an integer. For reasons of convergence, we prefer to use a second-order, *central difference* approximation for the first partial derivative:

$$\frac{\partial a}{\partial r} \approx \frac{a(r + \Delta r, z) - a(r - \Delta r, z)}{2\,\Delta r} \tag{8.17}$$

which is seen to involve the $r + \Delta r$ and $r - \Delta r$ points. For the second radial derivative we use

$$\frac{\partial^2 a}{\partial r^2} \approx \frac{a(r + \Delta r, z) - 2a(r, z) + a(r - \Delta r, z)}{\Delta r^2} \tag{8.18}$$

The approximations for the radial derivatives are substituted into the governing PDE, Equation (8.12), to give

$$\frac{\partial a}{\partial z} = Aa(r + \Delta r, z) + Ba(r, z) + Ca(r - \Delta r, z) + \mathscr{R}_A/V_z(r) \tag{8.19}$$

where

$$\begin{aligned}
A &= \mathscr{D}_A[1/(2r\,\Delta r) + 1/\Delta r^2]/V_z(r) \\
B &= \mathscr{D}_A[-2/\Delta r^2]/V_z(r) \\
C &= \mathscr{D}_A[-1/(2r\,\Delta r) + 1/\Delta r^2]/V_z(r)
\end{aligned} \tag{8.20}$$

Equation (8.19) is identical to Equation (8.12) in the limit as $\Delta r \to 0$ and is a reasonable approximation to it for small but finite Δr. It can be rewritten in terms of the index variable i. For $i = 1, \ldots, I - 1$,

$$\frac{da(i, z)}{dz} = A(i)a(i + 1, z) + B(i)a(i, z) + C(i)a(i - 1, z) + \mathscr{R}_A/V_z(i) \tag{8.21}$$

In this formulation, the concentrations have been discretized and are now given by a set of ODEs—a typical member of the set being Equation (8.21), which applies for $i=1$ to $i=I-1$. As indicated by the notation in Equation (8.21), A, B, and C depend on i since, as shown by Equation (8.20), they depend on $r = i\Delta r$. Special forms, developed below, apply at the centerline where $i=0$ and at the wall where $i=I$.

Equation (8.12) becomes indeterminate at the centerline since both r and $\partial a/\partial r$ go to zero. Application of L'Hospital's rule gives a special form for $r=0$:

$$\frac{\partial a}{\partial z} = \frac{\mathscr{D}_A}{V_z(0)}\left[2\frac{\partial^2 a}{\partial z^2}\right] + \frac{\mathscr{R}_A}{V_z(0)} \quad \text{at} \quad r=0$$

Applying the difference approximation of Equation (8.18) and noting that $a(1, z) = a(-1, z)$ due to the assumed symmetry at the centerline gives

$$\frac{da}{dz} = A(0)\,a(1, z) + B(0)\,a(0, z) + (\mathscr{R}_A)_0/V_z(0) \quad \text{at} \quad r=0 \tag{8.22}$$

where

$$\begin{aligned} A(0) &= \mathscr{D}_A[4/\Delta r^2]/V_z(0) \\ B(0) &= \mathscr{D}_A[-4/\Delta r^2]/V_z(0) \end{aligned} \tag{8.23}$$

The concentration at the wall, $a(I)$, is found by applying the zero flux boundary condition, Equation (8.14). A simple way is to set $a(I) = a(I-1)$ since this gives a zero first derivative. However, this approximation to a first derivative converges only $\mathbf{O}(\Delta r)$ while all the other approximations converge $\mathbf{O}(\Delta r^2)$. A better way is to use

$$a_{new}(I) = \frac{4a_{new}(I-1) - a_{new}(I-2)}{3} \tag{8.24}$$

which converges $\mathbf{O}(\Delta r^2)$. This result comes from fitting $a(i)$ as a quadratic in i in the vicinity of the wall. The constants in the quadratic are found from the values of $a(I-1)$ and $a(I-2)$ and by forcing $\partial a/\partial r = 0$ at the wall. Alternatively, Equation (8.24) is obtained by using a second-order, forward difference approximation for the derivative at $r = R$. See Appendix 8.2.

Equations (8.21) and (8.22) constitute a set of simultaneous ODEs in the independent variable z. The dependent variables are the $a(i)$ terms. Each ODE is coupled to the adjacent ODEs; i.e., the equation for $a(i)$ contains $a(i-1)$ and $a(i+1)$. Equation (8.24) is a special, degenerate member of the set, and Equation (8.22) for $a(0)$ is also special because, due to symmetry, there is only one adjacent point, $a(1)$. The overall set may be solved by any desired method. Euler's method is discussed below and is illustrated in Example 8.5. There are a great variety of commercial and freeware packages available for solving simultaneous ODEs. Most of them even work. Packages designed for stiff equations are best. The stiffness arises from the fact that $V_z(i)$ becomes very small near the tube wall. There are also software packages that will handle the discretization automatically.

8.3.2 Euler's Method

Euler's method for solving the above set of ODEs uses a first-order, forward difference approximation in the z-direction, Equation (8.16). Substituting this into Equation (8.21) and solving for the forward point gives

$$a_{new}(i) = A(i)\Delta z a_{old}(i+1) + [1 + B(i)\Delta z]a_{old}(i) \\ + C(i)\Delta z a_{old}(i-1) + (\mathcal{R}_A)_i \Delta z/V_z(i) \quad \text{for} \quad i = 1 \text{ to } I-1 \quad (8.25)$$

where A, B, and C are given by Equation (8.20). The equation for the centerline is

$$a_{new}(0) = A(0)\Delta z a_{old}(1) + [1 + B(0)\Delta z]a_{old}(0) \\ + (\mathcal{R}_A)_0 \Delta z/V_z(0) \quad (8.26)$$

where A and B are given by Equation (8.23). The wall equation finishes the set:

$$a_{new}(I) = \frac{4a_{new}(I-1) - a_{new}(I-2)}{3} \quad (8.27)$$

Equations (8.25) through (8.27) allow concentrations to be calculated at the "new" axial position, $z + \Delta z$, given values at the "old" position, z. If there is no reaction, the new concentration is a weighted average of the old concentrations at three different radial positions, $r + \Delta r$, r, and $r - \Delta r$. In the absence of reaction, there is no change in the average composition, and any concentration fluctuations will gradually smooth out. When the reaction term is present, it is evaluated at the old ith point. Figure 8.2 shows a diagram of the computational scheme. The three circled points at axial position z are used to calculate the new value at the point $z + \Delta z$. The dotted lines in Figure 8.2 show how the radial position r can be changed to determine concentrations for the various values of i. The complete radial profile at $z + \Delta z$ can be found from knowledge of the profile at z. The profile at $z = 0$ is known from the inlet boundary condition, Equation (8.13). The marching-ahead procedure can be used to find the profile at $z = \Delta z$, and so on, repeating the procedure in a stepwise manner until the end of the tube is reached. Colloquially, this solution technique can be called marching ahead with a sideways shuffle. It is worth noting that the axial step size Δz can be changed as the calculation proceeds. This may be necessary if the velocity profile changes during the course of the reaction, as discussed in Section 8.7.

Equations (8.25), (8.26), and (8.27) use the dimensioned independent variables, r and z, but use of the dimensionless variables, $\imath$ and $\jmath$, is often preferred. See Equations (8.56), (8.57), and (8.58) for an example.

A marching-ahead solution to a parabolic partial differential equation is conceptually straightforward and directly analogous to the marching-ahead method we have used for solving ordinary differential equations. The difficulties associated with the numerical solution are the familiar ones of accuracy and stability.

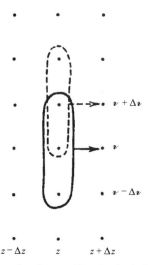

FIGURE 8.2 Computational template for marching-ahead solution.

8.3.3 Accuracy and Stability

The number of radial increments can be picked arbitrarily. A good approach is to begin with a small number, $I=4$, for debugging purposes. When the program is debugged, the value for I is successively doubled until a reasonable degree of accuracy is achieved or until computational times become excessive. If the latter occurs first, find a more sophisticated solution method or a faster computer.

Given a value for I and the corresponding value for Δr, it remains to determine Δz. The choice for Δz is not arbitrary but is constrained by stability considerations. One requirement is that the coefficients on the $a_{old}(i)$ and $a_{old}(0)$ terms in Equations (8.25) and (8.26) cannot be negative. Thus, the *numerical* (or *discretization*) *stability criterion* is

$$[1 + B(i)\Delta z] \geq 0 \quad \text{for} \quad i = 0 \text{ to } I - 1 \qquad (8.28)$$

where $B(i)$ is obtained from Equations (8.20) or (8.23). Since $B(i)$ varies with radial position—i.e., with i—the stability criterion should be checked at all values of i. Normal velocity profiles will have $V_z(R) = 0$ due to the zero-slip condition of hydrodynamics. For such profiles, the near-wall point, $r = R - \Delta r$, will generally give the most restrictive—i.e., smallest—value for Δz.

$$\Delta z_{max} = \frac{\Delta r^2 V_z(R - \Delta r)}{2 \mathscr{D}_A} \qquad (8.29)$$

This stability requirement is quite demanding. Superficially, it appears that Δz_{max} decreases as Δr^2, but $V_z(R - \Delta r)$ is also decreasing, in approximate proportion to Δr. The net effect is that Δz_{max} varies as Δr^3. Doubling the number of radial points will increase the number of axial points by a factor of 8 and will increase the computation time by a factor of 16. The net effect is that Δz quickly becomes so small that the convergence order of the ODE-solver ceases to be important.

Equation (8.29) provides no guarantee of stability. It is a necessary condition for stability that is imposed by the discretization scheme. Practical experience indicates that it is usually a sufficient condition as well, but exceptions exist when reaction rates (or heat-generation rates) become very high, as in regions near thermal runaway. There is a second, *physical stability* criterion that prevents excessively large changes in concentration or temperature. For example, Δa, the calculated change in the concentration of a component that is consumed by the reaction, must be smaller than a itself. Thus, there are two stability conditions imposed on Δz: numerical stability and physical stability. Violations of either stability criterion are usually easy to detect. The calculation blows up. Example 8.8 shows what happens when the numerical stability limit is violated.

Regarding accuracy, the finite difference approximations for the radial derivatives converge $O(\Delta r^2)$. The approximation for the axial derivative converges $O(\Delta z)$, but the stability criterion forces Δz to decrease at least as fast as Δr^2. Thus, the entire computation should converge $O(\Delta r^2)$. The proof of convergence requires that the computations be repeated for a series of successively smaller grid sizes.

8.3.4 The Trapezoidal Rule

The final step in the design calculations for a laminar flow reactor is determination of mixing-cup averages based on Equation (8.4). The trapezoidal rule is recommended for this numerical integration because it is easy to implement and because it converges $O(\Delta r^2)$ in keeping with the rest of the calculations.

For I equally sized increments in the radial direction, the general form for the trapezoidal rule is

$$\int_0^R F(r)dr \approx \Delta r \left[\frac{F(0)}{2} + \frac{F(I)}{2} + \sum_{i=1}^{I-1} F(i) \right] \tag{8.30}$$

For the case at hand,

$$F(r) = 2\pi r a(r) V_z(r) = 2\pi i \Delta r a(i) V_z(i) \tag{8.31}$$

Both $F(0)$ and $F(R)$ vanish for a velocity profile with zero slip at the wall. The mixing-cup average is determined when the integral of $F(r)$ is normalized by $Q = \pi R^2 \bar{u}$. There is merit in using the trapezoidal rule to calculate Q by integrating $dQ = 2\pi r V_z dr$. Errors tend to cancel when the ratio is taken.

The next few examples show the various numerical methods for a simple laminar flow reactor, gradually adding complications.

Example 8.2: An isothermal reactor with $L=2$ m, $R=0.01$ m is being used for a first-order reaction. The rate constant is $0.005\,\text{s}^{-1}$, and $\bar{u}=0.01\,\text{m/s}$. Estimate the outlet concentration, assuming piston flow.

Solution: For piston flow, $a_{out}=a_{in}\exp(-kL/\bar{u})$ and $Y=a_{in}/a_{out}=\exp(-1)=0.3679$.

Example 8.3: The reactor of Example 8.2 is actually in laminar flow with a parabolic velocity profile. Estimate the outlet concentration ignoring molecular diffusion.

Solution: Example 8.1 laid the groundwork for this case of laminar flow without diffusion. The mixing-cup average is

$$Y = \frac{a_{mix}(L)}{a_{in}} = \frac{\int_0^R 2\pi r V_z(r) \exp[-kL/V_z(r)]\,dr}{Q}$$

The following Excel macro illustrates the use of the trapezoidal rule for evaluating both the numerator and denominator in this equation.

```
DefDbl A-H, K-L, P-Z
DefLng I-J, M-O
Sub Exp8_3()

L=2
Ro=0.01
U=0.01
k=0.005
Itotal=2

For jj=1 To 8 'This outer loop varies the radial grid
                'size to test convergence

  Itotal=2 * Itotal
  dr=Ro/Itotal
  Range("A"& CStr(jj)).Select
  ActiveCell.FormulaR1C1=Itotal

  Fsum=0 'Set to F(0)/2+F(R)/2 for the general
          'trapezoidal rule
  Qsum=0 'Set to Q(0)/2+Q(R)/2 for the general
          'trapezoidal rule
  For i=1 To Itotal-1
    r=i * dr
    Vz=2 * U * (1-r^2/Ro^2)
```

```
        Q=r * Vz 'Factor of 2*Pi omitted since it will
                 'cancel in the ratio
        F=Q * Exp(-k * L/Vz)
        Fsum=Fsum+F * dr
        Qsum=Qsum+Q * dr
    Next i
    aout=Fsum/Qsum

    Range("B"& CStr(jj)).Select
    ActiveCell.FormulaR1C1=aout

Next jj

End Sub
```

The results are

I	a_{out}/a_{in}
4	0.46365
8	0.44737
16	0.44413
32	0.44344
64	0.44327
128	0.44322
256	0.44321
512	0.44321

The performance of the laminar flow reactor is appreciably worse than that of a PFR, but remains better than that of a CSTR (which gives $Y=0.5$ for $k\bar{t}=1$). The computed value of 0.4432 may be useful in validating more complicated codes that include diffusion.

Example 8.4: Suppose that the reactive component in the laminar flow reactor of Example 8.2 has a diffusivity of 5×10^{-9} m²/s. Calculate the minimum number of axial steps, J, needed for discretization stability when the radial increments are sized using $I=4, 8, 16, 32, 64$, and 128. Also, suggest some actual step sizes that would be reasonable to use.

Solution: Begin with $I=4$ so that $\Delta r = R/I = 0.0025$ m. The near-wall velocity occurs at $r = R - \Delta r = 0.0075$ m:

$$V_z = 2\bar{u}[1 - r^2/R^2] = 0.02[1 - 0.0075^2/(0.01)^2] = 0.00875 \text{ m/s}$$

$$\Delta z_{max} = \Delta r^2 V_z (R - \Delta r)/[2\mathscr{D}_A] = (0.0025)^2(0.00875)/2 \times 10^{-9} = 5.47 \text{ m}$$

$J_{min} = L/\Delta z_{max} = 2/5.47 = 0.3656$, but this must be rounded up to an integer. Thus, $J_{min}=1$ for $I=4$. Repeating the calculations for the other values of I gives

I	J_{min}	J_{used}
4	1	2
8	3	4
16	22	32
32	167	256
64	1322	2048
128	10527	16384

The third column represents choices for J that are used in the examples that follow. For $I=8$ and higher, they increase by a factor of 8 as I is doubled.

Example 8.5: Use the method of lines combined with Euler's method to determine the mixing-cup average outlet for the reactor of Example 8.4.

Solution: For a first-order reaction, we can arbitrarily set $a_{in}=1$ so that the results can be interpreted as the fraction unreacted. The choices for I and J determined in Example 8.4 will be used. The marching-ahead procedure uses Equations (8.25), (8.26), and (8.27) to calculate concentrations. The trapezoidal rule is used to calculate the mixing-cup average at the end of the reactor. The results are

I	J	a_{out}/a_{in}
4	1	0.37363
8	4	0.39941
16	32	0.42914
32	256	0.43165
64	2048	0.43175
128	16384	0.43171

These results were calculated using the following Excel macro:

```
DefDbl A-H, K-L, P-Z
DefLng I-J, M-O
Sub Fig8_1()
Dim aold(256), anew(256), Vz(256)
Dim A(256), B(256), C(256), D(256)

ain=1
Da=0.000000005
L=2
R=0.01
U=0.01
k=0.005
Itotal=2
```

```
For jj = 1 To 7  'This outer loop varies Itotal to check
                'convergence

    Itotal = 2 * Itotal
    If Itotal = 4 Then JTotal = 2
    If Itotal = 8 Then JTotal = 4
    If Itotal > 8 Then JTotal = 8 * JTotal
    dr = R/Itotal
    dz = L/JTotal

'Set constants in Equation 8.26
    A(0) = 4 * Da/dr ^ 2 * dz/2/U
    B(0) = - 4 * Da/dr ^ 2 * dz/2/U
    D(0) = - k * dz/2/U
    aold(0) = 1

'Set constants in Equation 8.25
    For i = 1 To Itotal - 1
        Vz(i) = 2 * U * (1 - (i * dr) ^ 2/R ^ 2)
        A(i) = Da * (1/(2 * dr ^ 2 * i) + 1/dr ^ 2) * dz/Vz(i)
        B(i) = Da * (-2/dr ^ 2) * dz/Vz(i)
        C(i) = Da * (-1/(2 * dr ^ 2 * i) + 1/dr ^ 2) * dz/Vz(i)
        D(i) = - k * dz/Vz(i)
        aold(i) = 1
    Next

'Set the initial conditions
    For i = 0 To Itotal
        aold(i) = ain
    Next

'March down the tube
    For j = 1 To JTotal
        anew(0) = A(0) * aold(1) + (1 + B(0)) * aold(0)
        + D(0) * aold(0)

'This is the sideways shuffle
    For i = 1 To Itotal - 1
        x = A(i) * aold(i + 1) + (1 + B(i)) * aold(i)
        anew(i) = x + C(i) * aold(i - 1) + D(i) * aold(i)
    Next j
        Next i
'Apply the wall boundary condition, Equation 8.27
```

```
  anew(Itotal) = 4 * anew(Itotal − 1)/3 − anew(Itotal − 2)/3
  'March a step forward
    For i = 0 To Itotal
      aold(i) = anew(i)
    Next i
    'Calculate the mixing cup average
    F = 0
    Q = 0
    For i = 1 To Itotal − 1
      F = F + 2 * dr * i * Vz(i) * anew(i)
      Q = Q + 2 * dr * i * Vz(i)
    Next i
    Y = F/Q

  'Output results for this mesh size
    Range("A"& CStr(jj)).Select
    ActiveCell.FormulaR1C1 = Itotal
    Range("B"& CStr(jj)).Select
    ActiveCell.FormulaR1C1 = JTotal
    Range("C"& CStr(jj)).Select
    ActiveCell.FormulaR1C1 = Y

  Next jj

  End Sub
```

Example 8.5 has $\mathscr{D}_A \bar{t}/R^2 = 0.01$. Since this is larger than 0.003, diffusion should have some effect according to Merrill and Hamrin. The diffusion-free result for $k\bar{t} = 1$ was found to be $Y = 0.4432$ in Example 8.3. The Example 8.5 result of 0.4317 is closer to piston flow, as expected.

8.3.5 Use of Dimensionless Variables

Example 8.5 used the natural, physical variables and the natural dimensions of the problem. A good case can be made for this practice. It is normal in engineering design since it tends to keep the physics of the design transparent and avoids errors, particularly when using physical property correlations. However, it is desirable to use dimensionless variables when results are being prepared for general use, as in a literature publication or when the calculations are so lengthy that rerunning them would be cumbersome. The usual approach in the chemical engineering literature is to introduce scaled, dimensionless independent variables quite early in the analysis of a problem.

The use of dimensionless variables will be illustrated using Equation (8.12) but with an added term for axial diffusion:

$$V_z(r)\frac{\partial a}{\partial z} = \mathscr{D}_A\left[\frac{\partial^2 a}{\partial z^2} + \frac{1}{r}\frac{\partial a}{\partial r} + \frac{\partial^2 a}{\partial r^2}\right] + \mathscr{R}_A \quad (8.32)$$

There are two independent variables, z and r. Both are lengths. They can be scaled separately using two different characteristic lengths or they can be scaled using a single characteristic length. We use two different lengths and define new variables $\mathscr{z} = z/L$ and $\mathscr{r} = r/R$ so that they both have a range from 0 to 1. Substituting the new variables into Equation (8.32) and doing some algebra gives

$$\frac{\partial a}{\partial \mathscr{z}} = \left(\frac{\mathscr{D}_A L}{R^2 V_z}\right)\left[\left(\frac{R^2}{L^2}\right)\frac{\partial^2 a}{\partial \mathscr{z}^2} + \frac{1}{\mathscr{r}}\frac{\partial a}{\partial \mathscr{r}} + \frac{\partial^2 a}{\partial \mathscr{r}^2}\right] + \mathscr{R}_A L/V_z \quad (8.33)$$

When expressed in the scaled variables, the $\partial^2 a/\partial \mathscr{z}^2$ and $\partial^2 a/\partial \mathscr{r}^2$ terms have the same magnitude, but the $\partial^2 a/\partial \mathscr{z}^2$ term is multiplied by a factor of R^2/L^2 that will not be larger than 0.01. Thus, this term, which corresponds to axial diffusion, may be neglected, consistent with the conclusion in Section 8.2.

The velocity profile is scaled by the mean velocity, $\bar{u}$, giving the dimensionless profile $\mathscr{V}_z(\mathscr{r}) = V_z(r)/\bar{u}$. To complete the conversion to dimensionless variables, the dependent variable, a, is divided by its nonzero inlet concentration. The dimensionless version of Equation (8.12) is

$$\mathscr{V}_z(\mathscr{r})\frac{\partial a^*}{\partial \mathscr{z}} = \left(\frac{\mathscr{D}_A \bar{t}}{R^2}\right)\left[\frac{1}{\mathscr{r}}\frac{\partial a^*}{\partial \mathscr{r}} + \frac{\partial^2 a^*}{\partial \mathscr{r}^2}\right] + \mathscr{R}_A \bar{t}/a_{in} \quad (8.34)$$

where $\bar{t} = L/\bar{u}$. Equation (8.34) contains the dimensionless number $\mathscr{D}_A \bar{t}/R^2$ that appears in Merrill and Hamrin's criterion, Equation (8.3), and a dimensionless reaction rate, $\mathscr{R}_A \bar{t}/a_{in}$. Merrill and Hamrin assumed a first-order reaction, $\mathscr{R}_A = -ka$, and calculated $a_{out} = a_{mix}(L)$ for various values of $\mathscr{D}_A \bar{t}/R^2$. They concluded that diffusion had a negligible effect on a_{out} when Equation (8.3) was satisfied.

The stability criterion, Equation (8.29), can be converted to dimensionless form. The result is

$$\Delta \mathscr{z}_{max} = 1/J_{min} = \frac{\Delta z_{max}}{L} = \frac{\Delta \mathscr{r}^2 \mathscr{V}_z(1-\Delta \mathscr{r})}{2[\mathscr{D}_A \bar{t}/R^2]} \quad (8.35)$$

and for the special case of a parabolic profile,

$$\Delta \mathscr{z}_{max} = \frac{\Delta \mathscr{r}^3[2-\Delta \mathscr{r}]}{2[\mathscr{D}_A \bar{t}/R^2]} \quad (8.36)$$

Example 8.6: Generalize Example 8.5 to determine the fraction unreacted for a first-order reaction in a laminar flow reactor as a function of the dimensionless groups $\mathscr{D}_A \bar{t}/R^2$ and $k\bar{t}$. Treat the case of a parabolic velocity profile.

Solution: The program of Example 8.5 can be used with minor modifications. Set U, R, and L all equal to 1. Then $\mathscr{D}_A \bar{t}/R^2$ will be equal to the value assigned to Da and $k\bar{t}$ will be equal to the value assigned to k. It is necessary to use the stability criterion to determine J. Example 8.5 had $\mathscr{D}_A \bar{t}/R^2 = 0.01$, and larger values for $\mathscr{D}_A \bar{t}/R^2$ require larger values for J.

Figure 8.1 includes a curve for laminar flow with $\mathscr{D}_A \bar{t}/R^2 = 0.1$. The performance of a laminar flow reactor with diffusion is intermediate between piston flow and laminar flow without diffusion, $\mathscr{D}_A \bar{t}/R^2 = 0$. Laminar flow reactors give better conversion than CSTRs, but do not generalize this result too far! It is restricted to a parabolic velocity profile. Laminar velocity profiles exist that, in the absence of diffusion, give reactor performance far worse than a CSTR.

Regardless of the shape of the velocity profile, radial diffusion will improve performance, and the case $\mathscr{D}_A \bar{t}/R^2 \to \infty$ corresponds to piston flow.

The thoughtful reader may wonder about a real reactor with a high level of radial diffusion. Won't there necessarily be a high level of axial diffusion as well and won't the limit of $\mathscr{D}_A \bar{t}/R^2 \to \infty$ really correspond to a CSTR rather than a PFR? The answer to this question is "yes, but...." The "but" is based on the restriction that $L/R > 16$. For reasonably long reactors, the effects of radial diffusion dominate those of axial diffusion until extremely high values of $\mathscr{D}_A \bar{t}/R^2$. If reactor performance is considered as a function of $\mathscr{D}_A \bar{t}/R^2$ (with $k\bar{t}$ fixed), there will be an interior maximum in performance as $\mathscr{D}_A \bar{t}/R^2 \to \infty$. This is the piston flow limit illustrated in Figure 8.3. There is another limit, that of a perfectly mixed flow reactor, which occurs at much higher values of $\mathscr{D}_A \bar{t}/R^2$ than those shown in Figure 8.3. The tools needed to quantify this idea are developed in Chapter 9. See Problem 9.11, but be warned that the computations are difficult and of limited utility.

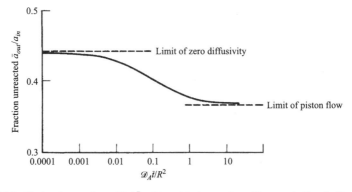

FIGURE 8.3 First-order reaction with $k\bar{t} = 1$ in a tubular reactor with a parabolic velocity profile.

8.4 SLIT FLOW AND RECTANGULAR COORDINATES

Results to this point have been confined to tubular reactors with circular cross sections. Tubes are an extremely practical geometry that is widely used for chemical reactors. Less common is slit flow such as occurs between closely spaced parallel plates, but practical heat exchangers and reactors do exist with this geometry. They are used when especially good mixing is needed within the cross section of the reactor. Using spiral-wound devices or stacked flat plates, it is practical to achieve slit heights as small as 0.003 m. This is far smaller than is feasible using a conventional, multitubular design.

Figure 8.4 illustrates pressure-driven flow between flat plates. The downstream direction is z. The cross-flow direction is y, with $y=0$ at the centerline and $y=\pm Y$ at the walls so that the channel height is $2Y$. Suppose the slit width (x-direction) is very large so that sidewall effects are negligible. The velocity profile for a laminar, Newtonian fluid of constant viscosity is

$$V_z(y) = 1.5\bar{u}\left[1 - \frac{y^2}{Y^2}\right] \tag{8.37}$$

The analog of Equation (8.12) in rectangular coordinates is

$$V_z(y)\frac{\partial a}{\partial z} = \mathscr{D}_A\left[\frac{\partial^2 a}{\partial y^2}\right] + \mathscr{R}_A \tag{8.38}$$

The boundary conditions are

$$\begin{aligned} a &= a_{in}(y) \quad \text{at} \quad z = 0 \\ \partial a/\partial y &= 0 \quad \text{at} \quad y = 0 \\ \partial a/\partial y &= 0 \quad \text{at} \quad y = \pm Y \end{aligned} \tag{8.39}$$

The zero slope boundary condition at $y=0$ assumes symmetry with respect to the centerline. The mathematics are then entirely analogous to those for the tubular geometries considered previously. Applying the method of lines gives

$$\frac{\partial a(y, z)}{\partial z} = Aa(y + \Delta y, z) + Ba(y, z) + C(y - \Delta y, z) + \frac{\mathscr{R}_A}{V_z(y)} \tag{8.40}$$

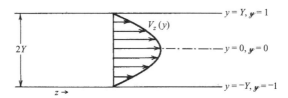

FIGURE 8.4 Pressure driven flow between parallel plates with both plates stationary.

$$A = \frac{\mathscr{D}_A}{V_z}\left[\frac{1}{\Delta y^2}\right]$$

$$B = \frac{\mathscr{D}_A}{V_z}\left[\frac{-2}{\Delta y^2}\right] \qquad (8.41)$$

$$C = \frac{\mathscr{D}_A}{V_z}\left[\frac{1}{\Delta y^2}\right] = A$$

With these revised definitions for A, B, and C, the marching-ahead equation for the interior points is identical to that for cylindrical coordinates, Equation (8.25). The centerline equation is no longer a special case except for the symmetry boundary condition that forces $a(-1) = a(1)$. The centerline equation is thus

$$a_{new}(0) = 2A(0)\,\Delta z\, a_{old}(1) + [1 + B(0)\,\Delta z]a_{old}(0) + (\mathscr{R}_A)_0 \Delta z / V_z(0) \qquad (8.42)$$

The wall boundary condition is unchanged, Equation (8.27).
The near-wall stability condition is

$$\Delta z_{max} = \frac{\Delta y^2 V_z(Y - \Delta y)}{2\mathscr{D}_A} \qquad (8.43)$$

Mixing-cup averages are calculated using

$$F(i) = a(i)V_z(i) \qquad (8.44)$$

instead of Equation (8.31), and Q can be obtained by integrating $dQ = V_z(y)dy$.

Example 8.7: Determine the flat-plate equivalent to Merrill and Hamrin's criterion.

Solution: Transform Equation (8.38) using the dimensionless independent variables $\mathscr{y} = z/L$ and $\mathscr{y} = y/Y$:

$$\mathscr{V}_z(\mathscr{y})\frac{\partial a}{\partial \mathscr{y}} = \left(\frac{\mathscr{D}_A \bar{t}}{Y^2}\right)\left[\frac{\partial^2 a}{\partial \mathscr{y}^2}\right] + \mathscr{R}_A \bar{t} \qquad (8.45)$$

Comparing this equation with Equation (8.34) shows that $\mathscr{D}_A \bar{t}/Y^2$ is the flat-plate counterpart of $\mathscr{D}_A \bar{t}/R^2$. We thus seek a value for $\mathscr{D}_A \bar{t}/Y^2$ below which diffusion has a negligible effect on the yield of a first-order reaction.

For comparison purposes, set $k\bar{t} = 1$ and compute a_{out}/a_{in} for the tubular case with $\mathscr{D}_A \bar{t}/R^2 = 0$ and with $\mathscr{D}_A \bar{t}/R^2 = 0.003$. The results using the programs in Examples 8.3 and 8.5 with $I = 128$ are 0.44321 and 0.43849, respectively. Thus, Merrill and Hamrin considered the difference between 0.44321 and 0.43849 to be negligible.

Turn now to the flat-plate geometry. The coefficients A, B, and C, and the mixing-cup averaging technique must be revised. This programming exercise is left to the reader. Run the modified program with $k\bar{t} = 1$ but without diffusion to give $a_{out}/a_{in} = 0.41890$ for $I = 128$ and $J = 16382$. The flat-plate geometry gives better performance than the tube. Why?

To ensure an apples-to-apples comparison, reduce $k\bar{t}$ until a_{out}/a_{in} matches the value of 0.44321 achieved in the tube. This is found to occur at $k\bar{t} = 0.9311$. Diffusion is now added until $a_{out}/a_{in} = 0.43849$ as in the case of a circular tube with $\mathscr{D}_A\bar{t}/R^2 = 0.003$. This is found to occur at about $\mathscr{D}_A\bar{t}/Y^2 = 0.008$. Thus, the flat-plate counterpart to the Merrill and Hamrin criterion is

$$\mathscr{D}_A\bar{t}/Y^2 < 0.008 \tag{8.46}$$

8.5 SPECIAL VELOCITY PROFILES

This section considers three special cases. The first is a flat velocity profile that can result from an extreme form of fluid rheology. The second is a linear profile that results from relative motion between adjacent solid surfaces. The third special case is for motionless mixers where the velocity profile is very complex, but its net effects can sometimes be approximated for reaction engineering purposes.

8.5.1 Flat Velocity Profiles

Flow in a Tube. Laminar flow with a flat velocity profile and slip at the walls can occur when a viscous fluid is strongly heated at the walls or is highly non-Newtonian. It is sometimes called *toothpaste flow*. If you have ever used Stripe® toothpaste, you will recognize that toothpaste flow is quite different than piston flow. Although $V_z(r) = \bar{u}$ and $\mathscr{V}_z(\iota) = 1$, there is little or no mixing in the radial direction, and what mixing there is occurs by diffusion. In this situation, the centerline is the critical location with respect to stability, and the stability criterion is

$$\Delta z_{max} = \frac{\Delta r^2 \bar{u}}{4\mathscr{D}_A} \tag{8.47}$$

and Δz_{max} varies as Δr^2. The flat velocity profile and Equation (8.47) apply to the packed-bed models treated in Chapter 9. The marching-ahead equations are unchanged from those presented in Section 8.3.1, although the coefficients must be evaluated using the flat profile.

Toothpaste flow is an extreme example of non-Newtonian flow. Problem 8.2 gives a more typical example. Molten polymers have velocity profiles that are flattened compared with the parabolic distribution. Calculations that assume a parabolic profile will be conservative in the sense that they will predict a lower conversion than would be predicted for the actual profile. The changes in velocity profile due to variations in temperature and composition are normally much more important than the fairly subtle effects due to non-Newtonian behavior.

Flow in a Slit. Turning to a slit geometry, a flat velocity profile gives the simplest possible solution using Euler's method. The stability limit is independent of y:

$$\Delta z_{max} = \frac{\Delta y^2 \bar{u}}{2 \mathscr{D}_A} \tag{8.48}$$

$$\Delta \tilde{y}_{max} = \frac{\Delta \tilde{y}^2}{2[\mathscr{D}_A \bar{t}/Y^2]}$$

The marching-ahead equation is also independent of y:

$$a_{new}(i) = A a_{old}(i+1) + (1 - 2A) a_{old}(i) + A a_{old}(i-1) + (\mathscr{R}_A)_i \bar{t} \Delta \tilde{y}_{max} \tag{8.49}$$

where

$$A = 0.5 \Delta \tilde{y} / \Delta \tilde{y}_{max} \tag{8.50}$$

Note that Equation (8.49) applies for every point except for $y = Y$ where the wall boundary condition is used, e.g., Equation (8.27). When $i = 0$, $a_{old}(-1) = a_{old}(+1)$.

Example 8.8: Explore conservation of mass, stability, and instability when the convective diffusion equation is solved using the method of lines combined with Euler's method.

Solution: These aspects of the solution technique can be demonstrated using Equation (8.49) as an algebraically simple example. Set $\mathscr{R}_A = 0$ and note that a uniform profile with $a_{old}(y) = a_{in}$ will propagate downstream as $a_{new}(y) = a_{in}$ so that mass is conserved. In the more general cases, such as Equation (8.25), $A + B + C = 0$ ensures that mass will be conserved.

According to Equation (8.50), the largest value for A that will give a stable solution is 0.5. With $A = 0.5$, Equation (8.49) becomes

$$a_{new}(i) = 0.5 a_{old}(i+1) + 0.5 a_{old}(i-1)$$

The use of this equation for a few axial steps within the interior region of the slit is illustrated below:

0	0	0	0	0	0	0	0	0.5
0	0	0	0	0	0	0	1	0
0	0	0	0	0	0	2	0	2.5
0	0	0	0	0	4	0	4	0
0	0	0	0	8	0	6	0	5
0	0	0	16	0	8	0	6	0
0	0	0	0	8	0	6	0	5
0	0	0	0	0	4	0	4	0
0	0	0	0	0	0	2	0	2.5
0	0	0	0	0	0	0	1	0
0	0	0	0	0	0	0	0	0.5

In this example, an initial steady-state solution with $a=0$ is propagated downstream. At the fourth axial position, the concentration in one cell is increased to 16. This can represent round-off error, a numerical blunder, or the injection of a tracer. Whatever the cause, the magnitude of the upset decreases at downstream points and gradually spreads out due to diffusion in the y-direction. The total quantity of injected material (16 in this case) remains constant. This is how a real system is expected to behave. The solution technique conserves mass and is stable.

Now consider a case where A violates the stability criterion. Pick $A=1$ to give

$$a_{new}(i) = a_{old}(i+1) - a_{old}(i) + a_{old}(i-1)$$

The solution now becomes

0	0	0	0	0	0	0	0	16
0	0	0	0	0	0	0	16	−80
0	0	0	0	0	0	16	−64	240
0	0	0	0	0	16	−48	160	−480
0	0	0	0	16	−32	96	−256	720
0	0	0	16	−16	48	−112	304	−816
0	0	0	0	16	−32	96	−256	720
0	0	0	0	0	16	−48	160	−480
0	0	0	0	0	0	16	−64	240
0	0	0	0	0	0	0	16	−80
0	0	0	0	0	0	0	0	16

This equation continues to conserve mass but is no longer stable. The original upset grows exponentially in magnitude and oscillates in sign. This marching-ahead scheme is clearly unstable in the presence of small blunders or round-off errors.

8.5.2 Flow Between Moving Flat Plates

Figure 8.5 shows another flow geometry for which rectangular coordinates are useful. The bottom plate is stationary but the top plate moves at velocity $2\bar{u}$.

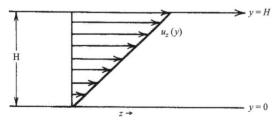

FIGURE 8.5 Drag flow between parallel plates with the upper plate in motion and no axial pressure drop.

The plates are separated by distance H, and the y-coordinate starts at the bottom plate. The velocity profile is linear:

$$V_z = \frac{2\bar{u}y}{H} \qquad (8.51)$$

This velocity profile is commonly called drag flow. It is used to model the flow of lubricant between sliding metal surfaces or the flow of polymer in extruders. A pressure-driven flow—typically in the opposite direction—is sometimes superimposed on the drag flow, but we will avoid this complication. Equation (8.51) also represents a limiting case of Couette flow (which is flow between coaxial cylinders, one of which is rotating) when the gap width is small. Equation (8.38) continues to govern convective diffusion in the flat-plate geometry, but the boundary conditions are different. The zero-flux condition applies at both walls, but there is no line of symmetry. Calculations must be made over the entire channel width and not just the half-width.

8.5.3 Motionless Mixers

Most motionless or static mixers consist of tubes or ducts in which stationary vanes (elements) have been installed to promote radial flow. There are many commercial types, some of which are shown in Figure 8.6. Similar results can be achieved in deep laminar flow by using a series of helically coiled tubes where the axis of each successive coil is at a 90° angle to the previous coil axis.[4] With enough static mixing elements or helical coils in series, piston flow can be approached. The flow geometry is complex and difficult to analyze. Velocity profiles, streamlines, and pressure drops can be computed using programs for *computational fluid dynamics* (CFD), such as Fluent®, but these computations have not yet become established and verified as design tools. The axial dispersion model discussed in Chapter 9 is one approach to data correlation. Another approach is to use Equation (8.12) for segments of the reactor but to periodically reinitialize the concentration profile. An empirical study[5] on Kenics-type static mixers found that four of the Kenics elements correspond to one zone of complete radial mixing. The computation is as follows:

1. Start with a uniform concentration profile, $a(z) = a_{in}$ at $z = 0$.
2. Solve Equation (8.12) using the methods described in this chapter and ignoring the presence of the mixing elements.
3. When an axial position corresponding to four mixing elements is reached, calculate the mixing-cup average composition a_{mix}.
4. Restart the solution of Equation (8.12) using a uniform concentration profile equal to the mixing-cup average, $a(z) = a_{mix}$.
5. Repeat Steps 2 through 4 until the end of the reactor is reached.

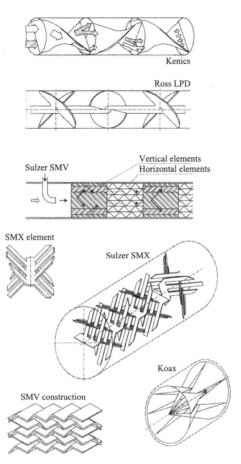

FIGURE 8.6 Commerical motionless mixers. (Drawing courtesy of Professor Pavel Ditl, Czech Technical University.)

This technique should give reasonable results for isothermal, first-order reactions. It and other modeling approaches are largely untested for complex and nonisothermal reactions.

8.6 CONVECTIVE DIFFUSION OF HEAT

Heat diffuses much like mass and is governed by similar equations. The temperature analog of Equation (8.12) is

$$V_z(r)\frac{\partial T}{\partial z} = \alpha_T\left[\frac{1}{r}\frac{\partial T}{\partial r} + \frac{\partial^2 T}{\partial r^2}\right] - \frac{\Delta H_R \mathcal{R}}{\rho C_P} \qquad (8.52)$$

where α_T is the thermal diffusivity and $\Delta H_R \mathscr{R}$ follows the summation convention of Equation (5.17). The units on thermal diffusivity are the same as those on molecular diffusivity, m²/s, but α_T will be several orders of magnitude larger than $\mathscr{D}_A$. The reason for this is that mass diffusion requires the actual displacement of molecules but heat can be transferred by vibrations between more-or-less stationary molecules or even between parts of a molecule as in a polymer chain. Note that $\alpha_T = \lambda/(\rho C_P)$, where λ is the thermal conductivity. Equation (8.52) assumes constant α_T and ρ. The assumption of constant density ignores expansion effects that can be significant in gases that are undergoing large pressure changes. Also ignored is viscous dissipation, which can be important in very high-viscosity fluids such as polymer melts. Standard texts on transport phenomena give the necessary embellishments of Equation (8.52).

The inlet and centerline boundary conditions associated with Equation (8.52) are similar to those used for mass transfer:

$$T = T_{in}(r) \quad \text{at} \quad z = 0 \qquad (8.53)$$

$$\partial T/\partial r = 0 \quad \text{at} \quad r = 0 \qquad (8.54)$$

The usual wall boundary condition is

$$T = T_{wall}(z) \quad \text{at} \quad r = R \qquad (8.55)$$

but the case of an insulated wall,

$$\partial T/\partial r = 0 \quad \text{at} \quad r = R$$

is occasionally used.

Equation (8.52) has the same form as Equation (8.12), and the solution techniques are essentially identical. Replace a with T, $\mathscr{D}_A$ with α_T, and $\mathscr{R}_A$ with $-\Delta H_R \mathscr{R}/(\rho C_P)$, and proceed as in Section 8.3.

The equations governing the convective diffusion of heat and mass are coupled through the temperature and composition dependence of the reaction rates. In the general case, Equation (8.52) is solved simultaneously with as many versions of Equation (8.12) as there are reactive components. The method of lines treats a single PDE as $I-1$ simultaneous ODEs. The general case has $N+1$ PDEs and thus is treated as $(N+1)(I-1)$ ODEs. Coding is easiest when the same axial step size is used for all the ODEs, but this step size must satisfy the most restrictive of the stability criteria. These criteria are given by Equation (8.29) for the various chemical species. The stability criterion for temperature is identical except that α_T replaces the molecular diffusivities and α_T is much larger, which leads to smaller step sizes. Thus, the step size for the overall program will be imposed by the stability requirement for the temperature equation. It may be that accurate results require very small axial steps and excessive computer time. Appendix 8.3 describes alternative finite difference approximations that eliminate the discretization stability condition. Algorithms exist where $\Delta z \sim \Delta r$ rather than $\Delta z \sim \Delta r^2$ (flat profile) or $\Delta z \sim \Delta r^3$ (parabolic profile) so that the number of computations increases by a factor of only 4 (rather than 8 or 16) when Δr is halved. The price for this is greater complexity in the individual calculations.

The equations governing convective diffusion of heat in rectangular-coordinate systems are directly analogous to those governing convective diffusion of mass. See Sections 8.4 and 8.5. The wall boundary condition is usually a specified temperature, and the stability criterion for the heat transfer equation is usually more demanding (smaller Δz_{max}) than that for mass transfer. Also, in slit flow problems, there is no requirement that the two walls be at the same temperature. When the wall temperatures are different, the marching-ahead equations must be applied to the entire slit width, and not just the half-width, since the temperature profiles (and the corresponding composition profiles) will not be symmetric about the centerline. There are no special equations for the centerline. Instead, the ordinary equation for an interior point e.g., Equation (8.40), is used throughout the interior with $a(y) \neq a(-y)$ and $T(y) \neq T(-y)$.

8.6.1 Dimensionless Equations for Heat Transfer

Transformation of the independent variables to dimensionless form uses $\mathscr{r} = r/R$ and $\mathscr{y} = z/L$. In most reactor design calculations, it is preferable to retain the dimensions on the dependent variable, temperature, to avoid confusion when calculating the Arrhenius temperature dependence and other temperature-dependent properties. The following set of marching-ahead equations are functionally equivalent to Equations (8.25)–(8.27) but are written in dimensionless form for a circular tube with temperature (still dimensioned) as the dependent variable. For the centerline,

$$T(0, \mathscr{y} + \Delta\mathscr{y}) = \left[1 - 4\left(\frac{\bar{u}\alpha_T \bar{t}}{\mathscr{V}_z(0)R^2}\right)\frac{\Delta\mathscr{y}}{\Delta\mathscr{r}^2}\right]T(0, \mathscr{y}) + 4\left(\frac{\bar{u}\alpha_T \bar{t}}{\mathscr{V}_z(0)R^2}\right)\frac{\Delta\mathscr{y}}{\Delta\mathscr{r}^2}T(\Delta\mathscr{r}, \mathscr{y})$$
$$- \frac{\Delta H_R \mathscr{R} \bar{t}\bar{u}}{\rho C_P \mathscr{V}_z(0)}\Delta\mathscr{y} \quad (8.56)$$

For interior points,

$$T(\mathscr{r}, \mathscr{y} + \Delta\mathscr{y}) = \left[1 - 2\left(\frac{\bar{u}\alpha_T \bar{t}}{\mathscr{V}_z(\mathscr{r})R^2}\right)\frac{\Delta\mathscr{y}}{\Delta\mathscr{r}^2}\right]T(\mathscr{r}, \mathscr{y})$$
$$+ \left(\frac{\bar{u}\alpha_T \bar{t}}{\mathscr{V}_z(\mathscr{r})R^2}\right)\frac{\Delta\mathscr{y}}{\Delta\mathscr{r}^2}\left[1 + \frac{\Delta\mathscr{r}}{2\mathscr{r}}\right]T(\mathscr{r} + \Delta\mathscr{r}, \mathscr{y})$$
$$+ \left(\frac{\bar{u}\alpha_T \bar{t}}{\mathscr{V}_z(\mathscr{r})R^2}\right)\frac{\Delta\mathscr{y}}{\Delta\mathscr{r}^2}\left[1 - \frac{\Delta\mathscr{r}}{2\mathscr{r}}\right]T(\mathscr{r} - \Delta\mathscr{r}, \mathscr{y}) - \frac{\Delta H_R \mathscr{R} \bar{t}\bar{u}}{\rho C_P \mathscr{V}_z(\mathscr{r})}\Delta\mathscr{y}$$
$$(8.57)$$

At the wall,

$$T_{wall}(\hat{y} + \Delta \hat{y}) = \frac{4T(1 - \Delta \hat{z}, \hat{y} + \Delta \hat{y}) - T(1 - 2\Delta \hat{z}, \hat{y} + \Delta \hat{y})}{3} \quad (8.58)$$

The more restrictive of the following stability criteria is used to calculate $\Delta \hat{y}_{max}$:

$$\frac{\Delta \hat{y}}{\Delta \hat{z}^2} \leq \frac{\mathscr{V}_z(\hat{z})R^2}{2\bar{u}\alpha_T \bar{t}} \qquad \Delta \hat{z} \leq \hat{z} \leq 1 - \Delta \hat{z} \quad (8.59)$$

$$\frac{\Delta \hat{y}}{\Delta \hat{z}^2} \leq \frac{\mathscr{V}_z(0)R^2}{4\bar{u}\alpha_T \bar{t}} \qquad \hat{z} = 0 \quad (8.60)$$

When the heat of reaction term is omitted, these equations govern laminar heat transfer in a tube. The case where T_{in} and T_{wall} are both constant and where the velocity profile is parabolic is known as the Graetz problem. An analytical solution to this linear problem dates from the 19th century but is hard to evaluate and is physically unrealistic. The smooth curve in Figure 8.7 corresponds to the analytical solution and the individual points correspond to a numerical solution found in Example 8.9. The numerical solution is easier to obtain but, of course, is no better at predicting the performance of a real heat exchanger. A major cause for the inaccuracy is the dependence of viscosity on temperature that causes changes in the velocity profile. Heating at the wall improves heat transfer while cooling hurts it. Empirical heat transfer

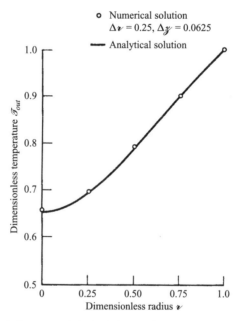

FIGURE 8.7 Numerical versus analytical solutions to the Graetz problem with $\alpha_T \bar{t}/R^2 = 0.4$.

correlations include a viscosity correction factor, e.g., the $(\mu_{bulk}/\mu_{wall})^{0.14}$ term in Equation (5.37). Section 8.7 takes a more fundamental approach by calculating $V_z(r)$ as it changes down the tube.

Example 8.9: Find the temperature distribution in a laminar flow, tubular heat exchanger having a uniform inlet temperature T_{in} and constant wall temperature T_{wall}. Ignore the temperature dependence of viscosity so that the velocity profile is parabolic everywhere in the reactor. Use $\alpha_T \bar{t}/R^2 = 0.4$ and report your results in terms of the dimensionless temperature

$$\mathscr{T} = (T - T_{in})/(T_{wall} - T_{in}) \tag{8.61}$$

Solution: A transformation to dimensionless temperatures can be useful to generalize results when physical properties are constant, and particularly when the reaction term is missing. The problem at hand is the classic Graetz problem and lends itself perfectly to the use of a dimensionless temperature. Equation (8.52) becomes

$$\mathscr{V}_z(\imath)\frac{\partial \mathscr{T}}{\partial \jmath} = \left(\frac{\alpha_T \bar{t}}{R^2}\right)\left[\frac{1}{\imath}\frac{\partial \mathscr{T}}{\partial \imath} + \frac{\partial^2 \mathscr{T}}{\partial \imath^2}\right] + \frac{\Delta H_R \mathscr{R}_A(\mathscr{T})\bar{t}}{a_{in}\rho C_p(T_{wall} - T_{in})} \tag{8.62}$$

but the heat of reaction term is dropped in the current problem. The dimensionless temperature ranges from $\mathscr{T} = 0$ at the inlet to $\mathscr{T} = 1$ at the walls. Since no heat is generated, $0 \leq \mathscr{T} \leq 1$ at every point in the heat exchanger. The dimensionless solution, $\mathscr{T}(\imath, \jmath)$, depends only on the value of $\alpha_T \bar{t}/R^2$ and is the same for all values of T_{in} and T_{wall}. The solution is easily calculated by the marching-ahead technique.

Use $\Delta \imath = 0.25$. The stability criterion at the near-wall position is obtained from Equation (8.36) with α_T replacing $\mathscr{D}_A$, or from Equation (8.59) evaluated at $\imath = 1 - \Delta \imath$. The result is

$$\Delta \jmath_{max} = \frac{\Delta \imath^2 (2\Delta \imath - \Delta \imath^2)}{(\alpha_T \bar{t}/R^2)} = 0.0684$$

which gives $J_{min} = 15$. Choose $J = 16$ so that $\Delta \jmath = 0.0625$.

The marching-ahead equations are obtained from Equations (8.56)–(8.58). At the centerline,

$$\mathscr{T}(0, \jmath + \Delta \jmath) = 0.2000\mathscr{T}(0, \jmath) + 0.8000\mathscr{T}(0.25, \jmath)$$

At the interior points,

$$\mathscr{T}(0.25, \jmath + \Delta \jmath) = 0.5733\mathscr{T}(0.25, \jmath) + 0.3200\mathscr{T}(0.50, \jmath) + 0.1067\mathscr{T}(0, \jmath)$$

$$\mathscr{T}(0.50, \jmath + \Delta \jmath) = 0.4667\mathscr{T}(0.50, \jmath) + 0.3333\mathscr{T}(0.75, \jmath) + 0.2000\mathscr{T}(0.25, \jmath)$$

$$\mathscr{T}(0.75, \jmath + \Delta \jmath) = 0.0857\mathscr{T}(0.75, \jmath) + 0.5333\mathscr{T}(1, \jmath) + 0.3810\mathscr{T}(0.5, \jmath)$$

At the wall,

$$\mathscr{T}(1, \hat{z} + \Delta\hat{z}) = 1.0$$

Note that the coefficients on temperatures sum to 1.0 in each equation. This is necessary because the asymptotic solution, $\hat{z} \gg 1$, must give $\mathscr{T} = 1$ for all $\hat{r}$. Had there been a heat of reaction, the coefficients would be unchanged but a generation term would be added to each equation.

The marching-ahead technique gives the following results for $\mathscr{T}$:

$\hat{r}$	$\hat{z} = 0$	$\hat{z} = 0.25$	$\hat{z} = 0.50$	$\hat{z} = 0.75$	$\hat{z} = 1.0$
0	0	0	0	0	1.0000
0.0625	0	0	0	0.5333	1.0000
0.1250	0	0	0.1778	0.5790	1.0000
0.1875	0	0.0569	0.2760	0.6507	1.0000
0.2500	0.0455	0.1209	0.3571	0.6942	1.0000
0.3125	0.1058	0.1884	0.4222	0.7289	1.0000
0.3750	0.1719	0.2544	0.4377	0.7567	1.0000
0.4375	0.2379	0.3171	0.5260	0.7802	1.0000
0.5000	0.3013	0.3755	0.5690	0.8006	1.0000
0.5625	0.3607	0.4295	0.6075	0.8187	1.0000
0.6250	0.4157	0.4791	0.6423	0.8349	1.0000
0.6875	0.4664	0.5246	0.6739	0.8496	1.0000
0.7500	0.5129	0.5661	0.7206	0.8629	1.0000
0.8125	0.5555	0.6041	0.7287	0.8749	1.0000
0.8750	0.5944	0.6388	0.7525	0.8859	1.0000
0.9375	0.6299	0.6705	0.7743	0.8960	1.0000
1.0000	0.6624	0.6994	0.7941	0.8960	1.0000

Figure 8.7 shows these results for $\hat{z} = 1$ and compares them with the analytical solution. The numerical approximation is quite good, even for a coarse grid with $I=4$ and $J=16$. This is the exception rather than the rule. Convergence should be tested using a finer grid size.

The results for $\hat{z} = 1$ give the outlet temperature distribution for a heat exchanger with $\alpha_T \bar{t}/R^2 = 0.4$. The results at $\hat{z} = 0.5$ give the outlet temperature distribution for a heat exchanger with $\alpha_T \bar{t}/R^2 = 0.2$. There is no reason to stop at $\hat{z} = 1.0$. Continue marching until $\hat{z} = 2$ and you will obtain the outlet temperature distribution for a heat exchanger with $\alpha_T \bar{t}/R^2 = 0.8$.

8.6.2 Optimal Wall Temperatures

The method of lines formulation for solving Equation (8.52) does not require that T_{wall} be constant, but allows $T_{wall}(z)$ to be an arbitrary function of axial position. A new value of T_{wall} may be used at each step in the calculations, just as a new $\Delta \hat{z}$ may be assigned at each step (subject to the stability criterion). The design engineer is thus free to pick a $T_{wall}(z)$ that optimizes reactor performance.

Reactor performance is an issue of selectivity, not of conversion. Otherwise, just push T_{wall} to its maximum possible value. Good selectivity results from an optimal trajectory of time versus temperature for all portions of the reacting fluid, but uniform treatment is difficult in laminar flow due to the large difference in residence time between the wall and centerline. No strategy for controlling the wall temperature can completely eliminate the resultant nonuniformity, but a good strategy for $T_{wall}(z)$ can mitigate the problem. With preheated feed, initial cooling at the wall can help compensate for long residence times. With cold feed, initial heating at the wall is needed to start the reaction, but a switch to cooling can be made at some downstream point. A good general approach to determining the optimal $T_{wall}(z)$ is to first find the best single wall temperature, then find the best two-zone strategy, the best three-zone strategy, and so on. The objective function for the optimization can be as simple as the mixing-cup outlet concentration of a desired intermediate. It can also be based on the concept of *thermal time distributions* introduced in Section 15.4.3.

Optimization requires that $\alpha_T \bar{t}/R^2$ have some reasonably high value so that the wall temperature has a significant influence on reactor performance. There is no requirement that $\mathscr{D}_A \bar{t}/R^2$ be large. Thus, the method can be used for polymer systems that have thermal diffusivities typical of organic liquids but low molecular diffusivities. The calculations needed to solve the optimization are much longer than those needed to solve the ODEs of Chapter 6, but they are still feasible on small computers.

8.7 RADIAL VARIATIONS IN VISCOSITY

Real fluids have viscosities that are functions of temperature and composition. This means that the viscosity will vary across the radius of a tubular reactor and that the velocity profile will be something other than parabolic. If the viscosity is lower near the wall, as in heating, the velocity profile will be flattened compared with the parabolic distribution. If the viscosity is higher near the wall, as in cooling, the velocity profile will be elongated. These phenomena can be important in laminar flow reactors, affecting performance and even operability. Generally speaking, a flattened velocity profile will improve performance by more closely approaching piston flow. Conversely, an elongated profile will usually hurt performance. This section gives a method for including the effects of variable viscosity in a reactor design problem. It is restricted to low Reynolds numbers, $Re < 100$, and is used mainly for reactions involving compounds with high molecular weights, such as greases, waxes, heavy oils, and synthetic polymers. It is usually possible to achieve turbulence with lower molecular weight compounds, and turbulence eliminates most of the problems associated with viscosity changes.

Variable viscosity in laminar tube flows is an example of the coupling of mass, energy, and momentum transport in a reactor design problem of practical significance. Elaborate computer codes are being devised that recognize this coupling in complex flow geometries. These codes are being verified and are becoming design tools for the reaction engineer. The present example is representative of a general class of single-phase, variable-viscosity, variable-density problems, yet it avoids undue complications in mathematical or numerical analysis.

Consider axisymmetric flow in a circular tube so that $V_\theta = 0$. Two additional assumptions are needed to treat the variable-viscosity problem in its simplest form:

1. The momentum of the fluid is negligible compared with viscous forces.
2. The radial velocity component V_r is negligible compared with the axial component V_z.

The first of these assumptions drops the momentum terms from the equations of motion, giving a situation known as *creeping flow*. This leaves V_r and V_z coupled through a pair of simultaneous, partial differential equations. The pair can be solved when circumstances warrant, but the second assumption allows much greater simplification. It allows V_z to be given by a single, ordinary differential equation:

$$0 = -\frac{dP}{dz} + \frac{1}{r}\frac{d}{dr}\left[\mu r \frac{dV_z}{dr}\right] \tag{8.63}$$

Note that pressure is treated as a function of z alone. This is consistent with the assumption of negligible V_r. Equation (8.63) is subject to the boundary conditions of radial symmetry, $dV_z/dr = 0$ at $r = 0$, and zero slip at the wall, $V_z = 0$ at $r = R$.

The key physical requirements for Equation (8.63) to hold are that the fluid be quite viscous, giving a low Reynolds number, and that the viscosity must change slowly in the axial direction, although it may change rapidly in the radial direction. In essence, Equation (8.63) postulates that the velocity profile $V_z(r)$ is in dynamic equilibrium with the radial viscosity profile $\mu(r)$. If $\mu(r)$ changes as a function of z, then $V_z(r)$ will change accordingly, always satisfying Equation (8.63). Any change in V_z will cause a change in V_r; but if the changes in $\mu(r)$ are slow enough, the radial velocity components will be small, and Equation (8.63) will remain a good approximation.

Solution of Equation (8.63) for the case of constant viscosity gives the parabolic velocity profile, Equation (8.1), and Poiseuille's equation for pressure drop, Equation (3.14). In the more general case of $\mu = \mu(r)$, the velocity profile and pressure drop are determined numerically.

The first step in developing the numerical method is to find a "formal" solution to Equation (8.63). Observe that Equation (8.63) is variable-separable:

$$r(dP/dz)dr = d[\mu r(dV_z/dr)]$$

This equation can be integrated twice. Note that dP/dz is a constant when integrating with respect to r. The constants of integration are found using the boundary conditions. The result is

$$V_z(r) = \frac{1}{2}\left[\frac{-dP}{dz}\right]\int_r^R \frac{r_1}{\mu}\,dr_1 \qquad (8.64)$$

where r_1 is a dummy variable of integration. Dummy variables are used to avoid confusion between the variable being integrated and the limits of the integration. In Equation (8.64), V_z is a function of the variable r that is the lower limit of the integral; V_z is not a function of r_1. The dummy variable is "integrated out" and the value of the integral would be the same if r_1 were replaced by any other symbol.

Equation (8.64) allows the shape of the velocity profile to be calculated (e.g., substitute $\mu = $ constant and see what happens), but the magnitude of the velocity depends on the yet unknown value for dP/dz. As is often the case in hydrodynamic calculations, pressure drops are determined through the use of a continuity equation. Here, the continuity equation takes the form of a constant mass flow rate down the tube:

$$W = \pi R^2 \bar{u}_{in}\bar{\rho}_{in} = \pi R^2 \bar{u}\bar{\rho} = \int_0^R 2\pi r \rho V_z\,dr \qquad (8.65)$$

Substituting Equation (8.64) into (8.65) allows $(-dP/dz)$ to be determined.

$$-\frac{dP}{dz} = \frac{W}{\pi \int_0^R \rho r \int_r^R (r_1/\mu)\,dr_1} = \frac{R^2 \bar{u}_{in}\bar{\rho}_{in}}{\int_0^R \rho r \int_r^R (r_1/\mu)\,dr_1} \qquad (8.66)$$

This is the local pressure gradient. It is assumed to vary slowly in the z-direction. The pressure at position z is

$$P = P_{in} + \int_0^z \left[\frac{dP}{dz}\right]dz \qquad (8.67)$$

Substituting Equation (8.66) into Equation (8.64) gives

$$V_z(r) = \frac{R^2 \bar{u}_{in}\bar{\rho}_{in}}{2}\frac{\int_r^R (r_1/\mu)\,dr_1}{\int_0^R \rho r \int_r^R (r_1/\mu)\,dr_1\,dr} \qquad (8.68)$$

A systematic method for combining the velocity and pressure calculations with the previous solutions techniques for composition and temperature starts with known values for all variables and proceeds as follows:

1. Take one axial step and compute new values for $a, b, \ldots, T$.
2. Use physical property correlations to estimate new values for μ and ρ.

3. Update $V_z(r)$ using Equation (8.68).
4. Calculate P at the new position using Equation (8.65).
5. Recalculate Δz_{max} using Equation (8.29) and change the actual Δz as required.
6. Repeat Steps 1–5 until $z = L$.

A numerical methodology for calculating $V_z(r)$ is developed in Example 8.10.

Example 8.10: Given tabulated data for $\mu(\imath)$ and $\rho(\imath)$, develop a numerical method for using Equation (8.68) to find the dimensionless velocity profile $\mathscr{V}_z(\imath) = V_z/\bar{u}$.

Solution: The numerical integration techniques require some care. The inlet to the reactor is usually assumed to have a flat viscosity profile and a parabolic velocity distribution. We would like the numerical integration to reproduce the parabolic distribution exactly when μ is constant. Otherwise, there will be an initial, fictitious change in $\mathscr{V}_z$ at the first axial increment. Define

$$G_1(\imath) = \int_{\imath}^{1} (\imath_1/\mu)\, d\imath_1$$

and

$$G_2 = \int_{0}^{1} (\rho/\bar{\rho}_{in})\imath G_1(\imath)\, d\imath$$

When μ is constant, the G_1 integrand is linear in $\imath$ and can be integrated exactly using the trapezoidal rule. The result of the G_1 integration is quadratic in $\imath$, and this is increased to cubic in $\imath$ in the G_2 integrand. Thus, G_2 cannot be integrated exactly with the trapezoidal rule or even Simpson's rule. There are many possible remedies to this problem, including just living with the error in G_2 since it will decrease $O(\Delta \imath^2)$. In the Basic program segment that follows, a correction of $\Delta \imath^3/8$ is added to G_2, so that the parabolic profile is reproduced exactly when μ is constant.

```
'Specify the number of radial increments, Itotal, and
'the values for visc(i) and rho(i) at each radial
'position. Also, the average density at the reactor
'inlet, rhoin, must be specified.

dr = 1/Itotal

'Use the trapezoidal rule to evaluate G1
G1(Itotal) = 0
```

```
For i=1 To Itotal
  m=Itotal-i
  G1(m) = G1(m+1) + dr^2/2*((m+1)/visc(m+1)
+          +m/visc(m))*dr
Next

'Now use it to evaluate G2
G2=0
For i=1 To Itotal-1
  G2=G2+i * dr * rho(i)/rhoin * G1(i) * dr
Next
G2=G2+rho(Itotal)/rhoin * G1(Itotal) * dr/2

'Apply a correction term to G2
G2=G2+dr ^ 3/8

'Calculate the velocity profile
For i=0 To Itotal
  Vz(i) =G1(i)/G2/2
Next i
```

The following is an example calculation where the viscosity varies by a factor of 50 across the tube, giving a significant elongation of the velocity profile compared with the parabolic case. The density was held constant in the calculations.

i	$\mu(i)$	Calculated $\mathscr{V}_z(i)$	Parabolic $\mathscr{V}_z(i)$
0	1.0	3.26	2.00
1	1.6	2.98	1.97
2	2.7	2.36	1.88
3	4.5	1.72	1.72
4	7.4	1.16	1.50
5	12.2	0.72	1.22
6	20.1	0.40	0.88
7	33.1	0.16	0.47
8	54.6	0.00	0.00

These results are plotted in Figure 8.8.

8.8 RADIAL VELOCITIES

The previous section gave a methodology for calculating $V_z(r)$ given $\mu(r)$ and $\rho(r)$. It will also be true that both μ and ρ will be functions of z. This will cause no difficulty provided the changes in the axial direction are slow.

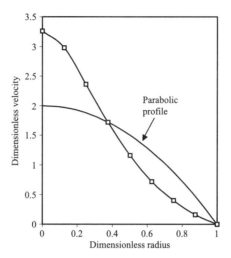

FIGURE 8.8 Elongated velocity profile resulting from a factor of 50 increase in viscosity across the tube radius.

The formulation of Equation (8.68) gives the fully developed velocity profile, $V_z(r)$, which corresponds to the local values of $\mu(r)$ and $\rho(r)$ without regard to upstream or downstream conditions. Changes in $V_z(r)$ must be gradual enough that the adjustment from one axial velocity profile to another requires only small velocities in the radial direction. We have assumed V_r to be small enough that it does not affect the equation of motion for V_z. This does not mean that V_r is zero. Instead, it can be calculated from the fluid continuity equation,

$$\partial(\rho V_z)/\partial z + (1/r)\partial(r\rho V_r)/\partial r = 0 \tag{8.69}$$

which is subject to the symmetry boundary condition that $V_r(0) = 0$. Equation (8.69) can be integrated to give

$$V_r = \frac{-1}{\rho r} \int_0^r r_1 \frac{\partial(\rho V_z)}{\partial z} dr_1 \tag{8.70}$$

Radial motion of fluid can have a significant, cumulative effect on the convective diffusion equations even when V_r has a negligible effect on the equation of motion for V_z. Thus, Equation (8.68) can give an accurate approximation for V_z even though Equations (8.12) and (8.52) need to be modified to account for radial convection. The extended versions of these equations are

$$V_z \frac{\partial a}{\partial z} + V_r \frac{\partial a}{\partial r} = \mathscr{D}_A \left[\frac{1}{r} \frac{\partial a}{\partial r} + \frac{\partial^2 a}{\partial r^2} \right] + \mathscr{R}_A \tag{8.71}$$

$$V_z \frac{\partial T}{\partial z} + V_r \frac{\partial T}{\partial r} = \alpha_T \left[\frac{1}{r} \frac{\partial T}{\partial r} + \frac{\partial^2 T}{\partial r^2} \right] - \frac{\Delta H_R \mathscr{R}}{\rho C_P} \qquad (8.72)$$

The boundary conditions are unchanged. The method of lines solution continues to use a second-order approximation for $\partial a/\partial r$ and merely adds a V_r term to the coefficients for the points at $r \pm \Delta r$.

The equivalent of radial flow for flat-plate geometries is V_y. The governing equations are similar to those for V_r. However, the various corrections for V_y are seldom necessary. The reason for this is that the distance Y is usually so small that diffusion in the y-direction tends to eliminate the composition and temperature differences that cause V_y. That is precisely why flat-plate geometries are used as chemical reactors and for laminar heat transfer.

It is sometimes interesting to calculate the paths followed by nondiffusive fluid elements as they flow through the reactor. These paths are called *streamlines* and are straight lines when the V_z profile does not change in the axial direction. The streamlines curve inward toward the center of the tube when the velocity profile elongates, as in cooling or polymerization. They curve outward when the velocity profile flattens, as in heating or depolymerization. Example 13.10 treats a case where they initially curve inward as the viscosity increases due to polymerization but later curve outward as the reaction goes to completion and diffusion mitigates the radial gradient in polymer concentration.

If desired, the streamlines can be calculated from

$$\int_0^{r_{in}} \rho r_1 V_z(r_1, 0) \, dr_1 = \int_0^r \rho r_1 V_z(r_1, z) \, dr_1 \qquad (8.73)$$

This mass balance equation shows that material that is initially at radial position r_{in} will move to radial position r for some downstream location, $z > 0$. A worked example of radial velocities and curved streamlines is given in Chapter 13, Example 13.10.

8.9 VARIABLE PHYSICAL PROPERTIES

The treatment of viscosity variations included the possibility of variable density. Equations (8.12) and (8.52) assumed constant density, constant $\mathscr{D}_A$, and constant α_T. We state here the appropriate generalizations of these equations to account for variable physical properties.

$$\frac{1}{A_c} \frac{\partial (A_c V_z a)}{\partial z} = \frac{\partial}{\partial z} \left[\mathscr{D}_A \frac{\partial a}{\partial z} \right] + \frac{1}{r} \frac{\partial}{\partial r} \left[\mathscr{D}_A r \frac{\partial a}{\partial r} \right] + \mathscr{R}_A \qquad (8.74)$$

304 CHEMICAL REACTOR DESIGN, OPTIMIZATION, AND SCALEUP

$$V_z\rho\frac{\partial H}{\partial z} = \frac{\partial}{\partial z}\left[\lambda\frac{\partial T}{\partial z}\right] + \frac{1}{r}\frac{\partial}{\partial r}\left[\lambda r\frac{\partial T}{\partial r}\right] - \Delta H_R \mathscr{R} \qquad (8.75)$$

For completeness, axial diffusion and variable cross-section terms were included in Equations (8.74) and (8.75). They are usually dropped. Also, the variations in $\mathscr{D}_A$ and λ are usually small enough that they can be brought outside the derivatives. The primary utility of these equations, compared with Equations (8.12) and (8.52), is for gas-phase reactions with a significant pressure drop.

8.10 SCALEUP OF LAMINAR FLOW REACTORS

Chapter 3 introduced the basic concepts of scaleup for tubular reactors. The theory developed in this chapter allows scaleup of laminar flow reactors on a more substantive basis. *Model-based scaleup* supposes that the reactor is reasonably well understood at the pilot scale and that a model of the proposed plant-scale reactor predicts performance that is acceptable, although possibly worse than that achieved in the pilot reactor. So be it. If you trust the model, go for it. The alternative is *blind scaleup*, where the pilot reactor produces good product and where the scaleup is based on general principles and high hopes. There are situations where blind scaleup is the best choice based on business considerations; but given your druthers, go for model-based scaleup.

Consider the scaleup of a small, tubular reactor in which diffusion of both mass and heat is important. As a practical matter, the same fluid, the same inlet temperature, and the same mean residence time will be used in the small and large reactors. Substitute fluids and *cold-flow models* are sometimes used to study the fluid mechanics of a reactor, but not the kinetics of the reaction.

The goal of a scaleup is to achieve similar product quality at a higher rate. The throughput scaleup factor is S. This determines the flow rate to the large system; and the requirement of constant $\bar{t}$ fixes the volume of the large system. For scaleup of flow in an open tube, the design engineer has two basic variables, R and T_{wall}. An exact scaleup requires that $\mathscr{D}_A \bar{t}/R^2$ and $\alpha_T \bar{t}/R^2$ be held constant, and the only way to do this is to keep the same tube diameter. Scaling in parallel is exact. Scaling in series may be exact and is generally conservative for incompressible fluids. See Section 3.2. Other forms of scaleup will be satisfactory only under special circumstances. One of these circumstances is isothermal laminar flow when $\mathscr{D}_A \bar{t}/R^2$ is small in the pilot reactor.

8.10.1 Isothermal Laminar Flow

Reactors in isothermal laminar flow are exactly scaleable using geometric similarity if diffusion is negligible in the pilot reactor. Converting Equation (8.2) to dimensionless form gives

$$\mathcal{V}_z(\imath)\frac{\partial a}{\partial \mathscr{y}} = \mathscr{R}_A \bar{t} \qquad (8.76)$$

The absolute reactor size as measured by R and L does not appear. Using the same feed composition and the same $\bar{t}$ in a geometrically similar reactor will give a geometrically similar composition distribution; i.e., the concentration at the point $(\imath, \mathscr{y})$ will be the same in the large and small reactors. Similarly, the viscosity profile will be the same when position is expressed in dimensionless form, and this leads to the same velocity profile, pressure drop, and mixing-cup average composition. These statements assume that diffusion really was negligible on the small scale and that the Reynolds number remains low in the large reactor. Blind scaleup will then give the same product from the large reactor as from the small. If diffusion was beneficial at the small scale, reactor performance will worsen upon scaleup. The Reynolds number may become too high upon scaleup for the creeping flow assumption of Section 8.7 to remain reasonable, but the probable consequence of a higher Reynolds number is improved performance at the cost of a somewhat higher pressure drop.

It may not be feasible to have an adequately low value for $\mathscr{D}_A \bar{t}/R^2$ and still scale using geometric similarity. Recall that reactor scaleups are done at constant $\bar{t}$. The problem is that the pilot reactor would require too high a flow rate and consume too much material when $\mathscr{D}_A \bar{t}/R^2$ is small enough (i.e., R is large enough) and L/R is large enough for reasonable scaleup. The choice is to devise a model-based scaleup. Model the pilot reactor using the actual value for $\mathscr{D}_A \bar{t}/R^2$. Confirm (and adjust) the model based on experimental measurements. Then model the large reactor using the appropriately reduced value for $\mathscr{D}_A \bar{t}/R^2$. If the predicted results are satisfactory, go for it. If the predictions are unsatisfactory, consider using motionless mixers in the large reactor. These devices lower the effective value for $\mathscr{D}_A \bar{t}/R^2$ by promoting radial mixing. The usual approach to scaling reactors that contain motionless mixers is to start with geometric similarity but to increase the number of mixing elements to compensate for the larger tube diameter. For mixers of the Kenics type, an extra element is needed each time the tube diameter is doubled.

8.10.2 Nonisothermal Laminar Flow

The temperature counterpart of $\mathscr{D}_A \bar{t}/R^2$ is $\alpha_T \bar{t}/R^2$; and if $\alpha_T \bar{t}/R^2$ is low enough, then the reactor will be adiabatic. Since $\alpha_T \gg \mathscr{D}_A$, the situation of an adiabatic, laminar flow reactor is rare. Should it occur, then $T(\imath, \mathscr{y})$ will be the same in the small and large reactors, and blind scaleup is possible. More commonly, $\alpha_T \bar{t}/R^2$ will be so large that radial diffusion of heat will be significant in the small reactor. The extent of radial diffusion will lessen upon scaleup, leading to the possibility of thermal runaway. If model-based scaleup predicts a reasonable outcome, go for it. Otherwise, consider scaling in series or parallel.

PROBLEMS

8.1. Polymerizations often give such high viscosities that laminar flow is inevitable. A typical monomer diffusivity in a polymerizing mixture is 1.0×10^{-10} m/s (the diffusivity of the polymer will be much lower). A pilot-scale reactor might have a radius of 1 cm. What is the maximum value for the mean residence time before molecular diffusion becomes important? What about a production-scale reactor with $R = 10$ cm?

8.2. The velocity profile for isothermal, laminar, non-Newtonian flow in a pipe can sometimes be approximated as

$$V_z = V_0[1 - (r/R)^{(\eta+1)/\eta}]$$

where η is called the flow index, or power law constant. The case $\eta = 1$ corresponds to a Newtonian fluid and gives a parabolic velocity profile. Find a_{out}/a_{in} for a first-order reaction given $k\bar{t} = 1.0$ and $\eta = 0.5$. Assume negligible diffusion.

8.3. Repeat Example 8.1 and obtain an analytical solution for the case of first-order reaction and pressure-driven flow between flat plates. Feel free to use software for the symbolic manipulations, but do substantiate your results.

8.4. Determine whether the sequence of a_{out}/a_{in} versus I in Example 8.5 is converging as expected. What is your prediction for the calculated value that would be obtained if the program is run with $I = 256$, and $J = 131{,}072$. Run the program to test your prediction.

8.5. Equation (8.4) defines the average concentration, a_{out}, of material flowing from the reactor. Omit the $V_z(r)$ term inside the integral and normalize by the cross-sectional area, $A_c = \pi R^2$, rather than the volumetric flow rate, Q. The result is the spatial average concentration $a_{spatial}$, and is what you would measure if the contents of the tube were frozen and a small disk of the material was cut out and analyzed. In-line devices for measuring concentration may measure $a_{spatial}$ rather than a_{out}. Is the difference important?
   (a) Calculate both averages for the case of a parabolic velocity profile and first-order reaction with $k\bar{t} = 1.0$.
   (b) Find the value of $k\bar{t}$ that maximizes the difference between these averages.

8.6. Determine the equivalent of Merrill and Hamrin's criterion for a tubular reactor when the reaction is:
   (a) Second order of the form $2A \to P$.
   (b) Half-order: $A \to P$, $\mathscr{R}_A = -ka^{1/2}$. Be sure to stop the reaction if the concentration of A drops to zero. It will go to zero at some locations in the reactor when $\mathscr{D}_A \bar{t}/R^2 = 0$. Does it still fall to zero when $\mathscr{D}_A \bar{t}/R^2$ is just large enough to affect a_{out}?

8.7. Consider an isothermal, laminar flow reactor with a parabolic velocity profile. Suppose an elementary, second-order reaction of the form $A + B \rightarrow P$ with rate $\mathscr{R} = kab$ is occurring with $ka_{in}\,\bar{t} = 2$. Assume $a_{in} = b_{in}$. Find a_{out}/a_{in} for the following cases:
(a) $\mathscr{D}_A\bar{t}/R^2 = \mathscr{D}_B\bar{t}/R^2 = 0.01$.
(b) $\mathscr{D}_A\bar{t}/R^2 = 0.01, \mathscr{D}_B\bar{t}/R^2 = 0.001$.

8.8. Which is better for isothermal chemical reactions, pressure driven flow or drag flow between flat plates? Assume laminar flow with first-order chemical reaction and compare systems with the same values for the slit width $(2Y = H)$, length, mean velocity, and reaction rate constant.

8.9. Free-radical polymerizations tend to be highly exothermic. The following data are representative of the thermal (i.e., spontaneous) polymerization of styrene:

$$\lambda = 0.13 \text{ J/(m·s·K)}$$
$$\mathscr{D}_A = 1.0 \times 10^{-9} \text{m}^2/\text{s}$$
$$\Delta H = -8 \times 10^4 \text{ J/g-mol}$$
$$C_P = 1.9 \times 10^3 \text{ J/(kg·K)}$$
$$\rho = 950 \text{ kg/m}^3$$
$$a_{in} = 9200 \text{ g-mol/m}^3$$
$$L = 7 \text{ m}$$
$$\bar{t} = 1\text{h}$$
$$k = 1.0 \times 10^{10} \exp(-10{,}000/T)\,\text{h}^{-1}$$
$$T_{in} = 120°\text{C}$$
$$T_{wall} = 120°\text{C}$$

Assume laminar flow and a parabolic velocity distribution. Calculate the temperature and composition profiles in the reactor. Start with $I = 4$ and double until your computer cries for mercy. Consider two cases: (a) $R = 0.01$ m; (b) $R = 0.20$ m.

8.10. Suppose the consecutive reactions $A \xrightarrow{k_I} B \xrightarrow{k_{II}} C$ are elementary with rate constants $k_I = 4.5 \times 10^{11} \exp(-10{,}000/T), \text{h}^{-1}$ and $k_{II} = 1.8 \times 10^{12} \exp(-12{,}000/T), \text{h}^{-1}$. The reactions are occurring in a tube in laminar flow with $a_{in} = 1$, $b_{in} = c_{in} = 0$. Both reactions are exothermic with $-\Delta H_I a_{in}/(\rho C_P) = -\Delta H_{II} a_{in}/(\rho C_P) = 50$ K. The reactor is operated with $\bar{t} = 1\text{h}$, $T_{in} = 400$ K, and $T_{wall} = 400$ K. Assume $\alpha_T \bar{t}/R^2 = 0.1$. Determine a_{out}, b_{out}, and c_{out} given
(a) $\mathscr{D}_A\bar{t}/R^2 = 0.01 \quad \mathscr{D}_B\bar{t}/R^2 = 0.01$
(b) $\mathscr{D}_A\bar{t}/R^2 = 0.01 \quad \mathscr{D}_B\bar{t}/R^2 = 0.001$

8.11. Determine the opposite of the Merrill and Hamlin criterion. That is, find the value of $\mathscr{D}_A\bar{t}/R^2$ above which a laminar flow reactor closely approximates a piston flow reactor for a first-order reaction. Make the comparison at $\bar{k t} = 1$.

8.12. An unreconstructed cgs'er messed up your viscosity correlation by reporting his results in centipoise rather than pascal seconds. How does this affect the sample velocity profile calculated in Example 8.10? What does the term "unreconstructed cgs'er" mean?

8.13. Suppose you are marching down the infamous tube and at step j have determined the temperature and composition at each radial point. A correlation is available to calculate viscosity, and it gives the results tabulated below. Assume constant density and $\mathbf{Re} = 0.1$. Determine the axial velocity profile. Plot your results and compare them with the parabolic distribution.

r/R	Isothermal μ	Cooling μ	Heating μ
1.000	1.0	54.6	0.018
0.875	1.0	33.1	0.030
0.750	1.0	20.1	0.050
0.625	1.0	12.2	0.082
0.500	1.0	7.4	0.135
0.375	1.0	4.5	0.223
0.250	1.0	2.7	0.368
0.125	1.0	1.6	0.607
0	1.0	1.0	1.000

8.14. Derive the equations necessary to calculate $V_z(y)$ given $\mu(y)$ for pressure-driven flow between flat plates.

8.15. The stated boundary condition associated with Equation (8.69) is that $V_r(0) = 0$. This is a symmetry condition consistent with the assumption that $V_\theta = 0$. There is also a zero-slip condition that $V_r(R) = 0$. Prove that both boundary conditions are satisfied by Equation (8.70). Are there boundary conditions on V_z? If so, what are they?

8.16. Stepwise condensation polymerizations can be modeled as a second-order reaction of the functional groups. Let a be the concentration of functional groups so that $\mathscr{R}_A = -ka^2$. The following viscosity relationship

$$\mu/\mu_0 = 1 + 100[1 - (a/a_{in})^3]$$

is reasonable for a condensation polymer in a solvent. Determine a_{out}/a_{in} for a laminar flow reactor with $\bar{k t} = 2$ and with negligible diffusion. Neglect the radial velocity component V_r.

8.17. Rework Problem 8.16 including the V_r; i.e., solve Equation (8.70). Plot the streamlines. See Example 13.10 for guidance.

REFERENCES

1. Merill, L. S., Jr. and Hamrin, C. E., Jr., "Conversion and temperature profiles for complex reactions in laminar and plug flow," *AIChE J.*, **16**, 194–198 (1970).
2. Nauman, E. B. and Buffham, B. A., *Mixing in Continuous Flow Systems*, Wiley, New York, 1983, pp. 31–33.
3. Nauman, E. B. and Savoca, J. T., "An engineering approach to an unsolved problem in multicomponent diffusion," *AIChE J.*, **47**, 1016–1021 (2001).
4. Nigam, K. D. P. and Saxena, A. K., "Coiled configuration for flow inversion and its effects on residence time distribution," *AIChE J.*, **30**, 363–368 (1984).
5. Nauman, E. B., "Reactions and residence time distributions in motionless mixers," *Can. J. Chem. Eng.*, **60**, 136–140 (1982).

SUGGESTIONS FOR FURTHER READING

The convective diffusion equations for mass and energy are given detailed treatments in most texts on transport phenomena. The classic reference is

Bird, R. B., Stewart, W. E., and Lightfoot, E. N., *Transport Phenomena*, Wiley, New York, 1960.

Practical applications to laminar flow reactors are still mainly in the research literature. The first good treatment of a variable-viscosity reactor is

Lynn, S. and Huff, J. E., "Polymerization in a Tubular Reactor," *AIChE J.*, **17**, 475–481 (1971).

A detailed model of an industrially important reaction, styrene polymerization, is given in

Wyman, C. E. and Carter, L. F., "A Numerical Model for Tubular Polymerization Reactors," *AIChE Symp. Ser.*, **72**, 1–16 (1976).

See also Chapter 13 of this book.

The appropriateness of neglecting radial flow in the axial momentum equation yet of retaining it in the convective diffusion equation is discussed in

McLaughlin, H. S., Mallikarjun, R., and Nauman, E. B., "The Effect of Radial Velocities on Laminar Flow, Tubular Reactor Models," *AIChE J.*, **32**, 419–425 (1986).

Gas-phase reactors are often in laminar flow but have such high diffusivities that radial concentration gradients tend to be unimportant. Combustion reactions are fast enough to be exceptions. See

Roesler, J. F., "An Experimental and Two-Dimensional Modeling Investigation of Combustion Chemistry in a Laminar Non-Plug-Flow Reactor," *Proc. 27th Symp. (Int.) Combust.*, **1**, 287–293 (1998).

The usefulness of your training in solving PDEs need not be limited to classic chemical engineering. For a potentially more remunerative application, see

Clewlow, L. and Strickland, C., *Implementing Derivatives Models*, Wiley, New York, 1998.

The derivatives are the financial type, e.g., option spreads. The methods used are implicit finite difference techniques. See Appendix 8.3.

APPENDIX 8.1: THE CONVECTIVE DIFFUSION EQUATION

This section derives a simple version of the convective diffusion equation, applicable to tubular reactors with a one-dimensional velocity profile $V_z(r)$. The starting point is Equation (1.4) applied to the differential volume element shown in Figure 8.9. The volume element is located at point (r, z) and is in the shape of a ring. Note that θ-dependence is ignored so that the results will not be applicable to systems with significant natural convection. Also, convection due to V_r is neglected. Component A is transported by radial and axial diffusion and by axial convection. The diffusive flux is governed by Fick's law.

The various terms needed for Equation (1.4) are

$$\text{Radial diffusion in} = -\mathscr{D}_A \left[\frac{\partial a}{\partial r}\right]_r [2\pi r \, \Delta z]$$

$$\text{Axial diffusion in} = -\mathscr{D}_A \left[\frac{\partial a}{\partial z}\right]_z [2\pi r \, \Delta r]$$

$$\text{Axial convection in} = V_z(z) a(z) [2\pi r \, \Delta r]$$

$$\text{Radial diffusion out} = -\mathscr{D}_A \left[\frac{\partial a}{\partial r}\right]_{r+\Delta r} [2\pi (r + \Delta r) \, \Delta z]$$

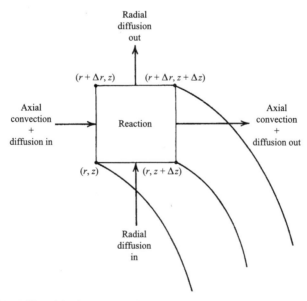

FIGURE 8.9 Differential volume element in cylindrical coordinates.

Axial diffusion out $= -\mathscr{D}_A \left[\dfrac{\partial a}{\partial z}\right]_{z+\Delta z} [2\pi r\, \Delta r]$

Axial convection out $= V_z(z) a(z+\Delta z)[2\pi r\, \Delta r]$

Formation of A by reaction $= \mathscr{R}_A [2\pi r\, \Delta r\, \Delta z]$

Accumulation $= \dfrac{\partial a}{\partial t}[2\pi r\, \Delta r\, \Delta z]$

Applying Equation (1.2), dividing everything by $[2\pi r\, \Delta r\, \Delta z]$, and rearranging gives

$$\dfrac{\partial a}{\partial t} + \dfrac{V_z(z+\Delta z)a(z+\Delta z) - V_z(z)a(z)}{\Delta z} = \dfrac{\mathscr{D}_A(\partial a/\partial z)_{z+\Delta z} - \mathscr{D}_A(\partial a/\partial z)_z}{\Delta z}$$
$$+ \dfrac{\mathscr{D}_A(\partial a/\partial r)_{r+\Delta r} - \mathscr{D}_A(\partial a/\partial r)_r}{\Delta r} + \mathscr{D}_A \dfrac{\partial a}{\partial r}\dfrac{1}{r} + \mathscr{R}_A$$

The limit is now taken as $\Delta z \to 0$ and $\Delta r \to 0$. The result is

$$\dfrac{\partial a}{\partial t} + \dfrac{\partial(V_z a)}{\partial z} = \dfrac{\partial(\mathscr{D}_A(\partial a/\partial z))}{\partial z} + \dfrac{\partial(\mathscr{D}_A(\partial a/\partial r))}{\partial r} + \dfrac{\mathscr{D}_A}{r}\dfrac{\partial a}{\partial r} + \mathscr{R}_A \qquad (8.77)$$

which is a more general version of Equation (8.12). Assume steady-state operation, V_z independent of z, and constant diffusivity to obtain Equation (8.12).

APPENDIX 8.2: FINITE DIFFERENCE APPROXIMATIONS

This section describes a number of finite difference approximations useful for solving second-order partial differential equations; that is, equations containing terms such as $\partial^2 f/\partial x^2$. The basic idea is to approximate f as a polynomial in x and then to differentiate the polynomial to obtain estimates for derivatives such as $\partial f/\partial x$ and $\partial^2 f/\partial x^2$. The polynomial approximation is a local one that applies to some region of space centered about point x. When the point changes, the polynomial approximation will change as well. We begin by fitting a quadratic to the three points shown below.

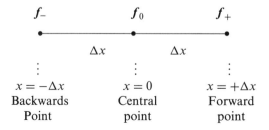

The quadratic has the form

$$f = A + Bx + Cx^2$$

Writing it for the three points gives

$$f_- = A - B\Delta x + C\Delta x^2$$
$$f_0 = A$$
$$f_+ = A + B\Delta x + C\Delta x^2$$

These equations are solved for A, B, and C to give

$$f = f_0 + \left(\frac{f_+ - f_-}{2\Delta x}\right)x + \left(\frac{f_+ - 2f_0 + f_-}{2\Delta x^2}\right)x^2$$

This is *a second-order approximation* and can be used to obtain derivatives up to the second. Differentiate to obtain

$$\frac{df}{dx} = \left(\frac{f_+ - f_-}{2\Delta x}\right) + \left(\frac{f_+ - 2f_0 + f_-}{\Delta x^2}\right)x$$

and

$$\frac{d^2f}{dx^2} = \frac{f_+ - 2f_0 + f_-}{\Delta x^2}$$

The value of the first derivative depends on the position at which it is evaluated. Setting $x = +\Delta x$ gives a *second-order, forward difference*:

$$\left[\frac{df}{dx}\right]_+ \approx \frac{3f_+ - 4f_0 + f_-}{2\Delta x}$$

Setting $x = 0$ gives a *second-order, central difference*:

$$\left[\frac{df}{dx}\right]_0 \approx \frac{f_+ - f_-}{2\Delta x}$$

Setting $x = \Delta x$ gives a *second-order, backward difference*:

$$\left[\frac{df}{dx}\right]_- \approx \frac{-f_+ + 4f_0 - 3f_-}{2\Delta x}$$

The second derivative is constant (independent of x) for this second-order approximation. We consider it to be a *central difference*:

$$\left[\frac{d^2f}{dx^2}\right]_0 \approx \frac{f_+ - 2f_0 + f_-}{\Delta x^2}$$

All higher derivatives are zero. Obviously, to obtain a nontrivial approximation to an nth derivative requires at least an nth-order polynomial. The various nontrivial derivatives obtained from an nth order polynomial will converge $O(\Delta x^n)$.

Example 8.11: Apply the various second-order approximations to $f = x\exp(x)$.

Solution: $f_+ = \Delta x \exp(\Delta x)$, $f_0 = 0$, $f_- = -\Delta x \exp(-\Delta x)$. The various derivative approximations are

$$\left[\frac{df}{dx}\right]_+ = \frac{3\exp(\Delta x) - \exp(-\Delta x)}{\Delta x}$$

$$\left[\frac{df}{dx}\right]_0 = \frac{\exp(\Delta x) + \exp(-\Delta x)}{2\Delta x}$$

$$\left[\frac{df}{dx}\right]_- = \frac{-\exp(\Delta x) + 3\exp(-\Delta x)}{2}$$

$$\left[\frac{d^2f}{dx^2}\right]_0 = \frac{\exp(\Delta x) - \exp(-\Delta x)}{\Delta x}$$

Evaluating them as a function of Δx gives

Δx	$\left[\dfrac{df}{dx}\right]_+$	Δ	$\left[\dfrac{df}{dx}\right]_0$	Δ	$\left[\dfrac{df}{dx}\right]_-$	Δ	$\left[\dfrac{d^2f}{dx^2}\right]$	Δ
1	3.893		1.543		−0.807		2.350	
		1.723		0.415		−0.892		0.266
1/2	2.170		1.128		0.805		2.084	
		0.633		0.096		−0.441		0.063
1/4	1.537		1.031		0.526		2.021	
		0.279		0.024		−0.231		0.016
1/8	1.258		1.008		0.757		2.005	
		0.131		0.006		−0.120		0.004
1/16	1.127		1.002		0.877		2.001	
		0.0064		0.002		0.0061		0.002
1/32	1.063		1.000		0.938		2.000	
∞	1		1		1		2	

It is apparent that the central difference approximations converge $O(\Delta x^2)$. The forward and backward approximations to the first derivative converge $O(\Delta x)$. This is because they are really approximating the derivatives at the points $x = \pm\Delta x$ rather than at $x = 0$.

For a *first-order approximation*, a straight line is fit between the points $x=0$ and x to get the *first-order, forward difference approximation*

$$\left[\frac{df}{dx}\right]_{+\Delta x/2} \approx \frac{f_+ - f_0}{\Delta x}$$

and between the points $x = -\Delta x$ and $x=0$ to get the *first-order, backward difference approximation*:

$$\left[\frac{df}{dx}\right]_{-\Delta x/2} \approx \frac{f_0 - f_-}{\Delta x}$$

These both converge $\mathbf{O}(\Delta x)$.

APPENDIX 8.3: IMPLICIT DIFFERENCING SCHEMES

The method of lines is called an *explicit* method because the "new" value $T(r, z + \Delta z)$ is given as an explicit function of the "old" values $T(r, z), T(r - \Delta r, z), \ldots$. See, for example, Equation (8.57). This explicit scheme is obtained by using a first-order, forward difference approximation for the axial derivative. See, for example, Equation (8.16). Other approximations for dT/dz are given in Appendix 8.2. These usually give rise to *implicit* methods where $T(r, z \pm \Delta z)$ is not found directly but is given as one member of a set of simultaneous algebraic equations. The simplest implicit scheme is known as *backward differencing* and is based on a first-order, backward difference approximation for $\partial T/\partial z$. Instead of Equation (8.57), we obtain

$$\left[1 - 2\left(\frac{\bar{u}\alpha_T \bar{t}}{V_z(\bar{t})R^2}\right)\frac{\Delta \bar{y}}{\Delta \bar{t}^2}\right]T(\bar{t}, \bar{y}) - \left(\frac{\bar{u}\alpha_T \bar{t}}{V_z(\bar{t})R^2}\right)\frac{\Delta \bar{y}}{\Delta \bar{t}^2}\left[1 + \frac{\Delta \bar{t}}{2\bar{t}}\right]T(\bar{t} + \Delta \bar{t}, \bar{y})$$

$$-\left(\frac{\bar{u}\alpha_T \bar{t}}{V_z(\bar{t})R^2}\right)\frac{\Delta \bar{y}}{\Delta \bar{t}^2}\left[1 - \frac{\Delta \bar{t}}{2\bar{t}}\right]T(\bar{t} - \Delta \bar{t}, \bar{y})$$

$$= T(\bar{t}, \bar{y} - \Delta \bar{y}) - \left[\frac{\Delta H_R \mathcal{R} \bar{t}\bar{u}}{\rho C_P V_z(\bar{t})}\right]_{\bar{y} - \Delta \bar{y}} \Delta \bar{y} \qquad (8.78)$$

Here, the temperatures on the left-hand side are the new, unknown values while that on the right is the previous, known value. Note that the heat sink/source term is evaluated at the previous location, $\bar{y} - \Delta \bar{y}$. The computational template is backwards from that shown in Figure 8.2, and Equation (8.78) cannot be solved directly since there are three unknowns. However, if a version of Equation (8.78) is written for every interior point and if appropriate special forms are written for the centerline and wall, then as many equations are obtained as there are unknown temperatures. The resulting algebraic equations are linear and can be solved by matrix inversion. The backward differencing scheme is stable for all Δz and Δy so that $I = 1/\Delta z$ and $J = 1/\Delta y$ can be picked independently. This avoids the need for extremely small Δy values that was encountered in Example 8.5. The method converges $\mathbf{O}(\Delta z^2, \Delta y)$.

Example 8.12: Use the backward differencing method to solve the heat transfer problem of Example 8.3. Select $\Delta z = 0.25$ and $\Delta y = 0.0625$.

$$\mathcal{T}(1, y) = 1.0$$

$$-0.5333\mathcal{T}(1, y) + 1.9143\mathcal{T}(0.75, y) - 0.3810\mathcal{T}(0.5, y) = \mathcal{T}(0.75, y - \Delta y)$$

$$-0.3333\mathcal{T}(0.75, y) + 1.5333\mathcal{T}(0.50, y) - 0.2000\mathcal{T}(0.25, y) = \mathcal{T}(0.50, y - \Delta y)$$

$$-0.3200\mathcal{T}(0.50, y) + 1.4267\mathcal{T}(0.25, y) - 0.1067\mathcal{T}(0, y) = \mathcal{T}(0.25, y - \Delta y)$$

$$-0.8000\mathcal{T}(0.25, y) - 1.800\mathcal{T}(0, y) = (0, y - \Delta y)$$

In matrix form

$$\begin{bmatrix} 1 & 0 & 0 & 0 & 0 \\ -0.5333 & 1.9143 & -0.3810 & 0 & 0 \\ 0 & -0.3333 & 1.5333 & -0.2000 & 0 \\ 0 & 0 & -0.3200 & 1.4267 & -0.1067 \\ 0 & 0 & 0 & -0.8000 & 1.8000 \end{bmatrix} \begin{bmatrix} \mathcal{T}(1, y) \\ \mathcal{T}(0.75, y) \\ \mathcal{T}(0.50, y) \\ \mathcal{T}(0.25, y) \\ \mathcal{T}(0, y) \end{bmatrix}$$

$$= \begin{bmatrix} 1 \\ \mathcal{T}(0.75, y - \Delta y) \\ \mathcal{T}(0.50, y - \Delta y) \\ \mathcal{T}(0.25, y - \Delta y) \\ \mathcal{T}(0, y - \Delta y) \end{bmatrix}$$

This system of equations is solved for each y, beginning with the inlet boundary:

$$\begin{bmatrix} 1 \\ \mathcal{T}(0.75, y - \Delta y) \\ \mathcal{T}(0.50, y - \Delta y) \\ \mathcal{T}(0.25, y - \Delta y) \\ \mathcal{T}(0, y - \Delta y) \end{bmatrix} = \begin{bmatrix} 1 \\ 0 \\ 0 \\ 0 \\ 0 \end{bmatrix}$$

Results are

$\hat{y}$	$\hat{z}=0$	$\hat{z}=0.25$	$\hat{z}=0.5$	$\hat{z}=0.75$	$\hat{z}=1$
0	0	0	0	0	1.0000
0.0625	0.0067	0.0152	0.0654	0.2916	1.0000
0.1250	0.0241	0.0458	0.1487	0.4605	1.0000
0.1875	0.0525	0.0880	0.2313	0.5652	1.0000
0.2500	0.0901	0.1372	0.3068	0.6349	1.0000
0.3125	0.1345	0.1901	0.3737	0.6846	1.0000
0.3740	0.1832	0.2440	0.4326	0.7223	1.0000
0.4375	0.2338	0.2972	0.4844	0.7523	1.0000
0.5000	0.2848	0.3486	0.5303	0.7771	1.0000
0.5625	0.3349	0.3975	0.5712	0.7982	1.0000
0.6250	0.3832	0.4437	0.6079	0.8166	1.0000
0.6875	0.4293	0.4869	0.6410	0.8327	1.0000
0.7500	0.4728	0.5271	0.6710	0.9471	1.0000
0.8125	0.5135	0.5645	0.6982	0.8601	1.0000
0.8750	0.5515	0.5991	0.7230	0.8718	1.0000
0.9375	0.5869	0.6311	0.7456	0.8824	1.0000
1.0000	0.6196	0.6606	0.7664	0.8921	1.0000

The backward differencing method requires the solution of $I+1$ simultaneous equations to find the radial temperature profile. It is *semi-implicit* since the solution is still marched-ahead in the axial direction. *Fully implicit* schemes exist where $(J+1)(I+1)$ equations are solved simultaneously, one for each grid point in the total system. Fully implicit schemes may be used for problems where axial diffusion or conduction is important so that second derivatives in the axial direction, $\partial^2 a/\partial z^2$ or $\partial^2 T/\partial z^2$, must be retained in the partial differential equation. An alternative approach for this case is the shooting method described in Chapter 9. When applied to partial differential equations, shooting methods are usually implemented using an implicit technique in the radial direction. This gives rise to a tridiagonal matrix that must be inverted at each step in axial marching. The *Thomas algorithm* is a simple and efficient way of performing this inversion. Some finite difference approximations combine forward and backward differencing. One of these, *Crank-Nicholson*, is widely used. It is semi-implicit, unconditionally stable (at least for the linear case), and converges $O(\Delta r^2, \Delta z^2)$.

CHAPTER 9
REAL TUBULAR REACTORS IN TURBULENT FLOW

The essence of reactor design is the combination of chemical kinetics with transport phenomena. The chemical kineticist, who can be a chemical engineer but by tradition is a physical chemist, is concerned with the interactions between molecules (and sometimes within molecules) in well-defined systems. By well-defined, we mean that all variables that affect the reaction can be controlled at uniform and measurable values. Chemical kinetic studies are usually conducted in small equipment where mixing and heat transfer are excellent and where the goal of having well-defined variables is realistic. Occasionally, the ideal conditions can be retained upon scaleup. Slow reactions in batch reactors or CSTRs are examples. More likely, scaleup to industrial conditions will involve fast reactions in large equipment where mixing and heat transfer limitations may emerge. Transport equations must be combined with the kinetic equations, and this is the realm of the chemical reaction engineer.

Chapter 8 combined transport with kinetics in the purest and most fundamental way. The flow fields were deterministic, time-invariant, and calculable. The reactor design equations were applied to simple geometries, such as circular tubes, and were based on intrinsic properties of the fluid, such as molecular diffusivity and viscosity. Such reactors do exist, particularly in polymerizations as discussed in Chapter 13, but they are less typical of industrial practice than the more complex reactors considered in this chapter.

The models of Chapter 9 contain at least one empirical parameter. This parameter is used to account for complex flow fields that are not deterministic, time-invariant, and calculable. We are specifically concerned with packed-bed reactors, turbulent-flow reactors, and static mixers (also known as motionless mixers). We begin with packed-bed reactors because they are ubiquitous within the petrochemical industry and because their mathematical treatment closely parallels that of the laminar flow reactors in Chapter 8.

9.1 PACKED-BED REACTORS

Packed-bed reactors are very widely used, particularly for solid-catalyzed heterogeneous reactions in which the packing serves as the catalyst. The velocity profile is quite complex in a packed-bed. When measured at a small distance from the surface of the packing, velocities are found to be approximately uniform except near the tube wall. Random packing gives more void space and thus higher velocities near the wall. The velocity profile is almost invariably modeled as being flat. This does not mean that the packed-bed is modeled as a piston flow reactor with negligible radial gradients in composition and temperature. Instead, radial mixing is limited in packed-bed reactors to the point that quite large differences can develop across the tube. Radial concentration and temperature profiles can be modeled using an effective radial diffusivity. Instead of Equation (8.12), we write

$$\bar{u}_s \frac{\partial a}{\partial z} = D_r \left[\frac{1}{r} \frac{\partial a}{\partial r} + \frac{\partial^2 a}{\partial r^2} \right] + \varepsilon \mathscr{R}_A \tag{9.1}$$

where D_r is a radial dispersion coefficient and ε is the void fraction. D_r has units of diffusivity, m^2/s. The major differences between this model and the convective diffusion equation used in Chapter 8 is that the velocity profile is now assumed to be flat and D_r is an empirically determined parameter instead of a molecular diffusivity. The value of D_r depends on factors such as the ratio of tube-to-packing diameters, the Reynolds number, and (at least at low Reynolds numbers) the physical properties of the fluid. Ordinarily, the same value for D_r is used for all reactants, finessing the problems of multicomponent diffusion and allowing the use of stoichiometry to eliminate Equation (9.1) for some of the components. Note that $\bar{u}_s$ in Equation (9.1) is the superficial velocity, this being the average velocity that would exist if the tube had no packing:

$$\bar{u}_s = \frac{Q}{A_c} = \frac{Q}{\pi R^2} \tag{9.2}$$

Note also that $\mathscr{R}_A$ is the reaction rate per fluid-phase volume and that $\varepsilon \mathscr{R}_A$ is the rate per total volume consisting of fluid plus packing. Except for the appearance of the void fraction ε, there is no overt sign that the reactor is a packed-bed. The reaction model is *pseudohomogeneous* and ignores the details of interactions between the packing and the fluid. These interactions are lumped into D_r and $\mathscr{R}_A$. The concentration a is the fluid-phase concentration, and the rate expression $\mathscr{R}_A(a, b)$ is based on fluid-phase concentrations. This approach is satisfactory when the reaction is truly homogeneous and the packing merely occupies space without participating in the reaction. For heterogeneous, solid-catalyzed reactions, the rate is presumably governed by surface concentrations, but the use of pseudohomogeneous kinetic expressions is nearly universal for the simple reason that the bulk concentrations can be measured while surface concentrations are not readily measurable. See Chapter 10 to understand the relationship between surface and bulk concentrations. We use Equation (9.1) for both homogeneous and heterogeneous reactions in packed-beds. The boundary conditions associated with Equation (9.1) are the same as those for Equation (8.12): a prescribed inlet concentration profile, $a_{in}(r)$, and zero gradients in concentration at the wall and centerline.

The temperature counterpart to Equation (9.1) is

$$\bar{u}_s \frac{\partial T}{\partial z} = E_r \left[\frac{1}{r} \frac{\partial T}{\partial r} + \frac{\partial^2 T}{\partial r^2} \right] - \frac{\varepsilon \Delta H_R \mathcal{R}}{\rho C_P} \qquad (9.3)$$

where E_r is an effective radial dispersion coefficient for heat and where $\Delta H_R \mathcal{R}$ has the usual interpretation as a sum. Two of the boundary conditions associated with Equation (9.3) are the ordinary ones of a prescribed inlet profile and a zero gradient at the centerline. The wall boundary condition has a form not previously encountered:

$$h_r[T(R) - T_{wall}] = -\lambda_r \, \partial T / \partial r \quad \text{at} \quad r = R \qquad (9.4)$$

It accounts for the especially high resistance to heat transfer that is observed near the wall in packed-bed reactors. Most of the heat transfer within a packed-bed is by fluid convection in the axial direction and by conduction through the solid packing in the radial direction. The high void fraction near the wall lowers the effective conductivity in that region. As in Section 8.6, T_{wall} is the inside temperature of the tube, but this may now be significantly different than the fluid temperature $T(R)$, just a short distance in from the wall. In essence, the system is modeled as if there were a thin, thermal boundary layer across which heat is transferred at a rate proportional to the temperature difference $[T(R)-T_{wall}]$. The proportionality constant is an empirical heat transfer coefficient, h_r. The left-hand side of Equation (9.4) gives the rate of heat transfer across the thermal boundary layer. At steady state, the heat transferred from the tube wall must equal the heat conducted and convected into the bed. Heat transfer within the bed is modeled using an effective thermal conductivity λ_r. The right-hand side of Equation (9.4) represents the conduction, and λ_r is an empirical constant.

It appears that the complete model for both mass and heat transfer contains four adjustable constants, D_r, E_r, h_r, and λ_r, but E_r and λ_r are constrained by the usual relationship between thermal diffusivity and thermal conductivity

$$E_r = \frac{\lambda_r}{\rho C_p} \qquad (9.5)$$

Thus, there are only three independent parameters. We take these to be D_r, h_r, and λ_r. Imperfect but generally useful correlations for these parameters are available. For a summary of published correlations and references to the original literature see Froment and Bischoff,[1] Dixon and Cresswell,[2] and Dixon.[3]

Figure 9.1 shows a correlation for D_r. The correlating variable is the particle Reynolds number, $\rho d_p \bar{u}_s / \mu$, where d_p is the diameter of the packing. The correlated variable is a dimensionless group known as the *Peclet number*, $(\mathbf{Pe})_\infty = (\bar{u}_s d_p / D_r)_\infty$, where the ∞ subscript denotes a tube with a large ratio of tube diameter to packing diameter, $d_t/d_p \gg 10$. Peclet numbers are commonly used in reactor design, and this chapter contains several varieties. All are dimensionless numbers formed by multiplying a velocity by a characteristic length and dividing by diffusivity. The Peclet number used to correlate data for packed-beds here in Section 9.1 has particle diameter, d_p, as the characteristic length and uses D_r as the diffusivity. The axial dispersion model discussed in Section 9.3 can also be applied to packed-beds, but the diffusivity is an axial diffusivity.

Many practical designs use packing with a diameter that is an appreciable fraction of the tube diameter. The following relationship is used to correct D_r for large packing:

$$\bar{u}_s d_p / D_r = \frac{(\bar{u}_s d_p / D_r)_\infty}{1 + 19.4(d_p/d_t)^2} \tag{9.6}$$

Shell-and-tube reactors may have $d_t/d_p = 3$ or even smaller. A value of 3 is seen to decrease $\bar{u}_s d_p / D_r$ by a factor of about 3. Reducing the tube diameter from $10d_p$ to $3d_p$ will increase D_r by a factor of about 10. Small tubes can thus have much better radial mixing than large tubes for two reasons: R is lower and D_r is higher.

The experimental results for $(\bar{u}d_p / D_r)_\infty$ in Figure 9.1 show a wide range of values at low Reynolds numbers. The physical properties of the fluid, and specifically its Schmidt number, $\mathbf{Sc} = \mu/(\rho \mathcal{D}_A)$, are important when the Reynolds number is low. Liquids will lie near the top of the range for $(\bar{u}_s d_p / D_r)_\infty$ and gases near the bottom. At high Reynolds numbers, hydrodynamics dominate, and the fluid properties become unimportant aside from their effect on Reynolds number. This is a fairly general phenomenon and is discussed further

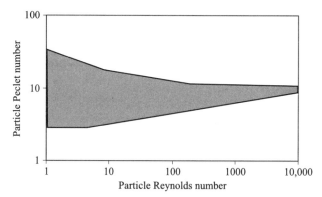

FIGURE 9.1 Existing data for the radial Peclet number in large-diameter packed beds, $(\mathbf{Pe})_\infty = (\bar{u}_s d_p / D_r)_\infty$ versus $\rho d_p \bar{u}_s / \mu$.

in Section 9.2. Figure 9.2 shows existing data for the effective thermal conductivity of packed beds. These data include both ceramic and metallic packings. More accurate results can be obtained from the semitheoretical predictions of Dixon and Cresswell.[2] Once λ_r is known, the wall heat transfer coefficient can be calculated from

$$\frac{h_r d_p}{\lambda_r} = \frac{3}{(\rho \bar{u}_s d_p / \mu)^{0.25}} \tag{9.7}$$

and E_r can be calculated from Equation (9.5). Thus, all model parameters can be estimated. The estimates require knowledge of only two system variables: the packing Reynolds number and the ratio of packing-to-tube diameters.

We turn now to the numerical solution of Equations (9.1) and (9.3). The solutions are necessarily simultaneous. Equation (9.1) is not needed for an isothermal reactor since, with a flat velocity profile and in the absence of a temperature profile, radial gradients in concentration do not arise and the model is equivalent to piston flow. Unmixed feed streams are an exception to this statement. By writing versions of Equation (9.1) for each component, we can model reactors with unmixed feed provided radial symmetry is preserved. Problem 9.1 describes a situation where this is possible.

The numerical techniques of Chapter 8 can be used for the simultaneous solution of Equation (9.3) and as many versions of Equation (9.1) as are necessary. The methods are unchanged except for the discretization stability criterion and the wall boundary condition. When the velocity profile is flat, the stability criterion is most demanding when at the centerline:

$$\Delta z_{max} = \frac{\Delta r^2 \bar{u}_s}{4 E_r} \tag{9.8}$$

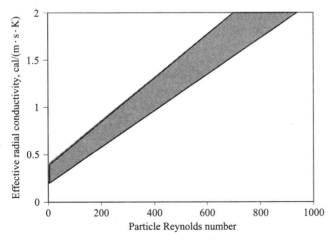

FIGURE 9.2 Existing data for the effective radial conductivity, λ_r.

or, in dimensionless form,

$$\Delta \hat{y}_{max} = \Delta \hat{z}^2 \frac{R^2}{4E_r t_s} \tag{9.9}$$

where $t_s = L/\bar{u}_s$. We have used E_r rather than D_r in the stability criterion because E_r will be larger.

Using a first-order approximation for the derivative in Equation (9.4), the wall boundary condition becomes

$$T(R, z) = \frac{h_r \Delta r T_{wall} + \lambda_r T(R - \Delta r, z)}{h_r \Delta r + \lambda_r} \tag{9.10}$$

A second-order approximation is preferred,

$$T(R, z) = \frac{2h_r \Delta r T_{wall} + 4\lambda_r T(R - \Delta r, z) - \lambda_r T(R - 2\Delta r, z)}{2h_r \Delta r + 3\lambda_r} \tag{9.11}$$

since it converges $O(\Delta r^2)$, as will the other derivative approximations. The computational templates for solving Equations (9.1) and (9.3) are similar to those used in Chapter 8. See Figure 8.2.

Example 9.1: The oxidation of o-xylene to phthalic anhydride is conducted in a multitubular reactor using air at approximately atmospheric pressure as the oxidant. Side reactions including complete oxidation are important but will be ignored in this example. The o-xylene concentration is low, $a_{in} = 44$ g/m^3, to stay under the explosive limit. Due to the large excess of oxygen, the reaction is pseudo-first-order in o-xylene concentration with $\ln(\varepsilon k) = 19.837 - 13.636/T$, where k is in s^{-1}. The tube is packed with 3-mm pellets consisting of V_2O_5 on potassium-promoted silica. The tube has an i.d. of 50 mm, is 5 m long, is operated with a superficial velocity of 1 m/s, and has a wall temperature of 640 K. Use $\rho = 1.29$ kg/m^3, $\mu = 3 \times 10^{-5}$ Pa·s, $C_P = 0.237$ cal/(g·K), and $\Delta H = -307$ kcal/mol. Assume $T_{in} = 640$ K. Use the two-dimensional, radial dispersion model to estimate the maximum temperature within the bed.

Solution: It is first necessary to estimate the parameters: D_r, E_r, h_r, and λ_r. The particle Reynolds number, $\rho d_p \bar{u}_s / \mu$, is 130, and Figure 9.1 gives $(\bar{u}_s d_p / D_r)_\infty \approx 10$. A small correction for d_p/d_t using Equation (9.6) gives $\bar{u}_s d_p / D_r = 8$ so that $D_r = 3.8 \times 10^{-4}$ m^2/s. Figure 9.2 gives $\lambda_r = 0.4$ cal/(m·s·K) so that $E_r = \lambda_r/(\rho C_P) = 1.3 \times 10^{-3}$ m^2/s. Equation 9.7 gives $h_r d_p/\lambda_r = 0.89$ so that $h_r = 120$ cal/(m^2·s).

The discretization stability criterion, Equation (9.9), gives $\Delta \hat{y}_{max} = 0.024 \Delta \hat{z}^2$. Pick $I = 5$, $\Delta \hat{z} = 0.2$, and $\Delta r = 0.005$ m. Then the stability criterion is satisfied by $J=1200$; $\Delta\mathscr{y} = 8.33 \times 10^{-4}$, and $\Delta z = 4.17 \times 10^{-3}$ m. The marching-ahead equation for concentration at the centerline is

$$a_{new}(0) = (1 - 4G_A)a(0) + 4G_A a(1) - \varepsilon k t_s a(0) \Delta\mathscr{y}$$

For the interior points,

$$a_{new}(i) = (1 - 2G_A)a(i) + G_A[1 + 0.5/I]a(i+1)$$
$$+ G_A[1 - 0.5/I]a(i-1) - \varepsilon k t_s a(i) \Delta\mathscr{y}$$

At the wall,

$$a_{new}(I) = (4/3)\, a(I-1) - (1/3)\, a(I-2)$$

where $G_A = (\Delta\mathscr{y}/\Delta\imath^2)(D_r t_s/R^2) = 0.0633$.

The equations for temperature are similar. At the centerline,

$$T_{new}(0) = (1 - 4G_T)T(0) + 4G_T T(1) - \left[\frac{-\Delta H_R a_{in}}{\rho C_P}\right]\varepsilon k t_s a(0) \Delta\mathscr{y}$$

For the interior points,

$$T_{new}(i) = (1 - 2G_T)T(i) + G_T[1 + 0.5/I]T(i+1) + G_T[1 - 0.5/I]T(i-1)$$
$$- \left[\frac{-\Delta H_R a_{in}}{\rho C_P}\right]\varepsilon k t_s a(i) \Delta\mathscr{y}$$

At the wall,

$$T_{new}(I) = \frac{2h_r R \Delta\imath T_{wall} + 4\lambda_r T(I-1) - \lambda_r T(I-2)}{2h_r R \Delta\imath + 3\lambda_r}$$

where $G_T = (\Delta\mathscr{y}/\Delta\imath^2)(E_r t_s/R^2) = 0.2167$ and where $[-\Delta H_R a_{in}/(\rho C_P)] = 417$ K is the adiabatic temperature rise for complete reaction. Solution of these equations shows the maximum temperature to be located on the centerline at an axial position of about 0.5 m down the tube. The maximum temperature is 661 K. Figure 9.3 shows the radial temperature and concentration profiles at the axial position of the maximum temperature. The example uses a low value for T_{in} so that the exotherm is quite modest. Under these conditions, the very crude grid, $I=5$, gives a fairly accurate solution. Industrial reactors tend to push the limits of catalyst degradation or undesired by-product production. Often, they are operated near a condition of thermal runaway where $d^2T/d^2z > 0$. Numerically accurate solutions will then require finer grids in the radial direction because of the large radial temperature gradients. There will also be large axial gradients, and physical stability of the computation may force the use of axial grids smaller than predicted by Equation (9.9).

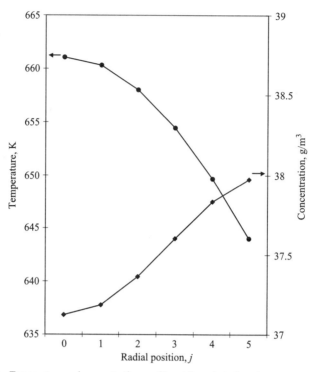

FIGURE 9.3 Temperature and concentration profiles at the point of maximum temperature for the packed-bed reactor of Example 9.1.

Example 9.2: Determine the value for T_{in} that will cause a thermal runaway in the packed tube of Example 9.1.

Solution: Table 9.1 shows the response of the system to a systematic variation in T_{in}. The calculations were carried out using $I = 64$. The solution with $T_{in} \approx 702$ represents a situation known as *parametric sensitivity*, where a small change in a parameter can cause a large change in the system response. Note that the 2 K change from $T_{in} = 690$ to $T_{in} = 692$ causes T_{max} to change by 6 K but the change from $T_{in} = 700$ to $T_{in} = 702$ causes T_{max} to change by 73 K. The axial temperature profile at the centerline is shown in Figure 9.4. There is a classic runaway with $d^2T/d^2z > 0$ for the illustrated case of $T_{in} = 704$ K. Figure 9.5 shows the variation of centerline and mixing-cup average concentrations with axial position. Note that the *o*-xylene is almost completely consumed at the centerline near the *hotspot*, but that the concentration subsequently increases due to radial mass transfer from outlying regions of the reactor.

The simplified reaction in Example 9.2 has the form A → B, and the runaway would be of no concern unless the temperature caused sintering or other

TABLE 9.1 Illustration of Parametric Sensitivity

T_{in}	T_{max}	a_{out}/a_{in}
690	719	0.595
692	725	0.576
694	733	0.551
696	744	0.513
698	762	0.446
700	823	0.275
702	896	0.135
704	930	0.083
706	953	0.057
708	971	0.042
710	987	0.032

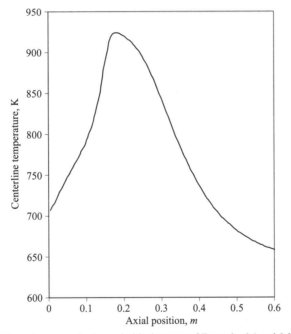

FIGURE 9.4 Thermal runaway in the packed-bed reactor of Examples 9.1 and 9.2; $T_{in} = 704$ K.

degradation of the catalyst. The real reaction has the form A → B → C, and the runaway would almost certainly provoke an undesired reaction B → C. See Problem 9.3. To maximize output of product B, it is typically desired to operate just below the value of T_{in} that would cause a runaway. As a practical matter, models using published parameter estimates are rarely accurate enough to allow a priori prediction of the best operating temperature. Instead, the

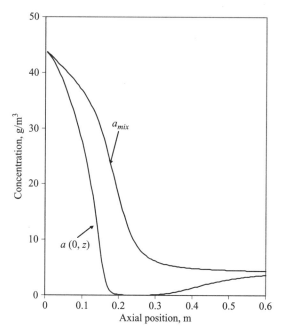

FIGURE 9.5 Reactant concentration profiles for a thermal runaway in the packed-bed reactor of Examples 9.1 and 9.2; $T_{in} = 704$ K.

models are used to guide experimentation and are tuned based on the experimental results.

Whenever there is an appreciable exotherm, scaleup of heterogeneous reactions is normally done in parallel using a multitubular reactor of the shell-and-tube type. The pilot reactor may consist of a single tube with the same packing, the same tube diameter, and the same tube length as intended for the full-scale reactor. The scaled-up reactor consists of hundreds or even thousands of these tubes in parallel. Such scaleup appears trivial, but there are occasional problems. See Cybulski et al.[4] One reason for the problems is that the packing is randomly dumped into the tubes, and random variations can lead to substantial differences in performance. This is a particular problem when d_t/d_p is small. One approach to minimizing the problem has been to use pilot reactors with at least three tubes in parallel. Thus, the scaleup is based on an average of three tubes instead of the possibly atypical performance of a single tube.

There is a general trend toward *structured packings* and *monoliths*, particularly in demanding applications such as automotive catalytic converters. In principle, the steady-state performance of such reactors can be modeled using Equations (9.1) and (9.3). However, the parameter estimates in Figures 9.1 and 9.2 and Equations (9.6)–(9.7) were developed for random packings, and even the boundary condition of Equation (9.4) may be inappropriate for monoliths or structured packings. Also, at least for automotive catalytic converters, the transient rather than steady-state performance of the reactor is of paramount importance. The transient terms $\partial a/\partial t$ and $\partial T/\partial t$ are easily added to Equations (9.1) and (9.3), but the results will mislead. These terms account for inventory changes in the gas phase but not changes in the amount of material absorbed on the solid surface. The surface inventory may be substantially larger than the gas-phase inventory, and a model that explicitly considers both phases is necessary for time-dependent calculations. This topic is briefly discussed in Section 10.6 and in Chapter 11.

It is also easy to add axial dispersion terms, $D_z(\partial^2 a/\partial z^2)$ and $E_z(\partial^2 T/\partial z^2)$. They convert the initial value problem into a two-point boundary value problem in the axial direction. Applying the method of lines gives a set of ODEs that can be solved using the reverse shooting method developed in Section 9.5. See also Appendix 8.3. However, axial dispersion is usually negligible compared with radial dispersion in packed-bed reactors. Perhaps more to the point, uncertainties in the value for D_r will usually overwhelm any possible contribution of D_z.

An important embellishment to the foregoing treatment of packed-bed reactors is to allow for temperature and concentration gradients within the catalyst pellets. The intrapellet diffusion of heat and mass is governed by differential equations that are about as complex as those governing the bulk properties of the bed. A set of simultaneous PDEs (ODEs if the pellets are spherical) must be solved to estimate the extent of reaction and conversion occurring within a single pellet. These local values are then substituted into Equations (9.1) and (9.3) so that we need to solve a set of PDEs that are embedded within a set of PDEs. The resulting system truly reflects the complexity of heterogeneous reactors and is an example of *multiscale modeling*. Practical solutions rarely go to this complexity. Most industrial reactors are designed on the basis of pseudohomogeneous models as in Equations (9.1) and (9.3), and the detailed catalyst behavior is described by the effectiveness factor defined in Chapter 10. In fact, radial gradients are sometimes neglected, even in single-tube calculations for multitubular designs. Such simplified models are anticonservative in the sense that the maximum temperatures are underestimated. At least the radial gradients within the bed should be calculated. Reasonable correlations for radial heat transfer now exist and should be used.

9.2 TURBULENT FLOW IN TUBES

Turbulent flow reactors are modeled quite differently from laminar flow reactors. In a turbulent flow field, nonzero velocity components exist in all three coordinate directions, and they fluctuate with time. Statistical methods must be used to obtain time average values for the various components and to characterize the instantaneous fluctuations about these averages. We divide the velocity into time average and fluctuating parts:

$$\mathbf{v} = \mathbf{\psi} + \mathbf{V} \tag{9.12}$$

where ψ represents the fluctuating velocity and $\mathbf{V}$ is the time average value:

$$\mathbf{V} = \lim_{t \to \infty} 1/t \int_0^t \mathbf{v}\, dt \qquad (9.13)$$

For turbulent flow in long, empty pipes, the time average velocities in the radial and tangential directions are zero since there is no net flow in these directions. The axial velocity component will have a nonzero time average profile $V_z(r)$. This profile is considerably flatter than the parabolic profile of laminar flow, but a profile nevertheless exists. The zero-slip boundary condition still applies and forces $V_z(R) = 0$. The time average velocity changes very rapidly near the tube wall. The region over which the change occurs is known as the *hydrodynamic boundary layer*. Sufficiently near the wall, flow in the boundary layer will be laminar, with attendant limitations on heat and mass transfer. Outside the boundary layer—meaning closer to the center of the tube—the time average velocity profile is approximately flat. Flow in this region is known as *core turbulence*. Here, the fluctuating velocity components are high and give rapid rates of heat and mass transfer in the radial direction. Thus, turbulent flow reactors are often modeled as having no composition or temperature gradients in the radial direction. This is not quite the same as assuming piston flow. At very high Reynolds numbers, the boundary layer thickness becomes small and a situation akin to piston flow is approached. At lower Reynolds numbers, a more sophisticated model may be needed.

To understand the new model, some concepts of turbulent mixing need to be considered. Suppose a small pulse of an ideal, nonreactive tracer is injected into a tube at the center. An *ideal tracer* is identical to the bulk fluid in terms of flow properties but is distinguishable in some nonflow aspect that is detectable with suitable instrumentation. Typical tracers are dyes, radioisotopes, and salt solutions. The first and most obvious thing that happens to the tracer is movement downstream at a rate equal to the time average axial velocity $\bar{u}$. If we are dealing with a stationary coordinate system (called an *Eulerian* coordinate system), the injected pulse just disappears downstream. Now, shift to a moving (*Lagrangian*) coordinate system that translates down the tube with the same velocity as the fluid. In this coordinate system, the center of the injected pulse remains stationary; but individual tracer particles spread out from the center due to the combined effects of molecular diffusion and the fluctuating velocity components. If the time average velocity profile were truly flat, the tracer concentration would soon become uniform in the radial and tangential directions, but would spread indefinitely in the axial direction. This kind of mixing has not been encountered in our previous discussions. Axial mixing is disallowed in the piston flow model and is usually neglected in laminar flow models. The models of Chapter 8 neglected molecular diffusion in the axial direction because axial concentration and temperature gradients are so much smaller than radial gradients. In turbulent flow, eddy diffusion due to the fluctuating velocity components dominates molecular diffusion, and the effective diffusivity is enhanced to the point of virtually eliminating the radial gradients and of causing possibly significant amounts of mixing in the axial direction. We seek a simple correction to piston flow that will account for this axial mixing and other small departures from ideality. A major use of this model is for isothermal reactions in turbulent, pipeline flows. However, the model that emerges is surprisingly versatile. It can be used for isothermal reactions in packed beds, whether laminar or turbulent, and in motionless mixers. It can also be extended to nonisothermal reactions.

9.3 THE AXIAL DISPERSION MODEL

A simple correction to piston flow is to add an axial diffusion term. The resulting equation remains an ODE and is known as the axial dispersion model:

$$\bar{u}\frac{da}{dz} = D\frac{d^2a}{dz^2} + \mathscr{R}_A \tag{9.14}$$

or in dimensionless form,

$$\frac{da}{d\mathscr{z}} = \frac{1}{\mathbf{Pe}}\frac{d^2a}{d\mathscr{z}^2} + \mathscr{R}_A \bar{t} \tag{9.15}$$

The parameter D is known as the *axial dispersion coefficient*, and the dimensionless number, $\mathbf{Pe} = \bar{u}L/D$, is the axial Peclet number. It is different than the Peclet number used in Section 9.1. Also, recall that the tube diameter is denoted by d_t. At high Reynolds numbers, D depends solely on fluctuating velocities in the axial direction. These fluctuating axial velocities cause mixing by a random process that is conceptually similar to molecular diffusion, except that the fluid elements being mixed are much larger than molecules. The same value for D is used for each component in a multicomponent system.

At lower Reynolds numbers, the axial velocity profile will not be flat; and it might seem that another correction must be added to Equation (9.14). It turns out, however, that Equation (9.14) remains a good model for real turbulent reactors (and even some laminar ones) given suitable values for D. The model lumps the combined effects of fluctuating velocity components, nonflat velocity profiles, and molecular diffusion into the single parameter D.

At a close level of scrutiny, real systems behave differently than predicted by the axial dispersion model; but the model is useful for many purposes. Values for **Pe** can be determined experimentally using transient experiments with nonreactive tracers. See Chapter 15. A correlation for D that combines experimental and theoretical results is shown in Figure 9.6. The dimensionless number, $\bar{u}d_t/D$, depends on the Reynolds number and on molecular diffusivity as measured by the Schmidt number, $\mathbf{Sc} = \mu/(\rho \mathscr{D}_A)$, but the dependence on **Sc** is weak for **Re** > 5000. As indicated in Figure 9.6, data for gases will lie near the top of the range and data for liquids will lie near the bottom. For high **Re**, $\bar{u}d_t/D = 5$ is a reasonable choice.

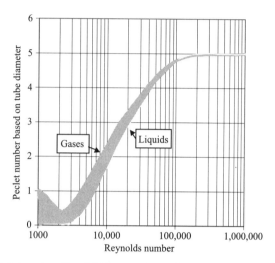

FIGURE 9.6 Peclet number, $\mathbf{Pe} = \bar{u}d_t/D$, versus Reynolds number, $\mathbf{Re} = \rho d_t \bar{u}/\mu$, for flow in an open tube.

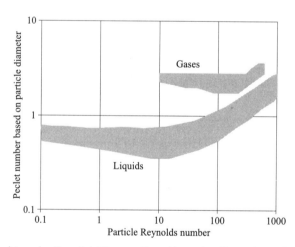

FIGURE 9.7 Peclet number $\mathbf{Pe} = \bar{u}_s d_p/D$, versus Reynolds number, $\mathbf{Re} = \rho d_p \bar{u}_s/\mu$ for packed beds.

The model can also be applied to packed-beds. Figure 9.7 illustrates the range of existing data.

9.3.1 The Danckwerts Boundary Conditions

The axial dispersion model has a long and honored history within chemical engineering. It was first used by Langmuir,[5] who also used the correct boundary conditions. These boundary conditions are quite subtle. Langmuir's work was forgotten, and it was many years before the correct boundary conditions were rediscovered by Danckwerts.[6]

The boundary conditions normally associated with Equation (9.14) are known as the Danckwerts or *closed* boundary conditions. They are obtained from mass balances across the inlet and outlet of the reactor. We suppose that the piping to and from the reactor is small and has a high **Re**. Thus, if we were to apply the axial dispersion model to the inlet and outlet streams, we would find $D_{in} = D_{out} = 0$, which is the definition of a closed system. See Figure 9.8. The flux in the inlet pipe is due solely to convection and has magnitude $Q_{in}a_{in}$. The flux just inside the reactor at location $z = 0+$ has two components. One component, $Q_{in}a(0+)$, is due to convection. The other component, $-DA_c[da/dz]_{0+}$, is due to diffusion (albeit eddy diffusion) from the relatively high concentrations at the inlet toward the lower concentrations within the reactor. The inflow to the plane at $z = 0$ must be matched by material leaving the plane at $z = 0+$ since no reaction occurs in a region that has no volume. Thus,

$$Q_{in}a_{in} = Q_{in}a(0+) - DA_c[da/dz]_{0+}$$

or

$$a_{in} = a(0+) - \frac{1}{\mathbf{Pe}}\left[\frac{da}{d\hat{z}}\right]_{0+}$$

(9.16)

is the inlet boundary condition for a closed system. The outlet boundary condition is obtained by a mass balance across a plane at $z = L$. We expect concentration to be a continuous function of z across the outlet plane so that $a(L+) = a(L-)$. Since $D_{out} = 0$, the balance gives $Q_{out}a(L-) = Q_{out}a(L+)$ and

$$\left[\frac{da}{dz}\right]_L = 0$$

or

$$\left[\frac{da}{d\hat{z}}\right]_1 = 0$$

(9.17)

as the outlet boundary condition for a closed system.

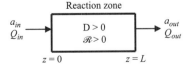

FIGURE 9.8 The axial dispersion model applied to a closed system.

These boundary conditions are really quite marvelous. Equation (9.16) predicts a discontinuity in concentration at the inlet to the reactor so that $a_{in} \neq a(0+)$ if $D>0$. This may seem counterintuitive until the behavior of a CSTR is recalled. At the inlet to a CSTR, the concentration goes immediately from a_{in} to a_{out}. The axial dispersion model behaves as a CSTR in the limit as $D \to \infty$. It behaves as a piston flow reactor, which has no inlet discontinuity, when $D=0$. For intermediate values of D, an inlet discontinuity in concentrations exists but is intermediate in size. The concentration $a(0+)$ results from backmixing between entering material and material downstream in the reactor. For a reactant, $a(0+) < a_{in}$.

The concentration is continuous at the reactor exit for all values of D and this forces the zero-slope condition of Equation (9.17). The zero-slope condition may also seem counterintuitive, but recall that CSTRs behave in the same way. The reaction stops so the concentration stops changing.

The marvelousness of the Danckwerts boundary conditions is further explored in Example 9.3, which treats open systems.

9.3.2 First-Order Reactions

Equation (9.14) is a linear ODE with constant coefficients. An analytical solution is possible when the reactor is isothermal and the reaction is first order. The general solution to Equation (9.14) with $\mathcal{R}_A = -ka$ is

$$a(z) = C_1 \exp\left[(1+p)\frac{\mathbf{Pe}}{2}\frac{z}{L}\right] + C_2 \exp\left[(1-p)\frac{\mathbf{Pe}}{2}\frac{z}{L}\right] \tag{9.18}$$

where $\mathbf{Pe} = \bar{u}L/d_t$ and

$$p = \sqrt{1 + \frac{4k\bar{t}}{\mathbf{Pe}}} \tag{9.19}$$

The constants C_1 and C_2 are evaluated using the boundary conditions, Equations (9.16) and (9.17). The outlet concentration is found by setting $z = L$. Algebra gives

$$\frac{a_{out}}{a_{in}} = \frac{4p \exp\left(\frac{\mathbf{Pe}}{2}\right)}{(1+p)^2 \exp\left(\frac{p\mathbf{Pe}}{2}\right) - (1-p)^2 \exp\left(\frac{-p\mathbf{Pe}}{2}\right)} \tag{9.20}$$

Conversions predicted from Equation (9.20) depend only on the values of $k\bar{t}$ and $\mathbf{Pe}$. The predicted conversions are smaller than those for piston flow but larger than those for perfect mixing. In fact,

$$\lim_{\mathbf{Pe} \to \infty} \frac{a_{out}}{a_{in}} = e^{-k\bar{t}} \tag{9.21}$$

so that the model approaches piston flow in the limit of high Peclet number (low D). Also,

$$\lim_{Pe \to 0} \frac{a_{out}}{a_{in}} = \frac{1}{1+k\bar{t}} \quad (9.22)$$

so that the axial dispersion model approaches perfect mixing in the limit of low Peclet number (high D). The model is thus universal in the sense that it spans the expected range of performance for well-designed real reactors. However, it should not be used, or be used only with caution, for **Pe** below about 8.

Example 9.3: Equation (9.20) was derived for a closed system. Repeat the derivation for an open system with $D_{in} > 0$ and $D_{out} > 0$ shown in Figure 9.9.

Solution: An open system extends from $-\infty$ to $+\infty$ as shown in Figure 9.9. The key to solving this problem is to note that the general solution, Equation (9.18), applies to each of the above regions; inlet, reaction zone, and outlet. If $k=0$ then $p=1$. Each of the equations contains two constants of integration. Thus, a total of six boundary conditions are required. They are

1. The far inlet boundary condition: $a = a_{in}$ at $z = -\infty$
2. Continuity of concentration at $z=0$: $a(0-) = a(0+)$
3. Continuity of flux at $z=0$: $Q_{in}a(0-) - A_c[da/dz]_{0-} = Q_{in}a(0+) - A_c[da/dz]_{0+}$
4. Continuity of concentration at $z=L$: $a(L-) = a(L+)$
5. Continuity of flux at $z=L$: $Q_{in}a(L-) - A_c[da/dz]_{L-} = Q_{in}a(L+) - A_c[da/dz]_{L+}$
6. The far outlet boundary condition: $a = a_{out}$ at $z = +\infty$

A substantial investment in algebra is needed to evaluate the six constants, but the result is remarkable. The exit concentration from an open system is identical to that from a closed system, Equation (9.20), and is thus independent of D_{in} and D_{out}! The physical basis for this result depends on the concentration profile, $a(z)$, for $z<0$. When $D=0$, the concentration is constant at a value if a_{in} until $z=0+$, when it suddenly plunges to $a(0+)$. When $D>0$, the concentration begins at a_{in} when $z=-\infty$ and gradually declines until it reaches exactly the same concentration, $a(0+)$, at exactly the same location, $z=0+$. For $z>0$, the open and closed systems have the same concentration profile and the same reactor performance.

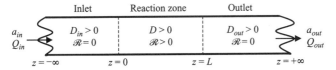

FIGURE 9.9 The axial dispersion model applied to an open system.

9.3.3 Utility of the Axial Dispersion Model

Chapters 8 and Section 9.1 gave preferred models for laminar flow and packed-bed reactors. The axial dispersion model can also be used for these reactors but is generally less accurate. Proper roles for the axial dispersion model are the following.

Turbulent Pipeline Flow. Turbulent pipeline flow is the original application of the axial dispersion model. For most kinetic schemes, piston flow predicts the highest possible conversion and selectivity. The axial dispersion model provides a less optimistic estimate, but the difference between the piston flow and axial dispersion models is usually small. For an open tube in well-developed turbulent flow, the assumption of piston flow is normally quite accurate unless the reaction is driven to near-completion. Figure 9.10 provides a quick means for estimating the effects of axial dispersion. The errors are percentages of the fraction unreacted. For a liquid at $\mathbf{Re} = 20{,}000$, Figure 9.6 gives $(\bar{u}d_t)/D \approx 3$ so that $\mathbf{Pe} \approx 3L/d_t$. For a reactor with $L/d_t = 33$, $\mathbf{Pe} \approx 100$, and Figure 9.10 shows that 1% error corresponds to $k\bar{t} \approx 1$. Thus, $a_{out}/a_{in} = 0.368$ for piston

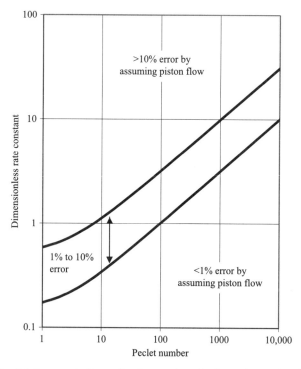

FIGURE 9.10 Relative error in the predicted conversion of a first-order reaction due to assuming piston flow rather than axial dispersion, $k\bar{t}$ versus $\mathbf{Pe}$.

flow and 0.364 for axial dispersion. Problem 9.5 gives an example where the correction for axial dispersion is much more significant. Such examples are the exception.

Isothermal Packed Beds. A packed reactor has a velocity profile that is nearly flat; and, for the usual case of uniform a_{in}, no concentration gradients will arise unless there is a radial temperature gradient. If there is no reaction exotherm (and if $T_{in} = T_{wall}$), the model of Section 9.1 degenerates to piston flow. This is overly optimistic for a real packed bed, and the axial dispersion model provides a correction. The correction will usually be small. Note that $\bar{u}$ should be replaced by $\bar{u}_s$ and that the void fraction ε should be inserted before the reaction term; e.g., $k\bar{t}$ becomes $\varepsilon k\bar{t}$ for reactions in a packed bed. Figure 9.7 gives $D\varepsilon/(\bar{u}_s d_p) \approx 2$ for moderate values of the particle Reynolds number. This gives $\mathbf{Pe} = \varepsilon L/(2d_p)$ or $\mathbf{Pe} \approx 300$ for the packed tube of Example 9.1. Again, the assumption of piston flow is quite reasonable unless the reaction goes to near-completion. It should be emphasized that the assumption of an isothermal reaction should be based on a small heat of reaction; e.g., as in a transesterification where the energy of a bond broken is approximately equal to that of a bond made or when inerts are present in large quantities. Calculate the adiabatic temperature rise. Sooner or later it will emerge upon scaleup.

Adiabatic Reactors. Like isothermal reactors, adiabatic reactors with a flat velocity profile will have no radial gradients in temperature or composition. There are axial gradients, and the axial dispersion model, including its extension to temperature in Section 9.4, can account for axial mixing. As a practical matter, it is difficult to build a small adiabatic reactor. Wall temperatures must be controlled to simulate the adiabatic temperature profile in the reactor, and guard heaters may be needed at the inlet and outlet to avoid losses by radiation. Even so, it is likely that uncertainties in the temperature profile will mask the relatively small effects of axial dispersion.

Laminar Pipeline Flows. The axial dispersion model can be used for laminar flow reactors if the reactor is so long that $\mathscr{D}_A \bar{t}/R^2 > 0.125$. With this high value for $\mathscr{D}_A \bar{t}/R^2$, the initial radial position of a molecule becomes unimportant. The molecule diffuses across the tube and samples many streamlines, some with high velocity and some with low velocity, during its stay in the reactor. It will travel with an average velocity near $\bar{u}$ and will emerge from the long reactor with a residence time close to $\bar{t}$. The axial dispersion model is a reasonable approximation for overall dispersion in a long, laminar flow reactor. The appropriate value for D is known from theory:

$$D = \mathscr{D}_A + \frac{\bar{u}^2 R^2}{48 \mathscr{D}_A} \tag{9.23}$$

As seen in Chapter 8, the stability criterion becomes quite demanding when $\mathscr{D}_A \bar{t}/R^2$ is large. The axial dispersion model may then be a useful alternative to solving Equation (8.12).

Motionless Mixers. These interesting devices consist of a tube or duct within which static elements are installed to promote radial flow. They are quite effective in promoting mixing in laminar flow systems, but their geometry is too complex to allow solution of the convective diffusion equation on a routine basis. The axial dispersion model may be useful for data correlations and scaleup when motionless mixers are used as reactors with premixed feed. A study on their use for homogeneous reactions in deep laminar flow, **Re** < 100, found that **Pe** ≈ 70 L, where L is the length in meters.[7] This dimensionally inconsistent result applies to 40-mm diameter Sulzer mixers of the SMX and SMV types. It obviously cannot be generalized. See also Fialova et al.[8] The lack of published data prevents *a priori* designs that utilize static mixers, but the axial dispersion model is a reasonable way to correlate pilot-plant data. Chapter 15 shows how **Pe** can be measured using inert tracers.

Static mixers are typically less effective in turbulent flow than an open tube when the comparison is made on the basis of constant pressure drop or capital cost. Whether laminar or turbulent, design correlations are generally lacking or else are vendor-proprietary and are rarely been subject to peer review.

9.4 NONISOTHERMAL AXIAL DISPERSION

The axial dispersion model is readily extended to nonisothermal reactors. The turbulent mixing that leads to flat concentration profiles will also give flat temperature profiles. An expression for the axial dispersion of heat can be written in direct analogy to Equation (9.14):

$$\bar{u}\frac{dT}{dz} = E\frac{d^2T}{dz^2} - \frac{2h}{\rho C_P}\frac{(T - T_{wall})}{R} - \frac{\Delta H_R \mathscr{R}}{\rho C_P} \quad (9.24)$$

where E is the axial dispersion coefficient for heat and where the usual summation conventions apply to $\Delta H_R \mathscr{R}$. For well-developed turbulence, the thermal Peclet number, $(\mathbf{Pe})_{thermal} = \bar{u}L/E$, should be identical to the mass Peclet number, $\mathbf{Pe} = \bar{u}L/D$. At lower Reynolds numbers, one would expect $\bar{u}L/E$ to depend on a thermal Schmidt number, $(\mathbf{Sc})_{thermal} = \mu/\rho\alpha_T = \mu C_P/\lambda$, which is more commonly called the Prandtl number. The inside heat transfer coefficient, h, can be estimated from standard correlations such as Equation (5.38).

The boundary conditions associated with Equation (9.24) are of the Danckwerts type:

$$Q_{in}T_{in} = Q_{in}T(0+) - EA_c[dT/dz]_{0+} \quad (9.25)$$

$$[dT/dz]_L = 0 \quad (9.26)$$

Correlations for E are not widely available. The more accurate model given in Section 9.1 is preferred for nonisothermal reactions in packed-beds. However, as discussed previously, this model degenerates to piston flow for an adiabatic reaction. The nonisothermal axial dispersion model is a conservative design methodology available for adiabatic reactions in packed beds and for nonisothermal reactions in turbulent pipeline flows. The fact that $E > D$ provides some basis for estimating E. Recognize that the axial dispersion model is a correction to what would otherwise be treated as piston flow. Thus, even setting $E = D$ should improve the accuracy of the predictions.

Only numerical solutions are possible when Equation (9.24) is solved simultaneously with Equation (9.14). This is true even for first-order reactions because of the intractable nonlinearity of the Arrhenius temperature dependence.

9.5 NUMERICAL SOLUTIONS TO TWO-POINT BOUNDARY VALUE PROBLEMS

The numerical solution of Equations (9.14) and (9.24) is more complicated than the solution of the first-order ODEs that govern piston flow or of the first-order ODEs that result from applying the method of lines to PDEs. The reason for the complication is the second derivative in the axial direction, d^2a/dz^2.

Apply finite difference approximations to Equation (9.15) using a backwards difference for $da/d\mathpzc{z}$ and a central difference for $d^2a/d\mathpzc{z}^2$. The result is

$$a_{j+1} = (2 + \text{Pe}\,\Delta\mathpzc{z})a_j - (1 + \text{Pe}\,\Delta\mathpzc{z})a_{j-1} - \text{Pe}\,\mathscr{R}_A \bar{t}\Delta\mathpzc{z}^2 \quad (9.27)$$

Thus, the value for the next, $j+1$, point requires knowledge of two previous points, j and $j-1$. To calculate a_2, we need to know both a_1 and a_0. The boundary conditions, Equations (9.16) and (9.17), give neither of these directly. In finite difference form, the inlet boundary condition is

$$a_1 = (1 + \text{Pe}\,\Delta\mathpzc{z})a_0 - \text{Pe}\,\Delta\mathpzc{z}\,a_{in} \quad (9.28)$$

where a_{in} is known. Thus, if we guess a_0, we can calculate a_1 using Equation (9.28) and we can then use Equation (9.27) to march down the tube. The outlet boundary condition is

$$a_{J+1} = a_J \quad (9.29)$$

where J is the number of steps in the axial direction. If Equation (9.29) is satisfied, the correct value for a_0 was guessed. Otherwise, guess a new a_0. This approach is known as forward shooting.

The forward shooting method seems straightforward but is troublesome to use. What we have done is to convert a two-point boundary value problem into an easier-to-solve initial value problem. Unfortunately, the conversion gives a numerical computation that is *ill-conditioned*. Extreme precision is needed at the inlet of the tube to get reasonable accuracy at the outlet. The phenomenon is akin to problems that arise in the numerical inversion of matrices and Laplace transforms.

Example 9.4: Use forward shooting to solve Equation (9.15) for a first-order reaction with $\mathbf{Pe} = 16$ and $k\bar{t} = 2$. Compare the result with the analytical solution, Equation (9.20).

Solution: Set $\Delta \dot{z} = 1/32$ so that $\mathbf{Pe}\Delta \dot{z} = 0.5$ and $\mathbf{Pe}\,k\bar{t}\,\Delta \dot{z}^2 = 0.03125$. Set $a_{in} = 1$ so that dimensionless or normalized concentrations are determined. Equation (9.27) becomes

$$a_{j+1} = 2.53125\, a_j - 1.5\, a_{j-1}$$

The computation is started using Equation (9.28):

$$a_1 = 1.5\, a_0 - 0.5$$

Results for a succession of guesses for a_0 give

a_0	a_{32}	a_{33}
0.90342	−20.8	−33.0
0.90343	0.93	1.37
0.903429	−1.24	−2.06
0.9034296	0.0630	0.0004
0.90342965	0.1715	0.1723
0.903429649	0.1693	0.1689
0.9034296493	0.1699	0.1699

The answer by the shooting method is $a_{out} = 0.17$. The analytical result is $a_{out} = 0.1640$. Note that the shooting method requires extreme precision on guesses for a_0 to obtain an answer of limited accuracy for a_{out}. Double precision is needed on a 16-bit computer. Better accuracy with the numerical approach can be achieved with a smaller step size or a more sophisticated integration routine such as Runge-Kutta, but better integration gives a more accurate value only after the right guess for a_0 is made. It does not eliminate the ill-conditioning inherent in forward shooting.

The best solution to such numerical difficulties is to change methods. Integration in the reverse direction eliminates most of the difficulty. Go back to Equation (9.15). Continue to use a second-order, central difference approximation for $d^2a/d\dot{z}^2$, but now use a first-order, forward difference approximation for $da/d\bar{z}$. Solve the resulting finite difference Equation for a_{j-1}:

$$a_{j-1} = (2 - \mathbf{Pe}\,\Delta\bar{z})a_j - (1 - \mathbf{Pe}\,\Delta\bar{z})a_{j+1} - \mathbf{Pe}\,\mathscr{R}_A\bar{t}\,\Delta\bar{z}^2 \tag{9.30}$$

The marching-ahead equation becomes a marching-backward equation. The method is called *reverse shooting*. The procedure is to guess $a_J = a_{out}$ and then to set $a_{J-1} = a_J$. The index j in Equation (9.30) begins at $J-2$ and is decremented by 1 until $j=0$ is reached. The reaction rate continues to be evaluated at the central, jth point. The test condition is whether a_{in} is correct when calculated using the inlet boundary condition

$$a_{in} = a_0 + \frac{a_0 - a_1}{\mathbf{Pe}\,\Delta\bar{z}} \tag{9.31}$$

Example 9.5: Repeat Example 9.4 using reverse shooting.

Solution: With $J = 32$, $\mathbf{Pe} = 16$, and $\bar{k}\bar{t} = 2$, Equation (9.30) gives

$$a_{j-1} = 1.53125 a_j - 0.5 a_{j+1}$$

Guess $a_{32} = a_{out}$ and then set $a_{31} = a_{32}$. Calculate a_j down to $j=0$. Then compare a_{in} with the value calculated using Equation (9.31) which, for this example, is just

$$a_{in} = 3 a_0 - 2 a_1$$

Some results are

a_{32}	a_{in}
0.16	1.0073
0.15	0.9444
0.159	1.0010
0.158	0.9947

Thus, we obtain $a_{out} = 0.159$ for a step size of $\Delta\bar{z} = 0.03125$. The ill-conditioning problem has been solved, but the solution remains inaccurate due to the simple integration scheme and the large step size.

The next example illustrates the use of reverse shooting in solving a problem in nonisothermal axial dispersion and shows how Runge-Kutta integration can be applied to second-order ODEs.

Example 9.6: Compare the nonisothermal axial dispersion model with piston flow for a first-order reaction in turbulent pipeline flow with $\mathbf{Re} = 10{,}000$. Pick the reaction parameters so that the reactor is at or near a region of thermal runaway.

Solution: The axial dispersion model requires the simultaneous solution of Equations (9.14) and (9.24). Piston flow is governed by the same equations except that $D = E = 0$. The following parameter values give rise to a near runaway:

$$k_0 \bar{t} = 2.0 \times 10^{11} \quad \text{(dimensionless)}$$

$$T_{act} = 10{,}000 \text{ K}$$

$$\frac{2h\bar{t}}{\rho C_P R} = 10 \quad \text{(dimensionless)}$$

$$-\frac{\Delta H_R a_{in}}{\rho C_P} = 200 \text{K}$$

$$T_{in} = T_{wall} = 373 \text{ K}$$

These parameters are enough to run the piston flow case. The solution is $a_{out}/a_{in} = 0.209$, $T_{out} = 376$ K, and $T_{max} = 403$ K occurring at $\dot{y} = 0.47$.

Turn now to the axial dispersion model. Plausible values for the dispersion coefficients at **Re** $= 10{,}000$ are

$$\frac{D}{\bar{u}d_t} = 0.45 \qquad \frac{D}{\bar{u}L} = 4.5$$

$$\frac{E}{\bar{u}d_t} = 0.60 \qquad \frac{E}{\bar{u}L} = 6.0$$

where we have assumed a low aspect ratio, $L/d_t = 10$, to magnify the effects of axial dispersion.

When the axial dispersion terms are present, $D > 0$ and $E > 0$, Equations (9.14) and (9.24) are second order. We will use reverse shooting and Runge-Kutta integration. The Runge-Kutta scheme (Appendix 2) applies only to first-order ODEs. To use it here, Equations (9.14) and (9.24) must be converted to an equivalent set of first-order ODEs. This can be done by defining two auxiliary variables:

$$a' = da/d\dot{y} \quad \text{and} \quad T' = dT/d\dot{y}$$

Then Equations (9.14) and (9.24) can be written as a set of four, first-order ODEs with boundary conditions as indicated below:

$$da/d\dot{y} = a' \qquad\qquad a = a_{out} \text{ at } \dot{y} = 1$$

$$\frac{da'}{d\dot{y}} = \frac{[a' + k_0 \bar{t} \exp(-T_{act}/T)a]}{(D/\bar{u}L)} \qquad a' = 0 \text{ at } z = 1$$

$dT/d\breve{y} = T'$ $\qquad T = T_{out}$ at $\breve{y} = 1$

$$\frac{dT'}{d\breve{y}} = \frac{\left[T' + \frac{2h\bar{t}}{\rho C_P R}(T - T_{wall}) - \left(\frac{-\Delta H_R a_{in}}{\rho C_P}\right) k_0 \bar{t} \exp(-T_{act}/T) a/a_{in}\right]}{(E/\bar{u}L)}$$

$T' = 0$ at $\breve{y} = 1$

There are four equations in four dependent variables, a, a', T, and T'. They can be integrated using the Runge-Kutta method as outlined in Appendix 2. Note that they are integrated in the reverse direction; e.g., $a_1 = a_0 - \mathscr{R}_A \Delta \breve{y}/2$, and similarly for a_2 and a_3 in Equations (2.47).

A double trial-and-error procedure is needed to determine a_0 and T_0. If done only once, this is probably best done by hand. This is the approach used in the sample program. Simultaneous satisfaction of the boundary conditions for concentration and temperature was aided by using an output response that combined the two errors. If repeated evaluations are necessary, a two-dimensional Newton's method can be used. Define

$$F(a_0, T_0) = a_0 - \frac{D}{UL} a'(0) - a_{in}$$

$$G(a_0, T_0) = T_0 - \frac{E}{\bar{u}L} T'(0) - T_{in}$$

and use the methodology of Appendix 4 to find a_0 and T_0 such that $F = G = 0$. The following is a comparison of results with and without axial dispersions:

	Piston flow	Axial dispersion
$D/\bar{u}L$	0	0.045
$E/\bar{u}L$	0	0.060
a_{out}/a_{in}	0.209	0.340
T_{out}	376 K	379 K
T_{max}	403 K	392 K
$\breve{y}(T_{max})$	0.47	0.37

A repetitious but straightforward Basic program for solving this axial dispersion problem follows:

```
tmeank0 = 200000000000#
Tact = 10000
h = 10
heat = 200
Tin = 373
Twall = 373
```

```
D = 0.045
E = 0.06
ain = 1
JJ = 32
dz = 1 / JJ

1 Input a0, T0
ap0 = 0
Tp0 = 0

For j = 1 To JJ
  Rp0 = RxRateP(a0, ap0, T0, Tp0)
  Sp0 = SourceP(a0, ap0, T0, Tp0)
  R0 = ap0
  S0 = Tp0

  a1 = a0-R0 * dz / 2
  T1 = T0-S0 * dz / 2
  ap1 = ap0-Rp0 * dz / 2
  Tp1 = Tp0-Sp0 * dz / 2

  Rp1 = RxRateP(a1, ap1, T1, Tp1)
  Sp1 = SourceP(a1, ap1, T1, Tp1)
  R1 = ap1
  S1 = Tp1

  a2 = a0-R1 * dz / 2
  T2 = T0-S1 * dz / 2
  ap2 = ap0-Rp1 * dz / 2
  Tp2 = Tp0-Sp1 * dz / 2

  Rp2 = RxRateP(a2, ap2, T2, Tp2)
  Sp2 = SourceP(a2, ap2, T2, Tp2)
  R2 = ap2
  S2 = Tp2

  a3 = a0-R2 * dz
  T3 = T0-S2 * dz
  ap3 = ap0-Rp2 * dz
  Tp3 = Tp0-Sp2 * dz

  Rp3 = RxRateP(a3, ap3, T3, Tp3)
  Sp3 = SourceP(a3, ap3, T3, Tp3)
  R3 = ap3
```

```
S3 = Tp3

a0 = a0-(R0 +2 * R1+2 * R2 +R3) / 6 * dz
T0 = T0-(S0+2 * S1+2 * S2 +S3) / 6 * dz
ap0 = ap0-(Rp0+2 * Rp1+2 * Rp2 +Rp3) / 6 * dz
Tp0 = Tp0-(Sp0+2 * Sp1+2 * Sp2 +Sp3) / 6 * dz

Next j

Atest = a0-D * ap0-ain
TTest = T0-E * Tp0-Tin
Ctest = Abs(Atest) +Abs(TTest) / 10
Print Ctest
GoTo 1 'Highly efficient code for a manual search even
        'though frowned upon by purists

Function RxRateP(a, ap, T, Tp)
  RxRateP = (ap+tmeank0 * Exp(-Tact / T) * a) / D
End Function

Function SourceP(a, ap, T, Tp)
  SourceP = (Tp+h * (T-Twall)+heat * RxRate(a, T)) / E
```

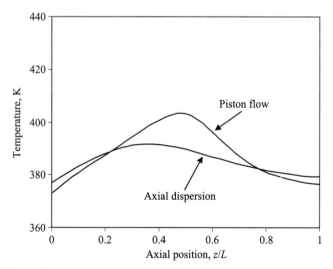

FIGURE 9.11 Comparison of piston flow and axial dispersion for the packed-bed reactor of Example 9.6; $T_{in} = T_{wall} = 373$ K.

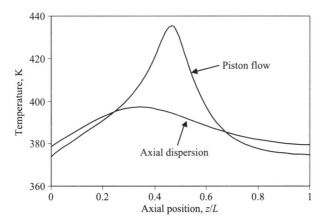

FIGURE 9.12 Comparison of piston flow and axial dispersion models at conditions near thermal runaway; $T_{in} = T_{wall} = 374$ K.

This example was chosen to be sensitive to axial dispersion but the effects are fairly modest. As expected, conversions are lower and the hotspots are colder when axial dispersion is considered. See Figure 9.11.

A more dramatic comparison of the piston flow and axial dispersion models is shown in Figure 9.12. Input parameters are the same as for Figure 9.11 except that T_{in} and T_{wall} were increased by 1 K. This is another example of parametric sensitivity. Compare Example 9.2.

Observe that the axial dispersion model provides a lower and thus *more* conservative estimate of conversion than does the piston flow model given the same values for the input parameters. There is a more subtle possibility. The model may show that it is possible to operate with *less* conservative values for some parameters—e.g., higher values for T_{in} and T_{wall}—without provoking adverse side reactions.

9.6 SCALEUP AND MODELING CONSIDERATIONS

Previous chapters have discussed how isothermal or adiabatic reactors can be scaled up. Nonisothermal reactors are more difficult. They can be scaled by maintaining the same tube diameter or by the modeling approach. The challenge is to increase tube diameter upon scaleup. This is rarely possible; and when it is possible, scaleup must be based on the modeling approach. If the predictions are satisfactory, and if you have confidence in the model, proceed with scaleup.

What models should be used, either for scaleup or to correlate pilot-plant data? Section 9.1 gives the preferred models for nonisothermal reactions in packed beds. These models have a reasonable experimental basis even though they use empirical parameters, D_r, h_r, and λ_r, to account for the packing and the complexity of the flow field. For laminar flow in open tubes, use the methods in Chapter 8. For highly turbulent flows in open tubes (with reasonable L/d_t ratios) use the axial dispersion model in both the isothermal and nonisothermal cases. The assumption $D = E$ will usually be safe, but do calculate how a PFR would perform. If there is a substantial difference between the PFR model and the axial dispersion model, understand the reason. For transitional flows, it is usually conservative to use the methods of Chapter 8 to calculate yields and selectivities but to assume turbulence for pressure-drop calculations.

PROBLEMS

9.1. A gas phase reaction, $A + B \xrightarrow{k}$ Products, is performed in a packed-bed reactor at essentially constant temperature and pressure. The following data are available: $d_t = 0.3$ m, $L = 8$ m, $\varepsilon = 0.5$, $D_r = 0.0005$ m^2/s, $\bar{u}_s = 0.25$ m/s, $a_{in} = b_{in}$. The current operation using premixed feed gives $Y = a_{out}/a_{in} = 0.02$. There is a safety concern about the premixing step. One proposal is to feed A and B separately. Component A would be fed into the base of the bed using a central tube with diameter 0.212 m and component B would be fed to the annulus between the central tube and the reactor wall. The two streams would mix and react only after they had entered the bed. The concentrations of the entering components would be increased by a factor of 2, but the bed-average concentrations and $\bar{u}_s$ would be unchanged. Determine the fraction unreacted that would result from the proposed modification.

9.2. Example 9.1 on the partial oxidation of o-xylene used a pseudo-first-order kinetic scheme. For this to be justified, the oxygen concentration must be approximately constant, which in turn requires low oxygen consumption and a low pressure drop. Are these assumptions reasonable for the reactor in Example 9.1? Specifically, estimate the total change in oxygen concentration given atmospheric discharge pressure and $a_{out} = 21$ g/m^3. Assume $\varepsilon = 0.4$.

9.3. Phthalic anhydride will, in the presence of the V_2O_5 catalyst of Example 9.1, undergo complete oxidation with $\Delta H_R = -760$ kcal/mol. Suppose the complete oxidation is pseudo-first-order in phthalic anhydride concentration and that $\ln(\varepsilon k_{II}) = 12.300 - 10,000/T$.

(a) To establish an upper limit on the yield of phthalic anhydride, pretend the reaction can be run isothermally. Determine yield as a function of temperature.

(b) To gain insight into the potential for a thermal runaway, calculate the adiabatic temperature rise if only the first oxidation goes to completion (i.e., $A \to B$) and if both the oxidation steps go to completion (i.e., $A \to B \to C$).

(c) Determine the value for T_{in} that will just cause a thermal runaway. This gives an upper limit on T_{in} for practical operation of the nonisothermal reactor. Take extra care to control error in your calculations.

(d) Based on the constraint found in part (c), determine the maximum value for the phthalic anhydride yield in the packed tube.

9.4. An alternative route to phthalic anhydride is the partial oxidation of naphthalene. The heat of reaction is -430 kcal/mol. This reaction can be performed using a promoted V_2O_5 catalyst on silica, much like that considered in Example 9.1. Suppose $\ln(\varepsilon k) = 31.6800 - 19{,}100/T$ for the naphthalene oxidation reaction and that the subsequent, complete oxidation of phthalic anhydride follows the kinetics of Problem 9.3. Suppose it is desired to use the same reactor as in Example 9.1 but with $a_{in} = 53$ g/m^3. Determine values for T_{in} and T_{wall} that maximize the output of phthalic anhydride from naphthalene.

9.5. Nerve gas is to be thermally decomposed by oxidation using a large excess of air in a 5-cm i.d. tubular reactor that is approximately isothermal at $620°$C. The entering concentration of nerve gas is 1% by volume. The outlet concentration must be less than 1 part in 10^{12} by volume. The observed half-life for the reaction is 0.2 s. How long should the tube be for an inlet velocity of 2 m/s? What will be the pressure drop given an atmospheric discharge?

9.6. Determine the yield of a second-order reaction in an isothermal tubular reactor governed by the axial dispersion model with $\mathbf{Pe} = 16$ and $a_{in} k \bar{t} = 2$.

9.7. Water at room temperature is flowing through a 1.0-in i.d. tubular reactor at $\mathbf{Re} = 1000$. What is the minimum tube length needed for the axial dispersion model to provide a reasonable estimate of reactor performance? What is the Peclet number at this minimum tube length? Why would anyone build such a reactor?

9.8. The marching equation for reverse shooting, Equation (9.24), was developed using a first-order, backward difference approximation for da/dz, even though a second-order approximation was necessary for d^2a/dz^2. Since the locations $j-1$, j, and $j+1$ are involved anyway, would it not be better to use a second-order, central difference approximation for da/dz?

(a) Would this allow convergence $O(\Delta z^2)$ for the reverse shooting method?

(b) Notwithstanding the theory, run a few values of J, differing by factors of 2, to experimentally confirm the orders of convergence for the two methods.

9.9. The piston flow model in Example 9.6 showed a thermal runaway when $T_{in} = T_{wall} = 374$. Will the axial dispersion model show a runaway (defined as $d^2T/dz^2 > 0$)? If so, at what value of $T_{in} = T_{wall}$?

9.10. Examples 9.1 and 9.2 used a distributed parameter model (simultaneous PDEs) for the phthalic anhydride reaction in a packed bed. Axial dispersion is a lumped parameter model (simultaneous ODEs) that can also be used for a packed bed. Apply the axial dispersion model to the phthalic reaction using D as determined from Figure 9.7 and $E = 1.33\ D$. Compare your results with those obtained in Examples 9.1 and 9.2.

9.11. Extend Figure 8.3 to the higher values of $\mathscr{D}_A \bar{t} / R^2$ needed to show an asymptotic approach to the performance of a CSTR. Assume $L/R = 16$. A partially implicit solution technique is suggested. See Appendix 8.3.

REFERENCES

1. Froment, G. F. and Bischoff, K. B., *Chemical Reaction Analysis and Design*, 2nd ed., Wiley, New York, 1990.
2. Dixon, A. G. and Cresswell, D. L., "Theoretical prediction of effective heat transfer parameters in packed beds," *AIChE J.*, **25**, 663–676 (1979).
3. Dixon, A.G., "An improved equation for the overall heat transfer coefficient in packed beds," *Chem. Eng. Process*, **35**, 323–331 (1996).
4. Cybulski, A., Eigenberger, G., and Stankiewicz, A., "Operational and structural nonidealities in modeling and design of multitubular catalytic reactors," *Ind. Eng. Chem. Res.* **36**, 3140–3148 (1997).
5. Langmuir, I., "The velocity of reactions in gases moving through heated vessels and the effect of convection and diffusion," *J. Am. Chem. Soc.*, **30**, 1742–1754 (1908).
6. Danckwerts, P. V., "Continuous flow systems: distribution of residence times," *Chem. Eng. Sci.*, **2**, 1–18 (1953).
7. Flaschel, E., Nguyen, K. T., and Renken, A., "Improvements of homogeneous tubular reactors operated at low Reynolds numbers," *Proc. 5th Eur. Conf. Mixing*, Jane Stansbury, Ed., 549–554 (1985).
8. Fialova, M., Redlich, K. H., and Winkler, K., "Axial dispersion of the liquid phase in vertical tubular contactors with static mixers," *Collect. Czech. Chem. Commun.*, **51**, 1925–1932 (1986).

SUGGESTIONS FOR FURTHER READING

The heat and mass transfer phenomena associated with packed-bed reactors are described in

Froment, G. F. and Bischoff, K. B., *Chemical Reaction Analysis and Design*, 2nd ed., Wiley, New York, 1990.

Correlations for heat transfer in packed-beds are still being developed. The current state of the art is represented by

Logtenberg, S. A., Nijemeisland, M., and Dixon, A. G., "Computational fluid dynamics simulations of fluid flow and heat transfer at the wall-particle contact points in a fixed-bed reactor," *Chem. Eng. Sci.* **54**, 2433–2439 (1999).

The more classic and time-tested work is

Dixon, A. G. and Cresswell, D. L., "Theoretical prediction of effective heat transfer parameters in packed beds," *AIChE J.*, **25**, 663–676 (1979).

A review article describing the occasionally pathological behavior of packed bed reactors is

Cybulski, A., Eigenberger, G., and Stankiewicz, A., "Operational and Structural Nonidealities in Modeling and Design of Multitubular Catalytic Reactors," *Ind. Eng. Chem. Res.*, **36**, 3140–3148 (1997).

Chapter 15 provides additional discussion of the axial dispersion model and of methods for measuring dispersion coefficients. A more advanced account is given in

Nauman, E. B. and Buffham, B. A., *Mixing in Continuous Flow Systems*, Wiley, New York, 1983.

Run an Internet search on static and motionless mixers to learn more about the utility of these devices, but be wary of the hype.

CHAPTER 10
HETEROGENEOUS CATALYSIS

The first eight chapters of this book treat homogeneous reactions. Chapter 9 provides models for packed-bed reactors, but the reaction kinetics are pseudohomogeneous so that the rate expressions are based on fluid-phase concentrations. There is a good reason for this. Fluid-phase concentrations are what can be measured. The fluid-phase concentrations at the outlet are what can be sold.

Chapter 10 begins a more detailed treatment of heterogeneous reactors. This chapter continues the use of pseudohomogeneous models for steady-state, packed-bed reactors, but derives expressions for the reaction rate that reflect the underlying kinetics of surface-catalyzed reactions. The kinetic models are *site-competition models* that apply to a variety of catalytic systems, including the enzymatic reactions treated in Chapter 12. Here in Chapter 10, the example system is a *solid-catalyzed gas reaction* that is typical of the traditional chemical industry. A few important examples are listed here:

- Ethylene is selectively oxidized to ethylene oxide using a silver-based catalyst in a fixed-bed reactor. Ethylene and oxygen are supplied from the gas phase and ethylene oxide is removed by it. The catalyst is stationary. Undesired, kinetically determined by-products include carbon monoxide and water. Ideally, a pure reactant is converted to one product with no by-products.
- Ethylbenzene is dehydrogenated in a fixed-bed reactor to give styrene. Hydrogen is produced as a stoichiometrically determined by-product. Undesired by-products including toluene, benzene, light hydrocarbons, coke, and additional hydrogen are kinetically determined. Ideally, a pure reactant is converted to one desired product and to hydrogen as the inevitable by-product.
- The final step in the methanol-to-gasoline process can be carried out in an adiabatic, fixed-bed reactor using a zeolite catalyst. A product mixture similar to ordinary gasoline is obtained. As is typical of polymerizations, a pure reactant is converted to a complex mixture of products.
- Catalytic reformers take linear alkanes, e.g., *n*-pentane, and produce branched alkanes, e.g., *i*-pentane. The catalyst is finely divided platinum on Si_2O_3.

Reforming is a common refinery reaction that begins with a complex mixture of reactants and produces a complex mixture of products.
- The catalytic converter on a car uses a precious-metal-based, solid catalyst, usually in the form of a monolith, to convert unburned hydrocarbons and carbon monoxide to carbon dioxide. Many different reactants are converted to two products: CO_2 and water.

Many more examples could be given. They all involve *interphase mass transfer* combined with chemical reaction. Gas-phase reactants are adsorbed onto a solid surface, react, and the products are desorbed. Most solid catalysts are supplied as cylindrical pellets with lengths and diameters in the range of 2–10 mm. More complex shapes and monoliths can be used when it is important to minimize pressure drop. The catalyst is *microporous* with pores ranging in diameter from a few angstroms to a few microns. The pores may have a bimodal distribution of sizes as illustrated in Figure 10.1. The internal surface area, accessible through the pores, is enormous, up to $2000 \, m^2$ per gram of catalyst. The internal surface area dwarfs the nominal, external area and accounts for most of the catalytic activity. The catalytic sites are atoms or molecules on the internal surface. The structural material of the catalyst particle is often an oxide, such as alumina (Al_2O_3) or (SiO_2). The structural material may provide the catalytic sites directly or it may *support* a more expensive substance, such as finely divided platinum. When heat transfer is important, the catalyst pellets are randomly packed in small-diameter (10–50 mm) tubes that are often quite long (2–10 m). A fluidized bed of small (50 µm) catalyst particles can also be used. If the adiabatic temperature change is small, the pellets are packed in large-diameter vessels. Annular flow reactors (see Figure 3.2) are used when it is important

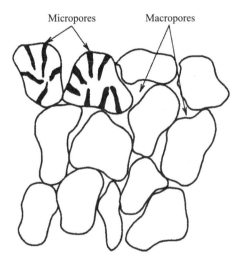

FIGURE 10.1 Diagram of bimodal catalyst pore structure.

to minimize the gas-phase pressure drop. Another approach is to flow the gas through the labyrinth of a monolithic catalyst, as in automobile exhaust systems. Regardless of the specific geometry used to contact the gas and the solid, all these schemes require a complex set of mass transfer and reaction steps, usually accompanied by heat transfer.

10.1 OVERVIEW OF TRANSPORT AND REACTION STEPS

Molecules enter the reactor with uniform concentrations a_{in} and leave with mixing-cup concentrations a_{out}. In between, they undergo the following steps:

1. Bulk transport of the reactants to the vicinity of a catalyst particle.
2. Mass transfer across a film resistance from the bulk gas phase to the external surface of the porous catalyst.
3. Transport of the reactants into the catalyst particle by diffusion through the pores.
4. Adsorption of reactant molecules onto the internal surface of the catalyst.
5. Reaction between adsorbed components on the catalytic surface.
6. Desorption of product molecules from the surface to the pores.
7. Diffusion of product molecules out of the pores to the external surface of the pellet.
8. Mass transfer of the products across a film resistance into the bulk gas phase.
9. Bulk transport of products to the reactor outlet.

All these steps can influence the overall reaction rate. The reactor models of Chapter 9 are used to predict the bulk, gas-phase concentrations of reactants and products at point (r, z) in the reactor. They directly model only Steps 1 and 9, and the effects of Steps 2 through 8 are lumped into the pseudohomogeneous rate expression, $\mathscr{R}(a, b, \ldots)$, where $a, b, \ldots$ are the bulk, gas-phase concentrations. The overall reaction mechanism is complex, and the rate expression is necessarily empirical. Heterogeneous catalysis remains an experimental science. The techniques of this chapter are useful to interpret experimental results. Their predictive value is limited.

The goal at this point is to examine Steps 2 through 8 in more detail so that the pseudohomogeneous reaction rate can reflect the mechanisms occurring within or on the catalyst. We seek a quantitative understanding of Steps 2 through 8 with a view toward improving the design of the catalyst and the catalytic reactor. The approach is to model the steps individually and then to couple them together. The modeling assumes that the system is at steady state. The coupling is based on the fact that each of Steps 2 through 8 must occur at the same rate in a steady-state system and that this rate, when expressed as moles per volume of gas phase per time, must equal the reaction rates in Steps 1 and 9.

10.2 GOVERNING EQUATIONS FOR TRANSPORT AND REACTION

Consider an observed reaction of the form $A + B \rightarrow P + Q$ occurring in a packed-bed reactor.

Step 1. Refer to Figure 10.2. The entering gas is transported to point (r, z) in the reactor and reacts, with rate $\varepsilon \mathscr{R}$. Equation (9.1) governs the combination of bulk transport and pseudohomogeneous reaction. We repeat it here:

$$\bar{u}_s \frac{\partial a}{\partial z} = D_r \left[\frac{1}{r} \frac{\partial a}{\partial r} + \frac{\partial^2 a}{\partial r^2} \right] + \varepsilon \mathscr{R}_A \tag{10.1}$$

The initial and boundary conditions are given in Chapter 9. The present treatment does not change the results given in Chapter 9, but instead provides a rational basis for using pseudohomogeneous kinetics for a solid-catalyzed reaction. The axial dispersion model in Chapter 9, again with pseudohomogeneous kinetics, is an alternative to Equation (10.1) that can be used when the radial temperature gradients are small.

Step 2. Reactant A in the gas phase at position (r, z) has concentration $a(r, z)$. It is transported across a film resistance and has concentration $a_s(r, z)$

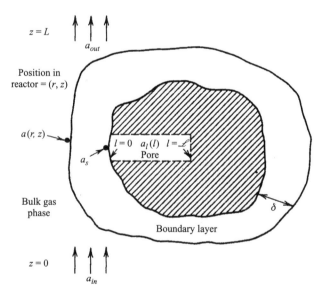

FIGURE 10.2 Illustration of pore and film resistances in a catalyst particle.

at the external surface of the catalyst pellet located at point (r, z). The detailed geometry of the gas and solid phases is ignored, so that both phases can exist at the same spatial location. The bulk and surface concentrations at location (r, z) are related through a mass transfer coefficient. The steady-state flux across the interface must be equal to the reaction rate. Thus, for component A,

$$\mathscr{R}_A = k_s A_s (a_s - a) \tag{10.2}$$

where k_s is a mass transfer coefficient and A_s is the external surface area of catalyst per unit volume of the gas phase. The units of k_s are moles per time per area per concentration driving force. These units simplify to length per time. The units of A_s are area per volume so that the product, $k_s A_s$, has dimensions of reciprocal time.

Step 3. Transport within a catalyst pore is usually modeled as a one-dimensional diffusion process. The pore is assumed to be straight and to have length $\mathscr{L}$. The concentration inside the pore is $a_l = a_l(l, r, z)$ where l is the position inside the pore measured from the external surface of the catalyst particle. See Figure 10.2. There is no convection inside the pore, and the diameter of the pore is assumed to be so small that there are no concentration gradients in the radial direction. The governing equation is an ODE.

$$0 = \mathscr{D}_A \frac{d^2 a_l}{dl^2} + \mathscr{R}_A \tag{10.3}$$

The solution to this equation, which is detailed in Section 10.4.1, gives the concentration at position l down a pore that has its mouth located at position (r, z) in the reactor. The reaction rate in Equation (10.3) remains based on the bulk gas-phase volume, not on the comparatively small volume inside the pore.

Step 4. A reactant molecule is adsorbed onto the internal surface of the catalyst. The adsorption step is modeled as an elementary reaction, the simplest version of which is

$$\text{A(gas)} + \text{S(solid)} \xrightarrow{k_a} \text{AS(solid)} \qquad \mathscr{R} = k_a a_l [\text{S}] \tag{10.4}$$

This kinetic relationship provides the necessary link between the gas-phase concentration a_l and the concentration of A in its adsorbed form, which is denoted as [AS]. The units for *surface concentration* are moles per unit area of catalyst surface. S denotes a catalytically active site on the surface, also with units of moles per area of catalyst surface.

Step 5. A surface reaction occurs between adsorbed species. The prototypical reaction is

$$\text{AS} + \text{BS} \xrightarrow{k_R} \text{PS} + \text{QS} \qquad \mathscr{R} = k_R [\text{AS}][\text{BS}] \tag{10.5}$$

where the product molecules, P and Q, are formed as adsorbed species. The surface reaction provides the link between reactant concentrations and product concentrations.

Step 6. The products are desorbed to give the gas-phase concentrations p_l and q_l. The desorption mechanism is written as

$$\text{PS(solid)} \xrightarrow{k_d} \text{P(gas)} + \text{S(solid)} \qquad \mathscr{R} = k_d[\text{PS}] \tag{10.6}$$

The catalytic sites, S, consumed in Step 4 are released in Step 6.

Step 7. Product species diffuse outward through the pores, the governing equations being similar to those used for the inward diffusion of reactants:

$$0 = \mathscr{D}_P \frac{d^2 p_l}{dl^2} + \mathscr{R}_P \tag{10.7}$$

The product molecules emerge from the interior of the catalyst at the same location (r, z) that the reactants entered.

Step 8. Product species diffuse across the fluid boundary layer at the external surface of the catalyst:

$$\mathscr{R}_P = k_s A_s (p_s - p) \tag{10.8}$$

Nominally, the value of $k_s A_s$ might be different for the different species. In practice, the difference is ignored.

Step 9. Product species generated at location (r, z) are transported to the reactor outlet. The governing equation is

$$\bar{u}_s \frac{\partial p}{\partial z} = D_r \left[\frac{1}{r} \frac{\partial p}{\partial r} + \frac{\partial^2 p}{\partial r^2} \right] + \varepsilon \mathscr{R}_P \tag{10.9}$$

Steps 1 through 9 constitute a model for heterogeneous catalysis in a fixed-bed reactor. There are many variations, particularly for Steps 4 through 6. For example, the Eley-Rideal mechanism described in Problem 10.4 envisions an adsorbed molecule reacting directly with a molecule in the gas phase. Other models contemplate a mixture of surface sites that can have different catalytic activity. For example, the platinum and the alumina used for hydrocarbon reforming may catalyze different reactions. Alternative models lead to rate expressions that differ in the details, but the functional forms for the rate expressions are usually similar.

10.3 INTRINSIC KINETICS

It is possible to eliminate the mass transfer resistances in Steps 2, 3, 7, and 8 by grinding the catalyst to a fine powder and exposing it to a high-velocity gas stream. The concentrations of reactants immediately adjacent to the catalytic surface are then equal to the concentrations in the bulk gas phase. The resulting kinetics are known as *intrinsic kinetics* since they are intrinsic to the catalyst surface and not to the design of the pores, or the pellets, or the reactor.

Most research in heterogeneous catalysis is concerned with the measurement, understanding, and modification of intrinsic kinetics.

When the mass transfer resistances are eliminated, the various gas-phase concentrations become equal: $a_l(l, r, z) = a_s(r, z) = a(r, z)$. The very small particle size means that heat transfer resistances are minimized so that the catalyst particles are isothermal. The recycle reactor of Figure 4.2 is an excellent means for measuring the intrinsic kinetics of a finely ground catalyst. At high recycle rates, the system behaves as a CSTR. It is sometimes called a *gradientless reactor* since there are no composition and temperature gradients in the catalyst bed or in a catalyst particle.

10.3.1 Intrinsic Rate Expressions from Equality of Rates

Suppose a gradientless reactor is used to obtain intrinsic rate data for a catalytic reaction. Gas-phase concentrations are measured, and the data are fit to a rate expression using the methods of Chapter 7. The rate expression can be arbitrary:

$$R = k a^m b^n p^r q^s \qquad (10.10)$$

As discussed in Chapter 7, this form can provide a good fit of the data if the reaction is not too close to equilibrium. However, most reaction engineers prefer a mechanistically based rate expression. This section describes how to obtain plausible functional forms for $\mathcal{R}$ based on simple models of the surface reactions and on the observation that all the rates in Steps 2 through 8 must be equal at steady state. Thus, the rate of transfer across the film resistance equals the rate of diffusion into a pore equals the rate of adsorption equals the rate of reaction equals the rate of desorption, and so on. This rate is the pseudohomogeneous rate shown in Steps 1 and 9.

Example 10.1: Consider the heterogeneously catalyzed reaction A → P. Derive a plausible form for the intrinsic kinetics. The goal is to determine a form for the reaction rate $\mathcal{R}$ that depends only on gas-phase concentrations.

Solution: Under the assumption of intrinsic kinetics, all mass transfer steps are eliminated, and the reaction rate is determined by Steps 4–6. The simplest possible version of Steps 4–6 treats them all as elementary, irreversible reactions:

$$A(gas) + S(solid) \xrightarrow{k_a} AS(solid) \qquad \mathcal{R} = k_a a_l [S]$$
$$AS(solid) \xrightarrow{k_R} PS(solid) \qquad \mathcal{R} = k_R [AS]$$
$$PS(solid) \xrightarrow{k_d} P(gas) + S(solid) \qquad \mathcal{R} = k_d [PS]$$

The reaction rates must be equal at steady state. Thus,

$$\mathcal{R} = k_a a_l [S] = k_R [AS] = k_d [PS]$$

A site balance ties these equations together:

$$S_0 = [S] + [AS] + [PS]$$

The site balance specifies that the number of empty plus occupied sites is a constant, S_0. Equality of the reaction rates plus the site balance gives four independent equations. Combining them allows a solution for $\mathscr{R}$ while eliminating the surface concentrations [S], [AS], and [PS]. Substitute the various reaction rates into the site balance to obtain

$$S_0 = \frac{\mathscr{R}}{k_a a_l} + \frac{\mathscr{R}}{k_R} + \frac{\mathscr{R}}{k_d}$$

But $a_l = a$ for intrinsic kinetics. Making this substitution and solving for $\mathscr{R}$ gives

$$\mathscr{R} = \frac{S_0 a}{\dfrac{1}{k_a} + \left(\dfrac{1}{k_R} + \dfrac{1}{k_d}\right)a} = \frac{S_0 k_a a}{1 + \left(\dfrac{k_a}{k_R} + \dfrac{k_a}{k_d}\right)a} \qquad (10.11)$$

Redefining constants gives

$$\mathscr{R} = \frac{ka}{1 + k_A a} \qquad (10.12)$$

Equation (10.12) is the simplest—and most generally useful—model that reflects heterogeneous catalysis. The active sites S are fixed in number, and the gas-phase molecules of component A compete for them. When the gas-phase concentration of component A is low, the $k_A a$ term in Equation (10.12) is small, and the reaction is first order in a. When a is large, all the active sites are occupied, and the reaction rate reaches a saturation value of k/k_A. The constant in the denominator, k_A, is formed from ratios of rate constants. This makes it less sensitive to temperature than k, which is a normal rate constant.

The form of Equation (10.12) is widely used for multiphase reactions. The same model, with slightly different physical interpretations, is used for enzyme catalysis and cell growth. See Chapter 12.

Example 10.2: Repeat Example 10.1 but now assume that each of Steps 4–6 is reversible.

Solution: The elementary reaction steps of adsorption, reaction, and desorption are now reversible. From this point on, we will set $a_l = a$, $p_l = P$, and so on, since the intrinsic kinetics are desired. The relationships between a_l, a_s, and a are addressed using an effectiveness factor in Section 10.4. The various reaction steps are

$$A(\text{gas}) + S(\text{solid}) \underset{k_a^-}{\overset{k_a^+}{\rightleftarrows}} AS(\text{solid}) \qquad \mathscr{R} = k_a^+ a[S] - k_a^-[AS]$$

$$\text{AS(solid)} \underset{k_R^-}{\overset{k_R^+}{\rightleftharpoons}} \text{PS(solid)} \qquad \mathscr{R} = k_R^+[\text{AS}] - k_R^-[\text{PS}]$$

$$\text{PS(solid)} \underset{k_d^-}{\overset{k_d^+}{\rightleftharpoons}} \text{P(gas)} + \text{S(solid)} \qquad \mathscr{R} = k_d^+[\text{PS}] - k_d^- p[\text{S}]$$

As in Example 10.1, the rates must all be equal at steady state:

$$\mathscr{R} = k_a^+ a[\text{S}] - k_a^-[\text{AS}] = k_R^+[\text{AS}] - k_R^-[\text{PS}] = k_d^+[\text{PS}] - k_d^- p[\text{S}]$$

The site balance is the same as in Example 10.1:

$$S_0 = [\text{S}] + [\text{AS}] + [\text{PS}]$$

As in Example 10.1, equality of the reaction rates plus the site balance gives four independent equations. Combining them allows a solution for $\mathscr{R}$ while eliminating the surface concentrations [S], [AS], and [PS]. After much algebra and a redefinition of constants,

$$\mathscr{R} = \frac{k_f a - k_r p}{1 + k_A a + k_P p} \tag{10.13}$$

Problem 10.1 gives the result before the redefinition of constants.

The numerator of Equation (10.13) is the expected form for a reversible, first-order reaction. The denominator shows that the reaction rate is retarded by all species that are adsorbed. This reflects competition for sites. Inerts can also compete for sites. Thus, the version of Equation (10.13) that applies when adsorbable inerts are present is

$$\mathscr{R} = \frac{k_f a - k_r p}{1 + k_A a + k_P p + k_I i} \tag{10.14}$$

where i is the gas-phase concentration of inerts. The inerts may be intentionally added or they may be undesired contaminants. When they are contaminants, their effect on the reaction rate represents a form of deactivation, in this case *reversible deactivation*, that ceases when the contaminant is removed from the feed.

Examples (10.1) and (10.2) used the fact that Steps 4, 5, and 6 must all proceed at the same rate. This matching of rates must always be true, and, as illustrated in the foregoing examples, can be used to derive expressions for the intrinsic reaction kinetics. There is another concept with a time-honored tradition in chemical engineering that should be recognized. It is the concept of *rate-determining step* or *rate-controlling step*.

10.3.2 Models Based on a Rate-Controlling Step

The idea is that a single step, say adsorption, may be so much slower than the other steps (e.g., reaction and desorption) that it determines the overall reaction rate. The concept of rate-determining step has been widely employed in the literature, starting with Hougen and Watson.[1] The advantage of this approach is that it generates kinetic models with less algebra than the equal rate approach. It has the disadvantage of giving less general models that may also mislead the unwary experimentalist into thinking that surface mechanisms can be unambiguously determined from steady-state experiments. This is rarely possible.

Irreversible Unimolecular Reactions. Consider the irreversible catalytic reaction A → P of Example 10.1. There are three kinetic steps: adsorption of A, the surface reaction, and desorption of P. All three of these steps must occur at exactly the same rate, but the relative magnitudes of the three rate constants, k_a, k_R, and k_d, determine the concentration of surface species. Suppose that k_a is much smaller than the other two rate constants. Then the surface sites will be mostly unoccupied so that $[S] \approx S_0$. Adsorption is the rate-controlling step. As soon as a molecule of A is absorbed it reacts to P, which is then quickly desorbed. If, on the other hand, the reaction step is slow, the entire surface will be saturated with A waiting to react, $[AS] \approx S_0$, and the surface reaction is rate-controlling. Finally, it may be that k_d is small. Then the surface will be saturated with P waiting to desorb, $[PS] \approx S_0$, and desorption is rate-controlling. The corresponding forms for the overall rate are:

Adsorption is rate-controlling, $\mathscr{R} = k_a S_0 a$ (first order in A)

Surface reaction is rate-controlling, $\mathscr{R} = k_R S_0$ (zero order in A)

Adsorption is rate-controlling, $\mathscr{R} = k_d S_0$ (zero order in A)

These results can be confirmed by taking the appropriate limits on the rate constants in Equation 10.11.

Reversible Unimolecular Reactions. The intrinsic reaction steps in heterogeneously catalyzed reactions are usually reversible. The various limiting cases can be found by taking limits before redefining the constants, e.g., take limits on Equation (10.11), not Equation (10.12). However, a more direct route is to assume that the fast steps achieve equilibrium before deriving the counterpart to Equation (10.11).

Example 10.3: Suppose that adsorption is much slower than surface reaction or desorption for the heterogeneously catalyzed reaction A $\rightleftharpoons$ P. Deduce the functional form for the pseudohomogeneous, intrinsic kinetics.

Solution: The adsorption step is slow, reversible, and rate-controlling. Its equation remains

$$A(gas) + S(solid) \underset{k_a^-}{\overset{k_a^+}{\rightleftharpoons}} AS(solid) \qquad \mathscr{R} = k_a^+ a[S] - k_a^-[AS]$$

The reaction and desorption steps are assumed to be so fast compared with adsorption that they achieve equilibrium:

$$\frac{[PS]}{[AS]} = \frac{k_R^+}{k_R^-} = K_R$$

$$\frac{p[S]}{[PS]} = \frac{k_d^+}{k_d^-} = K_d$$

The site balance is unchanged from Examples 10.1 and 10.2:

$$S_0 = [S] + [AS] + [PS]$$

There are enough equations to eliminate the surface equations from the reaction rate. After redefinition of constants,

$$\mathscr{R} = \frac{k_f a - k_r p}{1 + k_P p} \tag{10.15}$$

When the adsorption step determines the rate, component A no longer retards the reaction. Any A that is adsorbed will quickly react, and the concentration of [AS] sites will be low. Note that the desorption step is now treated as being reversible. Thus, any P in the gas phase will retard the reaction even if the surface reaction is irreversible, $k_r = 0$.

Example 10.4: Repeat Example 10.3, assuming now that the surface reaction controls the rate.

Solution: Appropriate equations for the adsorption, reaction, and desorption steps are

$$\frac{[AS]}{a[S]} = \frac{k_a^+}{k_a^-} = K_a$$

$$\mathscr{R} = k_R^+[AS] - k_R^-[PS]$$

$$\frac{p[S]}{[PS]} = \frac{k_d^+}{k_d^-} = K_d$$

The site balance is unchanged. Elimination of [S], [AS], and [PS] gives

$$\mathscr{R} = \frac{S_0[k_R^+ K_a K_d a - k_R^- p]}{K_d + K_a K_d a + p} = \frac{k_f a - k_r p}{1 + k_A a + k_P p} \tag{10.16}$$

This result is experimentally indistinguishable from the general form, Equation (10.12), derived in Example 10.1 using the equality of rates method. Thus, assuming a particular step to be rate-controlling may not lead to any simplification of the intrinsic rate expression. Furthermore, when a simplified form such as Equation (10.15) is experimentally determined, it does not necessarily justify the assumptions used to derive the simplified form. Other models may lead to the same form.

Bimolecular Reactions. Models of surface-catalyzed reactions involving two gas-phase reactants can be derived using either the equal rates method or the method of rate-controlling steps. The latter technique is algebraically simpler and serves to illustrate general principles.

Example 10.5: Derive a model of the Hougen and Watson type for the overall reaction $2A \to P$, assuming that the surface reaction is the rate-determining step.

Solution: A plausible mechanism for the observed reaction is

$$A + S \rightleftharpoons AS$$
$$2AS \to PS$$
$$PS \rightleftharpoons P + S$$

The adsorption and desorption steps achieve equilibrium:

$$\frac{[AS]}{a[S]} = \frac{k_a^+}{k_a^-} = K_a$$

$$\frac{p[S]}{[PS]} = \frac{k_d^+}{k_d^-} = K_d$$

The rate for the reaction step is

$$\mathscr{R} = k_R[AS]^2$$

The site balance is identical to those in previous examples.

$$S_0 = [S] + [AS] + [PS]$$

Elimination of the surface concentrations gives

$$\mathscr{R} = \frac{S_0^2 K_a^2 k_R a^2}{(1 + K_A a + p/K_d)^2} = \frac{k a^2}{(1 + k_A a + k_P p)^2} \qquad (10.17)$$

The retardation due to adsorption appears as a square because two catalytic sites are involved. It is likely that the reaction, a dimerization, requires that the reacting [AS] groups must be physically adjacent for reaction to occur. The square in the denominator is intended to account for this.

Examples of Hougen-Watson kinetic models, which are also called Langmuir-Hinshelwood models, can be derived for a great variety of assumed surface mechanisms. See Butt[2] and *Perry's Handbook* (see "Suggestions for Further reading" in Chapter 5) for collections of the many possible models. The models usually have numerators that are the same as would be expected for a homogeneous reaction. The denominators reveal the heterogeneous nature of the reactions. They come in almost endless varieties, but all reflect competition for the catalytic sites by the adsorbable species.

10.3.3 Recommended Models

Enthusiastic theoreticians have created far too many models of solid-catalyzed gas reactions. As suggested earlier, it is difficult to distinguish between them—given any reasonable experimental program—and essentially impossible to distinguish between them if the experimental program is confined to steady-state measurements. Recall the rampaging elephants of Section 7.1. Recall also that Equation (10.10) provides a reasonable fit with no more than $N+1$ adjustable constants, where N is the number of components including adsorbable inerts. For a solid-catalyzed reaction, the following form is recommended:

$$\mathscr{R} = \frac{\text{Rate expression for a homogeneous reaction}}{(\text{Site competition term})^{n=1 \text{ or } 2}} \qquad (10.18)$$

where $n=1$ if the reaction is unimolecular and $n=2$ if it is bimolecular. The numerator should contain only one rate constant plus an equilibrium constant if the reaction is reversible. See Section 7.2.2.

For reactions of the form $A \rightleftharpoons P$, the recommended rate expression is

$$\mathscr{R} = \frac{k(a - p/K_{kinetic})}{1 + k_A a + k_P p + k_I i} \qquad (10.19)$$

For reactions of the form $A \rightleftharpoons P + Q$, the recommended rate expression is

$$\mathscr{R} = \frac{k(a - pq/K_{kinetic})}{1 + k_A a + k_P p + k_Q q + k_I i} \qquad (10.20)$$

For reactions of the form $A + B \rightleftharpoons P$, the recommended rate expression is

$$\mathscr{R} = \frac{k(ab - p/K_{kinetic})}{(1 + k_A a + k_B b + k_P p + k_I i)^2} \qquad (10.21)$$

For reactions of the form A + B ⇌ P + Q, the recommended rate expression is

$$\mathcal{R} = \frac{k(ab - pq/K_{kinetic})}{(1 + k_A a + k_B b + k_P p + k_Q q + k_I i)^2} \quad (10.22)$$

Each of these expressions has a primary rate constant k and an equilibrium constant that can be ignored if the reaction is essentially irreversible. The primary rate constant is usually fit to an Arrhenius temperature dependence while the temperature dependence of the equilibrium constant is determined using the methods of Section 7.2. The recommended kinetic expressions also contain an adsorption constant for each reactive component plus a lumped constant for adsorbable inerts. These denominator constants can be fit to an Arrhenius temperature dependence as well, but the activations energies are usually small since the denominator constants are ratios of rate constants and the individual activation energies will tend to cancel. The usual range of temperature measurements is small enough that the denominator constants can be regarded as independent of temperature. The power of 2 in the denominator when the forward reaction is bimolecular is somewhat arbitrary. The same quality fit can usually be achieved using $n = 1$ with different values for the adsorption constants. Proper fitting of the adsorption constants demands an extensive experimental program where the gas-phase concentrations of reactants and products are varied over a wide range.

10.4 EFFECTIVENESS FACTORS

Few fixed-bed reactors operate in a region where the intrinsic kinetics are applicable. The particles are usually large to minimize pressure drop, and this means that diffusion within the pores, Steps 3 and 7, can limit the reaction rate. Also, the superficial fluid velocity may be low enough that the external film resistances of Steps 2 and 8 become important. A method is needed to estimate actual reaction rates given the intrinsic kinetics and operating conditions within the reactor. The usual approach is to define the *effectiveness factor* as

$$\eta = \frac{\text{Actual reaction rate}}{\text{Rate predicted from intrinsic kinetics}} \quad (10.23)$$

and then to correlate η with the operating conditions in the reactor.

The global design equation used for Steps 1 and 9 is modified to include the effectiveness factor:

$$\bar{u}_s \frac{\partial a}{\partial z} = D_r \left[\frac{1}{r} \frac{\partial a}{\partial r} + \frac{\partial^2 a}{\partial r^2} \right] + \varepsilon \eta \mathcal{R}_A \quad (10.24)$$

where $\mathcal{R}_A$ now represents the intrinsic kinetics. Suppose the intrinsic kinetics are known and that η has been determined as a function of local operating conditions in the reactor. Then Steps 2 through 8 can be ignored. Their effects are built into the effectiveness factor and the intrinsic kinetics, and the reactor design methods of Chapter 9 can be applied with no changes other than using $\eta \mathscr{R}_A$ as the rate expression.

What is needed at this point is a correlation or other means for estimating η at every point in the reactor. This may be done empirically; for example, by running a single tube of what ultimately will be a multitubular reactor. However, some progress has been made in determining η from first principles. We outline the salient results achieved to date.

10.4.1 Pore Diffusion

The most important mass transfer limitation is diffusion in the micropores of the catalyst. A simplified model of pore diffusion treats the pores as long, narrow cylinders of length $\mathscr{L}$. The narrowness allows radial gradients to be neglected so that concentrations depend only on the distance l from the mouth of the pore. Equation (10.3) governs diffusion within the pore. The boundary condition at the mouth of the pore is

$$a_l = a_s \quad \text{at} \quad l = 0$$

The other boundary condition is

$$\frac{da_l}{dl} = 0 \quad \text{at} \quad l = \mathscr{L}$$

An analytical solution is possible when the reaction is first order; e.g., a reaction of the form A → P with adsorption as the rate-controlling step. Then Equation (10.3) becomes

$$0 = \mathscr{D}_A \frac{d^2 a_l}{dl^2} - k a_l$$

Solution subject to the boundary conditions gives

$$\frac{a_l}{a_s} = \frac{\exp(-2\mathscr{L}\sqrt{k/\mathscr{D}_A})\exp(l\sqrt{k/\mathscr{D}_A}) + \exp(-l\sqrt{k/\mathscr{D}_A})}{1 + \exp(-2\mathscr{L}\sqrt{k/\mathscr{D}_A})}$$

This gives the concentration profile inside the pore, $a_l(l)$. The total rate of reaction within a pore can be found using the principle of equal rates. The reaction rate within a pore must equal the rate at which reactant molecules enter the pore. Molecules enter by diffusion. The flux of reactants molecules diffusing into a pore of diameter d_{pore} equals the reaction rate. Thus,

$$\text{Actual rate} = \mathscr{R}_A = (\pi d_{pore}^2/4)\left[-\mathscr{D}_A \frac{da_l}{dl}\right]_{l=0}$$

$$= (\pi d_{pore}^2/4) a_s \mathscr{D}_A \sqrt{k/\mathscr{D}_A} \left[\frac{1 - \exp(-2\mathscr{L}\sqrt{k/\mathscr{D}_A})}{1 + \exp(-2\mathscr{L}\sqrt{k/\mathscr{D}_A})}\right]$$

is the actual rate as affected by pore diffusion. If there were no diffusion limitation inside the pore, the entire volume of the pore would have concentration a_s and the intrinsic rate would apply:

$$\text{Intrinsic rate} = (\pi d_{pore}^2/4)\,\mathscr{L}ka_s$$

The ratio of actual rate to intrinsic rate is the effectiveness factor:

$$\eta = \frac{\mathscr{D}_A}{k\mathscr{L}}\sqrt{k/\mathscr{D}_A}\left[\frac{1-\exp(-2\mathscr{L}\sqrt{k/\mathscr{D}_A})}{1+\exp(-2\mathscr{L}\sqrt{k/\mathscr{D}_A})}\right] = \frac{\tanh(\mathscr{L}\sqrt{k/\mathscr{D}_A})}{\mathscr{L}\sqrt{k/\mathscr{D}_A}} \quad (10.25)$$

It depends only on $\mathscr{L}\sqrt{k/\mathscr{D}_A}$, which is a dimensionless group known as the *Thiele modulus*. The Thiele modulus can be measured experimentally by comparing actual rates to intrinsic rates. It can also be predicted from first principles given an estimate of the pore length $\mathscr{L}$. Note that the pore radius does not enter the calculations (although the effective diffusivity will be affected by the pore radius when d_{pore} is less than about 100 nm).

Example 10.6: A commercial process for the dehydrogenation of ethylbenzene uses 3-mm spherical catalyst particles. The rate constant is $15\,\text{s}^{-1}$, and the diffusivity of ethylbenzene in steam is $4\times10^{-5}\,\text{m}^2/\text{s}$ under reaction conditions. Assume that the pore diameter is large enough that this bulk diffusivity applies. Determine a likely lower bound for the isothermal effectiveness factor.

Solution: The lowest η corresponds to the largest value for $\mathscr{L}$. Suppose $\mathscr{L} = R_p = 1.5\,\text{mm}$. Then

$$\mathscr{L}\sqrt{k/\mathscr{D}_A} = 0.92$$

and

$$\eta = \frac{\tanh(0.92)}{0.92} = 0.79$$

Many theoretical embellishments have been made to the basic model of pore diffusion as presented here. Effectiveness factors have been derived for reaction orders other than first and for Hougen and Watson kinetics. These require a numerical solution of Equation (10.3). Shape and tortuosity factors have been introduced to treat pores that have geometries other than the idealized cylinders considered here. The *Knudsen diffusivity* or a combination of Knudsen and bulk diffusivities has been used for very small pores. While these studies have theoretical importance and may help explain some observations, they are not yet developed well enough for predictive use. Our knowledge of the internal structure of a porous catalyst is still rather rudimentary and imposes a basic limitation on theoretical predictions. We will give a brief account of Knudsen diffusion.

In bulk diffusion, the predominant interaction of molecules is with other molecules in the fluid phase. This is the ordinary kind of diffusion, and the corresponding diffusivity is denoted as $\mathscr{D}_A$. At low gas densities in small-diameter pores, the mean free path of molecules may become comparable to the pore diameter. Then, the predominant interaction is with the walls of the pore, and diffusion within a pore is governed by the Knudsen diffusivity, $\mathscr{D}_K$. This diffusivity is predicted by the kinetic theory of gases to be

$$\mathscr{D}_K = \frac{d_{pore}}{3}\sqrt{\frac{8R_g T}{\pi M_A}} \qquad (10.26)$$

where M_A is the molecular weight of the diffusing species.

Example 10.7: Repeat Example 10.6, assuming a pore diameter of $20\,\text{nm} = 2 \times 10^{-8}\,\text{m}$. The reaction temperature is 625°C.

Solution:

$$\mathscr{D}_K = \frac{2 \times 10^{-8}}{3}\sqrt{\frac{8}{\pi} \cdot \frac{8.314\,\text{J}}{\text{mol}\cdot\text{K}} \cdot \frac{898\,\text{K}}{0.106\,\text{kg/mol}}} = 3 \times 10^{-6}\,\text{m}^2/\text{s}$$

This is an order of magnitude less than the bulk diffusivity, and

$$\mathscr{L}\sqrt{k/\mathscr{D}_K} = 3.5$$

$$\eta = \frac{\tanh(3.5)}{3.5} = 0.29$$

Example 10.8: How fine would you have to grind the ethylbenzene catalyst for laboratory kinetic studies to give the intrinsic kinetics? Assume the small pore diameter of Example 10.7.

Solution: Take $\eta = 0.98$ as an adequate approach to the intrinsic kinetics. Setting this value for η and solving Equation (10.25) for $\mathscr{L}\sqrt{k/\mathscr{D}_K}$ gives $\mathscr{L}\sqrt{k/\mathscr{D}_K} = 0.248$. Suppose $k = 15\,\text{s}^{-1}$ and $\mathscr{D}_K = 3 \times 10^{-6}\,\text{m}^2/\text{s}$. Then $\mathscr{L} = 1.1 \times 10^{-4}\,\text{m} = 110\,\mu\text{m}$.

The value for $\mathscr{L}$ is conservatively interpreted as the particle diameter. This is a perfectly feasible size for use in a laboratory reactor. Due to pressure-drop limitations, it is too small for a full-scale packed bed. However, even smaller catalyst particles, $d_p \approx 50\,\mu\text{m}$, are used in fluidized-bed reactors. For such small particles we can assume $\eta = 1$, even for the 3-nm pore diameters found in some cracking catalysts.

When the Knudsen and bulk diffusivities are significantly different, η is determined by the smaller of the two. The pore diameters for most commercial catalysts are in the range 1–100 nm. At a typical operating temperature of about 700 K, this gives Knudsen diffusivities in the range of 10^{-6}–10^{-8} m²/s. Bulk diffusivities at atmospheric pressure will usually be in the range of 10^{-4}–10^{-6} m²/s. The Knudsen diffusivity is independent of pressure but the bulk diffusivity varies approximately as P^{-1}. Thus, Knudsen diffusion will determine η at low to moderate pressures, but the bulk diffusivity can be limiting at high pressures. When the two diffusivities are commensurate, the combined effect is actually worse than either acting alone. The following equation is adequate for most purposes:

$$\frac{1}{\mathscr{D}_{eff}} = \frac{1}{\mathscr{D}_K} + \frac{1}{\mathscr{D}_A} \tag{10.27}$$

A more rigorous result together with theoretical justification, has been given by Rothfield.[3]

10.4.2 Film Mass Transfer

The concentration of gas over the active catalyst surface at location l in a pore is $a_l(l)$. The pore diffusion model of Section 10.4.1 linked concentrations within the pore to the concentration at the pore mouth, a_s. The film resistance between the external surface of the catalyst (i.e., at the mouths of the pore) and the concentration in the bulk gas phase is frequently small. Thus, $a_s \approx a$, and the effectiveness factor depends only on diffusion within the particle. However, situations exist where the film resistance also makes a contribution to η so that Steps 2 and 8 must be considered. This contribution can be determined using the principle of equal rates; i.e., the overall reaction rate equals the rate of mass transfer across the stagnant film at the external surface of the particle. Assume A is consumed by a first-order reaction. The results of the previous section give the overall reaction rate as a function of the concentration at the external surface, a_s.

$$\mathscr{R}_A = \text{Actual rate} = -\left[\frac{\tanh(\mathscr{L}\sqrt{k/\mathscr{D}_A})}{\mathscr{L}\sqrt{k/\mathscr{D}_A}}\right]ka_s \tag{10.28}$$

The overall effectiveness factor for the first-order reaction is defined using the bulk gas concentration a.

$$\mathscr{R}_A = -\eta k a \tag{10.29}$$

The concentrations a_s and a are coupled by the rate of mass transfer across the film:

$$\mathscr{R}_A = k_s A_s (a_s - a) \tag{10.30}$$

Equations (10.28), (10.29), and (10.30) are combined to eliminate a_s and to obtain an expression for $\mathscr{R}_A$ based on the bulk, gas-phase concentration a.

The result has the form of Equation (10.29) with

$$\eta = \frac{k_s A_s \tanh(\mathscr{L}\sqrt{k/\mathscr{D}_A})}{k_s A_s \mathscr{L}\sqrt{k/\mathscr{D}_A} + k \tanh(\mathscr{L}\sqrt{k/\mathscr{D}_A})} = \frac{k_s A_s \eta_0}{k_s A_s + k\eta_0} \quad (10.31)$$

where η_0 is the effectiveness factor ignoring film resistance; i.e., η_0 is given by Equation (10.25).

Equation (10.29) is the appropriate reaction rate to use in global models such as Equation (10.1). The reaction rate would be $-ka$ if there were no mass transfer resistance. The effectiveness factor η accounts for pore diffusion and film resistance so that the effective rate is $-\eta ka$.

Typically, the film resistance is important only when the internal pore resistance is also important. If the Thiele modulus is small, the film resistance will usually be negligible. This idea is explored in Problems 10.11 and 10.12.

Reactions other than first order can be treated numerically, but a priori predictions of effectiveness factors are rarely possible, even for the simple cases considered here. The approach of Examples 10.6 through 10.8 can sometimes be used to estimate whether mass transfer resistances are important. When mass transfer is important, effectiveness factors are determined experimentally.

10.4.3 Nonisothermal Effectiveness

Catalyst pellets often operate with internal temperatures that are substantially different from the bulk gas temperature. Large heats of reaction and the low thermal conductivities typical of catalyst supports make temperature gradients likely in all but the finely ground powders used for intrinsic kinetic studies. There may also be a film resistance to heat transfer at the external surface of the catalyst.

The definition of the effectiveness factor, Equation (10.23), is unchanged, but an exothermic reaction can have reaction rates inside the pellet that are higher than would be predicted using the bulk gas temperature. Thus, $\eta > 1$ is expected for exothermic reactions in the absence of mass diffusion limitations. (The case $\eta > 1$ is also possible for some isothermal reactions with weird kinetics.) With systems that have low thermal conductivities but high molecular diffusivities and high heats of reaction, the actual rate can be an order of magnitude higher than the intrinsic rate. Thus, $\eta \gg 1$ is theoretically possible for exothermic reactions. When mass transfer limitations do emerge, concentrations will be lower inside the pellet than outside. The decreased concentration may have a larger effect on the rate than the increased temperature. Thus, $\eta < 1$ is certainly possible for an exothermic reaction. For an endothermic reaction, $\eta < 1$ is expected, except perhaps for some esoteric kinetic schemes.

The theory of nonisothermal effectiveness is sufficiently well advanced to allow order-of-magnitude estimates for η. The analysis requires simultaneous solutions for the concentration and temperature profiles within a pellet. The solutions are necessarily numerical. Solutions are feasible for actual pellet shapes (such as cylinders) but are significantly easier for spherical pellets since this allows a one-dimensional form for the energy equation:

$$\lambda_{eff}\left(\frac{d^2T}{dr_p^2}+\frac{2}{r_p}\frac{dT}{dr_p}\right)=-\Delta H_R \mathcal{R} \qquad (10.32)$$

where r_p is the radial coordinate within a pellet and λ_{eff} is the effective thermal conductivity for the pellet. The boundary conditions associated with Equation (10.32) are $T=T_s$ at the external surface and $dT/dr_p=0$ at the center of the pellet. Equation (10.32) must be solved simultaneously with component balance equations. The pore diffusion model of Section 10.4.1 is inappropriate for this purpose. Instead, we use a model for effective diffusion that is directly compatible with the heat transfer model. This model is

$$\mathcal{D}_{eff}\left(\frac{d^2a}{dr_p^2}+\frac{2}{r_p}\frac{da}{dr_p}\right)=\mathcal{R}_A \qquad (10.33)$$

subject to the boundary conditions that $a=a_s$ at the external surface and that $da/dr_p=0$ at the center. Equation (10.33) is obviously consistent with Equation (10.32). Numerical solutions to these simultaneous equations have been given by Weisz and Hicks[4] for the case of a first-order, irreversible reaction. The solution for η depends on three dimensionless groups: an Arrhenius number, $E/(R_g T_s)$, which is the ratio of the activation temperature to the temperature at the external surface of the particle; a heat generation number,

$$\beta=\frac{-\Delta H_R \mathcal{D}_{eff} a_s}{\lambda_{eff} T_s}$$

and a modified Thiele modulus,

$$\frac{d_p}{2}\sqrt{\frac{k}{\mathcal{D}_{eff}}}$$

Figure 10.3 shows results for an Arrhenius number of 20. With plausible estimates for λ_{eff} and $\mathcal{D}_{eff}$, the magnitude of η can be calculated. For the special case of $\Delta H_R=0$ (i.e., $\beta=0$), Equation (10.33) is an alternative to the pore diffusion model for isothermal effectiveness. It predicts rather different results. For example, suppose $(d_p/2)\sqrt{k/\mathcal{D}_{eff}}=\mathcal{L}\sqrt{k/\mathcal{D}_A}=1$. Then Equation (10.25) gives $\eta=0.76$ while the solution of Equation (10.33) (see Figure 10.3 with $\beta=0$) gives $\eta=0.99$. The lesson from this is that $\mathcal{D}_{eff}$ and $\mathcal{D}_A$ are fundamentally different quantities and have different values for the same physical system.

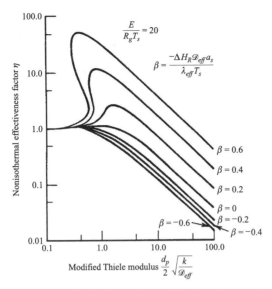

FIGURE 10.3 Nonisothermal effectiveness factors for first-order reactions in spherical pellets. (Adapted from Weisz, P. B. and Hicks, J. S., *Chem. Eng. Sci.*, **17**, 265 (1962).)

10.4.4 Deactivation

The definition of effectiveness factor, Equation (10.23), can be expanded to account for deactivation processes that decrease the activity of a catalyst as a function of time. In this context, the intrinsic kinetics in Equation (10.23) should be determined using a new, freshly prepared catalytic surface. The activity of the surface, and thus the actual rate of reaction, will change with time so that $\eta = \eta(\theta)$ where θ is the time the catalyst has been *on stream*. It is necessary to consider deactivation processes in reactor designs since they can have a marked effect on process economics and even operability. Deactivation is usually classified as being either physical or chemical in nature, but this classification is somewhat arbitrary. *Physical deactivation* includes blocking of pores by entrained solids, loss of active sites due to agglomeration (*site sintering*), closure of pores by internal collapse (*support sintering*), and the reversible loss of active sites by physical adsorption of impurities. *Chemical deactivation* includes the irreversible loss of active sites through chemisorption of impurities, loss of sites due to coking, and pore blockage due to coking.

Some deactivation processes are reversible. Deactivation by physical adsorption occurs whenever there is a gas-phase impurity that is below its critical point. It can be reversed by eliminating the impurity from the feed stream. This form of deactivation is better modeled using a site-competition model that includes the impurities—e.g., any of Equations (10.18)–(10.21)—rather than using the effectiveness factor. Water may be included in the reaction mixture so that the water–gas shift reaction will minimize the formation of coke. Off-line decoking can be done by high-temperature oxidation. Deactivation by chemisorption requires a chemical treatment to remove the chemically bonded poison from the catalytic surface and it can be modeled using a time-dependent effective factor. Except for physical adsorption, catalytic activity can never be restored completely. Even with continuous reactivation, the catalyst will gradually decline in activity and have a finite life that must be considered in the overall process economics. A change of catalyst for a large reactor can cost millions of dollars.

The catalyst used in fluidized-bed catalytic crackers is deactivated by the coking that occurs during one pass through the reactor—a matter of seconds. More stable catalysts can last for years, with the gradual decline in activity being accommodated by a gradual increase in reaction temperature. The effectiveness factor may decrease by a factor of 5 before the catalyst is changed. The selectivity of the reaction will usually drop during the course of the run. The decision to change the catalyst is usually based on this loss of selectivity, but it can be based on a temperature limitation imposed by materials of construction.

Some deactivation processes lower the number of active sites S_0. Others add mass transfer resistances. In either case, they cause a reduction in the reaction rate that is reflected in a time-dependent effectiveness factor:

$$[\text{Actual rate}] = \eta(\theta) \, [\text{Intrinsic rate of fresh catalyst}] \quad (10.34)$$

Some progress has been made in developing theoretical expressions for $\eta(\theta)$ for deactivation processes such as coking. Deactivation by loss of active sites can be modeled as a chemical reaction proceeding in parallel with the main reaction. It may be substantially independent of the main reaction. Site sintering, for example, will depend mainly on the reaction temperature. It is normally modeled as a second-order reaction:

$$\frac{dS_0}{d\theta} = -k_S S_0^2$$

The number of active sites is a multiplicative factor in the rate of the main reaction. See for example Equations (10.11) and (10.16). Thus, the decline in reaction rate can be modeled using a time-dependent effectiveness. A reasonable functional form for the time-dependent effectiveness factor is

$$\eta(\theta) = \frac{\eta_{fresh}}{1 + S_0 k_S \theta} = \frac{\eta_{fresh}}{1 + k_D \theta} \quad (10.35)$$

which results from the second-order model of site sintering. An alternative to Equation (10.35) is a first-order model:

$$\eta(\theta) = \eta_{fresh} \exp(-k_D \theta) \quad (10.36)$$

It is necessary to determine $\eta(\theta)$ under reaction conditions, and a life test should be included in any catalyst development effort. The data from this test will allow η to be fitted as a function of time on stream, θ. Equations (10.35) and (10.36) can obviously be used to model deactivation processes other than site sintering, and k_D can be regarded as an empirical constant with units of reciprocal time.

Typically, $(k_D)^{-1} \gg \bar{t}$, and the global design equation, e.g., Equation (10.1), can be solved independently of the deactivation process. The packed-bed reactor will operate at an approximate steady state that will track the slowly changing catalyst activity. The rate at which this slow change occurs dictates the process design. If the catalyst deactivates in minutes, continuous regeneration is necessary and fixed-bed reactors are infeasible. Fluidized-bed and other moving-solids reactors are used to continuously circulate the catalyst between reaction and regeneration zones. If the catalyst degenerates in days, fixed-bed reactors are usually regenerated in situ. Typically, several packed beds in parallel are operated in a swing-cycle between reaction and regeneration modes. If the catalyst degenerates in years, the spent catalyst is usually returned to the vendor for regeneration or recovery of valuable components such as precious metals.

Do not infer from the above discussion that all the catalyst in a fixed bed ages at the same rate. This is not usually true. Instead, the time-dependent effectiveness factor will vary from point to point in the reactor. The deactivation rate constant k_D will be a function of temperature. It is usually fit to an Arrhenius temperature dependence. For chemical deactivation by chemisorption or coking, deactivation will normally be much higher at the inlet to the bed. In extreme cases, a sharp deactivation front will travel down the bed. Behind the front, the catalyst is deactivated so that there is little or no conversion. At the front, the conversion rises sharply and becomes nearly complete over a short distance. The catalyst ahead of the front does nothing, but remains active, until the front advances to it. When the front reaches the end of the bed, the entire catalyst charge is regenerated or replaced.

10.5 EXPERIMENTAL DETERMINATION OF INTRINSIC KINETICS

A CSTR is the preferred method for measuring intrinsic kinetics. The finely ground catalyst is packed into a short, fixed bed within a recycle reactor such as that shown in Figure 4.2. Alternatively, the catalyst is put in a mesh cage and rotated at high speed so that the catalyst and cage act as the agitator of a CSTR. The reaction rate for component A is calculated as if the reaction were homogeneous:

$$\mathscr{R}_A = \frac{a_{out} - a_{in}}{\bar{t}} = \frac{a_{out} - a_{in}}{\varepsilon V / Q_{out}} \qquad (10.37)$$

The mass of the catalyst does not appear. However, physical intuition or the S_0 terms in Equations (10.11) and (10.16) suggest that doubling the amount of catalyst should double the reaction rate. How are rate data taken on a CSTR translated to a packed-bed reactor or even to another CSTR operating with a different catalyst density?

Homogeneous reactions occur in the fluid phase, and the volume available for reaction is εV. Solid-catalyzed reactions occur on the catalyst surface, and area available for the reaction is $V\rho_c a_c$, where V is the total reactor volume (i.e., gas plus catalyst), ρ_c is the average density of catalyst in the reactor (i.e., mass of catalyst per total reactor volume), and a_c is the surface area per mass of catalyst. The pseudohomogeneous reaction rate calculated using Equation (10.37) is multiplied by εV to get the rate of formation of component A in moles per time. The equivalent heterogeneous rate is based on the catalyst surface area and is multiplied by $V\rho_c a_c$ to obtain the rate of formation of component A in moles per time. Setting the two rates equal gives

$$\varepsilon V \mathcal{R}_{homogeneous} = V\rho_c a_c \mathcal{R}_{heterogeneous}$$

The void fraction should be the *total void fraction* including the pore volume. We now distinguish ε_{total} from the *superficial void fraction* used in the Ergun equation and in the packed-bed correlations of Chapter 9. The pore volume is accessible to gas molecules and can constitute a substantial fraction of the gas-phase volume. It is included in reaction rate calculations through the use of the total void fraction. The superficial void fraction ignores the pore volume. It is the appropriate parameter for the hydrodynamic calculations because fluid velocities go to zero at the external surface of the catalyst particles. The pore volume is accessible by diffusion, not bulk flow.

The homogeneous and heterogeneous rates are related by

$$\mathcal{R}_{homogeneous} = \frac{\rho_c a_c}{\varepsilon_{total}} \mathcal{R}_{heterogeneous} \qquad (10.38)$$

and Equation (10.1) should be written as

$$\bar{u}_s \frac{\partial a}{\partial z} = D_r \left[\frac{1}{r} \frac{\partial a}{\partial r} + \frac{\partial^2 a}{\partial r^2} \right] + \varepsilon_{total}\, \eta(\theta)[\mathcal{R}_A]_{intrinsic\ pseudohomogeneous} \qquad (10.39)$$

However, the intrinsic pseudohomogeneous rate used in Equation (10.39) is not identical to the rate determined from the CSTR measurements since the catalyst density will be different. The correction procedure is

1. Calculate $\mathcal{R}_A$ from the CSTR data using Equation (10.37).
2. Divide by the stoichiometric coefficient for component A, ν_A, to obtain $\mathcal{R}_{homogeneous}$ for the CSTR.
3. Use Equation (10.38) to calculate $\mathcal{R}_{heterogeneous}$ using the CSTR values for ρ_c, a_c, and ε_{total}.
4. Determine ρ_c, a_c, and ε_{total} for the packed bed.
5. Use Equation (10.38) again, now determining $\mathcal{R}_{homogeneous}$ for the packed bed.

Alternatively, write the global design equation as

$$\bar{u}_s \frac{\partial a}{\partial z} = D_r \left[\frac{1}{r} \frac{\partial a}{\partial r} + \frac{\partial^2 a}{\partial r^2} \right] + \eta(\theta) \rho_c a_c [\mathcal{R}_A]_{heterogeneous} \qquad (10.40)$$

Note that $[\mathcal{R}_A]_{heterogeneous}$ has units of mol/(m$^2\cdot$s) but remains a function of gas-phase concentrations. The composite term of Chapter 9 and Equation (10.1) is $\varepsilon_{total} \mathcal{R}_A = \rho_c a_c [\mathcal{R}_A]_{heterogeneous}$.

Example 10.9: A recycle reactor containing 101 g of catalyst is used in an experimental study. The catalyst is packed into a segment of the reactor having a volume of 125 cm^3. The recycle lines and pump have an additional volume of 150 cm^3. The particle density of the catalyst is 1.12 g/cm^3, its internal void fraction is 0.505, and its surface area is 400 m^2/g. A gas mixture is fed to the system at 150 cm^3/s. The inlet concentration of reactant A is 1.6 mol/m^3. The outlet concentration of reactant A is 0.4 mol/m^3. Determine the intrinsic pseudohomogeneous reaction rate, the rate per unit mass of catalyst, and the rate per unit surface area of catalyst. The reaction is A $\to$ P so $\nu_A = 1$.

Solution: The gas-phase volume, $\varepsilon_{total} V$, is the entire reactor except for the volume taken up by mechanical parts and by the skeleton of the catalyst particle:

$$\varepsilon_{total} V = 125 + 150 - \frac{101}{1.12}(1 - 0.505) = 230 \text{ cm}^3$$

$$\varepsilon_{total} = 230/275 = 0.836$$

$$\bar{t} = \frac{230 \text{ cm}^3}{150 \text{ cm}^3/\text{s}} = 1.53 \text{ s}$$

The intrinsic pseudohomogeneous rate as determined using the CSTR is

$$[\mathcal{R}_A]_{homogeneous} = \mathcal{R}_A \frac{0.4 - 1.6}{1.53} = -0.783 \text{ mol}/(\text{m}^3 \cdot \text{s}^1)$$

The average density of the catalyst is

$$\rho_c = \frac{101}{275} = 0.367 \text{ g/cm}^3 = 367 \text{ kg/m}^3$$

and

$$[\mathcal{R}_A]_{catalyst\ mass} \frac{\varepsilon_{total} \mathcal{R}_A}{\rho_c} = -1.78 \times 10^{-3} \text{ mol}/(\text{kg}\cdot\text{s})$$

$$[\mathcal{R}_A]_{heterogeneous} = [\mathcal{R}_A]_{surface\ area} \frac{\varepsilon_{total} \mathcal{R}_A}{\rho_c a_c}$$

$$= -4.45 \times 10^{-9} \text{ mol}/(\text{m}^2 \cdot \text{s})$$

Example 10.9 used two different definitions of the catalyst density and at least two more definitions are in common usage. The value $\rho_c = 367$ refers to the *reactor average density*. It is quite low in the example because so much of the reactor volume is empty. Normally, the reactor would be packed almost completely, and the reactor average density would approach the *bulk density*. The bulk density is what would be measured if the catalyst were dumped into a large container and gently shaken. The bulk density is not stated in the example, but it would be about 800 kg/m^3 for the catalyst pellets prior to grinding. The catalyst will pack to something less than the bulk density in a small-diameter tube. The *pellet density* in the example is 1120 kg/m^3. It is the mass of a catalyst pellet divided by the external volume of the pellet. The final density is the *skeletal density* of the pellet. It is the density of the solid support and equals $1120/(1 - 0.505) = 2260 \text{ kg/m}^3$ for the example catalyst. The various densities fall in the order

$$\rho_c < \rho_{bulk} < \rho_{pellet} < \rho_{skeletal}$$

Example 10.10: Suppose the reaction in Example 10.9 is first order. Determine the pseudohomogeneous rate constant, the rate constant based on catalyst mass, and the rate constant based on catalyst surface area.

Solution: Since $\mathscr{R}_A = -k a_{out}$ for a CSTR, the rates in the previous example are just divided by the appropriate exit concentrations to obtain k. The ordinary, gas-phase concentration is used for the pseudohomogeneous rate:

$$k_{homogeneous} = 0.783/0.4 = 1.96 \text{ s}^{-1}$$

The reactant concentration per unit mass is used for the rate based on catalyst mass:

$$[a_{out}]_{catalyst\ mass} = \frac{\varepsilon_{total} a_{out}}{\rho_c} = 9.11 \times 10^{-4} \text{ mol/kg}$$

$$k_{catalyst\ mass} = \frac{[\mathscr{R}_A]_{catalyst\ mass}}{[a_{out}]_{catalyst\ mass}} = 1.96 \text{ s}^{-1}$$

Similarly,

$$k_{surface\ area} = \frac{[\mathscr{R}_A]_{surface\ area}}{[a_{out}]_{surface\ area}} = \frac{[\mathscr{R}_A]_{surface\ area}}{\frac{\varepsilon_{total} a_{out}}{\rho_c a_c}} = \frac{4.45 \times 10^{-9}}{2.28 \times 10^{-9}} = 1.96 \text{ s}^{-1}$$

Example 10.11: The piping in the recycle reactor of Example 10.9 has been revised to lower the recycle line and pump volume to 100 cm^3. What effect will this have on the exit concentration of component A if all other conditions are held constant?

Solution: The catalyst charge is unchanged. If the reaction is truly heterogeneous and there are no mass transfer resistances, the rate of reaction of component A should be unchanged. More specifically, the pseudohomogeneous rate for the CSTR will change since the gas-phase volume and residence time changes, but the heterogeneous rate should be the same.

$$\varepsilon_{total} V = 125 + 100 - \frac{101}{1.12}(1 - 0.505) = 180 \, \text{cm}^3$$

$$\varepsilon_{total} = 180/225 = 0.800$$

$$\rho_c = \frac{101}{225} = 0.449 \, \text{g/cm}^3 = 449 \, \text{kg/m}^3$$

$$\bar{t} = \frac{180 \, \text{cm}^3}{150 \, \text{cm}^3/\text{s}} = 1.20 \, \text{s}$$

Assume

$$[\mathscr{R}_A]_{catalyst\,mass} = \frac{\varepsilon_{total} \mathscr{R}_A}{\rho_c} = -1.78 \times 10^{-3} \, \text{mol/(kg·s)}$$

as in Example 10.9, and convert to a pseudohomogeneous rate for the CSTR:

$$[\mathscr{R}_A]_{homogeneous} = \frac{\rho_c [\mathscr{R}_A]_{catalyst\,mass}}{\varepsilon_{total}} = -0.999 \, \text{mol/(m}^3\text{·s)}$$

Now assume $a_{in} = 1.6 \, \text{mol/m}^3$ and use Equation (10.38) to find a_{out}. The result is

$$a_{out} = a_{in} + \bar{t} [\mathscr{R}_A]_{homogeneous} = 0.401 \, \text{mol/m}^3$$

This is identical within rounding error to the exit concentration in Example 10.9.

It is a good idea to run the laboratory reactor without catalyst to check for homogeneous reactions. However, this method does not work when the homogeneous reaction involves reactants that do not occur in the feed but are created by a heterogeneous reaction. It then becomes important to maintain the same ratio of free volume to catalyst volume in the laboratory reactor used for intrinsic kinetic studies as in the pilot or production reactors.

10.6 UNSTEADY OPERATION AND SURFACE INVENTORIES

The global design equations for packed beds—e.g., Equations (10.1), (10.9), (10.39), and (10.40)—all have a similar limitation to that of the axial dispersion model treated in Chapter 9. They all assume steady-state operation. Adding an accumulation term, $\partial a/\partial t$ accounts for the change in the gas-phase inventory of component A but not for the surface inventory of A in the adsorbed form. The adsorbed inventory can be a large multiple of the gas-phase inventory.

Example 10.12: Estimate the surface inventory of component A for the catalytic CSTR in Example 10.9. Assume that the surface reaction is rate-controlling and that A is the only adsorbed species. Suppose A is a moderately large molecule that occupies a site that is 1 nm by 1 nm.

Solution: If the surface reaction controls the overall rate, all the active sites will be occupied. Assume that the entire surface is active so that it will be covered with a monolayer of A molecules. The surface area is $101(400) = 40,400 \, \text{m}^2$ (4 ha or 10 U.S. acres!).

$$\text{Adsorbed A} = \frac{40,400}{10^{-18}} = 4 \times 10^{22} \text{ molecules} = 0.067 \, \text{mol}$$

$$\text{Gas phase A} = (0.4 \, \text{mol/m}^3)(230 \times 10^{-6} \, \text{m}^3) = 9.2 \times 10^{-5} \, \text{mol}$$

Thus, the surface contains 700 times more A than the gas phase.

Example 10.13: How long will it take the reactor in Example 10.9 to achieve steady state?

Solution: The surface is estimated to contain 0.067 mol of A in adsorbed form. The inlet gas contains 1.6 mol of A per cubic meter and is flowing at $150 \, \text{cm}^3/\text{s}$ so that A is entering the reactor at a rate of $0.00024 \, \text{mol/s}$. Five minutes are needed to supply the surface if all the incoming gas were adsorbed. Fifteen to thirty minutes would be a reasonable startup time. Recall that the reactor has a gas-phase residence time V/Q_{out} of only 1.5 s! The residence time of the adsorbed species is 700 times larger than the average (nonadsorbed) molecule.

Anyone wishing to model the startup transient of a packed-bed reactor or to explore the possible benefits of periodic operation on selectivity should consider whether surface inventories are significant. The above examples show that they certainly can be. The unsteady versions of, say, Equation (10.1) should be supplemented with separate component balances for the adsorbed species. It may also be necessary to write separate energy balances for the gas and solid phases. Chapter 11 gives the general methodology for treating the component and energy balances in multiphase reactors, but implementation requires considerable sophistication.

PROBLEMS

10.1. The precursor result from which Equation (10.13) was obtained is

$$\mathscr{R} = \frac{S_0[(k_a^+ k_R^+ k_d^+)a - (k_a^- k_R^- k_d^-)p]}{(k_a^- k_R^- + k_a^- k_d^+ + k_R^+ k_d^+) + k_a^+(k_R^+ + k_R^- + k_d^+)a + k_d^-(k_a^- + k_R^+ + k_R^-)p}$$

Take the limit as k_a^+ and k_a^- both approach zero with a fixed ratio between them, $k_a^+/k_a^- = K_a$. The magnitude of $\mathscr{R}$ decreases and the functional form changes. Explain the physical basis for these changes.

10.2. What happens to $\mathscr{R}$ in Problem 10.1 when desorption is rate controlling?

10.3. Repeat Examples 10.3 and 10.4 for the case where desorption is rate controlling.

10.4. The Eley-Rideal mechanism for gas–solid heterogeneous catalysis envisions reaction between a molecule adsorbed on the solid surface and one that is still in the gas phase. Consider a reaction of the form

$$A + B \rightarrow P$$

There are two logical possibilities for the reaction mechanism:
(a) $A(gas) + BS(solid) \rightarrow PS(solid)$
(b) $A(gas) + BS(solid) \rightarrow P(gas) + S(solid)$

Determine the form of the pseudohomogeneous, intrinsic kinetics for each of these cases. Assume that the surface reaction step, as shown above, is rate limiting.

10.5. The ethylbenzene dehydrogenation catalyst of Example 3.1 has a first-order rate constant of 3.752 s^{-1} at 700°C. How does this compare with the catalyst used by Wenner and Dybdal.[5] They reported

$$k = 12{,}600\exp(-19{,}800/T)$$

where $\mathscr{R} = kP_{EB}$ with k in pound-moles per hour per atmosphere per pound of catalyst and T in degrees Rankine. P_{EB} is the partial pressure of ethylbenzene in atmospheres. The bulk density of the catalyst is 90 lb/ft^3 and the void fraction is 0.4.

10.6. An observed, gas–solid-catalyzed reaction is $A + B \rightarrow P$. Suppose the surface mechanism is

$$A + S \rightleftharpoons AS \qquad \frac{[AS]}{a[S]} = K_I$$

$$B + S \rightleftharpoons BS \qquad \frac{[BS]}{b[S]} = K_{II}$$

$$AS + BS \overset{k}{\rightleftharpoons} PS + S \qquad \mathscr{R} = k[AS][BS]$$

$$PS \rightleftharpoons P + S \qquad \frac{p[S]}{[PS]} = K_{IV}$$

Determine the functional form of the rate equation.

10.7. The following surface mechanism has been evoked to explain an observed reaction:

$$A_2 + 2S \rightleftharpoons 2AS \quad (I)$$

$$B + S \rightleftharpoons BS \quad (II)$$
$$AS + BS \rightleftharpoons CS + DS \quad (III)$$
$$CS \rightleftharpoons C + S \quad (IV)$$
$$DS \rightleftharpoons D + S \quad (V)$$

(a) What is the observed reaction?
(b) Develop a Hougen and Watson kinetic model, assuming Reaction (III) is rate controlling.

10.8. Repeat Problem 10.7, assuming that Reaction (I) is rate controlling.

10.9. The catalytic hydrogenation of butyraldehyde to butanol:

$$H_2 + C_3H_7\overset{\overset{O}{\|}}{C}H \rightleftharpoons C_3H_7CH_2OH$$

has a reported[6] rate equation of the form

$$\mathscr{R} = \frac{k(P_{H_2}P_B - P_{BOH}/K_{eq})}{(1 + K_1 P_{H_2} + K_2 P_{BAL} + K_3 P_{BOH})^2}$$

where P_{H_2}, P_{BAL}, and P_{BOH} are the partial pressures of hydrogen, butyraldehyde, and butanol, respectively.

(a) Develop a surface reaction model to rationalize the observed form of the kinetics.
(b) Is K_{eq} the thermodynamic or kinetic equilibrium constant? Is it a function of pressure?

10.10. Bimolecular reactions are sometimes catalyzed using two different metals dispersed on a common support. A mechanism might be

$$A + S1 \rightleftharpoons AS1$$
$$B + S2 \rightleftharpoons BS2$$
$$AS1 + BS2 \rightleftharpoons PS1 + QS2$$
$$PS2 \rightleftharpoons P + S2$$
$$QS2 \rightleftharpoons Q + S2$$

Derive a Hougen and Watson kinetic model, assuming that the surface reaction is rate-controlling.

10.11. Refer to Equation (10.31) and consider a catalyst pellet (not finely ground) for which $\eta_0 \approx 1$. What will be the value of η when film resistance is included? The obvious answer to this question is also the solution to Problem 10.12, but go one step beyond the obvious and ask what is likely to be the magnitude of k when $\eta_0 \approx 1$. What does this imply for η?

10.12. Consider a nonporous catalyst particle where the active surface is all external. There is obviously no pore resistance, but a film resistance to mass transfer can still exist. Determine the isothermal effectiveness factor for first-order kinetics.
Hint: The actual reaction rate is $\mathscr{R}_A = -ka_s$.

10.13. A platinum catalyst supported on Al_2O_3 is used for the oxidation of sulfur dioxide:

$$SO_2 + \tfrac{1}{2}O_2 \rightarrow SO_3 \qquad \Delta H_R = -95\,\text{kJ/mol}$$

The catalyst consists of 3-mm pellets that pack to a bulk density of 1350 kg/m³ and $\varepsilon = 0.5$. Mercury porosimetry has found $R_{pore} = 5\,\text{nm}$. The feed mixture to a differential reactor consisted of 5 mol% SO_2 and 95 mol% air. The following initial rate data were obtained at atmospheric pressure:

T, K	$\mathscr{R}$, mol/h (per g, catalyst)
653	0.031
673	0.053
693	0.078
713	0.107

Do an order-of-magnitude calculation for the nonisothermal effectiveness factor.
Hint: Use the pore model to estimate an isothermal effectiveness factor and obtain $\mathscr{D}_{eff}$ from that. Assume $\lambda_{eff} = 0.15\,\text{J/(m·s·K)}$.

10.14. Suppose that catalyst pellets in the shape of right-circular cylinders have a measured effectiveness factor of η when used in a packed-bed reactor for a first-order reaction. In an effort to increase catalyst activity, it is proposed to use a pellet with a central hole of radius $R_h < R_p$. Determine the best value for R_h/R_p based on an effective diffusivity model similar to Equation (10.33). Assume isothermal operation; ignore any diffusion limitations in the central hole, and assume that the ends of the cylinder are sealed to diffusion. You may assume that k, R_p, and $\mathscr{D}_{eff}$ are known.
Hints: First convince yourself that there is an optimal solution by considering the limiting cases of η near zero, where a large hole can almost double the catalyst activity, and of η near 1, where any hole hurts because it removes catalyst mass. Then convert Equation (10.33) to the form appropriate to an infinitely long cylinder. Brush up on your Bessel functions or trust your symbolic manipulator if you go for an analytical solution. Figuring out how to best display the results is part of the problem.

10.15. Work Problem 10.14 using the pore diffusion model rather than the effective diffusivity model.

10.16. Charge the reactor with the optimized pellets from Problem 10.14 or 10.15. What does it do to the value for $\eta(\theta)\rho_c a_c [\mathcal{R}_A]_{heterogeneous} = \eta(\theta)\varepsilon_{total} [\mathcal{R}_A]_{homogeneous}$ used to model the reactor? If you have not worked Problem 10.14 or 10.15, assume the new pellet increases the reaction rate per pellet by a factor of 1.5 when $R_h/R_p = 0.5$.

REFERENCES

1. Hougan, O. A. and Watson, K. M., *Chemical Process Principles*, Vol. III, Wiley, New York, 1947.
2. Butt, J. B. *Reaction Kinetics and Reactor Design*, Prentice-Hall, New York, 1980.
3. Rothfield, L. B., "Gaseous counterdiffusion in catalyst pellets," *AIChE J.*, **9**, 19–21 (1963).
4. Weisz, P. B. and Hicks, J. S., "The behaviour of porous catalyst particles in view of internal mass and heat diffusion effects," *Chem. Eng. Sci.*, **17**, 265–275 (1962).
5. Wenner, R. R. and Dybdal, F. C., "Catalytic dehydrogenation of ethylbenzene," *Chem. Eng. Prog.*, **44**, 275–286 (1948).
6. Cropley, J. B., Burgess, L. M., and Loke, R. A., "Butyraldehyde hydrogenation: a case study in process design," *CHEMTECH*, **14**(6), 374–380 (June 1984).

SUGGESTIONS FOR FURTHER READING

A comprehensive treatment of heterogeneous catalysis is given by

Thomas, J. M., and Thomas, J. W., *Principles and Practice of Heterogeneous Catalysis*, Wiley, New York, 1996.

All forms of catalysis are described in

Cornils, B. et al., *Catalysis from A to Z: A Concise Encyclopedia*, Wiley, New York, 2000.

A chemical engineering classic still in print is

Smith, J. M., *Chemical Engineering Kinetics*, 3rd ed., McGraw-Hill, New York, 1981.

A recent second edition of another standard text that emphasizes gas–solid reactions is

Butt, J. B., *Reaction Kinetics and Reactor Design*, 2nd ed., Marcel Dekker, New York, 2000.

Interactions between thermodynamics and kinetics are sometimes given short shrift. The following article will help remedy this omission:

Shinnar, R. and Feng, C. A., "Structure of complex catalytic reactions: thermodynamic constraints in kinetic modeling and catalyst evaluation," *I & EC Fundam.*, **24**, 153–170 (1985).

For a recent review of catalyst deactivation, see

Worstell, J. H., Doll, M. J., and Worstell, J. H., "What's causing your catalyst to decay?," *Chem. Eng. Prog.*, **96**(9), 59–64 (September 2000).

CHAPTER 11
MULTIPHASE REACTORS

The packed-bed reactors discussed in Chapters 9 and 10 are multiphase reactors, but the solid phase is stationary, and convective flow occurs only through the fluid phase. The reaction kinetics are pseudohomogeneous, and components balances are written only for the fluid phase.

Chapter 11 treats reactors where mass and component balances are needed for at least two phases and where there is interphase mass transfer. Most examples have two fluid phases, typically gas–liquid. Reaction is usually confined to one phase, although the general formulation allows reaction in any phase. A third phase, when present, is usually solid and usually catalytic. The solid phase may be either mobile or stationary. Some example systems are shown in Table 11.1.

When two or more phases are present, it is rarely possible to design a reactor on a strictly first-principles basis. Rather than starting with the mass, energy, and momentum transport equations, as was done for the laminar flow systems in Chapter 8, we tend to use simplified flow models with empirical correlations for mass transfer coefficients and interfacial areas. The approach is conceptually similar to that used for friction factors and heat transfer coefficients in turbulent flow systems. It usually provides an adequate basis for design and scaleup, although extra care must be taken that the correlations are appropriate.

Multiphase reactors can be batch, fed-batch, or continuous. Most of the design equations derived in this chapter are general and apply to any of the operating modes. Unsteady operation of nominally continuous processes is treated in Chapter 14.

11.1 GAS–LIQUID AND LIQUID–LIQUID REACTORS

After specifying the phases involved in the reaction, it is necessary to specify the contacting regimes. The ideal contacting regimes for reactors with two fluid phases are:

1. Both phases are perfectly mixed.
2. One phase is perfectly mixed and the other phase is in piston flow.

TABLE 11.1 Examples of Multiphase Reactors

Reaction	First phase	Second phase	Third phase
Phenol alkylation	Phenol	Gaseous alkenes	None
Refinery alkylation	Liquid alkanes (e.g., isobutane)	Gaseous alkenes (e.g., 1-butene)	HF or H_2SO_4
Aerobic fermentation	Water+organic carbon source	Air	Bacteria
Anaerobic fermentation	Water+sugar	Carbon dioxide	Yeast
Fluidized catalytic cracking (FCC)	Heavy oils, $C_{18}+$	Fluidized catalyst particles	None
Trickle-bed hydrocracking	Refinery residues, $C_{30}+$	Hydrogen	Stationary catalyst particles

3. The phases are in countercurrent piston flow.
4. The phases are in cocurrent piston flow.

These simple situations can be embellished. For example, the axial dispersion model can be applied to the piston flow elements. However, uncertainties in reaction rates and mass transfer coefficients are likely to mask secondary effects such as axial dispersion.

11.1.1 Two-Phase Stirred Tank Reactors

Stirred tanks are often used for gas–liquid reactions. The usual geometry is for the liquid to enter at the top of the reactor and to leave at the bottom. The gas enters through a sparge ring underneath the impeller and leaves through the vapor space at the top of the reactor. A simple but effective way of modeling this and many similar situations is to assume perfect mixing within each phase.

Figure 11.1 gives a conceptual view of a *two-phase, continuous-flow stirred tank reactor*, i.e., a two-phase CSTR. For convenience, we refer to one phase as being liquid and to the other as being gas, but the mixing and contacting scheme shown in Figure 11.1 can also apply to liquid–liquid systems. It can even apply to two gas phases separated by a membrane. Both phases are assumed to be internally homogeneous. They contact each other through an interface that has area A_i, with units of area per unit volume. The total interfacial area is $A_i V$, where $V = V_l + V_g$ is the *working volume* in the reactor as measured under operating conditions. The working volume ends at the top of the liquid level and is measured while gas is being added (i.e., the *gassed* condition). The working volume excludes the vapor space at the top of the reactor. The *gas-phase holdup* is the ratio V_g/V and the *liquid-phase holdup* is the ratio V_l/V.

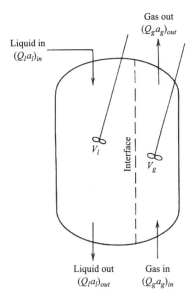

FIGURE 11.1 A two-phase, continuous-flow stirred tank reactor.

The interfacial area $A_i V$ usually excludes the contact area between the vapor space and the liquid at the top of the reactor. The justification for this is that most gas–liquid reactors have gas bubbles as a dispersed phase. This gives a much larger interfacial area than the nominal contact area at the top of the reactor. There are exceptions—e.g., polyester reactors where by-product water is removed only through the nominal interface at the top of the reactor—but these are old and inefficient designs. This nominal area scales as $S^{2/3}$ while the contact area with a dispersed phase can scale as S.

Mass Transfer Rates. Mass transfer occurs across the interface. The rate of mass transfer is proportional to the interfacial area and the concentration driving force. Suppose component A is being transferred from the gas to the liquid. The concentration of A in the gas phase is a_g and the concentration of A in the liquid phase is a_l. Both concentrations have units of moles per cubic meter; however they are not directly comparable because they are in different phases. This fact makes mass transfer more difficult than heat transfer since the temperature is the temperature regardless of what phase it is measured in, and the driving force for heat transfer across an interface is just the temperature difference $T_g - T_l$. For mass transfer, the driving force is not $a_g - a_l$. Instead, one of the concentrations must be converted to its equivalent value in the other phase.

The conversion is carried out using the equilibrium relationship between the gas- and liquid-phase concentrations. Usual practice is to assume Henry's law. Thus, the gas-phase concentration that is equivalent to a_l is $K_H a_l$, where K_H is

Henry's law constant. The overall driving force for mass transfer is $a_g - K_H a_l$, and the rate of mass transfer across the interface is

$$\text{Mass transfer rate} = K_g A_i V(a_g - K_H a_l) \tag{11.1}$$

where K_g is the overall mass transfer coefficient based on the equivalent gas-phase driving force, $a_g - K_H a_l$; A_i is the *interfacial area per unit volume* of the reactor; and V is the reactor volume. Experimental measurements and literature correlations usually give the composite quantity $K_g A_i$ rather than the individual variables.

Equation (11.1) bases the driving force on concentrations that have SI units of mol/m^3. Corresponding units for the composite quantity $K_g A_i$ are s^{-1}, and K_g by itself has units of velocity, m/s. These units are appropriate when the focus is on reactor design since the reaction rate also depends on concentration. However, the mass transfer literature frequently bases the liquid-phase driving force on mole fractions and the gas-phase driving force on mole fractions or partial pressures. This leads to units for $K_g A_i$ such as mol m^{-3} s^{-1} (gas-phase mole fraction)$^{-1}$ or mol/(m$^3 \cdot$s$\cdot$Pa). Example 11.9 includes a conversion of such units into SI units. The SI units are m/s for K_g, m^{-1} for A_i, s^{-1} for $K_g A_i$, and mol/s for $K_g A_i V(a_g - k_H a_l)$. The quantity $K_g(a_g - k_H a_l)$ has units of flux, mol/(m$\cdot$s). Quantities involving K_i have identical units to those involving K_g.

Equation (11.1) replaces the liquid-phase concentration with an equivalent gas-phase concentration. It is obviously possible to do it the other way, replacing the gas-phase concentration with an equivalent liquid-phase concentration. Then

$$\text{Mass transfer rate} = K_l A_i V(a_g / K_H - a_l) \tag{11.2}$$

Equations (11.1) and (11.2) must predict the same rate. This gives

$$K_l = K_H K_g \tag{11.3}$$

Henry's law constant is dimensionless when a_g and a_l are in mol/m^3, but conventional units for K_H are atmospheres or torr per mole fraction. Thus, the gas-phase concentration is expressed in terms of its partial pressure and the liquid-phase concentration is expressed as a mole fraction.

The mass transfer coefficients, K_g and K_l, are overall coefficients analogous to an overall heat transfer coefficient, but the analogy between heat and mass transfer breaks down for mass transfer across a phase boundary. Temperature has a common measure, so that thermal equilibrium is reached when the two phases have the same temperature. Compositional equilibrium is achieved at different values for the phase compositions. The equilibrium concentrations are related, not by equality, as for temperature, but by proportionality through an equilibrium relationship. This proportionality constant can be the *Henry's law constant* K_H, but there is no guarantee that Henry's law will apply over the necessary concentration range. More generally, K_H is a function of composition and temperature that serves as a (local) proportionality constant between the gas- and liquid-phase concentrations.

When K_H is a function of composition, the concept of overall mass transfer coefficient stops being useful. Instead, the overall resistance to mass transfer is divided between two film resistances, one for each phase. This is done by assuming that equilibrium is achieved at the interface. The equilibrium values are related by a function having the form of Henry's law:

$$a_g^* = K_H a_l^* \tag{11.4}$$

but it is possible that $K_H = K_H(a_l)$ varies with composition. The driving forces for mass transfer can now be expressed using concentrations within a single phase. The rate of mass transfer across the interface is

$$\text{Mass transfer rate} = k_g A_i V(a_g - a_g^*) = k_l A_i V(a_l^* - a_l) \tag{11.5}$$

Here, k_g and k_l are the gas-side and liquid-side mass transfer coefficients. Their units are identical to those for K_g and K_l, m/s. Like the overall coefficients, they are usually measured and reported as the composite quantities $k_g A_i$ and $k_l A_i$ with SI units of s^{-1}.

Equations (11.4) and (11.5) can be combined to find the interfacial concentrations:

$$a_l^* = \frac{a_g^*}{K_H^*} = \frac{k_g a_g + k_l a_l}{k_g K_H^* + k_l} \tag{11.6}$$

where K_H^* denotes the local value for the Henry's law constant at the interface. The usual case for a gas–liquid system is $k_g K_H^* \gg k_l$ so that the liquid-side resistance is controlling. Then $a_g^* \approx a_g$ and

$$a_l^* \approx \frac{a_g}{K_H^*} \tag{11.7}$$

If Henry's law holds throughout the composition range, then K_H will have the same value in all of these equations. Furthermore, all the mass transfer rates in these equations are equal. Algebra gives

$$K_l = K_H K_g = \frac{1}{1/k_l + 1/(K_H k_g)} \tag{11.8}$$

which is reminiscent of the equation for the overall heat transfer coefficient between a tube and a jacket, except that there is no wall resistance. A *membrane reactor* has the mass transfer equivalent to a wall resistance.

Example 11.1: Suppose an aerobic fermentation is being conducted in an agitated vessel at 38°C and that $k_l A_i = 0.1\,s^{-1}$ and $k_g A_i = 20\,s^{-1}$ have been measured for the mass transfer of oxygen from air at atmospheric pressure. Determine $K_l A_i$ and $K_g A_i$.

Solution: A standard reference gives $K_H = 4.04 \times 10^7$ mm of mercury per mole fraction for oxygen in water at 38°C (standard references are notoriously reluctant to change units). Thus, an oxygen partial pressure of 160 torr will be in equilibrium with water containing 3.95×10^{-6} mole fraction oxygen. The corresponding oxygen concentrations work out to be 10.7 mol/m^3 for the gas phase and 0.219 mol/m^3 for the liquid phase. Thus, the dimensionless value for Henry's law constant is 48.9. Substitute this and the given values for $k_l A_i$ and $k_g A_i$ into Equation (11.6) to obtain $K_l A_i = 0.1\,\text{s}^{-1}$ and $K_g A_i = 0.002\,\text{s}^{-1}$. The calculation also shows that the gas-side resistance is negligible compared with the liquid-side resistance. This is typical, except when the gas being transferred reacts very rapidly in the liquid. See "Enhancement Factors" in Section 11.1.5.

Example 11.2: What happens if the air in Example 11.1 is replaced with pure oxygen?

Solution: With pure oxygen, the composition is uniform in the gas phase so that $a_g^* = a_g$. There is no resistance to mass transfer in a phase consisting of a pure component. However, the gas-side resistance is already negligible in Example 11.1. Thus, the conversion to pure oxygen has no effect on $K_l A_i$ and $K_g A_i$ as calculated in that example. Going to pure oxygen will increase a_g^* and a_l^* each by a factor of about 5. What this does to the mass transfer rate depends on the response of the organism to the increased oxygen supply. If the organism's growth was strictly limited by oxygen—as it is in some wastewater treatment processes—the oxygen transfer rate could increase by a factor of 5, and the dissolved oxygen concentration a_l would remain low. If the fermentation was limited by something else, there might be no effect except that the dissolved oxygen content would increase by a factor of 5.

Membrane Reactors. Consider the two-phase stirred tank shown in Figure 11.1 but suppose there is a membrane separating the phases. The equilibrium relationship of Equation (11.4) no longer holds. Instead, the mass transfer rate across the interface is given by

$$\text{Mass transfer rate} = D_A \frac{a_g^*/K_H - a_l^*}{\Delta x} \qquad (11.9)$$

where a_g^*/K_H is the liquid-equivalent concentration on the gas side of the membrane, a_l^* is the concentration on the liquid side, and D_A is the effective diffusivity in the membrane based on the liquid-phase driving force. With these revised definitions for a_g^* and a_l^*, Equation (11.5) continues to hold and gives the same mass transfer rate as Equation (11.9). The value for D_A is very dependent on the species since membranes can be quite selective.

Equation (11.9) does not require K_H to be constant throughout the range of compositions in the reactor; but if it is constant, the overall mass transfer coefficient is

$$K_l = \cfrac{1}{\cfrac{1}{k_l} + \left(\cfrac{\Delta x}{D_A}\right) + \left(\cfrac{1}{K_H k_g}\right)} \qquad (11.10)$$

Recall that the designations "liquid" and "gas" are arbitrary. The same equations can be used for liquid–liquid and gas–gas membrane reactors. Values for the effective diffusivity D_A depend on which driving force is used in the calculations. Equation (11.9) uses the liquid-phase driving force. Multiply by K_H to find the equivalent value for D_A based on a gas-phase driving force. Experimental values for D_A depend strongly on the species being transported and on which phase is present inside the membrane. Many membranes are strongly asymmetric so that the phase inside the membrane is the phase in contact with the side of the membrane that has the larger pores.

Phase Balances for Components. Material balances can be written for each phase. For the general case of unsteady operation and variable physical properties, the liquid-phase balance is

$$\frac{d(V_l a_l)}{dt} = (Q_l a_l)_{in} + k_l A_i V(a_l^* - a_l) + V_l (\mathscr{R}_A)_l - (Q_l a_l)_{out} \qquad (11.11)$$

Note that the accumulation and reaction terms are based on the volume of the liquid phase but that the mass transfer term is based on the working volume, $V = V_l + V_g$. The gas-phase balance is

$$\frac{d(V_g a_g)}{dt} = (Q_g a_g)_{in} - k_g A_i V(a_g^* - a_g) + V_g (\mathscr{R}_A)_g - (Q_g a_g)_{out} \qquad (11.12)$$

These component balances are conceptually identical to a component balance written for a homogeneous system, Equation (1.6), but there is now a source term due to mass transfer across the interface. There are two equations (ODEs) and two primary unknowns, a_g and a_l. The concentrations at the interface, a_l^* and a_g^*, are also unknown but can be found using the equilibrium relationship, Equation (11.4), and the equality of transfer rates, Equation (11.5). For membrane reactors, Equation (11.9) replaces Equation (11.4). Solution is possible whether or not K_H is constant, but the case where it is constant allows a_l^* and a_g^* to be eliminated directly

$$\frac{d(V_l a_l)}{dt} = (Q_l a_l)_{in} + K_l A_i V(a_g/K_H - a_l) + V_l (\mathscr{R}_A)_l - (Q_l a_l)_{out} \qquad (11.13)$$

$$\frac{d(V_g a_g)}{dt} = (Q_g a_g)_{in} - K_l A_i V(a_g/K_H - a_l) + V_g (\mathscr{R}_A)_g - (Q_g a_g)_{out} \qquad (11.14)$$

We have elected to use the overall mass transfer coefficient K_l, which is based on the equivalent liquid-phase driving force $a_g/K_H - a_l$, but this choice was arbitrary so that K_g and $a_g - K_H a_l$ could have been used instead. Equations (11.13) and (11.14) can be used for membrane reactors when K_l is given by Equation (11.10).

Operating Modes. The component and mass balances are quite general and apply to any operating mode; e.g., batch, semibatch, or steady state. Table 11.2 gives examples for the various modes.

The flow terms are dropped for batch reactors, but mass transfer between the phases can still occur. The long-time solutions to Equations (11.11) and (11.12) give the equilibrium concentrations and volumes for the two phases. This can be considered an application of the method of false transients wherein an equilibrium problem is solved using simultaneous ODEs rather than simultaneous algebraic equations. Long-time solutions should be computed even for batch reactions that do not go to completion since this provides a test of the reactor model. Does the equilibrium solution make sense? The general design problem for a batch reactor allows V, V_g, V_l, and A_i to vary during the course of the reaction. Like the variable-volume batch reactions considered in Section 2.6, solutions can be quite complex. In two-phase systems, two equations of state may be needed, one for each phase.

Fed-batch operation with the liquid charged initially followed by continuous gas sparging is quite common for gas–liquid reactions. Set $(Q_{in})_l = (Q_{out})_l = 0$ and $d(V_g a_g)/dt = 0$. Typically, $(\mathscr{R}_A)_g = 0$ as well. Do not set $(Q_{out})_g = 0$ as it is usually necessary to remove inerts (e.g., nitrogen in an aerobic fermentation or saturates for an alkylation). This form of semibatch reaction is useful when the gas has a low solubility in the liquid. The equipment can be a conventional stirred tank with a sparge ring as shown in Figure 11.2. When heat transfer is important, much of the reactor volume can be provided by a heat exchanger in a recycle loop. Figure 11.3 illustrates a semibatch reactor used for phenol alkylation. The tank in that process acts as a collection vessel and has no agitator. Gas–liquid mixing is provided by a proprietary mixing nozzle, and the pump provides general circulation.

TABLE 11.2 Two-Phase Reactions in a Stirred Tank Reactor

Mode	Example
Batch	A condensation reaction where a product precipitates as a solid
Semibatch, reactant addition	A batch alkylation where a gaseous alkene is continuously charged
Semibatch, product removal	An aerobic fermentation where by-product CO_2 is allowed to escape
Continuous	Bleaching of paper pulp with Cl_2 or ClO_2

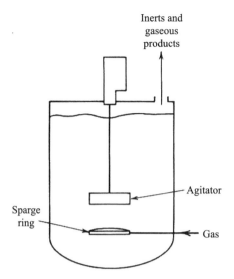

FIGURE 11.2 Mechanically agitated vessel with gas sparging.

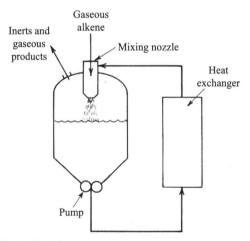

FIGURE 11.3 Semibatch alkylation reactor.

Semibatch or fully continuous operation with continuous removal of a by-product gas is also common. It is an important technique for relieving an equilibrium limitation, e.g., by-product water in an esterification. The pressure in the vapor space can be reduced or a dry, inert gas can be sparged to increase A_i and lower a_g^*, thereby increasing mass transfer and lowering a_l so that the forward reaction can proceed.

The time derivatives are dropped for steady-state, continuous flow, although the method of false transients may still be convenient for solving Equations (11.11) and (11.12) (or, for variable K_H, Equations (11.9) and (11.10) together with the appropriate auxiliary equations). The general case is somewhat less complicated than for two-phase batch reactions since system parameters such as V, V_g, V_l, and A_i will have steady-state values. Still, a realistic solution can be quite complicated.

In the relatively simple examples that follow, the system parameters are assumed to be known. Finding them may require a sophisticated effort in modeling and measurement. Measurement techniques are discussed in Section 11.1.2.

Example 11.3: The gas supply for the stirred tank in Examples 11.1 and 11.2 is suddenly changed from air to pure oxygen. How long does it take for the dissolved oxygen concentration to reach its new value? Assume the tank is operated in a batch mode with respect to water. Ignore any evaporation of water and any reaction involving oxygen. Ignore any changes in volumetric flow rate of the gas and assume that V_l and V_g remain constant.

Solution: The initial liquid-phase concentration of oxygen is $0.219 \, \text{mol/m}^3$ as in Example 11.1. The final oxygen concentration will be $1.05 \, \text{mol/m}^3$. The phase balances, Equations (11.11) and (11.12), govern the dynamic response. The flow and reaction terms are dropped from the liquid phase balance to give

$$\frac{da_l}{dt} = \frac{k_l A_i}{[V_l/V]}(a_l^* - a_l)$$

As a first approximation, suppose that the concentration of oxygen in the gas phase changes instantly from 20.9% oxygen to 100% oxygen. Then a_l^* will change instantly from 0.219 to $1.05 \, \text{mol/m}^3$, and the gas-phase balance is not required. The parameter $k_l A_i = 0.1 \, \text{s}^{-1}$ was specified in Example 11.1 so the only unknown parameter is the liquid holdup, V_l/V. A typical value for a mechanically agitated tank is 0.9. The liquid-phase balance becomes

$$\frac{da_l}{dt} = \frac{0.1}{[0.9]}(a_l^* - a_l)$$

subject to the initial condition that $a_l = 0.219$ at $t = 0$. Solution gives

$$a_l = 1.05 - 0.83 \exp(-0.111 t)$$

The 95% response time is 27 s.

Example 11.4: The assumption in Example 11.3 that the gas composition changes instantly is suspicious. Remedy this defect in the analysis.

Solution: The liquid-phase balance is unchanged except that a_l^* now varies with time, $a_l = a_l(t)$. The gas-phase balance comes from Equation (11.12). When simplified for the present case it becomes

$$\frac{da_g}{dt} = \frac{Q_g}{V_g}(a_g)_{in} - \frac{k_g A_i}{[V_g/V]}(a_g^* - a_g) - \frac{Q_g}{V_g}a_g$$

This introduces the gas-phase residence time V_g/Q_g as a new parameter. It also introduces an ambiguity regarding the term $k_g A_i(a_g^* - a_g)$. There is no resistance to mass transfer within a pure component so $k_g A_i \to \infty$ and $a_g^* - a_g \to 0$. Thus, $k_g A_i(a_g^* - a_g)$ is an indeterminate form of the $\infty \times 0$ variety. Its value must continue to equal the rate at which oxygen is transferred into the liquid phase. Equation (11.5) remains true and the pair of simultaneous ODEs become

$$\frac{da_l}{dt} = \frac{k_l A_i}{[V_l/V]}(a_l^* - a_l) = \frac{k_l A_i}{[V_l/V]}(a_g/K_H - a_l)$$

$$\frac{da_g}{dt} = \frac{Q_g}{V_g}(a_g)_{in} - \frac{k_l A_i}{[V_g/V]}(a_g/K_H - a_l) - \frac{Q_g}{V_g}a_g$$

The initial conditions are $a_l = 0.219$ and $a_g = 10.7$ mol/m^3 at $t = 0$. An analytical solution is possible but messy. The solution depends on the values of $k_l A_i$ and V_l/V (as in Example 11.3) and on the value of V_g/Q_g. In essence, Example 11.3 assumed $V_g/Q_g \approx 0$ so that the gas-phase concentration quickly responded to the change in inlet concentration.

Example 11.5: Suppose that the liquid phase in a gas–liquid CSTR contains a catalyst for the first-order reaction of a compound supplied from the gas phase. The reaction is

$$A \to \text{Products} \quad \mathscr{R} = ka_l$$

The reactor is operated in the semibatch mode with component A being sparged into the stirred tank. Unreacted A and the reaction products leave through the gas phase so that the mass of liquid remains constant. To the extent that these assumptions are true and the catalyst does not deactivate, a pseudo-steady-state can be achieved. Find $(a_g)_{out}$. Assume that Henry's law is valid throughout the composition range and ignore any changes in the gas density.

Solution: Both phases are assumed to be perfectly mixed so that $(a_g)_{out}$ is just a_g. Equation (11.13) provides the material balance for component A in the liquid phase:

$$0 = K_l A_i V(a_g/K_H - a_l) - V_l k a_l$$

The gas-phase balance comes from Equation (11.14):

$$0 = (Q_g a_g)_{in} - K_l A_i V(a_g/K_H - a_l) - (Q_g a_g)_{out}$$

There are two equations and two unknowns. Eliminating a_l gives

$$(a_g)_{out} = a_g = \frac{(a_g)_{in}}{1 + \left(\dfrac{V_l k(K_l A_i V)}{Q K_H (K_l A_i V + V_l k)}\right)}$$

where Q denotes the inlet and outlet gas flow rates since the gas density is constant.

Complete conversion of component A requires a high mass transfer rate, $K_l A_i \to \infty$, combined with a high reaction rate, $k \to \infty$. If $k \to \infty$ while $K_l A_i$ remains finite, the reaction is *mass transfer limited*. Some of the entering gas will not be absorbed and thus will not react. This situation is sometimes called *bypassing*.

Example 11.6: Suppose the liquid-phase reaction is

$$A + B \to \text{Products} \qquad \mathscr{R} = k a_l b_l$$

where A is a sparingly soluble gas that is sparged continuously and where the minor component B is charged to the vessel initially. Assume that component B is nonvolatile and that the gas-side film resistance is negligible. Determine $a_l(t)$ and $b_l(t)$.

Solution: Example 11.5 treats a system that could operate indefinitely since the liquid phase serves only as a catalyst. The present example is more realistic since the liquid phase is depleted and the reaction eventually ends. The assumption that the gas-side resistance is negligible is equivalent to assuming that $a_g^* = a_g$ throughout the course of the reaction. Equilibrium at the interface then fixes $a_l^* = a_g/K_H$ at all times. Dropping the flow and accumulation terms in the balance for the liquid phase, i.e., Equation (11.11), gives

$$0 = k_l A_i V(a_g/K_H - a_l) - V_l k a_l b_l$$

Solving for a_l,

$$a_l = \frac{k_l A_i V a_g / K_H}{k_l A_i V + V_l k b}$$

Note that a_l will gradually increase during the course of the reaction and will reach its saturation value, a_g/K_H, when B is depleted. Dropping the accumulation term for $a_l(t)$ represents a form of the pseudo-steady hypothesis. Since component B is not transferred between phases, its material balance has the usual form for a batch reactor:

$$\frac{db_l}{dt} = -k a_l b_l = -k \left[\frac{k_l A_i V a_g / K_H}{k_l A_i V + V_l k b}\right] b_l = \frac{-k' b_l}{1 + k_B b_l}$$

MULTIPHASE REACTORS   **393**

The appearance of this "heterogeneous" form for the rate expression reflects the presence of a mass transfer step in series with the reaction step. If the parameter values are known, this ODE for $b_l(t)$ can be integrated subject to the initial condition that $b_l = (b_l)_0$ at $t = 0$. The result can then be used to find $a_l(t)$.

The question arises as to how V, V_g, V_l, and A_i might vary during the course of the reaction. The problem statement does not give the necessary information to determine this. The reader is encouraged to create and solve some plausible scenarios, one of which allows V, V_g, V_l, and A_i to remain approximately constant.

Example 11.7: Carbon dioxide is sometimes removed from natural gas by reactive absorption in a tray column. The absorbent, typically an amine, is fed to the top of the column and gas is fed at the bottom. Liquid and gas flow patterns are similar to those in a distillation column with gas rising, liquid falling, and gas–liquid contacting occurring on the trays. Develop a model for a multitray CO_2 scrubber assuming that individual trays behave as two-phase, stirred tank reactors.

Solution: The liquid-phase reaction has the form $C + A \rightarrow P$, which we assume to be elementary with rate constant k. Suppose there are J trays in the column and that they are numbered starting from the bottom. Figure 11.4 shows a typical tray and indicates the notation. Since the

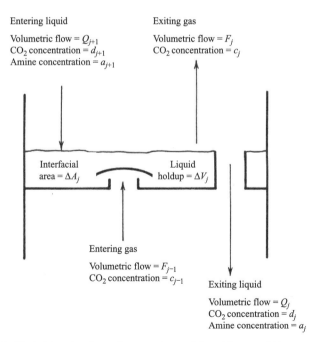

FIGURE 11.4 Typical tray in a tray column reactor used for acid-gas scrubbing.

column has many trays (typically 20 or more), composition changes on each tray are small, and it is reasonable to assume perfect mixing within each phase on an individual tray. The system is at steady state, and there is no reaction in the gas phase. The gas-phase balance for CO_2 is given by Equation (11.12), which simplifies to

$$0 = F_{j-1}c_{j-1} - k_g \Delta A_i(c_j - c_j^*) - F_j c_j$$

where ΔA_i is the interfacial area per tray. For the dissolved but unreacted CO_2 in the liquid phase, Equation (11.11) becomes

$$0 = Q_{j+1}d_{j+1} + k_l \Delta A_i(d_j^* - d_j) - \Delta V k a_j d_j - Q_j d_j$$

The liquid-phase balance for the amine is

$$0 = Q_{j+1}a_{j+1} - \Delta V k a_j d_j - Q_j a_j$$

In the current notation, Equations (11.6) and (11.7) give

$$d_j^* = \frac{k_g c_j + k_l d_j}{k_g K_H^* + k_l}$$

$$c_j^* = \frac{k_g c_j + k_l d_j}{k_g + k_l/K_H^*}$$

so that the interfacial calculations can be made (although by an iterative process when K_H is not constant). The inputs to the jth tray are c_{j-1}, d_{j+1}, a_{j+1}, F_{j-1}, and Q_{j+1}. Suppose these are known. Then there are three equations and three compositional unknowns: c_j, d_j, and a_j. There are also two unknown flow rates, F_j and Q_j, which may change significantly from the top to the bottom of the column. Stagewise values for these can be calculated from equations of state and overall mass balances for the two phases. Most gas scrubbing systems are designed to remove acid gases such as H_2S and SO_2 in addition to CO_2. Additionally, the heats of absorption can be significant so that energy balances may be needed as well. The overall computation can be quite complex, involving upwards of three nonlinear equations per tray so that hundreds of simultaneous equations must be solved simultaneously. This can be done using the multidimensional Newton's method described in Appendix 4. An alternative approach is to guess the composition of, say, the exiting liquid stream at the bottom of the column. With this initial guess, a sequential, tray-by-tray calculation is possible that involves simultaneous solution of only the basic set of five to ten equations per tray. This approach is conceptually similar to the shooting method described in Section 9.5. It presents similar difficulties due to numerical ill-conditioning for which there is no easy remedy. The implicit scheme is usually preferred.

Example 11.7 hints at the complications that are possible in reactive gas absorption. Gas absorption is an important unit operation that has been the subject of extensive research and development. Large, proprietary computer codes are available for purchase, and process simulation tools such as Aspen can do the job. However, as shown in Example 11.8, simple but useful approximations are sometimes possible.

Example 11.8: With highly reactive absorbents, the mass transfer resistance in the gas phase can be controlling. Determine the number of trays needed to reduce the CO_2 concentration in a methane stream from 5% to 100 ppm (by volume), assuming the liquid mass transfer and reaction steps are fast. A 0.9-m diameter column is to be operated at 8 atm and 50°C with a gas feed rate of $0.2 \, m^3/s$. The trays are bubble caps operated with a 0.1-m liquid level. Literature correlations suggest $k_g = 0.002 \, m/s$ and $\Delta A_i = 20 \, m^2$ per square meter of tray area.

Solution: Ideal gas behavior is a reasonable approximation for the feed stream. The inlet concentrations are $287 \, mol/m^3$ of methane and $15 \, mol/m^3$ of carbon dioxide. The column pressure drop is mainly due to the liquid head on the trays and will be negligible compared with 8 atm unless there are an enormous number of trays. Thus, the gas flow rate F will be approximately constant for the column as a whole. With fast reaction and a controlling gas-side resistance, $c_j^* = 0$. The gas-phase balance gives everything that is necessary to solve the problem:

$$0 = Fc_{j+1} - Fc_j - k_g \Delta A_i c_j$$

and

$$\frac{c_{j-1}}{c_j} = 1 + \frac{k_g \Delta A_i}{F} = 1.13$$

For the column at a whole,

$$\frac{c_0}{c_J} = (1.13)^J = \frac{0.05}{0.0001}$$

Solution gives $J = 51$ trays. The indicated separation appears feasible in a bubble-cap column although the design engineer should not be content with the glib assumption of negligible liquid-side resistance.

Overall and Phase Balances for Mass. The examples so far in Chapter 11 were designed to be simple yet show some essential features of gas–liquid reactors. Only component balances for the phases, Equations (11.11) and (11.12), have been used. They are reasonably rigorous, but they do not provide guidance regarding how the various operating parameters can be determined. This is done in Section 11.1.2. Also, total mass balances must supplement the component balances in order to handle the general case of semibatch operation where parameters such as V, V_g, V_l, and A_i can all vary with time.

An overall mass balance is written for the system as a whole. Interphase mass transfer does not appear in the system mass balance since gains in one phase exactly equal losses in the other. The net result is conceptually identical to Equation (1.3), but there are now two inlets and two outlets and the total inventory is summed over both phases. The result is

$$\frac{d(V_g \rho_g + V_l \rho_l)}{dt} = (Q_{in}\rho_{in})_l + (Q_{in}\rho_{in})_g - (Q_{out}\rho_{out})_l - (Q_{out}\rho_{out})_g \quad (11.15)$$

The phase mass balances are more complicated since the mass in a phase can grow or wane due to interphase mass transfer. The phase balances are

$$\frac{d(V_l \rho_l)}{dt} = (Q_{in}\rho_{in})_l - (Q_{out}\rho_{out})_l + \sum_{Components} M_A k_l A_i V(a_l^* - a_l) \quad (11.16)$$

$$\frac{d(V_g \rho_g)}{dt} = (Q_{in}\rho_{in})_g - (Q_{out}\rho_{out})_g - \sum_{Components} M_A k_g A_i V(a_g - a_g^*) \quad (11.17)$$

where M_A is the mass per mole (kg/mol) for component A. Equation (11.15) holds for each component so that Equations (11.16) and (11.17) sum to Equation (11.15). The film coefficients should depend on eddy diffusion more than molecular diffusion and, like the axial dispersion coefficient in Chapter 9, should be approximately the same for all chemical species. An exception to this statement occurs when a component is rapidly consumed by chemical reaction. Another exception is membrane reactors, where the membrane may be quite selective; i.e., D_A can vary greatly between species.

Energy Balances. When the reaction temperatures are unknown, two heat balances are also needed:

$$\frac{d(V_l \rho_l H_l)}{dt} = (Q_l \rho_l H_l)_{in} + UA_{ext}(T_{ext} - T_l)$$
$$+ h_i A_i V(T_g - T_l) - V_l(\Delta H_R \mathscr{R})_l - (Q_l \rho_l H_l)_{out} \quad (11.18)$$

and

$$\frac{d(V_g \rho_g H_g)}{dt} = (Q_g \rho_g H_g)_{in} - h_i A_i V(T_g - T_l)$$
$$- V_g(\Delta H_R \mathscr{R})_g - (Q_g \rho_g H_g)_{out} \quad (11.19)$$

These equations allow for interphase heat transfer with h_i as the heat transfer coefficient. As written, they envision heat transfer to the environment to occur only through the liquid phase since it is normally the continuous phase.

The component, mass, and energy balances constitute the complete set of design equations for a two-phase reactor with perfect mixing in each phase, although they must be supplemented by physical property relationships and parameter correlations. Solutions for the single-phase versions of these equations occupied most of Chapters 2, 4, and 5, The reader will appreciate that these equations are rarely used in their full form, even for single-phase systems. A full solution for the two-phase case may require an extensive modeling effort, but an approximate solution can provide substantial insight and can motivate a more substantial study.

11.1.2 Measurement of Mass Transfer Coefficients

Fundamental theory is insufficient to predict mass transfer coefficients and liquid-phase holdup. This section describes experimental methods for determining them and gives typical values. A great many correlations are available in the literature, but stirred tanks have many design variations. Liquid-phase properties can have a large effect on the results. Rheology (e.g., non-Newtonian behavior) is important for fermentations and polymerizations. Correlations may be based on small-scale experiments using well-defined conditions, but clean and dirty fluids behave differently due to the effects of interfacial agents on bubble coalescence. Most industrial processes, and fermentations in particular, are dirty almost by definition. Suspended solids retard mass transfer. An additive intended to solve one problem can cause another; e.g., antifoaming agents can reduce mass transfer. Because of this complexity, great care must be taken in using any literature correlation. To understand the complexity, browse through the appropriate sections of *Perry's Handbook* (see "Suggestions for Further Reading," Chapter 5), but rejoice that measurement techniques for $k_l A_i$ and phase holdup are relatively simple if the liquid-phase reaction is not too fast. See also the typical values in Table 11.3.

Transient Techniques for Nonreactive Systems. This technique is used to measure oxygen transfer rates in a stirred tank that is batched with respect to the liquid phase. The results can usually be used for CSTRs since a moderate liquid throughput will have little effect on $k_l A_i$ or holdup. The composition of the liquid phase should approximate that intended for actual operation, but the reaction or oxygen must be suppressed. Holdup is determined from the height of the liquid with and without gas flow. The mass transfer coefficient is determined by a sudden step change in gas composition similar to that in Example 11.3. The usual approach is to sparge air into the system until the liquid phase saturates with respect to oxygen. The air supply is suddenly replaced with nitrogen at the same volumetric flow rate, and a dissolved oxygen meter is used to monitor the oxygen content in the liquid phase. The analysis is similar to that in Example 11.4, but slightly simpler because $(a_g)_{in} = 0$. The phase balances are

$$\frac{da_l}{dt} = \frac{k_l A_i}{[V_l/V]}(a_g/K_H - a_l) \qquad (11.20)$$

$$\frac{da_g}{dt} = -\frac{k_l A_i}{[V_g/V]}(a_g/K_H - a_l) - \frac{Q_g}{V_g}a_g \quad (11.21)$$

These equations contain only one unknown parameter, $k_l A_i$. Assume values for it and solve Equations (11.20) and (11.21) simultaneously. Compare the calculated results with the experimental measurements using nonlinear least-squares analysis as in Equation (7.8). This is the preferred, modern approach, but the precomputer literature relied on computationally simpler methods for fitting $k_l A_i$.

Equations (11.20) and (11.21) are linear, first-order ODEs with coefficients that are assumed constant. The equations can be combined to give a second-order ODE in a_l:

$$\frac{d^2 a_l}{dt^2} + \left(\frac{k_l A_i V}{V_l} + \frac{k_l A_i K_H}{V_g} + \frac{Q_g}{V_g}\right)\frac{da_l}{dt} + \frac{k_l A_i V Q_g}{V_l V_g} a_l = 0 \quad (11.22)$$

This equation can be used to estimate $k_l A_i$ from an experimental $a_l(t)$ curve in at least three ways. They are:

1. *Initial second-derivative method:* At $t = 0$, $da_l/dt = 0$. Therefore,

$$\left[\frac{1}{a_l}\frac{d^2 a_l}{dt^2}\right]_0 = -\frac{k_l A_i V Q_g}{V_l V_g} \quad (11.23)$$

so that $k_l A_i$ can be calculated assuming that the other parameters are known. This method suffers the obvious difficulty of measuring a second derivative.

2. *The inflection point method:* At the inflection point, $d^2 a_l/dt^2 = 0$ and

$$\left[\frac{1}{a_l}\frac{da_l}{dt}\right]_{inflect} = \left[\frac{d \ln a_l}{dt}\right]_{inflect} = \frac{-k_l A_i V Q_g}{k_l A_i V V_g + k_l A_i V V_l K_H + V_l Q_g} \quad (11.24)$$

This method is reasonable in terms of accuracy and allows short experiments.

3. *The asymptotic method:* For most systems, $V_l/(k_l A_i) \gg V_g/Q_g$. Then at long times,

$$a_l(t) = \exp\left[\frac{-k_l A_i V t}{V_l}\right] \quad (11.25)$$

When applicable, this method is the least demanding in terms of experimental accuracy. It is merely necessary to estimate the slope of what should be a straight line when $\ln a_l$ is plotted versus t. By comparison, the inflection point method requires estimating the slope at an earlier time before it is constant.

Measurements Using Liquid-Phase Reactions. Liquid-phase reactions, and the oxidation of sodium sulfite to sodium sulfate in particular, are sometimes used to determine $k_l A_i$. As for the transient method, the system is batch with respect to the liquid phase. Pure oxygen is sparged into the vessel. A pseudo-steady-state results. There is no gas outlet, and the inlet flow rate is adjusted so that the vessel pressure remains constant. Under these circumstances, the inlet flow rate equals the mass transfer rate. Equations (11.5) and (11.12) are combined to give a particularly simple result:

$$Q_g a_g = k_l A_i V(a_l^* - 0) = k_l A_i V a_l^* = k_l A_i V a_g / K_H$$

or

$$k_l A_i = Q_g K_H / V \qquad (11.26)$$

Example 11.9: An article in the literature reports the absorption rate of pure oxygen into a sodium sulfite solution at 20°C using an agitated stirred tank having a liquid depth of 3 ft. A large excess of oxygen was continuously injected into the tank through a sparge ring located just under the agitator. The liquid reaction (sulfite oxidation) was semibatch, but there was sufficient sodium sulfite present so that the dissolved oxygen concentration was approximately zero throughout the experiment. The oxygen consumption was measured using gas flow rates. For a particular set of operating conditions, the result was reported as $K_g A_i = 0.04$ lb·mol/(h·ft³·atm). What was actually measured and what is its value in SI units?

Solution: The experimental conditions are consistent with Equation (11.26) so that $k_l A_i$ was measured. The experimental result was reported as $K_g A_i$ because the overall mass transfer coefficient was based on the equivalent gas-phase driving force expressed in partial pressure units rather than concentration units. Because a pure gas was sparged, $k_g \to \infty$ and $K_l = k_l$. Equation (11.3) relates K_g and K_l through Henry's law constant.

The overall driving force for mass transfer is $\Delta P = P_g - P_l$, where P_l is the concentration of oxygen in the liquid phase expressed as an equivalent partial pressure. For the experimental conditions, $P_l \approx 0$ due to the fast, liquid-phase reaction. The oxygen pressure on the gas side varies due to the liquid head. Assume that the pressure at the top of the tank was 1 atm. Then $P_g = 0.975$ atm since the vapor pressure of water at 20°C should be subtracted. At the bottom of the tank, $P_g = 1.0635$ atm. The logarithmic mean is appropriate: $\Delta P = 1.018$ atm. Thus, the transfer rate was

$$K_g A_i \Delta P = 0.04 \times 1.018 = 0.0407 \text{ lb·mol}/(\text{ft}^3 \cdot \text{h}) = 0.181 \text{mol}/(\text{m}^3 \cdot \text{s})$$

The same rate must be found when the rate is expressed as $k_l A_i a_l^*$, where a_l^* is the solubility of oxygen in water at 20°C and 1.018 atm. Henry's law gives $a_l^* = 1.46 \text{ mol/m}^3$. Thus,

$$k_l A_i = \frac{0.181}{1.46} = 0.124 \text{ s}^{-1}$$

Typical Values. Table 11.3 shows typical parameter values for mechanically agitated tanks and other gas–liquid contacting devices. Not shown are values for $k_g A_i$ since these are usually so large that they have no influence on the mass transfer rate.

Actual Flow Patterns. The assumption of a well-mixed liquid phase is reasonable by the criteria used for single-phase stirred tanks. The same assumption is normally made for the gas phase, but with far less justification. Since the gas phase is dispersed and coalescence is retarded in dirty systems, one might prefer a *segregated flow* model in which gas bubbles circulate as discrete entities. There is a conceptual difference between segregated flow and perfect mixing when the gas contains inerts, as when air is used as an oxygen source, since depletion of the reactive component can be appreciable for bubbles that have remained in the system for a long time. It makes less difference when the reactive gas is pure, although old bubbles will shrink and eventually disappear. There is no practical effect in steady-state operation as long as the liquid phase is well mixed and reaction is confined to the liquid phase. The liquid will be continually contacting the entire population of bubbles, and the measured mass transfer coefficients reflect the average contents of the bubbles with respect to size and gas concentration. Similarly, the fact that the pressure varies as a function of position in the reactor makes no real difference provided that the mass transfer measurements are made on the full-scale vessel. The subtleties of bubble circulation and coalescence would be important if, for example, a second order reaction occurred in the gas phase. *Population balance methods* must be used when the fate of individual bubbles is important. These are briefly discussed in Section 11.5 and are applied to the distribution of residence times in Chapter 15.

TABLE 11.3 Typical Operating Ranges for Gas–Liquid Contacting Devices

Device	Liquid holdup, V_l/V	$k_l A_i$, s^{-1}
Mechanically agitated tanks	0.9	0.02–0.2
Tray columns	0.15	0.01–0.05
Packed columns	0.05	0.005–0.02
Bubble columns	0.95	0.005–0.0

Source: Data from Harnby, N., Edward, M. R., and Nienow, A. W., *Mixing in the Process Industries*, Butterworths, London, 1985.

11.1.3 Fluid–Fluid Contacting in Piston Flow

Table 11.4 lists reactors used for systems with two fluid phases. The gas–liquid case is typical, but most of these reactors can be used for liquid–liquid systems as well. Stirred tanks and packed columns are also used for three-phase systems where the third phase is a catalytic solid. The equipment listed in Table 11.4 is also used for separation processes, but our interest is on reactions and on steady-state, continuous flow.

Stirred tanks are modeled assuming that both phases are well mixed. Tray columns are usually modeled as well mixed on each tray so that the overall column is modeled as a series of two-phase, stirred tanks. (Distillation trays with tray efficiencies greater than 100% have some progressive flow within a tray.) When reaction is confined to a single, well-mixed phase, the flow regime for the other phase makes little difference; but when the reacting phase approximates piston flow, the flow regime in the other phase must be considered. The important cases are where both phases approximate piston flow, either countercurrent or cocurrent.

For simplicity of discussion and notation, we will refer to one phase as being liquid and the other phase as being gas. The gas phase flows upward in the $+z$-direction. The liquid phase may flow upward (cocurrent) or downward. A steady-state but otherwise general component balance gives

$$\pm \frac{d(A_l \bar{u}_l a_l)}{dz} = A_l (\mathscr{R}_A)_l + k_l A_i' (a_l^* - a_l) \tag{11.27}$$

where A_l is the cross-sectional area of the liquid phase, $\bar{u}_l$ is its velocity, and A_i' is the interfacial area per unit height of the column. The plus sign on the derivative

TABLE 11.4 Typical Flow and Mixing Regimes for Gas–Liquid Reactors

Type of reactor	Liquid phase	Gas phase
Stirred tank with sparger	Continuous, well mixed	Discontinuous, but usually assumed well mixed
Rotating-disk and pulsed columns	Continuous, piston flow	Dispersed, piston flow
Bubble columns	Continuous, piston flow	Dispersed, piston flow
Packed columns	Continuous or trickle, piston flow	Continuous, trickle, or dispersed; piston flow
Tray columns	Continuous and well mixed on an individual tray	Discontinuous but often assumed well mixed on an individual tray
Spray towers	Discontinuous, piston flow	Continuous, typically well mixed

in Equation (11.27) is used for cocurrent flow; the minus sign is used for countercurrent flow. Many gas–liquid contactors operate in countercurrent flow. An example is the rotating-disk column shown in Figure 11.5. The deep-shaft fermentor shown in Figure 11.6 is an exception, as is the trickle-bed reactor discussed in Section 11.2. The gas-phase material balance is

$$\frac{d(A_g \bar{u}_g a_g)}{dz} = A_g (\mathscr{R}_A)_g - k_g A_i'(a_g^* - a_g) \qquad (11.28)$$

The mass transfer equations, Equations (11.1)–(11.10), remain valid when A_i' replaces A_i. Equations (11.27) and (11.28) contain one independent variable, z, and two dependent variables, a_l and a_g. There are also two auxiliary variables, the interfacial compositions a_l^* and a_g^*. They can be determined using Equations (11.5) and (11.6) (with A_i' replacing A). The general case regards K_H in Equation (11.4) as a function of composition. When Henry's law applies throughout the composition range, overall coefficients can be used instead of the individual film coefficients. This allows immediate elimination of the interface compositions:

$$\pm \frac{d(A_l \bar{u}_l a_l)}{dz} = A_l (\mathscr{R}_A)_l + K_l A_i'(a_g/K_H - a_l) \qquad (11.29)$$

$$\frac{d(A_g \bar{u}_g a_g)}{dz} = A_g (\mathscr{R}_A)_g - K_l A_i'(a_g/K_H - a_l) \qquad (11.30)$$

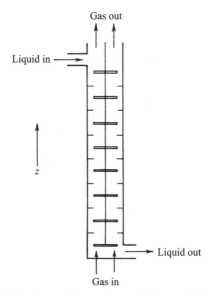

FIGURE 11.5 Rotating-disk column with countercurrent downward flow of a liquid and upward flow of a gas or lower-density liquid.

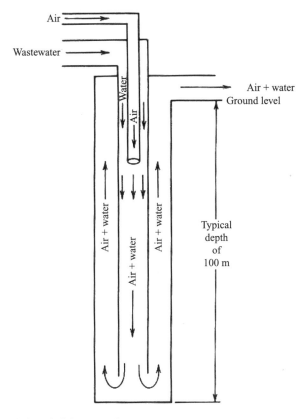

FIGURE 11.6 A deep-shaft fermentor for wastewater treatment.

All the parameters in these equations will be functions of z. Bubbles will grow as they rise in the column because of the lower hydrostatic head, and a_g will decrease even for a pure gas. Bubble coalescence and breakup may be important. Depletion or enrichment of the gas phase because of reaction and mass transfer may be important. The *air-lift* provided to the upward liquid flow may augment or even replace a conventional pump for the liquid phase. These effects are important in a device such as a deep-shaft fermentor. Bubbles breaking the surface will be large due to the pressure change and may be depleted in oxygen so that little or no mass transfer occurs in the upper part of the column. In the lower part, oxygen partial pressures of 2 atm (total pressure of 10 atm) will give a very high driving force for mass transfer. A rigorous analysis of a deep-shaft fermentor or for any two-phase, tubular reactor is a difficult problem in fluid mechanics. Most literature studies have made simplifying assumptions such as constant A_l and A_i. Table 11.5 lists a number of possible simplifications. To some extent, this table should be read in the negative as the assumptions are unlikely to be satisfied in real gas–liquid systems. They

404 CHEMICAL REACTOR DESIGN, OPTIMIZATION, AND SCALEUP

TABLE 11.5 Simplifying Assumptions for Gas–Liquid Reactors

Assumption	Possible rationale
P is constant	Short column or column operated at high pressure or with the liquid phase discontinuous
a_g and a_g^* are constant	P is constant and the reactant gas is pure or is fed at a high rate
$A_l \bar{u}_l \rho_l$ and $A_g \bar{u}_g \rho_g$ are constant	Constant mass flow in each phase; i.e., negligible net mass transfer
A_l and A_g are constant	Consistent with constant pressure and negligible net mass transfer
$k_l A_i$ is constant	Redispersion of gas compensates for coalescence and pressure effects; negligible depletion of the reactive component or else a high level of inerts are present in the gas phase
$\left\lvert \dfrac{dP}{dz} \right\rvert = \rho_l g$	Gas-lift is negligible so that the liquid exerts its normal static head

are more likely to be satisfied in a liquid–liquid system. Examples 11.10 and 11.11 avoid the complications by assuming that all operating parameters are independent of position.

Example 11.10: Determine phase concentrations for a liquid–liquid reaction in a packed-bed reactor. The reactive component is dilute in both phases. It enters the reactor in one phase but undergoes a pseudo-first-order reaction in the other phase. All parameters are constant.

Solution: The phase in which reaction occurs will be denoted by the subscript l, and the other phase will be denoted by the subscript g. Henry's law constant will be replaced by a *liquid–liquid partition coefficient*, but will still be denoted by K_H. Then the system is governed by Equations (11.29) and (11.30) with $(\mathscr{R}_A)_l = -ka_l$ and $(\mathscr{R}_A)_g = 0$. The initial conditions are $(a_l)_{in} = 0$ and $(a_g)_{in} = a_{in}$.

The governing equations are

$$\pm \frac{da_l}{dz} = \frac{-ka_l}{\bar{u}_l} + \frac{K_l A_i'}{A_l \bar{u}_l}(a_g/K_H - a_l)$$

$$\frac{da_g}{dz} = -\frac{K_l A_i'}{A_g \bar{u}_g}(a_g/K_H - a_l)$$

The initial conditions are $a_g = a_{in}$ at $z = 0$, $a_l = 0$ at $z = 0$ for cocurrent flow, and $a_l = 0$ at $z = L$ for countercurrent flow.

The ODEs are linear with constant coefficients. They can be converted to a single, second order ODE, much like Equation (11.22), if an analytical solution is desired. A numerical solution is easier and better illustrates what is necessary for anything but the simplest problem. Convert the independent variable to dimensionless form, $\mathcal{z} = z/L$. Then

$$\pm \frac{da_l}{d\mathcal{z}} = -\frac{kLa_l}{\bar{u}_l} + \frac{K_l A_i' L}{A_l \bar{u}_l}(a_g/K_H - a_l)$$

$$\frac{da_g}{d\mathcal{z}} = -\frac{K_l A_i' L}{A_g \bar{u}_g}(a_g/K_H - a_l)$$

Solutions for the cocurrent and countercurrent cases are displayed in Figure 11.7. The countercurrent case requires calculations of the shooting type where values for $(a_l)_{out} = a_l (z = 0)$ are guessed until the initial condition that $(a_l)_{in} = 0$ is satisfied. Normalized concentrations with $a_{in} = 1$ can be used.

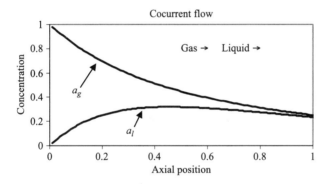

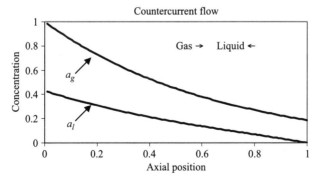

FIGURE 11.7 A pseudo-first-order reaction in one phase with reactant supplied from the other phase. See Example 11.10.

The solution is governed by four dimensionless constants. The values used for Figure 11.7 are

$$\frac{kL}{\bar{u}_l} = 2 \quad \frac{K_l A_i' L}{A_l \bar{u}_l} = 1 \quad \frac{K_l A_i' L}{A_g \bar{u}_g} = 1 \quad K_H = 0.5$$

Example 11.11: Assume the pseudo-first-order reaction in Example 11.10 was $A + B \rightarrow C$ with component B present in great excess. Now suppose that B is confined to the l phase and is present in limited supply, $(b_l)_{in} = b_{in}$. Determine the phase concentrations in the reactor of Example 11.11.

Solution: There are now three ODEs to be solved. They are

$$\pm \frac{da_l}{d\mathcal{z}} = -\frac{kLa_l b_l}{\bar{u}_l} + \frac{K_l A' L_i}{A_l \bar{u}_l}(a_g/K_H - a_l)$$

$$\frac{da_g}{d\mathcal{z}} = -\frac{K_l A_i' L}{A_g \bar{u}_g}(a_g/K_H - a_l)$$

$$\pm \frac{db_l}{d\mathcal{z}} = -\frac{kLa_l b_l}{\bar{u}_l}$$

Numerical solutions using the parameter values of Example 11.10 and $b_{in} = 1$ are shown in Figure 11.8. The countercurrent case now requires guesses for both for $(a_l)_{out} = a_l(z=0)$ and $(b_l)_{out} = b_l(z=0)$ to satisfy the initial conditions that $(a_l)_{in} = 0$ and $(b_l)_{in} = b_{in}$ are satisfied.

11.1.4 Other Mixing Combinations

Piston Flow in Contact with a CSTR. A liquid-phase reaction in a spray tower is conceptually similar to the transpired-wall reactors in Section 3.3. The liquid drops are in piston flow but absorb components from a well-mixed gas phase. The rate of absorption is a function of z as it can be in a transpired-wall reactor. The component balance for the piston flow phase is

$$\frac{d(A_l \bar{u}_l a_l)}{dz} = A_l(\mathcal{R}_A)_l + k_l A_i'(a_l^* - a_l) \qquad (11.31)$$

The component balance for the CSTR phase is

$$0 = (Q_g a_g)_{in} - \int_0^L k_g A_i'(a_g^* - a_g)\, dz + V_g(\mathcal{R}_A)_g - (Q_g)_{out} a_g \qquad (11.32)$$

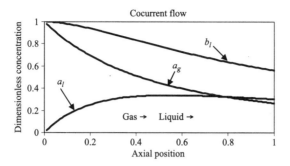

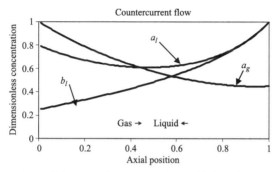

FIGURE 11.8 A + B → C in a two-phase reactor with A fed from the nonreacting phase and B from the reacting phase. See Example 11.11.

There are four unknowns: $a_g = (a_g)_{out}$ which is independent of z; and a_g^*, a_l^*, and a_l, which will generally vary in the z-direction. Equations (11.6) and (11.7) can be used to calculate the interfacial concentrations, a_g^* and a_l^*, if a_g and a_l are known. A numerical solution for the general case begins with a guess for a_g. This allows Equation(11.31) to be integrated so that a_g^*, a_l^*, and a_l are all calculated as functions of z. The results for a_g^* are substituted into Equation (11.32) to check the assumed value for a_g. Analytical solutions are possible for a few special cases.

Example 11.12: Solve Equations (11.31) and (11.32) for the simple case of constant parameters and a pseudo-first-order reaction occurring in the liquid phase of a component supplied from the gas phase. The gas-phase film resistance is negligible. The inlet concentration of the reactive component is a_{in}.

Solution: Note that $K_l = k_l$ when the gas-side resistance is negligible. Then Equation (11.31) simplifies to

$$\frac{da_l}{dz} = -\frac{ka_l}{\bar{u}_l} + \frac{k_l A_i'}{A_l \bar{u}_l}(a_g/K_H - a_l) = \left[\frac{k_l A_i'}{A_l \bar{u}_l K_H}\right]a_g - \left[\frac{k}{\bar{u}_l} + \frac{k_l A_i'}{A_l \bar{u}_l}\right]a_l$$

Integrate this ODE, subject to the initial condition that $a_l=0$ at $z=0$. The result is

$$a_l = \frac{a_g k_l A'_i}{K_H[kA_l + k_l A'_i]}\left[1 - \exp\left(-\frac{[kA_l + k_l A'_i]}{A_l \bar{u}_l}z\right)\right] = a_g C_1[1 - \exp(-C_2 z)]$$

where C_1 and C_2 are constants. This result gives the concentration profile in the liquid phase, but the gas phase concentration is still unknown. The component balance for the gas phase is necessary. Equation (11.32) simplifies to

$$0 = Q_g a_{in} - \int_0^L k_g A'_i(a_g^* - a_g)\, dz - Q_g a_g$$

The mass transfer term in this equation is indeterminate since $k_l \to \infty$ and $a_g^* - a_g \to 0$. The indeterminacy is overcome by using Equation (11.5). Thus,

$$0 = Q_g a_{in} - \int_0^L k_l A'_i(a_g/K_H - a_l)\, dz$$

or

$$a_g = a_{in} - \int_0^L \frac{k_l A'_i}{Q_g}(a_g/K_H - a_l)\, dz$$

$$= a_{in} - a_g \frac{k_l A'_i}{\bar{u}_g A_g}\int_0^L \{1/K_H - C_1[1 - \exp(-C_2 z)]\}\, dz$$

Integrating and solving for a_g gives

$$a_g = \frac{a_{in}}{1 + \dfrac{k_l A'_i}{\bar{u}_g A_g}\{L/K_H - C_1 L + (C_1/C_2)[1 - \exp(-C_2 L)]\}}$$

$$a_g = a_{in} \Bigg/ \left\{1 + \frac{k_l A'_i L}{\bar{u}_g A_g}\left\{1/K_H - \frac{k_l A'_i}{K_H[kA_l + k_l A'_i]} + \frac{A_l \bar{u} k_l A'_i}{K_H[kA_l + k_l A'_i]^2 L}\right.\right.$$
$$\left.\left.\times \left[1 - \exp\left(-\frac{[kA_l + k_l A'_i]}{A_l \bar{u}_l}L\right)\right]\right\}\right\}$$

With some algebra, the parameters used in this expression can all be related to four dimensionless groups used in Example 11.10:

$$\frac{kL}{\bar{u}_l}, \quad \frac{K_l A'_i L}{A_l \bar{u}_l}, \quad \frac{K_l A'_i L}{A_g \bar{u}_g}, \quad K_H$$

Using the same numerical values as in Example 11.10 gives $C_1 = 2/3$, $C_2 L = 3$, $a_g = 0.393 a_{in}$, and $a_l = 0.262[1 - \exp(-3z/L)]$.

Axial Dispersion. Enthusiastic modelers sometimes add axial dispersion terms to their two-phase, piston flow models. The component balances are

$$\pm \frac{d(A_l \bar{u}_l a_l)}{dz} = A_l D_l \frac{d^2 a_l}{dz^2} + A_l (\mathscr{R}_A)_l + k_l A_i'(a_l^* - a_l) \tag{11.33}$$

$$\frac{d(A_g \bar{u}_g a_g)}{dz} = A_g D_g \frac{d^2 a_g}{dz^2} + A_g (\mathscr{R}_A)_g - k_g A_i'(a_g^* - a_g) \tag{11.34}$$

where D_l and D_g are the axial dispersion coefficients for the two phases. In principle, Equations (11.33) and (11.34) can include the entire range of flow in a phase, from perfect mixing (e.g., $D_l \to \infty$) to piston flow (e.g., $D_l \to 0$). In practice, the axial dispersion model is best suited to model small deviations from piston flow. Values for the dispersion coefficients can be measured using the tracer techniques described in Chapter 15. It is usually possible to find tracers that remain in one phase (e.g., nonvolatile liquids or sparingly soluble gases).

11.1.5 Prediction of Mass Transfer Coefficients

As mentioned in Section 11.1.2, fundamental theory is insufficient to predict mass transfer coefficients from first principles. However, existing results do provide a framework for interpreting and sometimes extrapolating experimental results.

Surface Renewal Theory. The *film model* for interphase mass transfer envisions a stagnant film of liquid adjacent to the interface. A similar film may also exist on the gas side. These hypothetical films act like membranes and cause diffusional resistances to mass transfer. The concentration on the gas side of the liquid film is a_l^*; that on the bulk liquid side is a_l; and concentrations within the film are governed by one-dimensional, steady-state diffusion:

$$\mathscr{D}_A \frac{d^2 a}{dx^2} = 0$$

This ODE is subject to the boundary conditions that $a_l = a_l^*$ at $x = 0$ and $a = a_l$ at $x = \delta$. The solution is

$$a(x) = a_l^* + (a_l - a_l^*) \frac{x}{\delta}$$

The flux through the film is given by

$$\text{Flux, mol/(m}^2 \cdot \text{s)} = -\mathscr{D}_A \frac{da}{dx} = \frac{\mathscr{D}_A}{\delta}(a_l^* - a_l) = k_l(a_l^* - a_l) \tag{11.35}$$

or

$$k_l = \frac{\mathscr{D}_A}{\delta} \qquad (11.36)$$

Equation (11.36) gives the central result of film theory; and, as is discussed in any good text on mass transfer, happens to be wrong. Experimental measurements show that k_l is proportional to $\sqrt{\mathscr{D}_A}$ rather than to $\mathscr{D}_A$, at least when the liquid phase is turbulent.

Two rather similar models have been devised to remedy the problems of simple film theory. Both the *penetration theory* of Higbie and the *surface renewal theory* of Danckwerts replace the idea of steady-state diffusion across a film with transient diffusion into a semi-infinite medium. We give here a brief account of surface renewal theory.

Surface renewal theory envisions a continuous exchange of material between the bulk fluid and the interface. Eddy diffusion brings material of uniform composition a_l to the interface and exposes it to the gas phase for a period of time t. The exposed fluid is then replaced with fresh fluid. Diffusion during the exposure period is governed by

$$\frac{\partial a}{\partial t} = \mathscr{D}_A \frac{\partial^2 a}{\partial x^2} \qquad (11.37)$$

This PDE is subject to the initial condition that $a = a_l$ at $t = 0$ and boundary conditions that $a = a_l^*$ at $x = 0$ and $a = a_l$ at $x = \infty$. The solution is differentiated to calculate the flux as in Equation (11.35). Unlike that result, however, the flux into the surface varies with the exposure time t, being high at first but gradually declining as the concentration gradient at $x = 0$ decreases. For short exposure times,

$$\text{Flux} = -\mathscr{D}_A \left[\frac{\partial a}{\partial x}\right]_{x=0} = (a_l^* - a_l)\sqrt{\frac{\mathscr{D}_A}{\pi t}}$$

This result gives the flux for a small portion of the surface that has been exposed for exactly t seconds. Other portions of the surface will have been exposed for different times and thus will have different instantaneous fluxes. To find the average flux, we need the differential distribution of exposure times, $f(t)$. Danckwerts assumed an exponential distribution:

$$f(t) = \frac{1}{\tau}\exp\left(-\frac{t}{\tau}\right)$$

where $f(t)\,dt$ is the fraction of the interfacial area that has been exposed from t to $t + dt$ seconds and τ is the mean exposure time. The average flux is

$$\int_0^\infty \left[\frac{1}{\tau}\exp\left(-\frac{t}{\tau}\right)\right](a_l^* - a_l)\sqrt{\frac{\mathscr{D}_A}{t}}\,dt = (a_l^* - a_l)\sqrt{\frac{\mathscr{D}_A}{\tau}} \qquad (11.38)$$

so that

$$k_l = \sqrt{\frac{\mathcal{D}_A}{\tau}} \qquad (11.39)$$

in agreement with experimental observations for turbulent systems.

Enhancement Factors. This section considers gas absorption with fast, liquid-phase reaction of the component being absorbed. Even a slow reaction in the bulk liquid can give $a_l \to 0$ since the volume of the bulk liquid can be quite large. Thus, the existence of any reaction will increase mass transfer by increasing the overall driving force, but a slow reaction does not change k_l as predicted by Equation (11.39). With a much faster reaction, however, component A will be significantly consumed as it diffuses into the turbulent eddy. The concentration gradient at the interface remains high, increasing the flux, and increasing the value of k_l. The magnitude of the effect can be predicted by adding a reaction term to Equation (11.37):

$$\frac{\partial a}{\partial t} = \mathcal{D}_A \frac{\partial^2 a}{\partial x^2} + \mathcal{R}_A \qquad (11.40)$$

The initial and boundary conditions are identical to those for Equation (11.37): $a = a_l^*$ at $x = 0$ and $a = a_l$ at $x = \infty$. For a first-order reaction, $\mathcal{R}_A = -ka$, and Equation (11.40) has an analytical solution. Using this solution to find the average flux gives

$$k_l = \sqrt{\frac{[1+(k\tau)^2]\mathcal{D}_A}{\tau}} = (k_l)_0 \sqrt{[1+(k\tau)^2]} = (k_l)_0 \mathscr{E} \qquad (11.41)$$

where $\mathscr{E}$ is the *enhancement factor* and $(k_l)_0$ denotes what the mass transfer coefficient would be if there were no reaction; i.e., as given by Equation (11.39). Paired measurements of k_l and $(k_l)_0$ have been used to estimate τ. It is typically on the order of 10^{-2} or 10^{-3} s so that a very fast reaction with rate constant $k > 10$ s^{-1} is needed to measurably enhance the mass transfer coefficient.

Any fast reaction can enhance mass transfer. Consider a very fast, second-order reaction between the gas-phase component A and a liquid component B. The concentration of B will quickly fall to zero in the vicinity of the freshly exposed surface; and a reaction plane, within which $b = 0$, will gradually move away from the surface. If components A and B have similar liquid-phase diffusivities, the enhancement factor is

$$\mathscr{E} = 1 + \frac{b_l}{a_l^*}$$

Since a_l^* is small for sparingly soluble gases, the enhancement factor can be quite large.

11.2 THREE-PHASE REACTORS

Some reactors involve three or even more phases. This section discusses the fairly specific situation of a gas phase, a liquid phase, and a solid phase.

11.2.1 Trickle-Bed Reactors

The solid is stationary, catalytic, and usually microporous. The liquid flows in a trickling regime where it wets the external surface of the catalyst but leaves substantial voidage available for the flow of gas. The usual industrial design is for concurrent, downward flow of both liquid and gas. Such reactors find wide use in the hydrogenation and hydrodesulfurization of heavy petroleum fractions. The goal is to simultaneous contact gas (e.g., hydrogen) and liquid (e.g., a heavy hydrocarbon) on a catalytic surface. The liquid phase supplies one of the reactants and also acts as a transfer medium between the gas phase and the solid phase. The design intent is for the liquid to wet the solid completely since any direct exposure of the solid to the gas phase would not contribute to the reaction.

Piston flow is a reasonable approximation for the liquid and gas phases. The design equations of Section 11.1.3 can be applied by adding an effective, pseudohomogeneous reaction rate for the liquid phase:

$$\pm \frac{d(A_l \bar{u}_l a_l)}{dz} = A_l[(\mathscr{R}_A)_l + \varepsilon \eta (\mathscr{R}_A)_s] + k_l A_i'(a_l^* - a_l) \quad (11.42)$$

$$\frac{d(A_g \bar{u}_g a_g)}{dz} = A_g (\mathscr{R}_A)_g - k_g A_i'(a_g^* - a_g) \quad (11.43)$$

Compare Equation (11.42) with Equation (9.1). The standard model for a two-phase, packed-bed reactor is a PDE that allows for radial dispersion. Most trickle-bed reactors have large diameters and operate adiabatically so that radial gradients do not arise. They are thus governed by ODEs. If a mixing term is required, the axial dispersion model can be used for one or both of the phases. See Equations (11.33) and (11.34).

The pseudohomogeneous reaction term in Equation (11.42) is analogous to that in Equation (9.1). We have explicitly included the effectiveness factor η to emphasis the heterogeneous nature of the catalytic reaction. The discussion in Section 10.5 on the measurement of intrinsic kinetics remains applicable, but it is now necessary to ensure that the liquid phase is saturated with the gas when the measurements are made. The void fraction ε is based on relative areas occupied by the liquid and solid phases. Thus,

$$\varepsilon = \frac{A_l}{A_l + A_s} \quad \text{where} \quad A_g + A_l + A_s = A_c \quad (11.44)$$

The effectiveness factor accounts for the diffusional resistances in the liquid-filled catalyst pores. It does not account for the mass transfer resistance between the liquid and gas phases. This is the job of the k_l and k_g terms.

Homogeneous, liquid-phase reactions may also be important in trickle beds, and a strictly homogeneous term has been included in Equation (11.42) to note this fact. There is usually no reaction in the gas phase. Normally, the gas phase merely supplies or removes the gaseous reactants (e.g., H_2 and H_2S in hydrodesulfurization).

The central difficulty in applying Equations (11.42) and (11.43) is the usual one of estimating parameters. Order-of-magnitude values for the liquid holdup and $k_l A'_i$ are given for packed beds in Table 11.3. Empirical correlations are unusually difficult for trickle beds. Vaporization of the liquid phase is common. From a formal viewpoint, this effect can be accounted for through the mass transfer term in Equation (11.42) and (11.43). In practice, results are specific to a particular chemical system and operating mode. Most models are proprietary.

11.2.2 Gas-Fed Slurry Reactors

These reactors contain suspended solid particles. A discontinuous gas phase is sparged into the reactor. Coal liquefaction is an example where the solid is consumed by the reaction. The three phases are hydrogen, a hydrocarbon-solvent/product mixture, and solid coal. Microbial cells immobilized on a particulate substrate are an example of a three-phase system where the slurried phase is catalytic. The liquid phase is water that contains the organic substrate. The gas phase supplies oxygen and removes carbon dioxide. The solid phase consists of microbial cells grown on the surface of a nonconsumable solid such as activated carbon.

A general model for a gas–liquid–solid reactor would have to consider homogeneous reactions occurring within the various phases and up to three sets of heterogeneous reactions: gas–liquid, gas–solid, and liquid–solid. Such a general treatment adds notational complexity without providing additional insight. When the solid acts only as a catalyst, pseudohomogeneous models can usually be used as in the trickle-bed case. Biochemical reactions are often treated in this manner using rate expressions similar in form to those for gas–solid heterogeneous catalysis. Noncatalytic fluid–solid reactions are more difficult since the *age* of individual particles may be important. This may require the use of *population balance models*. See Example 11.17 and Chapter 15. A slurry reactor with consumable solids can be self-classifying. Particles will stay in the reactor when they are large but will be entrained in the liquid stream as they decrease in size. Careful hydrodynamic design can lead to complete conversion of the solid phase.

11.3 MOVING SOLIDS REACTORS

Fixed-bed reactors are ideal for many solid-catalyzed gas reactions. The contacting of the solid by the gas tends to be quite uniform, and long contact times are possible. However, packed beds have severe heat transfer limitations, and scaleup must often be done using many small-diameter tubes in parallel rather than a single, large-diameter bed. Also, the large particle sizes needed to minimize pressure drop lead to diffusional resistances within the catalyst particles. If catalyst deactivation is rapid, the fixed-bed geometry may cause problems in regeneration. For gas–solid noncatalytic reactions, the solid particles may shrink or grow as the reaction proceeds. This too is not easily accommodated in a fixed bed.

Many types of gas–solid reactors have been designed to allow motion of the solid relative to the fixed walls of the reactors. This motion is desired for one of the following reasons:

1. To enhance heat transfer between the particle and the environment
2. To enable use of small particles
3. To enable continuous regeneration of catalyst particles
4. To facilitate continuous removal of ash and slag
5. To accommodate size changes of the particles concurrent with reaction

The particle motion can be accomplished by purely mechanical means—perhaps aided by gravity—as in rotary cement kilns and fireplace grates. Chemical engineers usually prefer designs where the particle motion is brought about through hydrodynamic forces that are generated by a fluid phase that also participates in the reaction. Such designs tend to be more controllable and scalable, although scalability can be a problem. Sophisticated pilot-plant and modeling efforts are usually necessary for any form of *fluidized-bed reactor*.

Fluidized-bed reactors have received attention from researchers that is disproportionate to their use in industry. The hydrodynamics of fluidization are interesting, and many aspects of fluidization can be studied in small equipment at room temperature and atmospheric pressure. This makes it well suited to academic research. A large number of fluidization regimes have been elucidated, and Figure 11.9 illustrates the more common regimes applicable to the catalyst particles typical of gas-fluidized beds. The particles are spherical with an average diameter of about 50 μm. The size distribution is fairly broad with 95 wt% in the range of 30–100 μm. The particles are microporous with a particle density just over $1 \, g/cm^3$, but with a skeletal density of about $2.5 \, g/cm^3$. They rest on a microporous distributor plate and, at low flow rates, form a packed bed. The pressure drop across the bed increases with increasing flow rate, and the drag on the particles becomes significant. When the pressure drop just equals the weight of the bed, the particles become suspended and mobile enough to be stirred mechanically; but relatively little solids motion is caused by just the gas flow. This is the state of *incipient fluidization* illustrated in Figure 11.9(b). The gas velocity at this point is called the *minimum fluidization velocity*, u_{min}.

As the gas flow is increased beyond u_{min}, the behavior of the bed depends on the density difference between the particles and the suspending fluid. If the

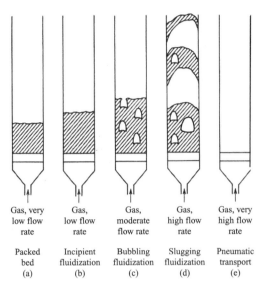

FIGURE 11.9 Fluidization regimes in a batch fluidized bed at low multiples of the minimum fluidization velocity.

density difference is small, as in a liquid-fluidized bed, the bed continues to expand and enters a region known as *particulate fluidization*. Similar behavior is less commonly observed in gas-fluidized beds, where it is known as *delayed bubbling*. More common is a sharp transition to the *bubbling regime* where pockets of gas form at the distributor plate and move upward through the bed. These voids are essentially free of solid particles and behave much like bubbles in a gas–liquid system. The voids grow as they rise in the bed due to pressure reduction, they can coalescence, and they provide a stirring mechanism for the suspended particles. Figure 11.9(c) illustrates the bubbling regime of fluidization that lasts until the superficial gas velocity is many times higher than u_{min}. At this point, the bubbles are so large that they span the reactor and cause slugging. Solids at the center of the bed are conveyed upward but rain down near the walls. Figure 11.9(d) depicts the *slugging* regime. Still higher gas velocities are used in industrial reactors, and the bed behaves in a smoother fashion. A regime called *turbulent fluidization* applies when the upper boundary of the bed is still reasonably well defined. Beyond this are *fast fluidization* and finally the *pneumatic transport* regime where the gas velocity exceeds the terminal velocity for even the larger particles. A batch fluidized bed operating even in the bubbling or faster regimes will elute some of the smaller particles, and it is normal practice to use a cyclone to return these particles to the bed. Internal baffles are sometimes used to promote more uniform contacting of the gas and solid.

11.3.1 Bubbling Fluidization

The dynamics of bubble formation and growth and of solids movement within the bubbling fluidized beds have been analyzed in great detail, and elaborate computer simulations have been developed for all regimes of fluidization. The reader must be referred to the specialized (and sometimes proprietary) literature for details on such models. Here we describe a fairly simple model that is applicable to the bubbling regime and that treats a catalytic fluidized bed much like a gas–liquid reactor. Pseudohomogeneous axial dispersion models are used for higher gas flow rates.

The bubbles play the role of the gas phase. The role of the liquid is played by an *emulsion phase* that consists of solid particles and suspending gas in a configuration similar to that at incipient fluidization. The quasi-phases are in cocurrent flow, with mass transfer between the phases and with a solid-catalyzed reaction occurring only in the emulsion phase. The downward flow of solids that occurs near the walls is not explicitly considered in this simplified model.

For the emulsion phase,

$$A_e u_e \frac{da_e}{dz} = A_e D_e \frac{d^2 a_e}{dz^2} + A_e \mathscr{R}_A + K_m A_i'(a_b - a_e) \quad (11.45)$$

where D_e is an axial dispersion coefficient, the kinetics are pseudohomogeneous with rate equal to that at incipient fluidization, and K_m is an overall mass transfer coefficient. Henry's law constant does not appear since, at equilibrium, the concentrations would be the same in each phase. Axial dispersion in the bubble phase is ignored. Thus,

$$A_b u_b \frac{da_b}{dz} = -K_m A_i'(a_b - a_e) \quad (11.46)$$

These equations are seen to be special cases of Equations (11.33) and (11.34). The exit concentration is averaged over both phases:

$$a_{out} = \frac{A_e u_e (a_e)_{out} + A_b u_b (a_b)_{out}}{A_e u_e + A_b u_b} \quad (11.47)$$

Values for the various parameters in these equations can be estimated from published correlations. See "Suggestions for Further Reading." It turns out, however, that bubbling fluidized beds do not perform particularly well as chemical reactors. At or near incipient fluidization, the reactor approximates piston flow. The small catalyst particles give effectiveness factors near 1, and the pressure drop—equal to the weight of the catalyst—is moderate. However, the catalyst particles are essentially quiescent so that heat transfer to the vessel walls is poor. At higher flow rates, the bubbles promote mixing in the emulsion phase and enhance heat transfer, but at the cost of increased axial dispersion.

The emulsion phase approaches the performance of a CSTR with its inherent lower yield for most reactions. To make matters worse, mass transfer between the emulsion and bubble phases becomes limiting to the point that some of the entering gas completely bypasses the catalytic emulsion phase. The system behaves like the reactor in Example 11.5.

11.3.2 Fast Fluidization

There are relatively few processes that use a fluidized catalyst. Those that do typically operate with gas velocities high enough to completely entrain the particles. The operating regime is called *fast fluidization* or *transport-line fluidization* and is illustrated in Figure 11.9(e). Elutriated particles are continuously collected in a cyclone and recycled back to the inlet of the reactor. The FCC process (fluidized catalytic cracking) used in many oil refineries is the most important example of this type of fluidized reactor. The catalyst deactivates so rapidly by coking that it survives one quick trip through a riser reactor. It is then collected, regenerated, and recycled back to the inlet of the cracking reactor. The regeneration step uses air to burn off the coke.

A well-defined bed of particles does not exist in the fast-fluidization regime. Instead, the particles are distributed more or less uniformly throughout the reactor. The two-phase model does not apply. Typically, the cracking reactor is described with a pseudohomogeneous, axial dispersion model. The maximum contact time in such a reactor is quite limited because of the low catalyst densities and high gas velocities that prevail in a fast-fluidized or transport-line reactor. Thus, the reaction must be fast, or low conversions must be acceptable. Also, the catalyst must be quite robust to minimize particle attrition.

11.3.3 Spouted Beds

A very different regime of fluidization is called *spouting*. Spouting can occur—and is usually undesirable—in a normal fluidized bed if the gas is introduced at localized points rather than being distributed evenly over the reactor cross section. See Figure 11.10(a). The velocity in the spout is high enough to entrain all particles, but they disengage in the low-velocity regions above the bed. This causes circulation of particles with upward movement in the spout but with motion that is generally downward in the bed. Contact times within the spout are quite short, and little reaction occurs there. Thus, the freely spouted bed in Figure 11.10(a) would show relatively low yields for a catalytic reaction. In Figure 11.10(b), however, the gas is forced to turn around and flow cocurrently with the downward-moving particles. The reaction environment in this region is close to that in a fixed-bed reactor, but the overall reactor is substantially better than a fixed-bed reactor in terms of fluid–particle heat transfer and heat transfer to the reactor walls. To a reasonable approximation, the reactor in

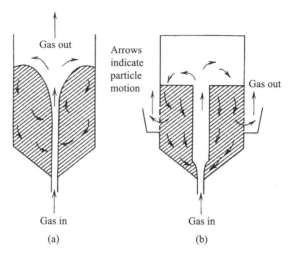

FIGURE 11.10 Spouted-bed reactor with (a) normal gas outlet; (b) side outlet.

Figure 11.10(b) can be modeled as a piston flow reactor with recycle. The fluid mechanics of spouting have been examined in detail so that model variables such as pressure drop, gas recycle rate, and solids circulation rate can be estimated. Spouted-bed reactors use relatively large particles. Particles of 1 mm (1000 μm) are typical, compared with 40–100 μm for most fluidizable catalysts.

The spouting regime of fluidization is used for the fluid–solid noncatalytic reactions, especially drying and combustion.

11.4 NONCATALYTIC FLUID–SOLID REACTIONS

Cases where a solid directly participates in an overall reaction include the burning of solid fuels, the decoking of cracking catalyst, the reduction of iron ore with hydrogen, and the purification of water in an ion-exchange bed. A unifying aspect of all these examples is that the solid participates directly, appears in the reaction stoichiometry, and will ultimately be consumed or exhausted. Often the size and shape of the fluid–solid interface will change as the reaction proceeds. Mass transfer resistances are frequently important, and the magnitude of these resistances may also change with the extent of reaction. The diversity of possible chemistries and physical phenomena is too great to allow comprehensive treatment. We necessarily take a limited view and refer the reader to the research literature on specific processes.

A glib generalization is that the design equations for noncatalytic fluid–solid reactors can be obtained by combining the intrinsic kinetics with the appropriate

TABLE 11.6 Examples of Fluid–Solid Noncatalytic Reactions

Particle geometry largely unaffected by reaction	Particle geometry strongly affected by reaction
Decoking of catalyst pellets	Combustion of coal
Ion-exchange reactions	Reduction of ore
Hydrogen storage in a metal lattice	Production of acetylene from CaC_2
Semiconductor doping	Semiconductor etching

transport equations. The experienced reader knows that this is not always possible, even for the solid-catalyzed reactions considered in Chapter 10, and is much more difficult when the solid participates in the reaction. The solid surface is undergoing change. See Table 11.6. Measurements usually require transient experiments. As a practical matter, the measurements will normally include mass transfer effects and are often made in pilot-scale equipment intended to simulate a full-scale reactor.

Consider a gas–solid reaction of the general form

$$\nu_A A + \nu_S S \rightarrow \text{Products}$$

Any of the following mass transfer resistances can be important:

Film diffusion: With a fast surface reaction on a nonporous particle, mass transfer limitations can arise in the fluid phase.

Pore diffusion: With porous particles, pore diffusion is likely to limit reaction rates at the internal surface.

Product layer diffusion: Many fluid-solid reactions generate ash or oxide layers that impede further reaction.

Sublimation: Some solids sublime before they react in the gas phase. Heat transfer can be the rate-limiting step.

Finally, of course, the surface reaction itself can be rate limiting.

A useful, semiempirical approach to noncatalytic surface reactions is to postulate a rate equation of the form

$$\mathscr{R}_A = \nu_A k a^n A_i \tag{11.48}$$

where $n=1$ or $n=0$ are typical choices. Equation (11.48) does not address the mechanism of the surface reaction but supposes that the rate will be proportional to the exposed area and perhaps to the concentration of A in the gas phase. The rate is specific to a given solid. The constant k and perhaps even n change if the composition of the solid changes.

The case of $n=1$ is a reasonable approximation for a great variety of cases, while $n=0$ covers another common situation where the reaction rate is limited by the disengagement of molecules from the surface. $\mathscr{R}_A$ has its usual interpretation as moles formed per unit volume of reactor per unit time when A_i is the surface area of the fluid–solid interface per unit volume of reactor. For single-particle experiments, A_i will be the surface area and $\mathscr{R}_A$ will be in moles reacted per unit time.

Example 11.13: Explore the suitability of Equation (11.48) for reflecting various forms of mass transfer and kinetic limitations.

Solution: We consider, in turn, each possibility as being rate limiting. Most of the cases use $n = 1$. For film diffusion control,

$$\mathscr{R}_A = k_g A_i (a_s - a)$$

where a_s is the gas-phase concentration of component A just above the reacting solid. When the film resistance is limiting, $a_s \approx 0$; and, with a redefinition of constants, the reaction rate has the form of Equation (11.48).

If pore diffusion is controlling, we repeat the effectiveness factor calculations in Chapter 10. Equation (10.29) has the form of Equation (11.48), and it includes both film resistance and pore diffusion.

Diffusion through a product layer can be treated like a film resistance. The surface concentration a_s is measured inside the ash layer at the unburned surface of the particle. If the ash thickness is constant and $a_s \approx 0$, then the rate has the form of Equation (11.48). The ash thickness will probably increase with time, and this will cause the rate constant applicable to a single particle to gradually decline with time.

Sublimation, dissolution, and ablative processes in general can be fit using $n = 0$ in Equation (11.48). The actual reaction, if there is one, occurs in the fluid phase with kinetics independent of Equation (11.48).

A strict kinetic limitation based on the gas-phase reactant can be modeled using a variable value for n although experience shows that a first order rate expressions with $n = 1$ often provides an excellent fit to experimental data regardless of the underlying reaction mechanism. A site-competition model such as Equation (10.12) can also be used.

The analysis of fluid–solid reactions is easier when the particle geometry is independent of the extent of reaction. Table 11.6 lists some situations where this assumption is reasonable. However, even when the reaction geometry is fixed, moving boundary problems and sharp reaction fronts are the general rule for fluid–solid reactions. The next few examples explore this point.

Example 11.14: Model the movement of the reaction front in an ion-exchange column.

Solution: We suppose that the mass transfer and diffusion steps are fast compared with bulk transport by convection. This is the design intent for ion-exchange columns. The reaction front moves through the bed at a speed dependent only on the supply of fluid-phase reactants. Assuming piston flow in a constant-diameter column, the location of the reaction front is given by

$$z_R(t) = \frac{1}{C_A} \int_0^\theta a_{in}(t)\bar{u}(t)\, dt \qquad (11.49)$$

Here, C_A is the capacity of the ion-exchange resin measured in moles of A per unit volume. The integral in Equation (11.49) measures the amount of material supplied to the reactor since startup. *Breakthrough* occurs no later than $z_R = L$, when all the active sites in the ion-exchange resin are occupied. Breakthrough will occur earlier in a real bed due to axial dispersion in the bed or due to mass transfer or reaction rate limitations.

Example 11.15: Coke formation is a major cause of catalyst deactivation. Decoking is accomplished by periodic oxidations in air. Consider a microporous catalyst that has its internal surface covered with a uniform layer of coke. Suppose that the decoking reaction is stopped short of completion. What is the distribution of residual coke under the following circumstances:

(a) The oxidation is reaction rate limited?
(b) The oxidation is pore diffusion limited?

Solution: For part (a), oxygen has access to the entire internal surface. We expect uniform combustion and a gradual reduction in coke thickness throughout the catalyst pellet. If a completely clean surface is required for catalytic activity, partial decoking will achieve very little.

For part (b), the reaction is fast, and oxygen is consumed as soon as it contacts carbon. Thus, there are two zones in the pellet. The outer zone contains oxygen and no carbon. The inner zone contains carbon at its original thickness and no oxygen. The reaction is confined to a narrow front between the zones. The rate at which the front advances is determined by the rate of diffusion of oxygen and the extent of carbon loading in the pores. It can be modeled using an effective diffusivity model such as Equation (10.33). The diffusional resistance occurs in the carbon-free spherical shell that starts at the external surface and ends at the reaction front. The size of this shell increases with time so that the diffusional resistance increases and the reaction rate decreases. The interior temperature of the catalyst particle may substantially exceed the bulk temperature, but this does not increase the reaction rate because of the diffusion limitation. The high temperature does increase the risk of catalyst deactivation through sintering. Partial oxidations in the diffusion-controlled regime give partial restoration of catalyst activity since some of the surface is completely cleaned. Decoking has been studied extensively because of its importance to the chemical industry. The two cases considered in this example are known as the *uniform* and *shell progressive* models, respectively. For further details see Lee.[1]

Example 11.16: Model the consumption of a solid when the gas-phase reactant is available at constant concentration and the reaction products are gaseous.

Solution: With a constant, Equation (11.48) written for the solid component becomes

$$\mathscr{R}_S = v_S k a^n A_i = v_S k' A_i$$

Take the single particle viewpoint so that A_i is the surface area of the particle and $\mathscr{R}_S$ is in moles per hour. The volumetric consumption rate of solid is

$$\frac{dV_S}{dt} = \frac{v_S k' A_i}{(\rho_{molar})_S} \tag{11.50}$$

The ratio V_S/A_i is the *linear burn rate*. We suppose it to be constant. Thus,

$$\frac{dz_S}{dt} = \frac{v_S k'}{(\rho_{molar})_S} = k'' \tag{11.51}$$

The direction of the burn is normal to the surface at every point that is in contact with the gas. Thus, z_S is measured perpendicularly to the reaction front. It is best measured using a planar solid so that A_i is constant, but it can be inferred from short-term experiments on spherical particles or even from careful multiparticle experiments.

The concept of linear burning rate is not confined to the reaction of a gas with a solid. The fuses on fireworks are designed to burn at a constant linear rate. The flame front on solid rocket fuel progresses at a constant linear rate. Both examples have two reactants (a fuel and an oxidizer) premixed in the solid. Heat transfer limits the burning rate. These materials are merely fast burning. Unlike explosives, they not do propagate a sonic shockwave that initiates further reaction.

When solid particles are subject to noncatalytic reactions, the effects of the reaction on individual particles are derived and then the results are averaged to determine overall properties. The general techniques for this averaging are called *population balance methods*. They are important in mass transfer operations such as crystallization, drop coagulation, and drop breakup. Chapter 15 uses these methods to analyze the distribution of residence times in flow systems. The following example shows how the methods can be applied to a collection of solid particles undergoing a consumptive surface reaction.

Example 11.17: Uniformly sized spheres are fed to a CSTR where they undergo a reaction that consumes the surface at a constant rate of k'', in meters per second. What fraction of the initial population will survive the reactor and what will be the average size upon exiting the reactor?

Solution: The radius of a single particle decreases with time according to the equation

$$R(t) = R_0 - k''t \tag{11.52}$$

provided that $t < R_0/k''$. There is no unique answer for the average size and survival probability since the population statistics depend on the distribution of time spent in the reactor by the various particles. We consider two cases here.

The piston flow case assumes that the particles spend the same time in the reactor, $\bar{t}$, even though the fluid phase is well mixed. This case resembles the mass transfer situation of piston flow in contact with a CSTR as considered in Section 11.1.4. The particles leave the reactor with size $R_0 - k''\bar{t}$. None will survive if $\bar{t} > R_0/k''$. Note that $\bar{t}$ is the mean residence time of the solid particles, not that of the fluid phase.

The other case assumes that the fluid particles are well mixed. Specifically, assume that they have an exponential distribution of residence times so that

$$f(t) = \frac{1}{\bar{t}}\exp\left(-\frac{t}{\bar{t}}\right) \tag{11.53}$$

where $f(t)dt$ is the fraction of the particles remaining in the reactor for a time between t to $t+dt$ seconds. The reasonableness of this assumption for a phase within a CSTR will become apparent in Chapter 15. Again, $\bar{t}$ is the mean residence time for the particles. The fraction of particles that survive the reactor is equal to the fraction that has a residence time less $t_{max} = R_0/k''$:

$$\text{Fraction surviving} = \int_0^{t_{max}} f(t)\,dt \tag{11.54}$$

The average size of the surviving particles is obtained by weighting $R(t)$ by the differential distribution function and integrating over the range of possible times:

$$\bar{R} = \int_0^{t_{max}} R(t)f(t)\,dt \tag{11.55}$$

Equations (11.54) and (11.55) apply to any distribution of particle residence times provided the linear consumption rate is constant. They do not require that the fluid phase is perfectly mixed, only that the consumption rate is strictly controlled by the surface reaction. For the special case of an exponential distribution of residence times per Equation (11.53), some calculus gives

$$\text{Fraction surviving} = 1 - \exp(-t_{max}/\bar{t})$$

And the mean size of these surviving particles is

$$\bar{R} = (R_0 - k''\bar{t})[1 - \exp(-t_{max}/\bar{t})] + k'' t_{max} \exp(-t_{max}/\bar{t})$$

If $t_{max} = R_0/k'' = \bar{t}$, then 63% of the particles survive their sojourn through the reactor, compared with none for the piston flow case. The average size of the surviving particles is $0.37 R_0$.

11.5 REACTION ENGINEERING FOR NANOTECHNOLOGY

Nanotechnology refers to electrical, optical, and mechanical devices, sometimes with biological components, with sizes that range from a few hundred nanometers down to the size of individual molecules. It is a burgeoning field of diverse methodologies. This section highlights a few uses of chemical reactions to fabricate such devices.

11.5.1 Microelectronics

If reinvented today, microelectronics would be called nanoelectronics since sizes have been pushed well below 1 μm. The fabrication of modern electronic devices, such as large-scale integrated circuits, involves an elaborate sequence of chemical operations. A typical process starts with a wafer of high-purity silicon that has been cut from a single crystal. Electronic functionality is achieved by creating a multilayer structure in and on the surface of the wafer in a precise geometric pattern. The pattern is laid down by a process known as *photolithography* using the following sequence of steps:

1. The surface is coated with a polymer, typically by spin coating.
2. An image is formed on the surface using hard-UV or soft x-rays. If the polymer is a *photoresist*, it cross-links in those areas exposed to radiation. If the polymer is a *negative photoresist*, it degrades in those areas exposed to radiation.
3. A solvent removes the polymer that is not cross-linked (or that has been degraded) and thus exposes the underlying surface.
4. The freshly exposed surface is treated with a chemical agent, or *dopant*, to modify its electrical properties (e.g., to create transistors).

MULTIPHASE REACTORS

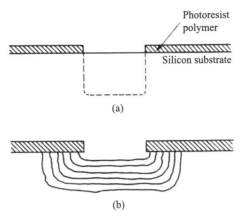

FIGURE 11.11 Etching of silicon wafer to create interconnections: (a) desired profile achievable by ion bombardment; (b) profile obtained by chemical etching.

5. The remaining polymer is removed using a more aggressive solvent.

This procedure, with minor variations, is repeated dozens of times in the manufacture of a semiconductor chip. The chemical treatment can be carried out using reagents in a liquid phase, but gas-phase treatment by a process known as *chemical vapor deposition* (*CVD*) has become more important as individual features in the integrated circuit become smaller.

Electrical connections are required between the various layers that are deposited on a wafer. One method for providing these connections is to pattern the surface using photolithography. Channels are then cut into the silicon substrate using *chemical etching*. A conductor such as copper will subsequently be deposited in the channels. Straight-sided channels that have a width equal to the opening in the photoresist are desired, but experience shows that the channel will undercut the photoresist by an appreciable distance. See Figure 11.11. The reason for this is found in Equation (11.51). The etchant is an aggressive chemical that reacts with the solid. The reaction proceeds in a direction everywhere normal to the existing surface so that the walls of the channel will be attacked at the same rate as the base. Since there are two walls being attacked simultaneously, the width-to-depth ratio for a deep channel is 2. This limitation and the smaller sizes of modern devices is leading to the replacement of chemical etching by more direct techniques such as ion beam etching.

From a reaction engineering viewpoint, semiconductor device fabrication is a sequence of semibatch reactions interspersed with mass transfer steps such as polymer dissolution and *physical vapor deposition* (e.g., vacuum metallizing and sputtering). Similar sequences are used to manufacture still experimental devices known as *NEMS* (for *nanoelectromechanical systems*).

11.5.2 Chemical Vapor Deposition

Chemical vapor deposition is distinguished from physical vapor deposition processes by the use of a chemical reaction, usually a decomposition, to create the chemical species that is deposited. An example important to the microelectronics industry is the formation of polycrystalline silicon by the decomposition of silane:

$$SiH_4 \rightarrow Si + 2H_2$$

The decomposition occurs on the surface and has an observed[2] rate of the form

$$\mathcal{R} = \frac{k[SiH_4]}{1 + k_A[SiH_4] + k_B[H_2]^{1/2}}$$

This form suggests a Hougen and Watson mechanism in which silane and hydrogen atoms occupy sites that must also be used by the silicon being deposited. The primary disposition reaction can be complemented by dopant reactions involving compounds such as AsH_3, PH_3, and B_2H_6, which deposit trace amounts of the dopant metals in the silicon lattice.

A similar reaction,

$$CH_4 \rightarrow C + 2H_2$$

is used in the diamond CVD process. The decomposition is accomplished using electrically heated filaments, microwave plasma discharge, or direct current arc discharge. Polycrystalline diamond is deposited as a thin, hard film.

Organic coatings are also possible. The classic example is the *paralene* process where a cyclic dimer of *p*-xylene is thermally decomposed at about 850°C to form *p*-xylene free radicals that polymerize into a conformal film when deposited on a solid surface. Other examples of polymerization from a deposited vapor have been developed, and advocates believe that this technology will replace spin coating of silicon wafers.

Manufacturing economics require that many devices be fabricated simultaneously in large reactors. Uniformity of treatment from point to point is extremely important, and the possibility of concentration gradients in the gas phase must be considered. For some reactor designs, standard models such as axial dispersion may be suitable for describing mixing in the gas phase. More typically, many vapor deposition reactors have such low L/R ratios that two-dimensional dispersion must be considered. A pseudo-steady model is

$$\bar{u}\frac{\partial a}{\partial z} = D_z \frac{\partial^2 a}{\partial z^2} + D_r \left(\frac{1}{r}\frac{\partial a}{\partial r} + \frac{\partial^2 a}{\partial r^2} \right) + \mathcal{R}_A \qquad (11.56)$$

This model has been applied to vacuum coaters where the material being vapor deposited is evaporated from one or more point sources. Note that D_z and D_r

are empirical parameters that account for both convection and diffusion. Rotary vacuum coaters avoid any dependence in the θ-direction by rotating the substrate as it is coated. Solution techniques for Equation (11.56) were outlined in Section 9.5.

11.5.3 Self-Assembly

The creation of chemical systems that arrange molecules in a specified form is touted as the area of chemical synthesis and reaction engineering that will provide the greatest advances in materials science and engineering. The touts are likely correct. This area is still too young to be treated in a systematic fashion. A general observation is that the continuum mechanical models used in this book and as the basis for essentially all engineering designs do not hold at the molecular scale. Statistical mechanical methods such as molecular dynamics, dissipative particle dynamics, and Ising-type models are being used as research tools. The conversion of these research tools into reliable design methodologies is a remaining challenge for the chemical reaction engineer.

11.6 SCALEUP OF MULTIPHASE REACTORS

The design equations presented in this chapter are independent of scale. The various parameters embedded in them are highly scale-dependent. The parameter estimates are almost entirely empirical, and this means that the reactor designer must depend on literature correlations if anything approaching a priori design is attempted. With few exceptions, the a priori design of a multiphase system is highly uncertain and rarely attempted. We turn instead to the problem of scaling up a multiphase pilot reactor. How can experiments on a small unit generate confidence in a proposed design?

11.6.1 Gas–Liquid Reactors

Small multiphase reactors are needed to estimate the reaction kinetics. The concept of intrinsic kinetics applies to gas–liquid reactors, but the elimination of all mass transfer resistances may not be possible, even at the small scale. Thus, the reaction kinetics may be confounded with mass transfer limitations. These confounded results need to be obtained using a reactor large enough that literature correlations for $k_l A_i$ can be applied. Direct measurement of $k_l A_i$ using one of the methods in Section 11.1.2 is highly desirable as well. There are numerous literature correlations for mass transfer coefficients and gas phase holdups in gas–liquid reactors. For stirred tanks, there are also correlations for agitator power in both the gassed and ungassed states. Compare the pilot-plant results with the values predicted from these correlations. If they agree and if the literature correlations have an experimental base that includes large equipment, a scaleup using the correlations is likely to succeed.

If $k_l A_i$ is known with good accuracy, it may be possible to back out the intrinsic kinetics using the methods of Section 7.1. Knowing the intrinsic kinetics may enable a scaleup where $k_l A_i (a_l^* - a_l)$ is different in the large and small units. However, it is better to adjust conditions in the pilot reactor so that they are identical to those expected in the larger reactor. Good pilot plants have this versatility. The new conditions may give suboptimal performance in the pilot unit but achievable performance in the full-scale reactor.

Mass transfer is generally improved in deep vessels because of the higher partial pressure of the gaseous component being transferred. The price for the improvement is the greater pumping energy needed to sparge the gas at the greater depth. Experiments run in the pilot unit at various liquid levels can be used to test this concept. The agitator speed should be varied over a wide range. If performance improves with increasing depth, scaleup to a large vessel should be reasonable. Another possible test method is to pressurize the vessel to increase the partial pressure of the component being transferred or to enrich the entering gas (e.g., by using, say, 30% oxygen rather than atmospheric oxygen). If performance improves, scaleup is likely to succeed. Alternatively, the tricks just mentioned may be designed into the larger unit as a means of overcoming any scaleup problems.

Does increased agitator speed improve performance in the pilot plant? If so, there is a potential scaleup problem. Installing a variable-speed drive with a somewhat over-sized motor can provide some scaleup insurance, the cost of which is apt to be minor compared with the cost of failure.

Example 11.18: Consider a gas-sparged CSTR with reaction occurring only in the liquid phase. Suppose a pilot-scale reactor gives a satisfactory product. Propose a scaleup to a larger vessel.

Solution: Ideally, the scaleup will maintain the same inlet concentrations for the two phases, the same relative flow rates and holdups for the two phases, and the same ratio of gas transferred to liquid throughput. It is also necessary to maintain a constant residence time in the liquid phase. It is simple to set the flow rates:

$$S = \left[\frac{Q_2}{Q_1}\right]_l = \left[\frac{Q_2}{Q_1}\right]_g \tag{11.57}$$

We would also like the following to be true, but their achievement is less direct than for the flow rates.

$$S = \left[\frac{V_2}{V_1}\right]_l = \left[\frac{V_2}{V_1}\right]_g = \frac{V_2}{V_1} \tag{11.58}$$

$$\left[\frac{Vk_lA_i(a_l^* - a_l)}{Q_l}\right]_2 = \left[\frac{Vk_lA_i(a_l^* - a_l)}{Q_l}\right]_1 \qquad (11.59)$$

We can operate at the required liquid volume—say, by putting the reactor on load cells—but the gas-phase volume and thus the total volume may change upon scaleup. Correlations are needed for the gas-phase holdup and for k_lA_i. A typical correlation for k_lA_i is that by Middleton:[3]

$$k_lA_i = C(P_g/V_l)^{0.7}(\bar{u}_s)_g^{0.6}$$

where C is a constant, P_g is the agitator power in the gassed condition, and $(\bar{u}_s)_g$ is the superficial gas velocity. It was obtained by experiments on only one agitator in one tank, but it is a place to start. Suppose the pilot-scale vessel is scaled using geometric symmetry. Then A_c scales as $S^{2/3}$ and $(\bar{u}_s)_g = Q/A_c$ scales as $S^{1/3}$. This fact imposes a limit on scaleup since scaling by too large a factor could blow the water out of the vessel. Consider $S = 64$ so that $(\bar{u}_s)_g$ increases by a factor of 4 upon scaleup. Can the pilot vessel accept a factor of 4 increase in the sparging rate and what does that do to the holdup? Alternatively, can the gas rate in the small vessel be lowered without too much impact on product quality? Suppose for the moment that experiments at the increased rate indicate no problem and that it causes only a minor increase in holdup. Then a geometrically similar scaleup will satisfy Equation (11.58), and Equation (11.59) will be satisfied if k_lA_i can be held constant upon scaleup. (Actually, a_g^* will increase upon scaleup due to the greater liquid head so k_lA_i can decrease, but this will usually be a small effect.) The Middleton correlation says that k_lA_i will be constant if

$$\left[\frac{(P_g/V_l)_2}{(P_g/V_l)_1}\right]^{0.7}\left[\frac{[(\bar{u}_s)_g]_2}{[(\bar{u}_s)_g]_1}\right]^{0.6} = 1$$

The factor of 4 increase in $(\bar{u}_s)_g$ allows $(P_g/V_l)_2$ to decrease to $0.3(P_g/V_l)_1$. This suggests that the installed horsepower for the full-scale plant would be about a third of that calculated for a conservative scaleup with constant power per volume. This will have a major impact on cost and is too large to ignore. The engineer can do any of the following:

1. Search the literature for correlations for k_lA_i and holdup to see if they predict a similar scaleup.

2. Operate the pilot plant over a wide range of agitator speeds and gas rates to confirm operability and to develop correlations for k_lA_i and holdup applicable to the particular geometry of the pilot reactor.

3. Contact mixing equipment vendors for a recommended scaleup. They have proprietary correlations and extensive experience on similar scaleups. They

will also guarantee success, although their liability will be limited to the value of the equipment and not for consequential damages.

Probably all three things should be done.

11.6.2 Gas–Moving-Solids Reactors

As mentioned in Section 11.3, fluidized-bed reactors are difficult to scale. One approach is to build a *cold-flow model* of the process. This is a unit in which the solids are fluidized to simulate the proposed plant, but at ambient temperature and with plain air as the fluidizing gas. The objective is to determine the gas and solid flow patterns. Experiments using both adsorbed and nonadsorbed tracers can be used in this determination. The nonadsorbed tracer determines the gas-phase residence time using the methods of Chapter 15. The adsorbed tracer also measures time spent on the solid surface, from which the *contact time distribution* can be estimated. See Section 15.4.2.

PROBLEMS

11.1. Henry's law constant, K_H, for carbon dioxide in water at 30°C is 11.1 10^7 mm of Hg per mole fraction. What is the dimensionless value for K_H so that a_l and a_g have the same units?

11.2. Complete Example 11.4 for the case where $V_g/Q_g = 20$ s. Solve the governing ODEs analytically or numerically as you prefer. How does this more rigorous approach change the 95% response time calculated in Example 11.3?

11.3. A reactive gas is slowly bubbled into a column of liquid. The bubbles are small, approximately spherical, and are well separated from each other. Assume Stokes' law and ignore the change in gas density due to elevation. The gas is pure and reacts in the liquid phase with first-order kinetics. Derive an expression for the size of the bubbles as a function of height in the column. Carefully specify any additional assumptions you need to make.

11.4. Example 11.5 treats a reaction that is catalyzed by a stagnant liquid phase. Find the outlet concentration of component A for the limiting case of high catalytic activity, $k \to \infty$. Repeat for the limiting case of high mass transfer, $k_l A_i \to \infty$.

11.5. Example 11.6 ignored the accumulation term for $a_l(t)$ in Equation (11.11). How does the result for a_l change if this term is retained? Consider only the asymptotic result as $t \to \infty$.

11.6. Confirm that the solutions in Examples 11.10 and 11.11 satisfy an overall material balance.

11.7. Find an analytical solution for the pair of ODEs in Example 11.10 for (a) the (easy) cocurrent case; (b) the countercurrent case.

11.8. The low-pressure chemical vapor deposition of silicon nitride on silicon involves two gaseous reactants: dichlorosilane and ammonia. The following reactions are believed to be important under typical conditions of $P = 1$ torr and $T = 1000\text{--}1200$ K:

$$\text{SiH}_2\text{Cl}_2 + \tfrac{4}{3}\text{NH}_3 \longrightarrow \tfrac{1}{3}\text{Si}_3\text{N}_4 + 2\text{HCl} + 2\text{H}_2 \tag{I}$$

$$\text{SiH}_2\text{Cl}_2 \rightleftarrows \text{SiCl}_2 + \text{H}_2 \tag{II}$$

$$\text{SiCl}_2 + \tfrac{4}{3}\text{NH}_3 \longrightarrow \tfrac{1}{3}\text{Si}_3\text{N}_4 + 2\text{HCl} + \text{H}_2 \tag{III}$$

$$2\text{NH}_3 \longrightarrow \text{N}_2 + 3\text{H}_2 \tag{IV}$$

Suppose the reactant gases are supplied continuously in large excess and flow past a single wafer of silicon. By performing multiple experiments, the growth of the nitride layer can be determined as a function of time and reactant concentrations. Develop an experimental program to determine rate expressions for Reactions (I) through (IV). Note that Reactions (I), (III), and (IV) are heterogeneous while Reaction (II) occurs in the gas phase. It is possible to include N_2, H_2, and HCl in the feed if this is useful. Exiting gas concentrations cannot be measured.

11.9. The shell progressive model in Example 11.15, part (b) envisions a mass transfer limitation. Is the limitation more likely to be based on oxygen diffusing in or on the combustion products diffusing out?

11.10. Determine the position of the reaction front in the diffusion-limited decoking of a spherical cracking catalyst.

Hint: Use a version of Equation (11.49) but correct for the spherical geometry and replace the convective flux with a diffusive flux.

11.11. Example 11.14 assumed piston flow when treating the moving-front phenomenon in an ion-exchange column. Expand the solution to include an axial dispersion term. How should breakthrough be defined in this case?

11.12. The transition from Equation (11.50) to Equation (11.51) seems to require the step that $dV_S/A_i = d[V_S/A_i] = dz_S$. This is not correct in general. Is the validity of Equation (11.51) limited to situations where A_i is actually constant?

11.13. An overly simplified model of fluidized-bed combustion treats the solid fuel as spherical particles freely suspended in upward-flowing gas. Suppose the particles react with zero-order kinetics and that there is no ash or oxide formation. It is desired that the particles be completely consumed by position $z = L$. This can be done in a column of constant diameter or in a column where the diameter increases or decreases with increasing height. Which approach is better with respect to minimizing the reactor volume? Develop a model that predicts the position of the particle as a function of time spent in the reactor. Ignore particle-to-particle interactions.

11.14. Suppose the pilot-scale stirred tank of Example 11.18 is at the ragged edge of acceptable operation so that $(\bar{u}_s)_g$ cannot be increased upon scaleup. Neither can Q_g/Q_l be decreased. What can be done to avoid a scaleup limitation? Your proposed solution should utilize the existing pilot reactor for experimental confirmation.

REFERENCES

1. Lee, H. H., *Homogeneous Reactor Design*, Butterworth, Boston, 1985.
2. Claassen, W. A. P., et. al., "The deposition of silicon from silane in a low-pressure hot-wall system," *J. Crystal Growth*, **57**, 259–266 (1982).
3. Middleton, J. C., "Gas–liquid dispersion and mixing," *Mixing in the Process Industries*, Harnby, N., Edwards, M. F., and A. W. Nienow, Eds., Butterworths, London, 1985, 322–355.

SUGGESTIONS FOR FURTHER READING

A general reference for stirred tank fans is

Tatterson, G. L., *Fluid Mixing and Gas Dispersion in Agitated Tanks*, McGraw-Hill, New York, 1991.

A review of most of the catalytic reactors discussed in this chapter can be found in

Dudukovic, M. P., "Trends in catalytic reaction engineering," *Catal. Today*, **48**, 5–15 (1999).

A comprehensive review of the complex hydrodynamic issues associated with three-phase reactors is given in

Mewes, D., Loser, T., Millies, M., "Modeling of two-phase flow in packings and monoliths," *Chem. Eng. Sci.*, **54**, 4729–4747 (1999).

The following review articles emphasize those forms of fluidization used in industrial reactors:

Bi, H. T., Ellis, N., Abba, I. A., and Grace, J. R., "A state-of-the-art review of gas–solid turbulent fluidization," *Chem. Eng. Sci.*, **55**, 4789–4825 (2000).

Marmo, L., Rovero, G., and Baldi, G., "Modeling of catalytic gas–solid fluidized bed reactors," *Catal. Today*, **52**, 235–247 (1999).

For a comprehensive treatment of spouted-bed reactors consult

Arkun, Y., Littman, H., and Morgan, M. H., III, "Modeling of spouted bed chemical reactors," *Encyclopedia of Fluid Mechanics,* Vol. 4, Solids and Gas–Solids, N. P. Cheremisinoff, Ed., 1089–1025, Gulf, Houston, TX, 1986.

An older but still useful summary of chemical reaction engineering in microelectronics is

Jensen, K. F., "Micro-reaction engineering applications of reaction engineering to processing of electronic and photonic materials," *Chem. Eng. Sci.*, **42**, 923–958 (1987).

The following book is both more modern and more specialized:

Hitchman, M. L. and Jensen, K. F., *Chemical Vapor Deposition: Principles and Applications*, Academic Press, New York, 1997.

Most books on self-assembly are collections of research papers. One that manages to get all the buzz words into the title is

Sauvage, J. P. and Hosseine, M. W., Eds., *Comprehensive Supramolecular Chemistry, Volume 9: Templating, Self-Assembly, and Self-Organization,* Pergamon, Oxford, 1999.

Literature references and the measurement of the contact time distribution in a large, cold-flow model of a gas-fluidized bed are reported in

Pustelnik, P. and Nauman, E. B., "Contact time distributions in a large fluidized bed," *AIChE J.,* **37**, 1589–1592 (1991).

CHAPTER 12
BIOCHEMICAL REACTION ENGINEERING

Biochemical engineering is a vibrant branch of chemical engineering with a significant current presence and even greater promise for the future. In terms of development, it can be compared with the petrochemical industry in the 1920 s. Despite its major potential, biochemical engineering has not yet been integrated into the standard undergraduate curriculum for chemical engineers. This means that most graduates lack an adequate background in biochemistry and molecular biology. This brief chapter will not remedy the deficiency. Instead, it introduces those aspects of biochemical reactor design that can be understood without detailed knowledge of the underlying science. A chemical engineer can make contributions to the field without becoming a biochemist or molecular biologist, just as chemical engineers with sometimes only rudimentary knowledge of organic chemistry made contributions to the petrochemical industry.

Proteins are key ingredients to life and to biochemistry. They are linear polymers of amino acids. The general formula for an amino acid is

$$H_2N-\underset{R}{\overset{COOH}{\underset{|}{C}}}-H$$

where R represents one of twenty different radicals found in nature. The amino and carboxy groups condense and eliminate water to form proteins. When proteins are formed by a living cell, the sequence of amino acids is dictated by the DNA within the cell. The term genetic engineering refers to manipulation of DNA to alter the recipe. Despite the name, genetic engineering is not an engineering discipline, but is a branch of molecular biology similar in spirit to organic synthesis. This chapter is not concerned with genetic engineering as such. Biochemical engineering (or sometimes agricultural engineering) comes later when genetically engineered organisms are to be grown in mass. Many current applications of biochemical engineering are based on naturally occurring organisms and biocatalysts. The products range from small, simple molecules such as methane and ethanol, to moderately complex compounds such as penicillin, to therapeutic proteins such as human growth factor, to whole cells such as yeast, and potentially to multicell aggregates such as skin. Some of these compounds—e.g., ethanol and penicillin—can be produced by traditional organic synthesis. Thus, a working distinction between biochemical reaction engineering and "ordinary" reaction engineering is the involvement of biocatalysts, specifically proteins having catalytic activity and known as *enzymes*.

12.1 ENZYME CATALYSIS

Enzymes are proteins that catalyze reactions. Thousands of enzymes have been classified and there is no clear limit as to the number that exists in nature or that can be created artificially. Enzymes have one or more catalytic sites that are similar in principle to the active sites on a solid catalyst that are discussed in Chapter 10, but there are major differences in the nature of the sites and in the nature of the reactions they catalyze. Mass transport to the active site of an enzyme is usually done in the liquid phase. Reaction rates in moles per volume per time are several orders of magnitude lower than rates typical of solid-catalyzed gas reactions. Optimal temperatures for enzymatic reactions span the range typical of living organisms, from about 4°C for cold-water fish, to about 40°C for birds and mammals, to over 100°C for thermophilic bacteria. Enzymatic reactions require very specific molecular orientations before they can proceed. As compensation for the lower reaction rates, enzymatic reactions are highly selective. They often require specific stereoisomers as the reactant (termed the *substrate* in the jargon of biochemistry) and can generate stereospecific products. Enzymes are subject to inhibition and deactivation like other forms of catalysis.

12.1.1 Michaelis-Menten and Similar Kinetics

Suppose the reaction S → P occurs using an enzyme as a catalyst. The following reaction mechanism is postulated:

$$S + E \rightleftarrows SE \qquad \frac{[SE]}{s[E]} = K$$

$$SE \longrightarrow P + E \qquad \mathscr{R} = k[SE]$$

where s denotes the substrate concentration, E denotes the active site, and SE denotes the adsorbed complex. This mechanism is somewhat different than that used for gas–solid catalysis since there is no explicit desorption step. In essence, product desorption is assumed to be instantaneous. The site balance is

$$[SE] + [E] = E_0$$

Substituting for [SE] and [E] gives

$$\frac{\mathscr{R}}{k} + \frac{\mathscr{R}}{ksK} = E_0$$

and

$$\mathscr{R} = \frac{E_0 k s}{1/K + s} = \frac{k's}{1 + k_S s} \quad (12.1)$$

which is the functional form expected when there is competition for active sites. Just as for gas–solid reactions, the reaction rate for a first-order reaction depends linearly on the amount of catalyst and hyperbolically on the reactant concentration. See Figure 12.1(a). Biochemists usually express Equation (12.1) as

$$\mathscr{R} = \frac{E_0 k s}{K_M + s} = \frac{\mathscr{R}_{max} s}{K_M + s} \quad (12.2)$$

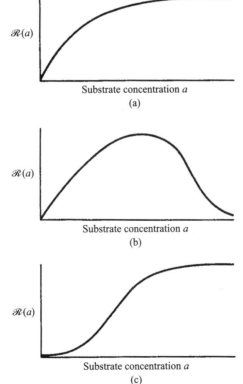

FIGURE 12.1 Effects of substrate (reactant) concentration on the rate of enzymatic reactions: (a) simple Michaelis-Menten kinetics; (b) substrate inhibition; (c) substrate activation.

where K_M is called the Michaelis constant. Either of Equations (12.1) and (12.2) contain two adjustable constants that must be found by fitting experimental data. Their form is equivalent to Hougen and Watson (or Langmuir-Hinshelwood) kinetics for the gas–solid reaction S → P with negligible adsorption, but the Michaelis-Menten equation was derived first.

Simple Michaelis-Menten kinetics exhibit the saturation behavior shown in Figure 12.1(a). Enzyme systems can also exhibit the more complex behavior shown in Figure 12.1(b) and 12.1(c). Figure 12.1(b) illustrates *substrate inhibition*, where high reactant concentrations lead to a decrease in reaction rate. The sigmoidal rate curve shown in Figure 12.1(c) illustrates *substrate activation*. These phenomena can be modeled with variants of Michaelis-Menten kinetics that involve two or more substrate molecules being adsorbed at a single site or enzyme molecules that have two or more interacting sites.

Example 12.1: Suppose an enzymatic reaction has the following mechanism:

$$S + E \rightleftharpoons SE \qquad \frac{[SE]}{s[E]} = K$$

$$S + SE \rightleftharpoons S^2E \qquad \frac{[S^2E]}{s[SE]} = K_2$$

$$SE \longrightarrow P + E \qquad \mathscr{R} = k[SE]$$

Determine the functional form of the rate equation.

Solution: The total concentration of active sites is

$$[S^2E] + [SE] + [E] = E_0$$

The two equilibrium relations and the rate expression allow the unknown surface concentrations $[S^2E]$, $[SE]$, and $[E]$ to be eliminated. The result is

$$\mathscr{R} = \frac{E_0 k s}{1/K + s + K_2 s^2} \qquad (12.3)$$

This equation gives $\mathscr{R}(0) = 0$, a maximum at $s = \sqrt{K_M/K_2}$, and $\mathscr{R}(\infty) = 0$. The assumed mechanism involves a first-order surface reaction with inhibition of the reaction if a second substrate molecule is adsorbed. A similar functional form for $\mathscr{R}(s)$ can be obtained by assuming a second-order, dual-site model. As in the case of gas–solid heterogeneous catalysis, it is not possible to verify reaction mechanisms simply by steady-state rate measurements.

Example 12.2: Suppose the reaction mechanism is

$$S + E \rightleftarrows SE \qquad \frac{[SE]}{s[E]} = K$$

$$S + E \rightleftarrows S^2E \qquad \frac{[S^2E]}{s[E]} = K_2$$

$$SE \longrightarrow P + E \qquad \mathscr{R}_I = k[SE]$$

$$S^2E \longrightarrow SE + P + E \qquad \mathscr{R}_{II} = k_2[S^2E]$$

Determine the functional form for $\mathscr{R}(s)$.

Solution: This case is different than any previously considered site models since product can be formed by two distinct reactions.
The overall rate is

$$\mathscr{R}_P = \mathscr{R}_I + \mathscr{R}_{II} = k[SE] + k_2[S^2E]$$

The site balance is the same as in Example 12.1. Eliminating the unknown surface concentrations gives

$$\mathscr{R}_P = \frac{E_0(ks + k_2K_2s^2)}{1/K + s + K_2s^2} \tag{12.4}$$

As written, this rate equation exhibits neither inhibition nor activation. However, the substrate inhibition of Example 12.1 occurs if $k_2 = 0$, and substrate activation occurs if $k = 0$.

Second-order enzymatic reactions require two adsorption events at the same site. For the reaction $A + B \rightarrow P$, there may be a compulsory order of adsorption (e.g., first A, then B) or the two reactants may adsorb in a random order. Different assumptions lead to slightly different kinetic expressions, but a general form with theoretical underpinnings is

$$\mathscr{R} = \frac{k'ab}{1 + k_A a + k_B + k_{AB}ab} \tag{12.5}$$

In summary, the simple Michaelis-Menten form of Equation (12.1) is usually sufficient for first-order reactions. It has two adjustable constants. Equation (12.4) is available for special cases where the reaction rate has an interior maximum or an inflection point. It has three adjustable constants after setting either $k_2 = 0$ (inhibition) or $k = 0$ (activation). These forms are consistent with two adsorptions of the reactant species. They each require three constants. The general form of Equation (12.4) has four constants, which is a little excessive for a first-order reaction. The general form for a reaction of the A + B → P type, Equation (12.5), also requires four constants, although it is possible that one or two of them will be zero.

The temperature dependence of enzymatic reactions is modeled with an Arrhenius form for the main rate constant k'. The practical range of operating temperatures is usually small, but the activation energies can be quite large. Temperature dependence of the inhibition constants can usually be ignored.

12.1.2 Inhibition, Activation, and Deactivation

Reactant molecules cause the substrate inhibition and activation discussed in Section 12.1.1. These effects and deactivation can also be caused by other molecules and by changes in environmental conditions.

Reversible inhibition ceases when the inhibiting molecule is removed from the system. The molecules can be eliminated from the feed in a flow system or from a batch reaction by a separation process such as dialysis. Two kinds of reversible inhibition are distinguished. *Competitive inhibition* occurs when an inhibitor molecule occupies a site before it is occupied by a substrate molecule. The assumed mechanism is

$$I + E \rightleftharpoons IE$$

Noncompetitive inhibition occurs when the inhibiting molecule is adsorbed after the substrate molecule has been absorbed. The assumed mechanism is

$$I + SE \rightleftharpoons ISE$$

The two forms of inhibition can occur together. Their combined effects are modeled by changing the denominator of the rate equation. For an irreversible, first-order reaction, a suitable rate equation is

$$\mathscr{R} = \frac{k's}{1 + k_S s + k_I i + k_{SI} si} \tag{12.6}$$

where $k_{SI} = 0$ for purely competitive inhibition and $k_I = 0$ for purely noncompetitive inhibition.

Some enzymes require *cofactors* to activate catalysis. Typical cofactors are metal atoms, ammonia, and small organic molecules that associate with the enzyme and help to structure the catalytic site. To conduct an enzymatic reaction, the necessary cofactors must be supplied along with the substrate and the enzyme. In cell metabolism, a variety of these cofactors act in conjunction with inhibitors to control the metabolic rate.

Deactivation generally refers to a change in the physical structure of the enzyme, often caused by an increase in temperature. Some of the amino acids in a protein chain are hydrophobic. Others are hydrophilic. Proteins in solution *fold* into elaborate but characteristic shapes to increase like-to-like interactions within the polymer and between the polymer and the solvent. The folded state is the *native or natural state*, and it is the state in which enzymes have their catalytic activity. At high temperatures, random thermal forces disrupt the folded chain and destroy the catalytic sites. Very high temperatures will cause coagulation or other structural and chemical changes. This leads to *irreversible deactivation*, and the proteins are *denatured*. Mammalian enzymes typically deactivate at temperatures above 45°C, but some enzymes continue to function at 100°C. Lower temperatures can cause unfolding without permanent damage to the enzyme. *Reversible deactivation* caused by relatively low temperatures or by short times at high temperatures can be reversed just by lowering the temperature.

Deactivation of both the reversible and irreversible varieties can have chemical causes such as pH changes. If the pH change is small, it can sometimes be modeled using an inhibition term in the Michaelis-Menten equation. Larger changes in pH can reversibly alter the configuration of the enzyme, and still larger changes may destroy it. Enzymes can be deactivated by other enzymes or by chemical agents such as ozone and chlorine.

12.1.3 Immobilized Enzymes

Some enzymes are cheap enough to be used in one-time applications such as washing clothes or removing blood stains from leather. Others are so expensive that they must be recycled or otherwise reused. One method to achieve this is put an ultrafilter, operating in the cross-flow mode, on the discharge stream of a continuous process. See Figure 12.2. Small molecules pass through the filter, but large molecules such as the enzyme are retained and recycled. The enzymes can also be entrapped within a porous solid, e.g., within the large-pore side of an asymmetric membrane. These are examples of *immobilization by physical entrapment*.

Immobilization by adsorption onto a surface such as activated carbon or to an ion-exchange resin gives a reversible and relatively weak bond, but this can be sufficient to increase the retention time in a flow system to acceptable levels. Recall Section 10.6 where it is shown that the residence time of an adsorbed species can be much larger than that of the mobile phase, in essence giving more time for catalysis.

Immobilization by chemical bonding gives strong, irreversible attachments to a solid support. The bonds are normally covalent but they can be electrostatic. Typical supports are functionalized glass and ceramic beads and fibers. Enzymes are sometimes cross-linked to form a gel. Occasionally, enzymes can be flocculated while retaining catalytic activity.

All these immobilization techniques run the risk of altering activity compared with the native enzyme. Improved activity is occasionally reported, but this is the exception. The immobilization techniques listed above are in approximate order of loss in activity. Physical entrapment normally causes no change. Adsorption will distort the shape of the molecule compared with the native

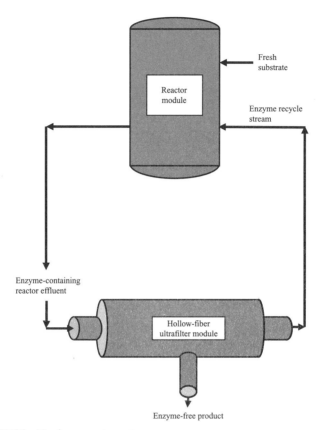

FIGURE 12.2 Membrane reactor system.

state. The effect of covalent bonding depends on the location of the bond relative to an active site. If remote from the site, it may have no effect. The chemical nature of the support can affect activity. Cross-linking requires two covalent attachments per enzyme molecule, and is thus likely to distort the shape of the enzyme to the point that catalytic activity is lost. Such distortions are even more likely, but not inevitable, for coagulated or flocculated enzymes. On the positive side, immobilization tends to stabilize enzymes against deactivation.

Immobilization can give rise to mass transfer limitations that do not occur for freely suspended enzymes in their native state. As a formality, these limitations can be incorporated into an effectiveness factor:

$$\eta = \frac{\text{Observed reaction rate}}{\text{Rate when enzyme is in its native state}} \qquad (12.7)$$

This definition recognizes that immobilization—e.g., at cellular membranes—is the native state for some enzymes. Although interesting mathematics are possible, effectiveness factors are measured experimentally, as is the case for the solid-catalyzed gas-phase reactions discussed in Chapter 10. Effectiveness factors greater than 1 are possible. For gas–solid reactions, these were caused by temperature gradients in the catalyst particle. This is rarely the case for an enzyme-catalyzed reaction since the rate of heat release is low. Occasionally, however, the spatial or chemical configuration of the site is enhanced by the artificial environment of the support.

12.1.4 Reactor Design for Enzyme Catalysis

When the product from a biochemical reactor is intended for use as a food or drug, the design process is subject to a set of government-mandated checks and balances to assure safe and effective finished products. The process must conform to a methodology known as Current Good Manufacturing Practice or CGMP (usually shortened to GMP). Subject to this requirement for special care and documentation, the design of a biochemical reactor is conceptually similar to that for ordinary chemical reactors.

Confined Enzymes in Steady-State Flow Reactors. The confinement can be accomplished using a membrane reactor as shown in Figure 12.2 or by a packed-bed reactor in which the enzymes are immobilized within the packing or by using a solid–liquid fluidized bed in which the enzymes are immobilized within the fluidized particles. All these geometries have a common feature that makes the reactor analysis relatively simple: no enzymes enter or leave the system during steady-state operation. Due to the high cost of enzymes, such reactors are a desirable way of performing enzyme catalysis.

The easiest reactor to analyze is a steady-state CSTR. Biochemists call it a *chemostat* because the chemistry within a CSTR is maintained in a static condition. Biochemists use the *dilution rate* to characterize the flow through a CSTR. The dilution rate is the reciprocal of the mean residence time.

Example 12.3: Suppose S → P according to first-order, Michaelis-Menten kinetics. Find s_{out} for a CSTR.

Solution: Most enzyme reactors use such high concentrations of water that the fluid density is constant. Applying Michaelis-Menten kinetics to the component balance for a steady-state CSTR gives

$$s_{in} - \frac{\bar{t}E_0 k s_{out}}{K_M + s_{out}} = s_{out}$$

Solving for s_{out} gives

$$s_{out} = 1/2\left[s_{in} - K_m - E_0 k\bar{t} - \sqrt{(s_{in} - K_M - E_0 k\bar{t})^2 + 4s_{in}K_M} \right]$$

The reverse solution, finding the value of $\bar{t}$ needed to achieve a desired value for s_{out}, is easier. Equation (1.54) gives the reverse solution for the general case where the reaction rate depends on s_{out} alone and density is constant. Applying Equation (1.54) to the present case gives

$$\bar{t} = \frac{(s_{in} - s_{out})(K_M + s_{out})}{E_0 k s_{out}}$$

Most biochemical reactors operate with dilute reactants so that they are nearly isothermal. This means that the packed-bed model of Section 9.1 is equivalent to piston flow. The axial dispersion model of Section 9.3 can be applied, but the correction to piston flow is usually small and requires a numerical solution if Michaelis-Menten kinetics are assumed.

Example 12.4: Suppose S → P according to first-order, Michaelis-Menten kinetics. Find s_{out} for a PFR.

Solution: The design equation for a PFR gives

$$\bar{u}\frac{ds}{dz} = \frac{-E_0 k s}{K_M + s}$$

subject to the initial condition that $s = s_{in}$ at $z = 0$. The solution is implicit in s_{out}:

$$s_{out} \exp\left[\frac{s_{out} - s_{in}}{K_M}\right] = s_{in} \exp\left[\frac{-E_0 k \bar{t}}{K_M}\right]$$

The corresponding solution for $\bar{t}$ is explicit:

$$\bar{t} = \frac{s_{in} - s_{out} + K_M \ln(s_{in}/s_{out})}{E_0 k}$$

Enzyme-Catalyzed Batch Reactions. Michaelis-Menten theory assumes equilibrium between occupied and unoccupied sites:

$$S + E \underset{k_r}{\overset{k_f}{\rightleftarrows}} SE \qquad \frac{[SE]}{s[E]} = \frac{k_f}{k_r} = K$$

If the enzyme charged to a batch reactor is pristine, some time will be required before equilibrium is reached. This time is usually short compared with the batch reaction time and can be ignored. Furthermore, $s_0 \gg E_0$ is usually true so that the depletion of substrate to establish the equilibrium is negligible. This means that Michaelis-Menten kinetics can be applied throughout the reaction cycle, and that the kinetic behavior of a batch reactor will be similar to that of a packed-bed PFR, as illustrated in Example 12.4. Simply replace $\bar{t}$ with t_{batch} to obtain the approximate result for a batch reactor.

This approximation is an example of the quasi-steady hypothesis discussed in Section 2.5.3.

Example 12.5: Formulate the governing equations for an enzyme-catalyzed batch reaction of the form S → P. The enzyme is pristine when charged to the reactor. Do not invoke the quasi-steady hypothesis.

Solution: The reactions are

$$S + E \rightarrow SE \qquad \mathscr{R}_I = k_f\left[s[E] - \frac{[SE]}{K}\right]$$

$$SE \rightarrow E + P \qquad \mathscr{R}_{II} = k[SE]$$

The corresponding ODEs for a constant-volume batch reaction are

$$\frac{ds}{dt} = -k_f\left[s[E] - \frac{[SE]}{K}\right] \qquad s = s_0 \text{ at } t = 0$$

$$\frac{d[E]}{dt} = -k_f\left[s[E] - \frac{[SE]}{K}\right] + k[SE] \qquad [E] = E_0 \text{ at } t = 0$$

$$\frac{d[SE]}{dt} = k_f\left[s[E] - \frac{[SE]}{K}\right] - k[SE] \qquad [SE] = 0 \text{ at } t = 0$$

$$\frac{dp}{dt} = k[SE] \qquad p = 0 \text{ at } t = 0$$

The initial condition for [SE] assumes that the enzyme was charged to the reactor in pristine condition. It makes no difference whether the enzyme is free or immobilized provided the reaction follows Michaelis-Menten kinetics.

Free Enzymes in Flow Reactors. Substitute $t = z/\bar{u}$ into the ODEs of Example 12.5. They then apply to a steady-state PFR that is fed with freely suspended, pristine enzyme. There is an initial distance down the reactor before the quasi-steady equilibrium is achieved between S in solution and S that is adsorbed on the enzyme. Under normal operating conditions, this distance will be short. Except for the loss of catalyst at the end of the reactor, the PFR will behave identically to the confined-enzyme case of Example 12.4. Unusual behavior will occur if k_f is small or if the substrate is very dilute so $s_{in} \approx E_{in}$. Then, the full equations in Example 12.5 should be (numerically) integrated.

The case for a CSTR is similar. Under normal operating conditions, the solution in Example 12.3 will apply to free enzymes as well as to confined enzymes. Like the PFR case, unusual behavior will occur if k_f is small or if the substrate is very dilute so $s_{in} \approx E_{in}$.

Example 12.6: Formulate the governing equations for an enzyme-catalyzed reaction of the form S → P in a CSTR. The enzyme is pristine when it enters the reactor. Do not invoke the quasi-steady hypothesis.

Solution: The reactions are the same as in Example 12.5. The steady-state performance of a CSTR is governed by algebraic equations, but time derivatives can be useful for finding the steady-state solution by the method of false transients. The governing equations are

$$V\frac{ds}{dt} = Qs_{in} - k_f\left[s[E] - \frac{[SE]}{K}\right]V - Qs$$

$$V\frac{d[E]}{dt} = QE_0 - Vk_f\left[s[E] - \frac{[SE]}{K}\right] + Vk[SE] - Q[E]$$

$$V\frac{d[SE]}{dt} = Vk_f\left[s[E] - \frac{[SE]}{K}\right] - Vk[SE] - Q[SE]$$

$$V\frac{dp}{dt} = k[SE] - Qp$$

It was assumed that $[SE]_{in} = p_{in} = 0$.

12.2 CELL CULTURE

Whole cells are grown for a variety of reasons. The cells may perform a desired transformation of the substrate, e.g., wastewater treatment; the cells themselves may be the desired produce, e.g., yeast production; or the cells may produce a desired product, e.g., penicillin. In the later case, the desired product may be excreted, as for the penicillin example, and recovered in relatively simple fashion. If the desired product is retained within the cell walls, it is necessary to *lyse* (rupture) the cells and recover the product from a complex mixture of cellular proteins. This approach is often needed for therapeutic proteins that are created by recombinant DNA technology. The resulting separation problem is one of the more challenging aspects of biochemical engineering. However, culture of the cells can be quite difficult experimentally and is even more demanding theoretically.

The easiest cells to grow are microbes that live independently in their natural environment. These include bacteria, yeasts, and molds. The hardest are the cells extracted from higher order plants and animals since they normally rely on complex interactions with other cells in the parent organism. Bacteria and yeasts are single-celled. Molds are multicelled but have relatively simple structures and nutritional requirements.

Microbial culture is also called *fermentation* and is most commonly a batch process (although oxygen may be supplied continuously). The chemistry of cell growth is extremely complex and, even in the simplest living cells, is only partially understood. The common gut bacterium, *Escherichia coli*, utilizes approximately 1000 enzymes. The nutrient mixture for cell growth must include a carbon source, which is typically a sugar such as glucose, and sources for nitrogen, phosphorus, and potassium; and a large variety of trace elements.

Oxygen may be required directly (*aerobic* fermentations) it or may be obtained from a carbohydrate, which also serves as the carbon source (*anaerobic* fermentations). Some bacteria, called *facultative anaerobes*, utilize molecular oxygen when available but can survive and grow without it. For others, called *strict anaerobes*, oxygen is a poison. Some microbes can be grown using *defined media*. This is a nutrient source consisting of simple sugars and salts in which all chemical components are identified as to composition and quantity. *Complex media* contain uncharacterized substances, typically proteins from natural sources.

Batch fermentation begins with an initial charge of cells called an *inoculum*. Growth of the desired cell mass usually occupies a substantial portion of the batch cycle and, conceptually at least, follows the curve illustrated in Figure 12.3. During the initial lag phase, cells are adjusting themselves to the new environment. Most microbes, and particularly bacteria, are extremely adaptable and can utilize a variety of carbon sources. However, one or more unique enzymes are generally required for each source. These are called *induced enzymes* and are manufactured by the cell in response to the new environment. The induction period is called the *lag phase* and will be short if the fermentation medium is similar to that used in culturing the inoculum. If the chemistry is dramatically different, appreciable cell death may occur before the surviving cells have retooled for the new environment.

Exponential growth occurs after cell metabolisms have adjusted and before a key nutrient becomes limiting or toxic products accumulate. In the *exponential growth phase*, the total cell mass will increase by a fixed percentage during each time interval, typically doubling every few hours. Ultimately, however, the

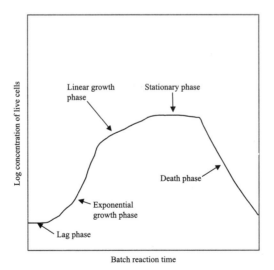

FIGURE 12.3 Idealized growth phases for a batch fermentation.

growth rate must slow and stop. A *linear growth phase* can occur in semibatch fermentations when some key reactant is supplied at a fixed rate. The oxygen supply may limit cell growth or the carbon source (substrate) may be fed to the system at a fixed rate. Ultimately, cell mass achieves a maximum and the culture enters a *stationary or maintenance phase*. A stationary population can be sustained using continuous culture techniques as described in Section 12.2.2, but the stationary phase is typically rather brief in batch fermentations. It is followed by a *death phase* where cells die or sporulate. The number of viable cells usually follows an exponential decay curve during this period. The cells can reduce their nutritional requirements in times of stress, and surviving cells will cannibalize the bodies of cells that lyse.

Models for batch culture can be constructed by assuming mechanisms for each phase of the cycle. These mechanisms must be reasonably complicated to account for a lag phase and for a prolonged stationary phase. *Unstructured models* treat the cells as a chemical entity that reacts with its environment. *Structured models* include some representation of the internal cell chemistry. *Metabolic models* focus on the energy-producing mechanisms within the cells.

12.2.1 Growth Dynamics

This section gives models for the rates of birth, growth, and death of cell populations. We seek models for (1) the rate at which biomass is created, (2) the rates at which substrates are consumed, (3) the rates at which products are generated, (4) the maintenance requirements for a static population, and (5) the death rate of cells. The emphasis is on unstructured models.

Biomass Production. Biomass is usually measured by dry weight of viable cells per unit volume X. We bypass the sometimes tricky problems associated with this measurement except to say that it is the province of the microbiologist and usually involves plate cultures and filtration followed by drying. Suppose there is one limiting nutrient S, and that all other nutrients are available in excess. Then the *Monod model* for growth is

$$\mathcal{R}_X = \mu X = \mu_{max} X \left[\frac{S}{K_S + S} \right] \qquad (12.8)$$

where μ is the specific growth rate for cell mass. Typical units are grams per liter per hour for $\mathcal{R}_X$, reciprocal hours for μ and μ_{max}, and grams per liter for S, X, and K_S. A typical value for μ_{max} under optimal conditions of temperature and pH is $1\,\text{h}^{-1}$. It the primary substrate is abundant, Equation 12.8 gives exponential growth;

$$X = X_0 e^{\mu_{max}(t-t_0)} \qquad (12.9)$$

where X_0 is the inoculum size and t_0 is the time when the induction phase ends. Equation (12.9) models induction as a pure time delay. The exponential growth phase ends when a substrate concentration becomes limiting.

It is possible for two or more substrates to become simultaneously limiting. Define a growth limitation factor, $G_i < 1$, for substrate i such that

$$\mathscr{R}_X = \mu X = \mu_{max} X G_1$$

when growth is limited by substrate 1. The factor G_1 is determined by growth experiments that manipulate the single variable S_1. Dual limitations can be modeled using a multiplicative form:

$$\mathscr{R}_X = \mu_{max} X G_1 G_2 \tag{12.10}$$

but this usually underestimates the growth rate. Another possibility is

$$\mathscr{R}_X = \mu_{max} X \, \text{MIN}[G_1, G_2] \tag{12.11}$$

but this usually overestimates the growth rate. Define the substrates so that $G_1 \leq G_2$. A compromise model is

$$\mathscr{R}_X = \mu_{max} X G_1 G_2^{\alpha_{12}} \tag{12.12}$$

where α_{12} is an empirical interaction parameter that is fit to data having at least one experiment with dual limitations.

Any substance present in great excess can inhibit growth or even cause death. Metabolic products are often toxic to the organism that produces them. Thus, a batch fermentation can be limited by accumulation of products as well as by depletion of the substrate. A simple model for growth in the presence of an inhibitor is

$$\mathscr{R}_X = \mu_{max} X \left[\frac{S}{K_S + S} \right] \left[1 - \frac{p}{p_{max}} \right]^m = \mu X G_S G_P \tag{12.13}$$

where p is the concentration of the inhibiting material, p_{max} is the value for p at which growth stops, and m is an empirical constant. Note that Equation (12.13) uses a multiplicative combination of the growth-limiting factors G_S and G_P. This may be overly pessimistic if G_S is appreciably less than 1. Equation (12.12) can be used if data are available to fit α_{12}.

Cell cultures can be inhibited by an excessive concentration of the substrate. One way to model substrate inhibition is to include an S^2 term in the denominator of the rate equation. See Equation (12.4).

Substrate Consumption. Consumption of substrates and generation of products can be described using empirical yield coefficients. Yields are usually based on the amount of limiting substrate that has been consumed. Thus, $Y_{P/S}$ denotes the mass of product produced per mass of substrate consumed, and $Y_{X/S}$ denotes the dry mass of cells produced per mass of substrate consumed. The value for $Y_{P/S}$ is not stoichiometrically determined and will vary within a growth cycle. When glucose is anaerobically fermented by a yeast, part of the carbon will appear as ethanol and coproduct CO_2, part will be incorporated into the cell mass, and minor amounts will appear as excreted by-products. A typical cell composition on a dry weight basis is about 50% carbon, 7% hydrogen, 12% nitrogen, 30% oxygen, and 1% other. The empirical formula $C_5H_7O_2N$ can be useful for back-of-the-envelope calculations. An upper limit on $Y_{X/S}$ of about 2 g of dry cell mass per gram of carbon in the substrate is predicted using this average composition. The actual value for $Y_{X/S}$ will be about half the upper limit, the balance of the carbon being used to generate energy. This missing half is accounted for by the $Y_{P/S}$ terms summed over all excreted products. An internally consistent model for cell growth closes the mass balance for carbon and other nutrients. The mass balance has the form

$$Y_{X/S} + \sum_{Products} Y_{P/S} = 1 \tag{12.14}$$

Equation (12.14) applies to all the elements that constitute a cell, but it is normally applied in the form of a carbon balance. The dominant terms in the balance are carbon in the cells, measured through $Y_{X/S}$, and carbon in the *primary metabolites*—e.g., ethanol and CO_2—measured through the $Y_{P/S}$ terms for these metabolites.

A convenient way to model the consumption of substrate is to divide it between consumption that is associated with the growth of new cells and consumption that is required to maintain existing cells:

$$\mathscr{R}_S = -\frac{\mathscr{R}_X}{Y_{X/S}} - M_S X \tag{12.15}$$

Here, M_S is the maintenance coefficient for substrate S. Typical units of M_S are grams of substrate per gram of dry cell mass per hour. Table 12.1 gives maintenance coefficients for various organisms and substrates. The maintenance

TABLE 12.1 Maintenance Coefficients for Various Organisms and Substrates

Organism	Substrate	M_S, h^{-1}	M_O, h^{-1}
Acetobacterium woodii	Lactate	0.07	Anaerobic
Aerobacter aerogenes	Citrate	0.06	0.05
Aerobacter aerogenes	Glucose	0.05	0.05
Aerobacter aerogenes	Glycerol	0.08	0.11
Saccharomyces cerevisiae	Glucose	0.02	0.02
Escherichia coli	Glucose	0.05	0.02
Penicillium chrysogenum	Glucose	0.02	0.02

Source: Data for *A. woodii* from Peters, V., Janssen, P. H., and Conrad, R. *FEMS Microbiol. Ecol.*, **26**, 317 (1998). Other data are from Roels, J. A. and N. W. F .Kossen, *Prog. Ind. Microbiology*, **14**, 95 (1978).

coefficient for oxygen is denoted M_O. It is seen to depend on both the organism and the carbon source.

Maintenance requirements exist for nitrogen and other elements (e.g., phosphorous). They are relatively small, but must be supplied to maintain a stationary cell population.

Excreted Products Balances. Most of the carbon going into a cell is converted to cell mass or to primary metabolic products. An aerobic fermentation may give CO_2 and water as the only products. The anaerobic fermentation of glucose by yeast normally gives 2 mol of ethanol and 1 mol of CO_2 for each mole of glucose. A lactic acid fermentation gives 2 mol of lactic acid per mole of glucose. Some bacteria have several metabolic pathways and can utilize a variety of five- and six-carbon sugars to produce ethanol, formic acid, acetic acid, lactic acid, succinic acid, CO_2, and H_2, depending on their environment. The specific mix of products, even for a fixed substrate and organism, depends on factors such as the phase of growth in the batch cycle, the pH of the medium, and whether or not molecular oxygen is available. For example, the fermentation of glucose by *Bacillus polymxya* gives a mixture of ethanol, acetic acid, lactic acid, and butanediol; but butanediol can be obtained in almost stoichiometric yield:

$$C_6H_6(OH)_6 + \tfrac{1}{2}O_2 \to C_4H_8(OH)_2 + 2CO_2 + H_2O$$

late in a batch cycle with an acidic medium and aeration. Occasionally, optimal growing conditions are suggested just by the overall stoichiometry and energetics of the metabolic reactions. For example, ethanol can be produced anaerobically: this reaction has a negative change in free energy so it can support microbial growth. An anaerobic fermentation for butanediol releases free hydrogen and has less favorable energetics. Thus, one might expect more alcohol than butanediol under strictly anaerobic conditions. With oxygen available, the butanediol route produces more energy and is favored, although, with enough oxygen, the reaction may go all the way to CO_2 and water. More detailed predictions require kinetic models of the actual metabolic pathways in a cell. The energy-producing pathways are now known to a large extent, and so structured, metabolic models are becoming possible.

Assume the product distribution is known and fixed. Then all products can be lumped into a single, equivalent product, P, which is stoichiometrically linked to the substrate. On a mass basis, carbon from the substrate must match the carbon that appears as product plus the carbon utilized in making biomass. When there is no change in the cell mass,

$$\mathscr{R}_P = -\mathscr{R}_S$$

The experimental quantity $Y_{X/S}$ includes both uses of the substrate. Define $\widehat{Y}_{X/S}$ as the theoretical yield of cell mass per mass of substrate if energy requirements are ignored, i.e., $\widehat{Y}_{X/S} \approx 2$ for carbon. Then product formation can be expressed in terms of cell mass and growth of cell mass:

$$\mathscr{R}_P = -\mathscr{R}_S - \frac{\mathscr{R}_X}{\widehat{Y}_{X/S}} = \frac{\mathscr{R}_X}{Y_{X/S}} - \frac{\mathscr{R}_X}{\widehat{Y}_{X/S}} + M_S X \qquad (12.16)$$

The microbiology literature distinguishes between products that are formed during the growth of cell mass (*growth-associated products*) and those that are made during the stationary phase (*non-growth-associated products*). Growth-associated products have a production rate proportional to $\mathscr{R}_X$, while the production rate for non-growth-associated products is proportional to the cell mass X. The primary metabolites are proportional to both $\mathscr{R}_X$ and X since they are generated during cell growth and during maintenance. This fact is reflected in Equation (12.16). Excreted enzymes (*extracellular enzymes*) and antibiotics are usually nongrowth-associated. They are sometimes called *secondary metabolites*.

Cell Death. Spontaneous death or sporulation of cells is commonly modeled as a first-order process. Equation (12.8) (or 12.12)) is modified to include a disappearance term:

$$\mathscr{R}_X = \mu X - k_d X = \mu_{max} X \left[\frac{S}{K_S + S} \right] - k_d X \qquad (12.17)$$

This model does not predict a stationary phase in a batch fermentation if k_d is constant. A nearly stationary phase can be modeled if k_d is made to depend on the accumulation of a toxic product.

If there is insufficient substrate for maintenance, the lysing of some cells to supply the maintenance requirements of others can be modeled as

$$\mathscr{R}_X = -k_d - Y_{X/S} M_X X \quad \text{for } S = 0 \qquad (12.18)$$

Here, the factor of $Y_{X/S}$ is merely a rough estimate of the amount of carbon that can be reutilized. The theoretical upper limit is $\widehat{Y}_{X/S}$.

Death kinetics are obviously important in chemical or thermal *sterilization*. The spores formed by some bacteria are the hardest to kill. Problem 12.3 gives data for a representative case.

12.2.2 Reactors for Freely Suspended Cells

The typical bioreactor is a two-phase stirred tank. It is a three-phase stirred tank if the cells are counted as a separate phase, but they are usually lumped with the aqueous phase that contains the microbes, dissolved nutrients, and soluble products. The gas phase supplies oxygen and removes by-product CO_2. The most common operating mode is batch with respect to biomass, batch or fed-batch with respect to nutrients, and fed-batch with respect to oxygen. Reactor aeration is discussed in Chapter 11. This present section concentrates on reaction models for the liquid phase.

Batch Reactors. The reaction rates presented in Section 12.2.1 can be combined to give a fairly comprehensive model for cell growth, substrate consumption, and product generation in a batch bioreactor. The model is necessarily of the unstructured variety and contains many empirical constants. Concentrations are measured on a weight basis with typical units of grams per liter or milligrams per liter. Such measures are sometimes denoted as w/v. A dimensionless weight per weight measure, denoted as w/w, is also common and is preferred for mathematical modeling.

Example 12.7: Develop a model for the anaerobic batch fermentation of glucose to ethanol and coproduct CO_2 using *Saccharomyces cerevisiae*. The starting mixture contains 10% glucose. The inoculum is 0.0005 w/w. Product inhibition stops cell growth at 14% ethanol. Assume $k_d = 0$ but include the cannibalization of cellular material beginning when the substrate is completely consumed.

Solution: The initial conditions for a batch reactor are $S = S_0$ and $X = X_0$ at $t = 0$. Equation (12.13) gives the cell growth when there is inhibition by the product:

$$\frac{dX}{dt} = \mathscr{R}_X = \mu_{max} X \left[\frac{S}{K_S + S}\right]\left[1 - \frac{p}{p_{max}}\right]^m \quad \text{at} \quad t = 0$$

Cannibalization according to Equation (12.18) begins when $S = 0$:

$$\frac{dX}{dt} = \mathscr{R}_X = -Y_{X/S} M_X X$$

Equation (12.15) governs substrate consumption. For a batch reactor with all the substrate charged initially,

$$\frac{dS}{dt} = -\frac{\mathscr{R}_X}{Y_{X/S}} - M_S X$$

Excreted products are governed by Equation (12.16):

$$\frac{dp}{dt} = \frac{\mathscr{R}_X}{Y_{X/S}} - \frac{\mathscr{R}_X}{\widehat{Y}_{X/S}} + M_S X \quad p = 0 \quad \text{at} \quad t = 0$$

As with most modeling efforts, the mathematical formulation is the easy part. Picking the right values from the literature or experiments is more work. An immediate task is to decide how to characterize the substrate and product concentrations. The balance equations for substrate and product apply to the carbon content. The glucose molecule contains 40% carbon by weight so S will be 0.4 times the glucose concentration, and $S_0 = 0.04$. Similarly,

p measures the carbon content of the excreted products. Two-thirds of the excreted carbon appears as ethanol, and ethanol is 52% carbon by weight. Thus, p will be $0.52/0.67 = 0.78$ times the ethanol concentration, and the value of p_{max} that corresponds to 14% ethanol is 0.109.

A typical value for μ_{max} is $0.5\,\text{h}^{-1}$. Use $K_S = 0.001$ on a carbon-equivalent, weight-per-weight basis (but see Problem 12.4). Assume $m = 1$.

Turning to the substrate balance, yeast cells contain about 50% carbon. The cell mass is measured as total dry weight, not just carbon. This gives $\hat{Y}_{X/S} = 2$ when S is measured as the carbon equivalent of glucose. A reasonable value for $Y_{X/S}$ is 1 so that half the carbon goes into biomass and half meets the associated energy requirements. The maintenance coefficient in carbon-equivalent units is $0.008\,\text{h}^{-1}$. Using these parameter estimates, the three simultaneous ODEs for $S > 0$, become

$$\frac{dX}{dt} = \mathcal{R}_X = 0.5X\left[\frac{S}{0.001 + S}\right]\left[1 - \frac{p}{0.109}\right] \quad X = 0.0005 \text{ at } t = 0$$

$$\frac{dS}{dt} = -\mathcal{R}_X - 0.008X \quad S_0 = 0.04 \text{ at } t = 0$$

$$\frac{dp}{dt} = \frac{\mathcal{R}_X}{2} + 0.008X \quad p = 0 \text{ at } t = 0$$

For $S = 0$,

$$\frac{dX}{dt} = -0.008X$$

$$\frac{dS}{dt} = \frac{dp}{dt} = 0$$

The solution is shown in Figure 12.4.

Most high-value fermentation products are made in batch. Vinification and brewing are examples where the desired products are excreted, and cell mass is undesired. High-end, nonexcreted products such as interferon and human insulin are produced in batch fermentations that typically use a genetically altered variety of *Escherichia coli*. The cells are harvested and lysed to obtain the desired product as part of a complex mixture. Viable whole cells are the desired product in the commercial, batch culture of baker's yeast (*Saccharomyces cerevisiae*) and in the mixed-strain culture (of a yeast, *Saccharomyces exiguus* with a bacterium, *Lactobacillus sanfrancisco*) used to make sourdough bread. These are all expensive, low-volume products for which batch fermentation is satisfactory. Continuous fermentation is attractive for high-volume, bulk chemicals such as fuel-grade ethanol.

Continuous Stirred Tanks Without Biomass Recycle. The chemostat without biomass recycle is a classic CSTR. The reactor is started in the batch mode.

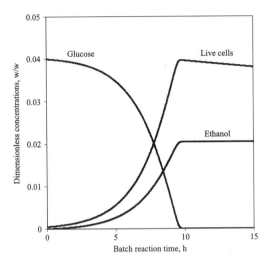

FIGURE 12.4 Simulation of a batch *Saccharomyces cerevisiae* fermentation.

Inlet and outlet flows are begun during the exponential growth phase. The input stream contains all necessary nutrients (except oxygen which is sparged continuously) but no cells. The outlet stream contains cells, unreacted nutrients, and fermentation products. If the flow rate is low enough, a steady state will be reached with the continuous production of cells being matched by their outflow. If the flow rate is too high, *washout* occurs, and the cell population is lost. This fact leads to a marvelous method for selecting the fastest-growing species from a mixed culture. Inoculate a batch reactor using wild microbes from dirt or other natural sources. Convert to continuous operation at a low flow rate, but gradually increase the flow rate. The last surviving species is the fastest growing (for the given substrate, pH, temperature, etc.).

Since $X_{in} = 0$, the steady-state cell balance for a CSTR is

$$\bar{t}\mathcal{R}_X = X_{out}$$

The growth rate in a CSTR has the general form $\mathcal{R}_X = \mu X_{out}$. Thus,

$$\bar{t}\mu X_{out} = X_{out}$$

This equation can always be satisfied with $X_{out} = 0$ so that the washout condition is always possible as a steady state. This steady state is achieved when there is no inoculum or when the flow rate is too high. A nontrivial solution with $X_{out} > 0$ requires that $\bar{t}\mu = 1$ or

$$\mu = 1/\bar{t} \tag{12.19}$$

Microbiologists would say that the growth rate equals the dilution rate at steady state.

The substrate balance is

$$S_{in} - \bar{t}\left(\frac{\mathcal{R}_X}{Y_{X/S}} + M_S X_{out}\right) = S_{out}$$

Substituting $\mathcal{R}_X = \mu X_{out}$ and solving for X_{out} gives

$$X_{out} = \frac{S_{in} - S_{out}}{\frac{\bar{t}\mu}{Y_{X/S}} + \bar{t}M_S} = \frac{[S_{in} - S_{out}]Y_{X/S}}{1 + \bar{t}Y_{X/S}M_S} \qquad (12.20)$$

A mass balance for product gives

$$p_{out} = p_{in} + \bar{t}\left[\frac{\mu X_{out}}{Y_{X/S}} - \frac{\mu X_{out}}{\widehat{Y}_{X/S}} + M_S X_{out}\right]$$

$$= p_{in} + \frac{X_{out}}{Y_{X/S}} - \frac{X_{out}}{\widehat{Y}_{X/S}} + \bar{t}M_S X_{out} \qquad (12.21)$$

Equations (12.20) and (12.21) apply to any functional form for the growth rate, μ. There are three unknowns in these equations: X_{out}, S_{out}, and p_{out}. A third equation is needed. It comes from substituting some specific functional form for μ into Equation (12.19). For example, if μ is given by Equation (12.13), the third equation is

$$\mu_{max}\left[\frac{S_{out}}{K_S + S_{out}}\right]\left[1 - \frac{p_{out}}{p_{max}}\right]^m = 1/\bar{t} \qquad (12.22)$$

Equations (12.20), (12.21), and (12.22) are solved simultaneously for X_{out}, S_{out}, and p_{out}.

Example 12.8: The batch reactor in Example 12.7 has been converted to a CSTR. Determine its steady-state performance at a mean residence time of 4 h. Ignore product inhibition.

Solution: Set $m=0$ to ignore production inhibition. Then Equation (12.22) can be solved for S_{out}:

$$S_{out} = \frac{K_S}{\bar{t}\mu_{max} - 1} \qquad (12.23)$$

The parameters used in Example 12.7 were $K_S = 0.001$, $\mu_{max} = 0.5$, and $\bar{t} = 4$. This gives $S_{out} = 0.001$ and $\mu = \mu_{max}/2 = 0.25$. These values and M_S are substituted into Equation (12.20) to give $X_{out} = 0.0378$. Equation (12.21) is then used to calculate $p_{out} = 0.0201$. A carbon balance confirms the results. Carbon enters as S_{in} and leaves as $S_{out} + X_{out}/2 + p_{out} = 0.001 + 0.0378/2 + 0.0201 = 0.040$.

It is clear from Equation (12.19) that $\bar{t}\mu_{max} > 1$ is necessary to avoid washout. However, it is not sufficient. The sufficient condition is found from the requirement that $S_{out} < S_{in}$ at steady state. This gives

$$\bar{t}\mu_{max} > 1 + \frac{K_S}{S_{in}} \qquad (12.24)$$

Operating near the washout point maximizes the production rate of cells. A feedback control system is needed to ensure that the limit is not exceeded. The easiest approach is to measure cell mass—e.g., by measuring turbidity—and to use the signal to control the flow rate. Figure 12.5 shows how cell mass varies as a function of $\bar{t}$ for the system of Examples 12.7 and 12.8. The minimum value for $\bar{t}$ is 2.05 h. Cell production is maximized at $\bar{t} = 2.37$ h.

Microbial kinetics can be quite complex. Multiple steady states are always possible, and oscillatory behavior is common, particularly when there are two or more microbial species in competition. The term *chemostat* can be quite misleading for a system that oscillates in the absence of a control system.

Continuous Stirred Tanks with Biomass Recycle. When the desired product is excreted, closing the system with respect to biomass offers a substantial reduction in the cost of nutrients. The idea is to force the cells into a sustained stationary or maintenance period where there is relatively little substrate used to grow biomass and where production of the desired product is maximized. One approach is to withhold some key nutrient so that cell growth is restricted, but to supply a carbon source and other components needed for the desired product. It is sometimes possible to maintain this state for weeks or months and to achieve high-volumetric productivities. There will be spontaneous cell loss (i.e., $k_d > 0$), and true steady-state operation requires continuous purging and makeup. The purge can be achieved by incomplete separation and recycle

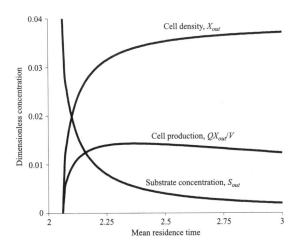

FIGURE 12.5 Production of *Saccharomyces cerevisiae* in a CSTR.

of the cells or by an explicit purge stream as shown in Figure 12.6. Cell makeup can be achieved by allowing some cell growth in the reactor. An alternative that is better in terms of raw material efficiency is to grow the makeup cells in a separate reactor that has nutrient levels optimized for growth rather than maintenance.

The *activated sludge* process for wastewater treatment uses recycle of live cells. The goal is to oxidize organics without generating too much sludge (i.e., biomass).

Piston Flow Bioreactors. There are several commercial examples of continuous-flow bioreactors that approximate piston flow. The deep-shaft fermentors described in Chapter 11 have been used for wastewater treatment when land is very expensive. Beer is sometimes brewed in an unstirred tower using a flocculating variety of yeast that settles to the bottom of the tower. Most plants for fuel-grade ethanol and lactic acid are continuous, and some designs approximate piston flow. Monod kinetics give low outlet concentrations of substrate even in a single CSTR because K_S is usually small. See Equation (12.23). However, batch kinetics give still lower values for S_{out}, and this can be important in cases like wastewater treatment. Wastewater-treatment lagoons are designed to have progressive flow from inlet to outlet. Sterilization by chlorine or ozone is best done in a piston flow reactor. Continuous bioreactors often use a tubular reactor or a flat-plate heat exchanger for *continuous thermal sterilization* of the substrate. The need to reduce viable organisms by many orders of magnitude

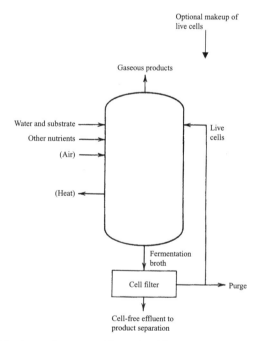

FIGURE 12.6 Continuous fermentor with recycle of live cells.

justifies use of the axial dispersion model. See Section 9.3. Note however that FDA requirements for the highest grade of biological purity (*water for injection* or WFI) requires sterilization by distillation or reverse osmosis.

12.2.3 Immobilized Cells

Like enzymes, whole cells are sometime immobilized by attachment to a surface or by entrapment within a carrier material. One motivation for this is similar to the motivation for using biomass recycle in a continuous process. The cells are grown under optimal conditions for cell growth but are used at conditions optimized for transformation of substrate. A great variety of reactor types have been proposed including packed beds, fluidized and spouted beds, and air-lift reactors. A semicommercial process for beer used an air-lift reactor to achieve reaction times of 1 day compared with 5–7 days for the normal batch process. Unfortunately, the beer suffered from a "mismatched flavour profile" that was attributed to mass transfer limitations.

There are few advantages to immobilizing cells that can be grown in free suspensions. However, some cells from multicelled plants and animals can be grown only when anchored to a surface and when interacting with adjacent cells. When the synthetic structure used for growth resembles the naturally occurring structure, the reaction is called *tissue culture* (as opposed to *cell culture*). The easier forms of tissue culture involve plants. The most common form, *meristem culture*, reproduces whole, genetically identical plants from tiny cuttings of a parent plant. This technique has been widely adopted by the nursery industry to mass-produce named varieties of ornamental plants. The more challenging forms of tissue culture involve cells of two or more types or animal cells grown simultaneously. Growth of complex structures like liver tissue or skin remains in the future.

PROBLEMS

12.1. It has been proposed that some enzymes exist in active and inactive forms that are in equilibrium. The active form binds substrate molecules for subsequent reaction while the inactive form does not. The overall reaction mechanism might be

$$E \rightleftarrows I \qquad \frac{[I]}{[E]} = K_I$$

$$S + E \rightleftarrows SE \qquad \frac{[SE]}{s[E]} = K$$

$$SE \longrightarrow P + E \qquad \mathscr{R} = k[SE]$$

Derive a kinetic model for this situation.

12.2. Set the time derivatives in Example 12.6 to zero to find the steady-state design equations for a CSTR with a Michaelis-Menten reaction. An analytical solution is possible. Find the solution and compare it with the solution in Example 12.3. Under what conditions does the quasi-steady solution in Example 12.3 become identical to the general solution in Example 12.6?

12.3. Wang et al.[1] report the death kinetics of *Bacillus stearothermophilus* spores using wet, thermal sterilization. Twenty minutes at 110°C reduces the viable count by a factor of 10^4. The activation temperature, E/R_g, is 34,200 K. How long will it take to deactivate by a factor of 10^7 at 120°C?

12.4. A literature value for the Monod constant for a *Saccharomyces cerevisiae* fermentation is $K_S = 25$ mg/liter.[2] How does this affect the simulation in Example 12.7?

12.5. A limiting case of Monod kinetics has $K_S = 0$ so that cell growth is zero order with respect to substrate concentration. Rework Example 12.7 for this situation, but do remember to stop cell growth when $S = 0$. Compare your results for X and p with those of Example 12.7. Make the comparison at the end of the exponential phase.

12.6. A simple way to model the lag phase is to suppose that the maximum growth rate μ_{max} evolves to its final value by a first-order rate process: $\mu_{max} = \mu_\infty [1 - \exp(-\alpha t)]$. Repeat Example 12.7 using $\alpha = 1\,\text{h}^{-1}$. Compare your results for X, S, and p with those of Example 12.7. Make the comparison at the end of the exponential phase.

12.7. Equation (12.17) postulates that spontaneous deaths occur throughout the batch cycle. This means that dX/dt is initially negative. Is it possible to lose the inoculum completely if the induction period is too long? Long induction periods correspond to small values of α in the lag phase model of Problem 12.6. Find the critical value for α at which the inoculum is lost.

12.8. Consider the CSTR with biomass recycle shown in Figure 12.6.
  (a) Determine a criterion for achieving a nontrivial steady state in a reactor that is closed with respect to biomass. Does it have anything to do with the residence time for cells? Does the criterion include the limiting case of no cell makeup, birth, death, or purge.
  (b) The purge stream means that the cells have a finite residence time in the reactor, although it will be much longer than for the liquid phase (i.e., water, excreted products, unreacted substrate). What is the mean residence time for cells? Does it depend on whether the makeup cells are fed to the reactor or whether they are born in the reactor?
  (c) How long does it take for the reactor to approximate the steady state found in part (a)?

REFERENCES

1. Wang, D. et al., *Fermentation and Enzyme Technology*, Wiley, New York, 1979.
2. Blanch, H. W. and Clark, D. S., *Biochemical Engineering*, Marcel Dekker, New York, 1996.

SUGGESTIONS FOR FURTHER READING

Specialized books on biochemical engineering include the following:

Blanch, H. W. and Clark, D. S., *Biochemical Engineering*, Marcel Dekker, New York, 1996.

Lee, J. M., *Biochemical Engineering*, Prentice Hall, New Jersey, 1992.

Bailey, J. F. and Ollis, D. F., *Biochemical Engineering Fundamentals*, 2nd ed., McGraw-Hill, New York, 1986.

CHAPTER 13
POLYMER REACTION ENGINEERING

Polymer reaction engineering is a specialized but important branch of chemical reaction engineering. The odds strongly favor the involvement of chemical engineers with polymers at some point in their career. The kinetics of polymerization reactions can be treated using the basic concepts of Chapters 1–5, but the chemistry and mathematics are more complicated than in the examples given there. The number of chemical species participating in a polymerization is potentially infinite, and the mathematical description of a batch polymerization requires an infinite set of differential equations. Analytical and numerical solutions are more difficult than for the small sets of equations dealt with thus far. Polymerizations also present some interesting mechanical problems in reactor design. Viscosity increases dramatically with molecular weight, and a polymer solution is typically 10^2–10^6 times more viscous than an ordinary liquid. Molecular diffusivities in polymer solutions are lowered by similar factors. Laminar flow is the rule, pressure drops are high, agitation is difficult, and heat and mass transfer limitations can be very severe. The polymer reaction engineer sometimes confronts these problems head on, but most often seeks to avoid them through clever reactor design.

13.1 POLYMERIZATION REACTIONS

Polymerization reactions are classified as being either *chain growth* or *step growth*. In *chain-growth polymerization*, a small molecule reacts with a growing polymer chain to form an incrementally longer chain:

$$M + P_l \rightarrow P_{l+1} \qquad l = 1, 2, \ldots \qquad (13.1)$$

where M represents the small molecule that is called the *monomer* and P_l denotes a polymer chain consisting of l monomer units that are chemically bonded. The chain may also contain a residual fragment of an *initiator* molecule that started the growth. A synonym for chain-growth polymerization is *addition polymerization*, so named because the monomer adds to the chain one unit at a time. The most important chain-growth polymers are *vinyl addition polymers* formed by the opening of a double bond. There is no by-product other than energy. Important examples are ethylene polymerization to form polyethylene and propylene polymerization to form polypropylene.

A quite different polymerization mechanism has two polymer molecules reacting together to form a larger new molecule:

$$P_l + P_m \to P_{l+m} \qquad l, m = 1, 2, \ldots \qquad (13.2)$$

The length increases in steps of size l and m that can be quite large, particularly near the end of the polymerization. This is called *step-growth polymerization* because the polymerization can occur in fairly large steps. A synonym is *condensation polymerization* because the reactants are condensed together and usually give a condensation by-product. The condensation reaction involves the breaking and making of covalent bonds of rather similar energy so that the heat of reaction is moderate. Important examples of condensation polymerizations are the reaction of phenol with formaldehyde to form phenolic resins and the reaction of terephthalic acid with ethylene glycol to form polyethylene terephthalate (PET). The condensation by-product is water in both these examples.

13.1.1 Step-Growth Polymerizations

Condensation polymers are often formed from two distinct monomers, each of which is difunctional. The monomers have the forms AMA and BNB where A and B are functional groups that react to couple the M and N units and form a condensation by-product, AB. M and N are the *mer* units that form the polymer chain. The first step in the polymerization forms dimer:

$$AMA + BNB \to AMNB + AB$$

Trimer can be formed in two ways, giving two different structures:

$$AMNB + AMA \to AMNMA + AB$$
$$AMNB + BNB \to BNMNB + AB$$

Tetramer can be formed in three ways, but the structures are identical:

$$AMNMA + BNB \to AMNMNB + AB$$
$$BNMNB + AMA \to AMNMNB + AB$$
$$2AMNB \to AMNMNB + AB$$

This example of step-growth polymerization has two monomers that can react with each other but not with themselves. It has only one dimer that can be self-reactive. This pattern continues indefinitely with two trimers, one self-reactive tetramer, and so on. Molecules with an odd number of mer units will come in two forms; A~A and B~B. Molecules with an even number have the form A~B.

Stoichiometry. Reactions of the form AMA + BNB → Polymer are known as *binary polycondensations*. The original concentrations of the A and B endgroups must be closely matched for the reaction to generate polymer that has a high molecular weight. Denote the concentrations of (unreacted) endgroups as [A] and [B], and suppose that A is the limiting endgroup. Then the stoichiometric ratio is defined as

$$S_{AB} = \frac{[A]_0}{[B]_0} \leq 1 \qquad (13.3)$$

If the reaction goes to completion, the polymer molecules will have the form B~B, and the final concentration of B endgroups will be $[B]_0 - [A]_0$. The *number average chain length* is

$$\bar{l}_N = \frac{\text{Monomers present initially}}{\text{Molecules present after polymerization}} = \frac{([B]_0 + [A]_0)/2}{([B]_0 - [A]_0)/2} = \frac{1 + S_{AB}}{1 - S_{AB}} \qquad (13.4)$$

The factor of 2 appears in this equation because each polymer molecule has two endgroups. A pesky factor of 2 haunts many polymer equations. It factors out in this example, but we are not always so lucky.

Example 13.1: Determine the stoichiometric requirements for achieving various degrees of polymerization for a binary polycondensation.

Solution: Table 13.1 shows results calculated using Equation (13.4). The stoichiometric requirement for a binary polycondensation is very demanding. High-molecular-weight polymer, say $\bar{l}_N > 100$, requires a weighing accuracy that is difficult to achieve in a flow system.

Example 13.1 shows one reason that binary polycondensations are usually performed in batch vessels with batch-weighing systems. Another reason is

TABLE 13 1 Number Average Chain Lengths for Binary Polycondensations Going to Completion

S_{AB}	$\bar{l}_N$
0.2	1.5
0.333	2
0.5	3
0.8	9
0.9	19
0.98	99
0.99	199
0.998	999

that some polycondensation reactions involve polyfunctional molecules that will cross-link and plug a continuous-flow reactor. An example is phenol, which is trifunctional when condensed with formaldehyde. It can react at two ortho locations and one para location to build an infinite, three-dimensional network. This may occur even when the stoichiometry is less than perfect. See Problem 13.3 for a specific example. In a batch polymerization, any cross-linked polymer is removed after each batch, while it can slowly accumulate and eventually plug a flow reactor.

Self-condensing monomers of the form AMB avoid a stoichiometry restriction. The idea is to synthesize the monomer using conditions where the A and B endgroups are not reactive but to polymerize the monomer under conditions where the endgroups do react. An important example is the polymerization of PET. A large excess of ethylene glycol is reacted with terephthalic acid to form a nominal trimer, diglycol terephthalate:

$$2HOCH_2CH_2OH + HO\overset{O}{\overset{\|}{C}}\Phi\overset{O}{\overset{\|}{C}}OOH \rightarrow HOCH_2CH_2O\overset{O}{\overset{\|}{C}}\Phi\overset{O}{\overset{\|}{C}}OCH_2CH_2OH$$

This new "monomer" is separated from the excess glycol and polymerized. The monomer has two hydroxy endgroups; but with catalysis and temperature, it will self-condense to give ethylene glycol as the by-product. The overall result is a one-to-one reaction of terephthalic acid with ethylene glycol, but a substantial amount of glycol is internally recycled.

Equilibrium Limitations. Self-condensing monomers have $S_{AB}=1$ and can therefore react to infinite molecular weight. Unfortunately, free lunches are rare. The cost of this lunch is that most self-condensations are reversible, so the condensation product must be removed for the reaction to proceed. The condensation of diglycol terephthalate gives high molecular polymer only if by-product ethylene glycol is removed. The condensation of lactic acid to poly(lactic acid) proceeds only if by-product water is removed. The normal technique for removal is to raise the temperature and lower the pressure so that the condensation product evaporates. The polymer industry calls this *devolatilization*. It can occur spontaneously if the condensation product is a gas like HCl, but requires substantial effort for high-boiling by-products such as water and ethylene glycol. Equilibrium limitations are avoided when the condensation product has negligible solubility in the reaction medium. An example is the formation of salt when the sodium salt of bisphenol A reacts with epichlorohydrin to form epoxy resins.

The reversibility of polycondensation reactions sometimes provides a method for polymer recycling. Reacting the polymer with the condensation product AB can quickly reduce the degree of polymerization to a self-condensing trimer. The trimer can be purified by distillation and repolymerized to give virgin polymer that is once again suitable for food-contact applications. Reacting PET with ethylene glycol is the glycolysis route to PET recycling.

Other Complications in Condensation Polymerizations. *Cyclization* occurs naturally in many condensation polymerizations. A polymer molecule of the form A~B reacts with itself to form a cyclic compound. For binary polycondensations, the number of mers in the ring will be even since only the even numbered chains are self-condensing. High solvent concentrations increase ring formation. Small molecules cannot form rings due to steric hindrance, and the reactive ends on large molecules are unlikely to meet due to statistics. Thus, a plot of cyclic oligomer concentration versus mer number will have a sharp lower limit, a maximum, and a gradually declining tail.

Solubility limits may arise when oligomers beyond a certain length precipitate from the reaction medium. Solubility generally decreases with increasing molecular weight so this phenomenon is fairly common. Precipitation generally stops the reaction for condensation polymers, and solubility limits can be used advantageously to prepare oligomers of a fixed molecular weight. Solubility limits can usually be avoided by adding a good solvent to the reaction medium.

Chain stoppers are unifunctional molecules of the form AX or BX, where the X moiety is nonreactive. They are used to stop the polymerization at a desired point or to stabilize the polymer chains by endcapping.

Random condensation copolymers can be formed by adding a third monomer to the reaction mix. For example, some 1,4-butanediol might replace some of the ethylene glycol in a PET polymerization. Suppose the three monomers are AMA, BNB, and BZB. The resulting polymer will have a structure such as

$$MNMNMNMZMNMNMZMNMZMNMNMNMN\ldots$$

It makes no difference in this example which endgroup is stoichiometrically limiting. The polymer is sometimes called a *terpolymer* because three monomers are involved, but the term copolymer is used inclusively for any polymer formed from more than one monomer. In the absence of BZB, AMA and BNB will polymerize to form a *strictly alternating copolymer*. When BZB is added, the polymer still alternates with respect to the M mers but is random with respect to the N and Z mers. Of course, the enthusiastic chemist might add some AYA or even some AWB to the mix. They will all happily copolymerize, albeit at different rates.

13.1.2 Chain-Growth Polymerizations

The most important example of an addition polymerization is the homopolymerization of a vinyl monomer. The general formula for a vinyl monomer is

$$\begin{array}{cc} R_1 & R_3 \\ | & | \\ C = C \\ | & | \\ R_2 & R_4 \end{array} \tag{13.5}$$

The most important vinyl monomers have a simpler form:

$$\begin{array}{c} CH_2 = CH \\ | \\ R \end{array} \qquad (13.6)$$

This monomer is ethylene when R is hydrogen, propylene when R is a methyl group, styrene when R is a benzene ring, and vinyl chloride when R is chlorine. The polymers formed from these four monomers account for the majority of all commercial plastics. The polymers come in great variety and are made by many different processes. All of the polymerizations share a characteristic that is extremely important from the viewpoint of reactor design. They are so energetic that control of the reaction exotherm is a key factor in all designs.

Thermal Effects in Addition Polymerizations. Table 13.2 shows the heats of reaction (per mole of monomer reacted) and nominal values of the adiabatic temperature rise for complete polymerization. The point made by Table 13.2 is clear even though the calculated values for ΔT_{adia} should not be taken literally for the vinyl addition polymers. All of these polymers have ceiling temperatures where polymerization stops. Some, like polyvinyl chloride, will dramatically decompose, but most will approach equilibrium between monomer and low-molecular-weight polymer. A controlled polymerization yielding high-molecular-weight polymer requires substantial removal of heat or operation at low conversions. Both approaches are used industrially.

The other entries in Table 13.2 show that heat removal is not a problem for most ring-opening and condensation polymerizations. Polycaprolactam (also called Nylon 6) is an addition polymer, but with rather similar bond energies for the monomer and the polymer. The reaction exotherm is small enough that large parts are made by essentially adiabatic reaction in a mold. An equilibrium between monomer and polymer does exist for polycaprolactam, but it occurs at commercially acceptable molecular weights.

TABLE 13.2 Typical Heats of Polymerization

Polymer	Polymerization type	ΔH (kJ/mol)	ΔT_{adia} (°C)
Polyethylene	Vinyl addition	95.0.	1610
Polyvinyl chloride	Vinyl addition	95.8	730
Polystyrene	Vinyl addition	69.9	320
Polymethyl methycralate	Vinyl addition	56.5	270
Polycaprolactam	Ring opening	15.9	68
Polysulfone	Condensation	25.1	24
Polycarbonate	Condensation	0	0
Polyethylene terephthalate	Condensation	0	0

Equilibrium between Monomer and Polymer. A monomer-with-polymer equilibrium is quite different from the polymer-with-condensation-product equilibrium discussed in Section 13.1.1. If the condensation product is removed from the reaction mixture, a condensation polymer increases in molecular weight. If the monomer is removed when it is in equilibrium with the polymer, the polymer depolymerizes to re-form the monomer. At temperatures suitable for long-term use, the equilibrium will be shifted toward stable polymer. However, at fabrication temperatures and at the high temperatures common in devolatilization, the production of monomer and low-molecular-weight oligomers can be significant.

Polymer Structure. The simple vinyl monomer of Equation (13.6) can be arranged in several ways along the polymer chain to give distinct materials with markedly different physical properties. The first consideration is whether the polymerization proceeds *head-to-head* or *head-to-tail*. The head-to-head arrangement has the pendant R groups on adjacent carbon atoms. It is quite uncommon. Most polymers are of the head-to-tail variety with the pendant groups separated by a carbon atom in the polymer backbone. The next structural distinction is *tacticity*. It is a stereochemical property that depends on the tetrahedral nature of carbon bonds. If the pendant R groups are all on the same side relative to the carbon–carbon bonds along the backbone, the polymer is *isotactic*. If they strictly alternate, the polymer is *syndiotactic*. If they alternate at random, the polymer is *atactic*. Polyethylene has no tacticity since the pendant group is hydrogen. Commercial polypropylene is isotactic. Commercial polystyrene is atactic although a syndiotactic variety has been introduced.

Diene-based polymers such as polybutadiene have other structural distinctions. The linear versions of these polymers have one residual double bond for each mer. When the double bonds are in the polymer chain, the cis and trans stereoisomers are possible. The double bonds can appear as pendant vinyl groups, which can then exhibit tacticity.

Most commercial polymers are substantially *linear*. They have a single chain of mers that forms the backbone of the molecule. *Side-chains* can occur and can have a major affect on physical properties. An elemental analysis of any polyolefin, (e.g., polyethylene, polypropylene, poly(1-butene), etc.) gives the same empirical formula, CH_2, and it is only the nature of the side-chains that distinguishes between the polyolefins. Polypropylene has methyl side-chains on every other carbon atom along the backbone. Side-chains at random locations are called *branches*. Branching and other polymer structures can be deduced using analytical techniques such as ^{13}C NMR.

Copolymerization e.g., of 1-butene or 1-hexene with ethylene, gives *short-chain branching*; e.g., the branches contain three or five carbon atoms. The random location of the side-chains lowers the crystallinity and density. *Long-chain branching* refers to branches that are similar in length to the polymer backbone and this type occurs in polyethylene manufactured using the high-pressure process. These branches also lower the crystallinity and density. *Star polymers* have three or more long chains originating from a common point. They are formed when the polymerization is initiated with a polyfunctional molecule.

The logical extension of branching is *cross-linking*, where the polymer becomes an immense, three-dimensional molecule. Originally distinct polymer chains are connected by chemical bonds or ionic attractions to two or more other chains. Lightly cross-linked materials swell in a solvent to form a gel but do not dissolve. Heavily cross-linked polymers can be as hard as a bowling ball. Intentional cross-linking is done after thermosetting polymers (e.g., phenolic and epoxy resins) and rubbers are molded into their final shapes. Unintentional cross-linking shows up as defects in polymer films (known as gels or fish eyes) or can shut down reactors.

Vinyl copolymers contain mers from two or more vinyl monomers. Most common are *random copolymers* that are formed when the monomers polymerize simultaneously. They can be made by most polymerization mechanisms. *Block copolymers* are formed by reacting one monomer to completion and then replacing it with a different monomer that continues to add to the same polymer chain. The polymerization of a *diblock copolymer* stops at this point. *Triblock and multiblock polymers* continue the polymerization with additional monomer depletion and replenishment steps. The polymer chain must retain its ability to grow throughout the process. This is possible for a few polymerization mechanisms that give *living polymers*.

13.2 MOLECULAR WEIGHT DISTRIBUTIONS

The molecular weight distribution (MWD) is of vital importance for polymers of all types. It determines the ease of manufacture, the ease of fabrication, and the end-use properties of the polymer. A proper kinetic description of a polymerization requires determination of the molecular weight distribution of the polymer in addition to the usual concepts of conversion and selectivity.

13.2.1 Distribution Functions and Moments

Theoretical molecular weight distributions are usually based on the chain length rather than the molecular weight. A multiplicative factor, molecular weight per mer, can be applied at the end of any calculation. The subtle differences due to endgroups are usually ignored. Let $c(l)$, where $l = 1, 2, \ldots$, be the concentration of polymer chains having length l. The discrete valued function $c(l)$ is sometimes called the molecular weight distribution, but it does not obey the rules of a probability distribution since it does not sum to 1. Instead,

$$\sum_{l=1}^{\infty} c(l) = c_{polymer} \qquad (13.7)$$

where $c_{polymer}$ is the total polymer concentration. A proper probability function is

$$f(l) = \frac{c(l)}{c_{polymer}} \quad \text{so that} \quad \sum_{l=1}^{\infty} f(l) = 1 \tag{13.8}$$

where $f(l)$ is the fraction of the total number of polymer molecules that has length l. Both $c(l)$ and $f(l)$ are defined only for integer values of l, where $l = 1, 2, \ldots$, but when $c(l)$ is measured experimentally, it is difficult to resolve individual oligomers beyond the first few, and $c(l)$ appears to be a continuous function of l. Integrals then replace the sums in Equations (13.7) and (13.8). When $f(l)$ is a continuous function of l, $f(l)dl$ gives the fraction of the molecules that have lengths in the range from l to $l + dl$. See Chapter 15 for a similar definition of a differential distribution function applicable to residence times. Here in Chapter 13, summations will be used rather than integrals.

The moments of the molecular weight distribution are defined as either

$$\mu_n = \sum_{l=1}^{\infty} l^n c(l) \quad \text{or} \quad \eta_n = \sum_{l=1}^{\infty} l^n f(l) \quad n = 0, 1, 2, \ldots \tag{13.9}$$

where $\mu_0 = c_{polymer}$ when $c(l)$ is used to define the moments and $\eta_0 = 1$ when $f(l)$ is used. Note that μ and η are not equivalent but differ by a factor of $c_{polymer}$. Statisticians use the η definition but denote them as μ. The polymer literature tends to use moments based on $c(l)$ and they will be used here in Chapter 13. The number average chain length (which is proportional to the number average molecular weight) is obtained from the first moment:

$$\bar{l}_N = \frac{\mu_1}{\mu_0} = \frac{\sum_{l=1}^{\infty} l c(l)}{c_{polymer}} = \sum_{l=1}^{\infty} l f(l) \tag{13.10}$$

An experimental determination of the molecular weight distribution conceptually sorts the polymer molecules into bins, with one bin for each degree of polymerization, $l = 1, 2, \ldots$. The contents of the bin are counted, and $f(l)$ is the fraction (by number) found in bin l. The bins could also be weighed to give a weight fraction in each bin. This defines the function $g(l)$. Since weight is proportional to chain length, the distribution by weight is

$$g(l) = \frac{lf(l)}{\bar{l}_N} \quad \text{and} \quad \sum_{l=1}^{\infty} g(l) = 1 \tag{13.11}$$

where the factor of $\bar{l}_N$ is included so that the probabilities sum to 1. The *weight average chain length* is

$$\bar{l}_W = \sum_{l=1}^{\infty} l g(l) = \frac{1}{\bar{l}_N} \sum_{l=1}^{\infty} l^2 f(l) = \frac{\mu_2}{\mu_1} \tag{13.12}$$

The ratio of weight-to-number average chain lengths is the **polydispersity**,

$$PD = \frac{\bar{l}_W}{\bar{l}_N} \geq 1 \tag{13.13}$$

This dimensionless number measures the breadth of the molecular weight distribution. It is 1 for a *monodisperse* population (e.g., for monomers before reaction) and is 2 for several common polymerization mechanisms.

13.2.2 Addition Rules for Molecular Weight

Suppose w_1 kilograms of a polymer with chain lengths $(\bar{l}_N)_1$ and $(\bar{l}_W)_1$ are mixed with w_2 kilograms of a polymer with chain lengths $(\bar{l}_N)_2$ and $(\bar{l}_W)_2$. Then the mixture has the following properties:

$$(\bar{l}_N)_{mix} = \frac{w_1 + w_2}{\dfrac{w_1}{(\bar{l}_N)_1} + \dfrac{w_2}{(\bar{l}_N)_2}} \tag{13.14}$$

$$(\bar{l}_W)_{mix} = \frac{w_1(\bar{l}_W)_1 + w_2(\bar{l}_W)_2}{w_1 + w_2} \tag{13.15}$$

These rules can be used in reactor design calculations when newly formed polymer is added to existing polymer.

13.2.3 Molecular Weight Measurements

When the full distribution is needed, it is measured by size-exclusion chromatography (also called gel permeation chromatography). This is a solution technique that requires dissolution of the polymer in a reasonable solvent such as tetrahydrofuran or tetrachlorlobenzene. For polymers that require exotic solvents or solution temperatures above about 150°C, a simple measurement of solution viscosity can be a useful surrogate for the actual molecular weight. The viscosity of the pure polymer (i.e., a polymer *melt* viscosity) can also be used. Such simplified techniques are often satisfactory for routine quality control, particularly for condensation polymers such as PET that vary in average molecular weight but usually have a polydispersity of 2.

Size-exclusion chromatography is primarily a research tool. The instrument is calibrated using polystyrene standards, and results are normally reported as *polystyrene-equivalent molecular weights,* not the actual molecular weights of the polymer being tested, which may or may not be known. Furthermore, *a low molecular weight cut-off* of 2000–3000 is usually employed. Thus, the measurements are more comparative than absolute. For theoretical calculations involving condensation polymers, everything is included, even the monomers.

For addition polymers, it is normal practice to exclude monomer from the calculations. Low-molecular-weight oligomers are usually excluded as well.

The detector installed on a size-exclusion chromatograph determines whether the measured molecular weight distribution is by number, by weight, or even by viscosity. Modern instruments include software to convert from one distribution to another. They also compute number average, weight average, and Z-average molecular weights, the last one being related to the third moment of the distribution. All these numbers are relative to the polystyrene standards.

13.3 KINETICS OF CONDENSATION POLYMERIZATIONS

The generic condensation polymerization begins with monomers AMA and BMB and produces molecules of the forms A~A, B~B, and A~B. Each step of the reaction generates a longer polymer by the step-growth mechanism of Equation (13.2) and produces 1 mol of condensation product AB.

13.3.1 Conversion

Suppose that the reactivity of the A and B endgroups is independent of the chains to which they are attached. This is a form of the *equal reactivity assumption* that is needed for almost all analytical solutions to polymer kinetic problems. If it is satisfied, we can ignore the details of the polymerization and just concentrate on the disappearance of the endgroups. For a batch system,

$$\frac{d[A]}{dt} = -k[A][B] = \frac{d[B]}{dt}$$

The initial condition for this ODE is based on Equation (13.3), $[A]_0 = S_{AB}[B]_0$. The solution is

$$X_A = 1 - \frac{(1 - S_{AB})}{\exp[(1 - S_{AB})[B]_0 kt] - S_{AB}} \qquad (13.16)$$

where X_A is the conversion of the limiting endgroup, A. The conversion depends on S_{AB} and on the dimensionless reaction time, $[B]_0 kt$. Perfect initial stoichiometry, as for a self-condensing monomer, gives an indeterminant form, to which L'Hospital's rule may be applied:

$$X_A = 1 - \frac{1}{1 + [B]_0 kt} \qquad \text{for} \qquad S_{AB} = 1 \qquad (13.17)$$

We see from Equations (13.16) and (13.17) that the conversion of endgroups obeys the kinetics of a simple, second-order reaction; the second-order reaction having perfect initial stoichiometry in the case of Equation (13.17).

13.3.2 Number and Weight Average Chain Lengths

Solutions for the general case are

$$\bar{l}_N = \frac{1 + S_{AB}}{1 + S_{AB} - 2X_A S_{AB}} \tag{13.18}$$

$$\bar{l}_W = \frac{(1 + S_{AB})(1 + X_A^2 S_{AB}) + 4X_A S_{AB}}{(1 + S_{AB})(1 - X_A^2 S_{AB})} \tag{13.19}$$

For zero conversion, $\bar{l}_N = \bar{l}_W = 1$ since only monomers are present initially. At high conversion, Equation (13.18) approaches Equation (13.4). The polydispersity for the complete conversion case is

$$PD = \frac{\bar{l}_W}{\bar{l}_N} = \frac{(1 + 6S_{AB} + S_{AB}^2)}{(1 + S_{AB})^2} \tag{13.20}$$

The polydispersity is 2 for perfect stoichiometry or self-condensing monomers, but Equation (13.4) shows that $\bar{l}_N \to \infty$ for this case. $PD = 2$ is an asymptotic value that applies exactly only in the limit of high molecular weight. However, PD closely approaches 2 at quite low chain lengths.

Example 13.2: Determine PD as a function of chain length for binary polycondensations that go to completion.

Solution: Equation (13.4) is used to relate $\bar{l}_N$ and S_{AB} at complete conversion. The polydispersity is then calculated using Equation (13.20). Some results are shown in Table 13.3. The polydispersity becomes experimentally indistinguishable from 2 at a chain length of about 10.

Example 13.3: The conversion of a self-condensing reaction can be limited to give polymers with finite lengths. How does the polydispersity of these polymers compare with those in Example 13.2 where the reaction went to completion with imperfect stoichiometry? Make the comparison at the same average chain length.

TABLE 13.3 Polydispersities for Binary Polycondensations

X_A	S_{AB}	$\bar{l}_N$	PD
1	0	1	1
1	0.2	1.5	1.5556
1	0.3333	2	1.7500
1	0.6667	5	1.9600
1	0.8182	10	1.9900
1	0.9048	20	1.9975
1	0.9608	50	1.9996
1	0.9802	100	1.9999

Solution: The number and weight average chain lengths for a self-condensing polymerization are obtained from Equation (13.18) and (13.19) by setting $S_{AB}=1$:

$$\bar{l}_N = \frac{1}{1-X_A} \tag{13.21}$$

$$\bar{l}_W = \frac{1+X_A}{1-X_A} \tag{13.22}$$

so that

$$PD = 1 + X_A \tag{13.23}$$

Table 13.4 tabulates results at the same value of $\bar{l}_N$ as in Table 13.3. The polydispersities are lower than when the same average chain length is prepared by a binary polycondensation going to completion. The stoichiometry-limited binary polycondensations have a higher polydispersity because the monomer in stoichiometric excess (the B monomer) is included in the calculations. This broadens the molecular weight distribution.

13.3.3 Molecular Weight Distribution Functions

The probability distribution for chain lengths in a binary polycondensation is

$$\begin{aligned} f(l) &= \frac{2(1-X_A S_{AB})(1-X_A)}{1+S_{AB}-2X_A S_{AB}} X_A^{l-1} S_{AB}^{l/2} & \text{if } l \text{ is even} \\ &= \frac{(1-X_A S_{AB})^2 + S_{AB}(1-X_A)^2}{1+S_{AB}-2X_A S_{AB}} X_A^{l-1} S_{AB}^{(l-1)/2} & \text{if } l \text{ is odd} \end{aligned} \tag{13.24}$$

Figures 13.1 and 13.2 illustrate these distributions by number and weight, respectively. The most abundant species by number is always the monomer, even for the case of perfect stoichiometry. The distribution by weight usually shows an interior maximum. Note that the even-numbered mers are missing because the A-type endgroups are reacted to completion.

If $S_{AB}=1$, Equation (13.24) reduces to a simple form:

$$f(l) = (1-X_A)X_A^{l-1} \qquad \text{for all } l \tag{13.25}$$

This is the famous *Flory distribution*. Here, it is expressed in terms of the conversion, but Equation (13.21) can be used to replace X_A with $\bar{l}_N$. The result is

$$f(l) = \frac{1}{\bar{l}_N}\left[1-\frac{1}{\bar{l}_N}\right]^{l-1} \tag{13.26}$$

TABLE 13.4 Polydispersities for Self-Condensing Polymerizations

X_A	S_{AB}	$\bar{l}_N$	PD
0	1	1	1
0.33	1	1.5	1.33
0.50	1	2	1.50
0.80	1	5	1.80
0.90	1	10	1.90
0.95	1	20	1.95
0.98	1	50	1.98
0.99	1	100	1.99

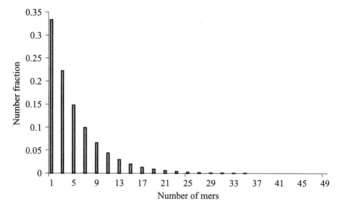

FIGURE 13.1 Molecular weight distribution by number fraction for a binary polycondensation going to completion with $\bar{l}_N = 5$ and $\bar{l}_W = 9.8$.

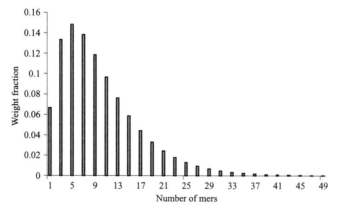

FIGURE 13.2 Molecular weight distribution by weight fraction for a binary polycondensation going to completion with $\bar{l}_N = 5$ and $\bar{l}_W = 9.8$.

The mean of this *most probable distribution* is (obviously) $\bar{l}_N$, and the polydispersity (PD) is

$$\text{PD} = 1 + X_A = 2 - \frac{1}{\bar{l}_N} \tag{13.27}$$

The equal reactivity assumption is sometimes wrong. Suppose the reaction in Equation (13.2) is elementary but that the rate constant, $k_{l,m}$, depends on l and m. The usual case is for the reactions involving the monomers and smaller oligomers to be relatively fast, but for the rate dependence to vanish for reasonably long chains. Then calculations based on equal reactivity are usually adequate if the desire is to make polymer of high molecular weight. If a detailed accounting is needed, brute force numerical calculations can be used. Numerical solutions for $\bar{l}_N$ of a few hundred were feasible in the early 1960s, and any reasonable kinetic scheme is now solvable in detail.

Example 13.4: Calculate the molecular weight distribution for a self-condensing polymerization with $k_{l,m} = k_0/(l+m)$. Stop the calculations when $\bar{l}_N = 5$. Compare the results with those of the Flory distribution.

Solution: Polymer of length l is formed by the reaction of any two molecules whose lengths sum to l. It is consumed when it reacts with any molecule. The batch rate equation governing the formation of polymer of length l is

$$\frac{dc_l}{dt} = \frac{k_0}{l}\sum_{m=1}^{l-1} c_m c_{l-m} - k_0 c_l \sum_{m=1}^{\infty} \frac{c_m}{l+m}$$

The following program segment illustrates the solution. The time scale is arbitrary. The code is computationally inefficient but straightforward. Execution times are trivial because $\bar{l}_N$ is small. Results are $\bar{l}_W = 7.04$ and PD = 1.41 compared with $\bar{l}_W = 9$ and PD = 1.8 for the equal-reactivity case. See Figure 13.3. Convergence was tested by comparing the equal-reactivity case—as calculated using the program—with the analytical result, Equation (13.27).

```
Dim c(100)
c(1) =1
xk=1
dt=0.001

Do
    u0 = 0
    u1 = 0
    u2 = 0
    For j = 1 To 99
```

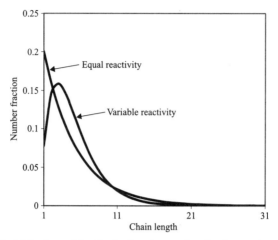

FIGURE 13.3 Molecular weight distributions by number for the equal-reactivity case and the variable-reactivity case of Example 13.4.

```
    Sum = 0
    For m = 1 To j - 1
        Sum = Sum + c(m) * c(j - m)
    Next m
    tot = 0
    For m = 1 To 99
        'tot = tot + xk * c(m) 'Equal reactivity case
        tot = tot + xk / (j + m) * c(m)
    Next m
    'c(j) = c(j) + (xk * Sum - c(j) * tot) * dt
    'Equal reactivity case
    c(j) = c(j) + (xk / j * Sum - c(j) * tot) * dt
    u0 = u0 + c(j)
    u1 = u1 + j * c(j)
    u2 = u2 + j * j * c(j)
    Next j
    xlength = u1 / u0
Loop While xlength < 5
```

13.4 KINETICS OF ADDITION POLYMERIZATIONS

Most addition polymerizations involve vinyl or diene monomers. The opening of a double bond can be catalyzed in several ways. Free-radical polymerization is the most common method for styrenic monomers, whereas coordination metal catalysis (Zigler-Natta and metallocene catalysis) is important for olefin polymerizations. The specific reaction mechanism may generate some catalyst residues, but there are no true coproducts. There are no stoichiometry requirements, and equilibrium limitations are usually unimportant so that quite long chains are formed; $\bar{l}_N > 500$ is typical of addition polymers.

The first step in an addition polymerization is *initiation* to form a polymer chain of length 1:

$$M + I \rightarrow IP_1 \tag{13.28}$$

The moiety denoted as I is the initiator group. It can be as simple as a free radical or as complicated as a transition metal atom bonded to organic ligands and located on a catalytic support. The next step in the polymerization is *propagation*, i.e., the repeated insertion of monomer units into the chain to create an incrementally longer chain

$$M + IP_l \rightarrow IP_{l+1} \qquad l = 1, 2, \ldots \tag{13.29}$$

The propagation reaction is a more mechanistic version of Equation (13.1) and accounts for most of the monomer consumption. The growth of the chain can be stopped by *chain transfer*, the simplest form of which is chain transfer to monomer:

$$M + IP_l \rightarrow IP_1 + P_l \qquad l = 1, 2, \ldots \tag{13.30}$$

Here, P_l is a finished polymer molecule, and IP_1 starts a new chain. The final step in the polymerization is *termination*. It deactivates the initiator group and gives a finished polymer molecule.

13.4.1 Living Polymers

A batch, anionic polymerization—e.g., of styrene catalyzed by butyl lithium—is among the simplest addition polymerizations to analyze. Butyl lithium is added to a solution of styrene monomer to form $Li^+Bu^-P_1$ as the first propagating species. The initiation step is fast and consumes I_0 molecules each of the initiator and of the monomer to give I_0 growing polymer chains. If the reaction is maintained at moderate temperatures, say $< 50°C$, and if there are no contaminants such as water, CO_2, or oxygen, then there are no chain transfer or termination reactions. The I_0 chains constitute *living polymer* molecules that continue to grow until all the monomer is consumed. The average chain length will be

$$\bar{l}_N(t \rightarrow \infty) = \frac{M_0}{I_0}$$

The chains share the same chemical environment and grow at the same average rate. If the final chain length is large, all the molecules will have more or less the same size and the resulting polymer will be approximately monodisperse. Furthermore, if a second monomer is added to the system after the first has been depleted, a block copolymer can be formed.

The present section analyzes the above concepts in detail. There are many different mathematical methods for analyzing molecular weight distributions. The *method of moments* is particularly easy when applied to a living polymer polymerization. Equation (13.30) shows the propagation reaction, each step of which consumes one monomer molecule. Assume equal reactivity. Then for a batch polymerization,

$$\frac{dM}{dt} = -k_p M \sum_{l=1}^{\infty} c_l = -k_p M \mu_0 = -k_p M I_0 \qquad (13.31)$$

where $\mu_0 = I_0$ is the total concentration of polymer molecules and is constant for living polymer. Equation (13.31) is subject to the initial condition that $M = M_0 - I_0$ at $t = 0$, zero time being just after the fast initiation step that consumes I_0 monomer molecules. The solution is

$$M = M(t) = (M_0 - I_0) \exp(-k_p I_0 t) \qquad (13.32)$$

Polymer of length l is formed when polymer of length $l - 1$ reacts with monomer and is consumed when it itself reacts with monomer:

$$\frac{dc_l}{dt} = k_p M c_{l-1} - k_p M c_l = k_p M (c_{l-1} - c_l) \qquad (13.33)$$

The time variable is transformed to incorporate M:

$$\frac{dc_l}{d\tau} = \frac{dc_l}{k_p M \, dt} = c_{l-1} - c_l \qquad (13.34)$$

where $c_0 = 0$ and $d\tau = k_p M \, dt$. Integrating Equation (13.32) gives

$$\tau = \frac{M_0 - I_0}{I_0} [1 - \exp(-k_p I_0 t)] = \frac{M_0 - I_0 - M}{I_0} \qquad (13.35)$$

Equation (13.34) is one member of an infinite set of ODEs. Add them all up to obtain

$$\sum_{l=1}^{\infty} \frac{dc_l}{d\tau} = \frac{d \sum_{l=1}^{\infty} c_l}{d\tau} = \frac{d\mu_0}{d\tau} = -c_1 + (c_1 - c_2) + (c_2 - c_3) + \cdots = 0$$

which is an ODE for μ_0. To obtain ODEs for μ_1 and μ_2, multiply Equation (13.34) by l and l^2 and then sum:

$$\sum_{l=1}^{\infty} l \frac{dc_l}{d\tau} = \frac{d \sum_{l=1}^{\infty} l c_l}{d\tau} = \frac{d\mu_1}{d\tau}$$

$$= -c_1 + 2(c_1 - c_2) + 3(c_2 - c_3) + \cdots = c_1 + c_2 + c_3 + \cdots = \mu_0$$

$$\sum_{l=1}^{\infty} l^2 \frac{dc_l}{d\tau} = \frac{d\sum_{l=1}^{\infty} l^2 c_l}{d\tau} = \frac{d\mu_2}{d\tau}$$

$$= -c_1 + 4(c_1 - c_2) + 9(c_2 - c_3) + \cdots = 3c_1 + 5c_2 + 7c_3 + \cdots$$

$$= \mu_0 + 2\mu_1$$

The infinite set of simultaneous ODEs has been transformed to a set of moment equations that can be solved sequentially. The first three members of the set are:

$$\frac{d\mu_0}{d\tau} = 0 \qquad \mu_0 = I_0 \quad \text{at} \quad \tau = 0$$

$$\frac{d\mu_1}{d\tau} = \mu_0 \qquad \mu_1 = I_0 \quad \text{at} \quad \tau = 0$$

$$\frac{d\mu_2}{d\tau} = \mu_0 + 2\mu_1 \qquad \mu_2 = I_0 \quad \text{at} \quad \tau = 0$$

An infinite set of moments is theoretically necessary to describe a molecular weight distribution; but as a practical matter, knowing moments 0, 1, and 2 is usually adequate. The initial condition for all the moments is $\mu_n = I_0$ at $\tau = 0$. Solution gives

$$\mu_0 = I_0$$
$$\mu_1 = I_0(\tau + 1)$$
$$\mu_2 = I_0(\tau^2 + 3\tau + 1)$$

From these, we find

$$\bar{l}_N = \frac{\mu_1}{\mu_0} = \tau + 1 \tag{13.36}$$

$$\bar{l}_W = \frac{\mu_2}{\mu_1} = \frac{\tau^2 + 3\tau + 1}{\tau + 1} \tag{13.37}$$

$$PD = \frac{\bar{l}_W}{\bar{l}_N} = \frac{\mu_0 \mu_2}{\mu_1^2} = \frac{\tau^2 + 3\tau + 1}{(\tau + 1)^2} \tag{13.38}$$

The predicted polydispersities are quite low, with a maximum value of 1.25 at $\bar{l}_N = 2$.

Anionic polymerizations make the molecular weight standards that are used to calibrate size-exclusion chromatographs. Equation (13.38) predicts $PD = 1.001$ at $\bar{l}_N = 1000$. Actual measurements give about 1.05. The difference is attributed to impurities in the feed that cause terminations and thus short chains. Also, the chromatograph has internal dispersion so that a truly monodisperse sample would show some spread. Even so, a PD of 1.05 is extremely narrow by polymer standards. This does not mean it is narrow in an absolute sense. Figure 13.4 shows the molecular weight distribution by number for a polymer that has a polydispersity of 1.05. The spread in molecular weights relative to the mean is appreciable.

13.4.2 Free-Radical Polymerizations

A typical free-radical polymerization involves all of the steps in addition polymerization: initiation, propagation, chain transfer, and termination. In contrast to anionic and cationic polymerization that can give living polymers, the lifetime of growing polymer chains is short under free-radical catalysis, typically on the order of milliseconds. The quasi-steady hypothesis discussed in Section 2.5.3 is usually applied in theoretical analyses because growing chains are terminated at approximately the same rate at which they are initiated. Although some monomers will initiate spontaneously upon heating (*thermal initiation*), a chemical initiator is usually used as the original source of free radicals. A typical initiator decomposes according to first-order kinetics to yield two free radicals:

$$I_2 \xrightarrow{k_i} 2f\, I\bullet \qquad \mathscr{R} = k_i[I_2]$$

The factor of f reflects the fact that some of the free radicals immediately recombine into stable molecules that do not initiate polymerization. A primary radical formed by the chemical initiator reacts with monomer to form a propagating radical that contains one monomer unit:

$$I\bullet + M \rightarrow IP_1\bullet$$

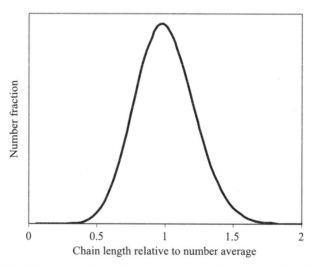

FIGURE 13.4 Molecular weight distribution for a polymer that has a polydispersity of 1.05.

This is the starting chain and is identical to the IP_1 species in Equation (13.28), The • notation explicitly shows the free-radical nature of the polymerization, and the moiety denoted by I represents a fragment from the chemical initiator; e.g. a butyl group when *t*-butyl peroxide is the initiator. The propagation reaction is

$$M + IP_l \cdot \rightarrow IP_{l+1} \cdot \quad l = 1, 2, \ldots \quad \mathscr{R}_p = k_p M \sum_{l=1}^{\infty} [IP_l \cdot] = k_p M P \cdot$$

where P• denotes the total concentration of growing chains. The constant value for k_p assumes equal reactivity.

Various forms of chain transfer are common. Chain transfer to monomer stops chain growth but initiates a new chain of length $l = 1$. It acts to limit the polymer molecular weight without reducing conversion. Chain transfer to polymer causes long-chain branching. Chain transfer to solvent limits molecular weight and may reduce conversion depending on the fate of the transferred free-radical. Chain transfer to a free radical scavenger such as *t*-butylcatechol inhibits polymerization.

In contrast to chain transfer, termination reactions destroy free radicals. Two mechanisms are considered. *Termination by combination* produces a single molecule of dead polymer:

$$IP_l \cdot + IP_m \cdot \rightarrow IP_{l+m}I \quad \mathscr{R} = k_c [P \cdot]^2$$

The free radicals combine to form a carbon-to-carbon bond and give a saturated polymer molecule with initiator fragments on both ends. *Termination by disproportionation* produces two polymer molecules, one of which will contain a double bond:

$$IP_l \cdot + IP_m \cdot \rightarrow IP_p + IP_q \quad \mathscr{R} = k_d [P \cdot]^2$$

where $l + m = p + q$.

Chain lifetimes are small and the concentration of free radicals is low. To a reasonable approximation, the system consists of unreacted monomer, unreacted initiator, and dead polymer. The quasi-steady hypothesis gives

$$\frac{dP \cdot}{dt} = fk_i[I_2] - (k_c + k_d)[P \cdot]^2 \approx 0$$

where it was assumed that the termination rate constants k_c and k_d are independent of chain length. Solving for [P•] and substituting into the propagation rate gives

$$\mathscr{R}_p = k_p M P \cdot = k_p M \sqrt{\frac{2fk_i[I_2]}{k_c + k_d}} \quad (13.39)$$

The propagation reaction accounts for nearly all the consumption of monomer. Thus, Equation (13.39) predicts that the polymerization rate will be first order in monomer concentration and half-order in initiator concentration. This is confirmed by experiments at low polymer concentrations, but can be violated at high polymer concentrations. The termination mechanisms require pairwise interactions between large molecules, and these become increasingly difficult at high polymer concentrations due to chain entanglements. The propagation reaction is less affected, and the net rate of polymerization can actually increase. The phenomenon of the rate increasing as M decreases is a form of autoacceleration known as the *gel effect*. It is particularly noticeable in the polymerization of methyl methacrylate. See Section 2.5.4 and Figure 2.5.

The ratio of propagation rate to termination rate gives the *dynamic chain length* of the growing polymer. Ignoring chain transfer,

$$(\bar{l}_N)_{live} = \frac{k_p M}{\sqrt{2f(k_c + k_d)k_i[I_2]}} \tag{13.40}$$

The dynamic chain length is the number average length of the growing chains before termination. The dead polymer will have the same average length as live polymer if termination is by disproportionation and will have twice this length if termination is by combination.

The growing polymer chains have the most probable distribution defined by Equation (13.26). Typically, $\bar{l}_N$ is large enough that $PD \approx 2$ for the growing chains. It remains 2 when termination occurs by disproportionation. Example 13.5 shows that the polydispersity drops to 1.5 for termination by pure combination. The addition rules of Section 13.2.2 can be applied to determine $1.5 < PD < 2$ for mixed-mode terminations, but disproportionation is the predominant form for commercial polymers.

Example 13.5: Determine the instantaneous distributions of chain lengths by number and weight before and after termination by combination. Apply the quasi-steady and equal reactivity assumptions to a batch polymerization with free-radical kinetics and chemical initiation.

Solution: The equal reactivity assumption says that k_p and k_c are independent of chain length. The quasi-steady hypothesis gives $d[R\bullet]/dt = 0$. Applying these to a material balance for growing chains of length l gives

$$k_p M[P_{l-1}\bullet] - k_p M[P_l\bullet] - k_c[P_l\bullet][P\bullet] = 0$$

where [P•] denotes the entire population of growing chains. Solving for $[P_l\bullet]$ gives

$$[P_l\bullet] = \frac{k_p M[P_{l-1}\bullet]}{k_p M + k_c[P\bullet]} = \omega[P_{l-1}\bullet]$$

where ω is a proportionality factor that is constant at any instant of time. Initiation creates species [P$_1$•], and propagation takes it from there. The relative concentrations of chain lengths are distributed according to

Chain length, l	Relative concentration, [P$_l$•]/[P$_1$•]
1	1
2	ω
3	ω^2
...	...

This is a geometric progression that sums to $(\bar{l}_N)_{live} = 1/(1-\omega)$. The fraction of the growing chains with length l is

$$f(l) = (1 - \omega)\omega^{l-1}$$

which is identical in form to Equation (13.25). Thus, the growing polymer chains have a Flory distribution. The dead polymer has a different and narrower distribution. It is narrower due to the averaging process that results when termination is by combination. An unusually long chain will usually combine with a shorter chain, giving a sum that is not too different from the average sum. We assume that termination occurs between pairs of growing chains chosen completely at random. Applying a theorem in probability theory shows that

$$\sum_{l=1}^{\infty} f_{dead}(l)e^{sl} = \left[\sum_{l=1}^{\infty} f(l)e^{sl}\right]^2$$

where $f_{dead}(l)$ and $f(l)$ are the probability distributions of the dead and living polymer, respectively. The exponentially weighted sums are known as *moment-generating functions*. They have the property that

$$\lim_{s \to \infty} \frac{d^n}{ds^n} \sum_{l=1}^{\infty} f_{dead}(l)e^{sl} = \frac{\mu_n}{\mu_0}$$

Note that $n = 1$ gives $\mu_1/\mu_0 = \bar{l}_N$, and $n = 2$ gives $\mu_2/\mu_0 = \bar{l}_N\bar{l}_W$ for the dead polymer. For the case at hand,

$$(\bar{l}_N)_{dead} = \frac{2}{1-\omega}, \quad (\bar{l}_W)_{dead} = \frac{2+\omega}{1-\omega}, \quad \text{and} \quad (PD)_{dead} = \frac{2+\omega}{2}$$

In the limit of long chain lengths, $\omega \to 1$, $(\bar{l}_N)_{dead} \to 2(\bar{l}_N)_{live}$, and $(\bar{l}_W)_{dead} \to 3(\bar{l}_N)_{live}$. Thus, the polydispersity has a limiting value of 1.5 for the assumed kinetic scheme.

Unlike the convention for condensation polymers, calculations of $\bar{l}_N$ for free-radical polymers exclude unreacted monomer. High-molecular weight polymer is formed at the beginning of a polymerization, and the molecular weight remains approximately constant as the reaction proceeds. Conversion in a free-radical polymerization means the conversion of monomer to high-molecular-weight polymer. A reaction mixture at 50% conversion may consist of 50% monomer and 50% polymer with a number average molecular weight of 100,000 ($\bar{l}_N \approx 1000$). The viscosity of the polymerizing mixture at 50% conversion might be 10,000 times higher than that of the pure monomer. In contrast, molecular weight increases gradually in a polycondensation, and really high molecular weights are not achieved until near the end of a batch reaction. Conversion in a polycondensation refers to consumption of the stoichiometrically limiting endgroup, and 50% conversion with $S_{AB}=1$ gives dimer. See Table 13.4. The viscosity of the reaction mixture at 50% conversion is perhaps twice that of the pure monomer(s). Figure 13.5 compares viscosity versus conversion curves for the two types of polymerizations. Living polymer systems behave more like condensation polymerizations since the molecular weight builds slowly. Transition metal catalyses behave like free-radical polymerizations with high-molecular-weight polymer created at the onset.

The above discussion has been based on conventional free-radical catalysis. There has been substantial research on long-lived free radicals that can give a living polymer without the severe cleanliness requirements of anionic polymerizations. Unfortunately, it has not yet had commercial success.

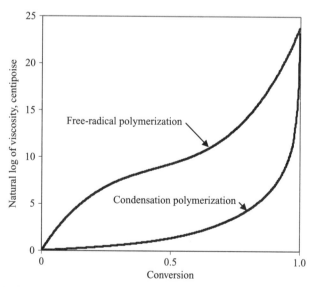

FIGURE 13.5 Representative profiles of viscosity versus conversion for free-radical and condensation polymerizations. The endpoint is 99.9% conversion with a number average chain length of 1000.

13.4.3 Transition Metal Catalysis

Ziegler-Natta catalysis enabled the polymerization of propylene and has subsequently come into wide use for the polymerization of olefins and dienes. The classic Ziegler-Natta catalyst is $TiCl_3$ plus $Al(C_2H_5)_3$, but there are variants that use other transition metals such as chromium and nickel. Polymerization is believed to occur by the repeated insertion of a double bond from the monomer into a previously formed metal-to-carbon bond. The homogeneous forms, where the polymerization occurs in a solvent for the polymer, give linear high density polyethylene (HDPE) of low polydispersity, $PD \approx 2$. Heterogeneous forms are used for gas-phase polymerization in a fluidized-bed reactor. The catalyst is typically supported on $MgCl_2$ and is introduced into the bed as a fine powder. These supported catalysts give high yields, high activities, and high stereospecificity for polypropylene, but they give polymer of undesirably high polydispersity, $PD > 10$. It is believed that the catalyst remains active and that chain transfer to monomer or to hydrogen controls the molecular weight. The polydispersity is attributed to large site-to-site variations in activity.

More recent inventions are the metallocene catalysts based on zirconium. They offer more uniform catalyst activity and can give a relatively narrow molecular weight distribution. More importantly, they offer better control over structure and copolymer composition distributions.

The chemical mechanisms of transition metal catalyses are complex. The dominant kinetic steps are propagation and chain transfer. There is no termination step for the polymer chains, but the catalytic sites can be activated and deactivated. The expected form for the propagation rate is

$$\mathcal{R}_p = k_p I M$$

where I is the concentration of active sites. All the catalytic sites are presumed active for homogeneous catalysis. However, cocatalysts such as $Al(C_2H_5)_3$ or methylaluminoxane (MAO) are required for the reaction to proceed, and their presence or absence will affect the value of I. Mass transfer limitations can arise in supported systems because the catalytic sites become buried under a deposited film of polymer. Monomer solubility and diffusion in the polymer are included in some models of the polymerization kinetics. A segregated model that follows the course of individual active sites is needed for gas-to-solid heterogeneous polymerizations.

13.4.4 Vinyl Copolymerizations

Vinyl monomers are often copolymerized, usually with free-radical or coordination metal catalysis, but occasionally by other mechanisms. Random copolymers are important items of commerce. The two monomers are present together in the reaction mixture and copolymerize to give more-or-less random arrangements of the monomers along the polymer chain.

Consider the polymerization of two vinyl monomers denoted by X and Y. Each propagation reaction can add either an X or a Y to the growing polymer chain, and it is unrealistic to assume that the monomers have equal reactivities. Furthermore, reaction probabilities can depend on the composition of the polymer chain already formed. We suppose that they depend only on the last member added to the chain. The growing chain to which an X-mer was last added is denoted as IX_n, and I denotes the catalytic site. There are four propagation reactions to consider:

$$IX_n + X \xrightarrow{k_{XX}} IX_{n+1} \qquad \mathscr{R} = k_{XX}[IX_n]x$$

$$IX_n + Y \xrightarrow{k_{XY}} IY_{n+1} \qquad \mathscr{R} = k_{XY}[IX_n]y$$

$$IY_n + X \xrightarrow{k_{YX}} IX_{n+1} \qquad \mathscr{R} = k_{YX}[IY_n]x$$

$$IY_n + Y \xrightarrow{k_{YY}} IY_{n+1} \qquad \mathscr{R} = k_{YY}[IY_n]y$$

The initiation and termination steps may also come in several varieties, but they will have little effect on overall chain composition provided the chains are long. The monomer consumption rates are

$$\mathscr{R}_X = -k_{XX}[IX_n]x - k_{YX}[IX_n]x$$

$$\mathscr{R}_Y = -k_{XY}[IX_n]y - k_{YY}[IY_n]y$$

The ratio of these propagation rates is the ratio in which the two monomers are incorporated into the polymer:

$$\frac{x_p}{y_p} = \frac{k_{XX}[IX_n]x + k_{YX}[IY_n]x}{k_{XY}[IX_n]y + k_{YY}[IY_n]y} = \frac{x}{y}\left[\frac{k_{XX}[IX_n] + k_{YX}[IY_n]}{k_{XY}[IX_n] + k_{YY}[IY_n]}\right]$$

This result can be simplified considerably by observing that, except for end effects, the number of transitions from X to Y along the polymer chain (i.e., structures like $\sim XY\sim$) must equal the number of transitions from Y to X (i.e., structures like $\sim YX\sim$). This requires that

$$k_{XY}[IX_n]y = k_{YX}[IY_n]x$$

Substitution gives the promised simplification:

$$\frac{x_p}{y_p} = \frac{x}{y}\left[\frac{r_X x + y}{x + r_Y y}\right] \tag{13.41}$$

where $r_X = k_{XX}/k_{XY}$ and $r_Y = k_{YY}/k_{YX}$ are known as *copolymer reactivity ratios*. The quantity x_p/y_p is the *instantaneous* ratio at which the two monomers are being incorporated into the polymer. It is usually different than the ratio of the monomers themselves, x/y. If $r_X > 1$, monomer X tends to homopolymerize so that sequences like ~XX~ will be favored. If $r_X < 1$, copolymerization is preferred, and sequences like ~XY~ will be common. Copolymer reactivity ratios are found by measuring the polymer compositions that result from polymerizations run at low conversion. Extensive compilations are available in the literature. Reactivity ratios are largely independent of temperature since they are ratios of rate constants that have similar activation energies. They are heavily dependent on the polymerization mechanism. Finding new catalysts that give desired values for r_X and r_Y is an active area of industrial research.

Example 13.6: The following data were obtained using low-conversion batch experiments on the bulk (solvent-free), free-radical copolymerization of styrene (X) and acrylonitrile (Y). Determine the copolymer reactivity ratios for this polymerization.

Mol% styrene in the monomer mixture	x/y	x_p/y_p	Mol% styrene in the instantaneous polymer
50	1.00	1.36	0.576
60	1.50	1.58	0.613
70	2.33	1.93	0.659
80	4.00	2.61	0.723
90	9.00	4.66	0.823

Solution: The conversion is low so that the polymer composition is given by Equation 13.41 with the monomer concentrations at the initial values. There are five data and only two unknowns, so that nonlinear regression is appropriate. The sum-of-squares to be minimized is

$$SS = \sum_{j=1}^{5} \left(\left[\frac{X_p}{Y_p} \right]_j - \frac{x_j}{y_j} \left[\frac{r_X x_j + y_j}{x_j + r_Y y_j} \right] \right)^2$$

Quantities with the subscript j are experimental data. Values for r_X and r_Y are determined using any convenient minimization technique. See Appendix 6. The results are $r_X = 0.41$ and $r_Y = 0.04$.

In a batch reactor, the relative monomer concentrations will change with time because the two monomers react at different rates. For polymerizations with a short chain life, the change in monomer concentration results in a *copolymer composition distribution* where polymer molecules formed early in the batch will have a different composition from molecules formed late in the batch. For living polymers, the drift in monomer composition causes a corresponding change down the growing chain. This phenomenon can be used advantageously to produce *tapered block copolymers*.

When uniform copolymers are desired, semibatch polymerization with continuous addition of the more reactive monomer is required. There is also a special, happenchance case where the monomer and polymer compositions are equal:

$$\frac{x_p}{y_p} = \frac{x}{y} \quad \text{if} \quad \left[\frac{r_X x + y}{x + r_Y y}\right] = 1 \qquad (13.42)$$

Equation (13.42) gives the condition that a *copolymer azeotrope* exists. The azeotropic composition is

$$\left[\frac{x}{y}\right]_{azeotrope} = \left[\frac{1 - r_Y}{1 - r_X}\right] \qquad (13.43)$$

which implies that either both r_X and $r_Y > 1$ or both r_X and $r_Y < 1$. The situation with r_X and $r_Y < 1$ is the more common. The styrene/acrylonitrile system with $r_X = 0.41$ and $r_Y = 0.04$ has an azeotrope at 62 mol% (76 wt%) styrene. Commercial polymer is manufactured with this composition. The narrow composition distribution gives superior color and clarity.

Example 13.7: A 50/50 (molar) mixture of styrene and acrylonitrile is batch polymerized by free-radical kinetics until 80% molar conversion of the monomers is achieved. Determine the copolymer composition distribution.

Solution: The solution to this problem does not require knowledge of the polymerization rate but only that polymerization somehow occurs to the specified extent. Suppose a small amount, dx, of monomer X is polymerized. The corresponding amount, dy, of monomer Y that copolymerizes is given by Equation (13.41). For a batch reaction,

$$\frac{dx}{dy} = \frac{x}{y}\left[\frac{r_X x + y}{x + r_Y y}\right] \qquad (13.44)$$

This equation can be integrated analytically, but a numerical solution is simpler. Set $x_0 = y_0 = 0.5$. Take small steps Δy and calculate the corresponding values for Δx until $x + y = 0.2$. Some results are given below.

Moles of monomer remaining	Mole fraction of styrene in the monomer	Instantaneous mole fraction of styrene in the polymer	Cumulative mole fraction of styrene in the polymer
1.0000	0.5000	0.5755	0.5755
0.8	0.4818	0.5698	0.5728
0.6	0.4538	0.5612	0.5693
0.4	0.4033	0.5467	0.5644
0.2	0.2743	0.5109	0.5564

This is a free-radical polymerization with short chain lives. The first molecules formed contain nearly 58 mol% styrene when there is only 50% styrene in the monomer mixture. The relative enrichment of styrene in the polymer depletes the concentration in the monomer mixture, and both the polymer and monomer concentrations drift lower as polymerization proceeds. If the reaction went to completion, the last 5% or so of the polymer would be substantially pure polyacrylonitrile.

The change in polymer composition as a result of monomer drift gives a *macroscopic composition distribution*. It can be eliminated by polymerizing at the azeotrope, by polymerizing to low conversion, by continuously adding the more reactive monomer to a semibatch reactor, or by polymerizing in a CSTR. The last method is usually preferred for polymers with short chain lives. There is also a *microscopic composition distribution* due to statistical fluctuations at the molecular level. Two molecules formed from the same monomer mixture will not have the same sequence down the chain nor exactly the same overall composition. The microscopic distribution follows binomial (coin tossing) statistics and will exist even if the polymer is manufactured at an azeotrope. Simulations using random numbers can determine the microscopic composition. For simulation of a single molecule in a macroscopically uniform sample that contains 60 mol% X and 40 mol% Y on average, flip a coin with a 60/40 bias $\bar{l}_N$ times. Do this for enough molecules that the statistics become clear. Figure 13.6 shows the combined macroscopic and microscopic distributions for the batch polymerization of Example 13.7.

Sequence length distributions are occasionally important. They measure the occurrences of structures like $\sim$YXY$\sim$, $\sim$YXXY$\sim$, and $\sim$YXXXY$\sim$ in a random copolymer. These can be calculated from the reactivity ratios and the polymer composition. See, for example, Ham.[1]

The copolymerization theory presented here assumes that the reactivity of a growing chain depends on the last mer added to the chain. Chemical theorists

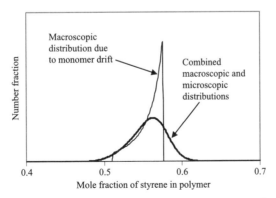

FIGURE 13.6 Copolymer composition distribution resulting from the batch polymerization of styrene and acrylonitrile.

point out that there can be *penultimate effects* where reactivity depends on the next-to-last (penultimate) mer as well as the last (ultimate) mer. Thus, ~XX would behave differently than ~YX. This is undoubtedly true, but the extra complexity seldom seems justified from a reaction engineering viewpoint.

13.5 POLYMERIZATION REACTORS

The properties of a polymer depend not only on its gross chemical composition but also on its molecular weight distribution, copolymer composition distribution, branch length distribution, and so on. The same monomer(s) can be converted to widely differing polymers depending on the polymerization mechanism and reactor type. This is an example of *product by process*, and no single product is best for all applications. Thus, there are several commercial varieties each of polyethylene, polystyrene, and polyvinyl chloride that are made by distinctly different processes.

Table 13.5 classifies polymerization reactors by the number and type of phases involved in the reaction. The most important consideration is whether the continuous phase contains a significant concentration of high-molecular-weight polymer. When it does, the reactor design must accommodate the high viscosities and low diffusivities typical of concentrated polymer solutions. It is the design of reactors for operation with a continuous polymer phase that most distinguishes polymer reaction engineering from chemical reaction engineering. When the polymer-rich phase is dispersed, the chemical kinetics and mass transfer steps may be very complex, but the qualitative aspects of the design resemble those of the heterogeneous reactors treated in Chapter 11.

A polymerization reactor will be heterogeneous whenever the polymer is insoluble in the mixture from which it was formed. This is a fairly common situation and gives rise to the precipitation, slurry, and gas-phase polymerizations listed in Table 13.5. If the polymer is soluble in its own monomers, a dispersed-phase polymerization requires the addition of a nonsolvent (typically water) together with appropriate interfacial agents. These extraneous materials require additional downstream purification and separation and may compromise product quality. *Bulk polymerizations* from pure, undiluted monomer are nominally preferred, although the use of a solvent to ease processing problems is quite common. For high-volume polymers, like high-volume chemicals, continuous operation is generally preferred over batch.

13.5.1 Stirred Tanks with a Continuous Polymer Phase

Continuous-flow stirred tank reactors are widely used for free-radical polymerizations. They have two main advantages: the solvent or monomer can be boiled to remove the heat of polymerization, and fairly narrow molecular weight and copolymer composition distributions can be achieved. Stirred tanks or

TABLE 13.5 Classification of Polymerization Reactors

Continuous phase	Dispersed phase	Type of polymerization	Example systems
Polymer dissolved in monomer	None	Homogeneous bulk polymerization	Poly(methyl methacrylate) in methyl methacrylate monomer
Polymer dissolved in solvent	None	Homogeneous solution polymerization	Polymerization of high-density polyethylene in hexane
Polymer in solution	Any (e.g., a condensation product or another polymer phase)	Heterogeneous bulk or solution polymerization	Salt precipitating from a condensation reaction. Prepolymerized rubber precipitating from a solution of polystyrene in styrene monomer
Water or other nonsolvent	Polymer or polymer in solution	Suspension, dispersion, or emulsion polymerization	Emulsion polymerization of a rubber latex. Suspension polymerization of expandable polystyrene
Liquid monomer	Polymer swollen with monomer	Precipitation or slurry polymerization	Polypropylene in a pool of liquid propylene
Gaseous monomer	Polymer	Gas-Phase polymerization	Fluidized-bed process for high-density polyethylene

functionally equivalent loop reactors are also used for the transition metal polymerization of olefins and dienes when the catalyst is homogeneous. CSTRs are used for the low-molecular-weight portions of condensation polymerizations when the stoichiometry requirements are not severe. Their principal disadvantage is that high conversions are difficult to achieve. They suffer from the usual problem of low average reaction rate. A second problem at high conversion is that the viscosity can be several orders of magnitude higher than the design limit for stirred tanks. Some practical designs use stirred tanks up to about 75% conversion, with further polymerization carried out in a tubular reactor.

Molecular Weight Distributions. The CSTRs produce the narrowest possible molecular weight distributions for fast chain growth, short chain lifetime polymerizations like free-radical and coordination metal catalysis. The mean residence time in the CSTR will be minutes to hours, and the chain lifetimes are fractions of a second. Any chain that initiates in the CSTR will finish its growth there. All the polymer molecules are formed under identical, well-mixed conditions and will have as narrow a molecular weight distribution (typically $PD \approx 2$) as is possible for the given kinetic scheme.

Condensation polymers and living addition polymers continue to grow throughout their stay in the reactor. All the chains in a CSTR grow in the same reaction environment and, subject to the equal reactivity assumption, grow at the same rate. The chain lengths at the reactor outlet will be proportional to the time spent in the reactor since initiation. As discussed in Chapter 15, the reaction times in a CSTR have an exponential distribution. This gives an exponential distribution of chain lengths. The most probable distribution approximates an exponential distribution whenever $\bar{l}_N$ is reasonably long. The polydispersity of a condensation polymer made in a CSTR will be approximately 2, as it is in a batch reactor. The same argument applies to a living polymer. It will also emerge from a CSTR with a polydispersity of 2, but this is dramatically different from the narrow molecular weight distribution possible in a batch reactor. CSTRs cannot be used to produce block copolymers or homopolymers with narrow molecular weight distributions. They can be used for most other polymers of commercial utility. For heterogeneous polymerizations with a suspended catalyst, they will not eliminate the problem in site-to-site variations in activity, but they will produce as narrow an MWD as is possible for the catalyst.

Example 13.8: Apply the method of moments to an anionic polymerization in a CSTR.

Solution: Assume that initiation occurs instantly as fresh monomer enters the reactor. A monomer balance for the CSTR gives

$$M_0 - I_0 - k\bar{t}MI_0 \sum_{i=1}^{\infty} c_i = M$$

or

$$M = \frac{M_0 - I_0}{1 + k_p \bar{t} I_0} \tag{13.45}$$

A balance on polymer of length l gives an infinite set of algebraic equations:

$$k_p \bar{t} c_{l-1} M - k_p \bar{t} c_l M = c_l$$

Sum these equations from $l = 1$ to ∞ to obtain

$$\mu_0 = I_0$$

Multiply them by l and l^2 and use the methods of Section 13.3.2 to obtain

$$\bar{l}_N = K^* + 1 \tag{13.46}$$

$$\bar{l}_W = \frac{2(K^*)^2 + 3K^* + 1}{K^* + 1} \tag{13.47}$$

where $K^* = k_p M \bar{t}$. Divide $\bar{l}_W$ by $\bar{l}_N$ and take the limit as $K^* \to \infty$ to show that $PD \to 2$ in the limit of high molecular weight.

An isothermal batch reactor produces a fairly narrow MWD, except possibly at high conversions. Conditions will vary from the start to the end of a batch reaction, and the molecular weight distribution will be somewhat broader than for a CSTR. The dynamic chain length of a free-radical polymer will drift because of changes in monomer concentration and initiation rate, although usually not by much when measured by the cumulative $\bar{l}_N$ and $\bar{l}_W$ of the finished polymer. When the catalyst is heterogeneous, an isothermal batch reactor will produce approximately the same result as for a CSTR, except possibly at high conversions. Of course, maintaining isothermal operation may not be easy.

Heat Transfer. Heat removal is the major issue for vinyl addition polymerizations, and CSTRs are well suited for this application. The homogeneous nucleation possible with boiling heat transfer allows easy scaleup at nearly isothermal conditions. In very large vessels, a temperature difference between the top and the bottom will arise due to the static head, but this has not been found to cause operational difficulties. The actual removal of heat occurs in the overhead condenser. There is sometimes a concern about fouling the condenser with polymer, but this problem can usually be overcome by good design and is not a problem when the polymer is soluble in the condensing liquid. Boiling heat transfer is routinely used for addition polymerizations. Temperature control by precooling the feed is occasionally used. A CSTR is obviously suited for precooled feeds while a PFR is not. Sensible heat transfer through the vessel walls or internal coils can be used in small reactors. Loop reactors, where most of the reactor volume is in the heat exchangers, are preferred for large reactors.

Most condensation polymers have negligible heats of reaction. See Table 13.2. Heat must be supplied to evaporate by-products such as water or ethylene glycol. An external heat exchanger is the best method for heating large reactors. Flashing the recycle stream as it enters the vessel also aids in devolatilization.

Copolymerizations. The uniform chemical environment of a CSTR makes it ideally suited for the production of copolymers. If the assumption of perfect mixing is justified, there will be no macroscopic composition distribution due to monomer drift, but the mixing time must remain short upon scaleup. See Sections 1.5 and 4.4. A real stirred tank or loop reactor will more closely approach perfect mixing when the circulation rate is high. The narrowing of the macroscopic composition distribution as a function of circulation rate can be modeled. When the macroscopic distribution becomes commensurate in breadth to the microscopic distribution, further inputs of power are unwarranted.

Laminar flow is almost inevitable in a polymer reactor when the polymer is soluble in the continuous phase. Molecular diffusivities are also low. This combination suggests that segregation is a possibility in a stirred tank reactor. See Chapter 15. A segregated stirred tank gives broader molecular weight distributions, higher conversions, and broader copolymer composition distributions than a CSTR that has good internal mixing. Given the appropriate rate expressions, calculations are quite feasible for the limiting cases of complete segregation and perfect mixing. These tend to be alarming since the copolymer composition distribution in a completely segregated CSTR will be even broader than that for a batch reactor. Fortunately, there is limited evidence of segregation in industrial reactors. It seems that segregation is an academic concept with little practical relevance.

13.5.2 Tubular Reactors with a Continuous Polymer Phase

Web-coating polymerizations (e.g., as used for photographic film and coated abrasives) literally achieve a piston flow reaction environment. Mechanically driven screw devices used as finishing reactors for PET closely approximate piston flow. Motionless mixers can do this as well. However, polymer reactors that closely approximate piston flow are the exception.

Tubular reactors are occasionally used for bulk, continuous polymerizations. A monomer or monomer mixture is introduced at one end of the tube and, if all goes well, a high-molecular-weight polymer emerges at the other. The classic example is the high-pressure polymerization of ethylene in a single tube approximately 1.75 inches in diameter and several miles long. Scaleup is done by adding another mile to the tube length. Some polystyrene plants use single-tube or shell-and-tube reactors as finishing reactors after the bulk of the polymer has been made. These commercial examples skirt three types of instability that can arise in tubular polymerizers:

Velocity profile elongation: Low fluid velocities near the tube wall give rise to high extents of polymerization, high viscosities, and yet lower velocities. The velocity profile elongates, possibly to the point of hydrodynamic instability.

Thermal runaway: Temperature control in a tubular polymerizer depends on convective diffusion of heat. This becomes difficult in a large-diameter tube, and temperatures may rise to a point where a thermal runaway becomes inevitable.

Tube-to-tube interactions: The problems of velocity profile elongation and thermal runaway can be eliminated by using a multitubular reactor with many small-diameter tubes in parallel. Unfortunately, this introduces another form of instability. Tubes may plug with polymer that cannot be displaced using the low-viscosity inlet fluid. Imagine a 1000-tube reactor with 999 plugged tubes!

Reacting to low conversions avoids most problems. The polyethylene example uses a single, small-diameter tube. This avoids thermal runaway and mitigates the velocity elongation problem. The polystyrene reactor avoids the multitubular stability problem by using a substantially polymerized, high-viscosity feed.

Figure 13.7 illustrates stability regimes for the thermally initiated polymerization of styrene for laminar flow in a single tube. Design and operating variables

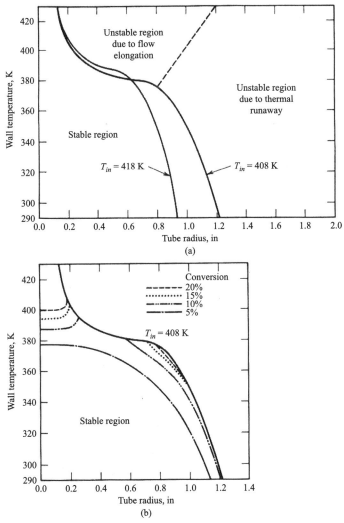

FIGURE 13.7 Performance of a laminar flow, tubular reactor for the bulk polymerization of styrene; $T_{in} = 35°C$ and $\bar{t} = 1$ h. (a) Stability regions. (b) Monomer-conversion within the stable region.

are the physical dimensions of the tube, the operating temperatures and pressures, and the flow rate. The aspect ratio L/R is unimportant provided it is reasonably large, and the operating pressure is unimportant unless it is so high that liquid compressibility becomes significant. Steady-state operation can be described by four variables: T_{in}, T_{wall}, R, and $\bar{t}$. Figure 13.7 shows the influence of two of these with the other two held constant at plausible values. Stable operation is easy to achieve in a capillary tube because mass and thermal diffusion remove gradients in the radial direction. Thus, the viscosity is uniform across the tube, and no hotspots develop. As the tube diameter is increased, mass and heat transfer become progressively more difficult. The mass diffusivity is several orders of magnitude lower than the thermal diffusivity, concentration gradients develop before thermal gradients, and velocity profile elongation emerges as the first limit on conversion. In still larger tubes, operation is limited by thermal stability. For the polystyrene example of Figure 13.7, a thermal runaway is predicted if the tube radius exceeds 1.25 in. This is true even if the tube wall were held at 0 K (pretending the model remains valid).

The polystyrene example in Figure 13.7 incorporates a detailed kinetic model of the polymerization. The model uses a reasonably sophisticated description of viscosity as a function of temperature, polymer concentration, and molecular weight. The convective diffusion equations for mass and heat are solved together with the effects of viscosity on the velocity profile. Assembling such a model requires weeks of effort. The following examples are simpler but illustrate design methodology for tubular polymerizers. We begin with a lumped parameter model for the polymerization of styrene

Example 13.9: Illustrate temperature and molecular weight changes in a tubular reactor by constructing a simple model of styrene polymerization in a tube.

Solution: The simplest model supposes piston flow in a tube of constant cross section and ignores the $a(d\bar{u}/dz)$ term in Equation (3.5), even though $\bar{u}$ will be allowed to vary. Thus,

$$\bar{u}(z)\frac{dM}{dz} = -\mathcal{R}_p$$

Equation (5.24) gives the energy balance:

$$\frac{dT}{dz} = \frac{-\Delta H_R \mathcal{R}_p}{\bar{u}\rho C_P} - \frac{2U}{\bar{u}\rho C_P R}(T - T_{ext})$$

The inside heat transfer coefficient is given by Equation (5.37):

$$hd_t/\lambda = 1.86 \text{Gz}^{1/3}\left(\frac{\mu_{bulk}}{\mu_{wall}}\right)^{0.14}$$

Assume that this is the controlling resistance so that $U=h$. A kinetic model is needed for R_p and for the instantaneous values of $\bar{l}_N$ and $\bar{l}_W$. The computer program in Appendix 13 includes values for physical properties and an expression for the polymerization kinetics. Cumulative values for the chain lengths are calculated as a function of position down the tube using

$$(\bar{l}_N)_{new} = \frac{w+dw}{\dfrac{w}{(\bar{l}_N)_{old}} + \dfrac{dw}{(\bar{l}_N)_{instant}}} \tag{13.48}$$

$$(\bar{l}_W)_{new} = \frac{w(\bar{l}_W)_{old} + dw(\bar{l}_W)_{instant}}{w+dw} \tag{13.49}$$

where w is the weight of polymer.

Predictions of this simplified tubular reactor model are in good agreement with experimental results for the shell-and-tube post-reactor used in some polystyrene processes. The code and sample results are given in Appendix 13. Predictions of thermal runaway for polymerization starting with pure styrene are in good agreement with those of rigorous models provided that hydrodynamic stability is not a factor. See Figure 13.8. The predicted runaway at a tube diameter of 0.061 m is in good agreement with the results shown in Figure 13.7.

The lumped parameter model of Example 13.9 takes no account of hydrodynamics and predicts stable operation in regions where the velocity profile is elongated to the point of instability. It also overestimates conversion in the stable regions. The next example illustrates the computations that are needed

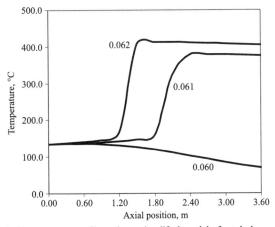

FIGURE 13.8 Temperature profiles using a simplified model of a tubular reactor with pure styrene feed; $T_{in} = 135°C$ and $T_{wall} = 20°C$. The parameter is the tube diameter in meters.

to include hydrodynamics in a tubular reactor calculation. The methodology follows that in Sections 8.7 and 8.8.

Example 13.10: Model the performance of a laminar flow, tubular reactor for the polymerization of a self-condensing monomer. Assume second-order kinetics based on endgroup concentrations with $a_{in}k\bar{t} = 1$ and a negligible heat of reaction. Ignore any condensation product. Assume that density is constant and viscosity varies with conversion as $\mu = \mu_0(100 - 99a/a_{in})$ where a is the concentration of remaining endgroups. Assume that an average value for the diffusion group, $\mathscr{D}_A\bar{t}/R^2 = 0.1$, applies to all endgroups. Include the radial velocity term in the convective diffusion equation and plot streamlines in the reactor.

Solution: The problem requires solution of the convective diffusion equation for mass but not for energy. Rewriting Equation (8.71) in dimensionless form gives

$$\mathscr{V}_z \frac{\partial a}{\partial \mathscr{y}} + \left(\frac{L}{R}\right)\mathscr{V}_r \frac{\partial a}{\partial \mathscr{r}} = \frac{\mathscr{D}_A \bar{t}}{R^2}\left[\frac{1}{\mathscr{r}}\frac{\partial a}{\partial \mathscr{r}} + \frac{\partial^2 a}{\partial \mathscr{r}^2}\right] + \bar{t}\mathscr{R}_p \qquad (13.50)$$

The propagation rate is assumed to be second order with respect to the endgroup concentration, $\mathscr{R}_p = ka^2$. The boundary conditions are a specified inlet concentration, zero flux at the wall, and symmetry at the centerline.

The dimensionless velocity component in the axial direction, $\mathscr{V}_z = V_z/\bar{u}$, is calculated using the method of Example 8.8. The component in the radial direction, $(L/R)\mathscr{V}_r$, is calculated using a dimensionless version of Equation (8.70):

$$\left(\frac{L}{R}\right)\mathscr{V}_r = \frac{-1}{\rho \mathscr{r}}\int_0^{\mathscr{r}} \mathscr{r}' \frac{\partial(\rho \mathscr{V}_z)}{\partial \mathscr{y}} d\mathscr{r}' \qquad (13.51)$$

A first-order difference approximation for the axial derivative, $\partial(\rho\mathscr{V}_z)/\partial\mathscr{y}$, is consistent with the first-order convergence of Euler's method. The convected-mean concentration is calculated from the dimensionless version of Equation (8.4):

$$a_{mix}(\mathscr{y}) = 2\int_0^1 a(\mathscr{r}, \mathscr{y})\mathscr{V}_z(\mathscr{r}, \mathscr{y})\mathscr{r}\, d\mathscr{r} \qquad (13.52)$$

The streamlines are calculated using a dimensionless version of Equation (8.73):

$$\int_0^{\mathscr{r}_{in}} \rho\mathscr{r}\,\mathscr{V}_z(\mathscr{r}, 0)d\mathscr{r} = \int_0^{\mathscr{r}} \rho\mathscr{r}'\,\mathscr{V}_z(\mathscr{r}', \mathscr{y})d\mathscr{r}' \qquad (13.53)$$

Values for z_{in} are selected and the integral on the left-hand side of Equation (13.52) is evaluated. Downstream, evaluation of the right-hand side proceeds from $z = 0$ up to some value $z' = z$ where the value of the integral exactly matches the original value on the left-hand side. The streamline has moved from z_{in} to z. The density terms will cancel out here and elsewhere in this example.

The overall solution is based on the method of lines discussed in Chapter 8. The resulting ODEs can then be solved by any convenient method. Appendix 13.2 gives an Excel macro that solves the ODEs using Euler's method. Figure 13.9 shows the behavior of the streamlines.

Example 13.10 treats an isothermal polymerization, but adding Equation (8.53) to include temperature dependence causes no special problems. Modern computers can easily solve small sets of simultaneous PDEs. Very large sets, say the 10,000 or so that are needed to describe individual species in a polymerization going to high molecular weight, are still expensive to solve. A key to easy computability in polymer reactor design is *lumping*, where various molecular species are combined and treated as a single entity. Example 13.10 uses the endgroup concentration to form a single lump containing all the molecules. Some models of polycondensations treat the first 20 or so oligomers exactly and then use one or more lumps to characterize the rest of the population. Models for chain-growth polymers are mainly of the single-lump variety. A more sophisticated treatment is needed to account for the dependence of molecular diffusivity on chain length, but a satisfactory design methodology has not yet emerged. Lumping is not confined to polymerizations. Comprehensive models of oil refineries use lumping to reduce the number of tracked species from about 5000 to about 1000.

13.5.3 Suspending-Phase Polymerizations

Many polymerizations use a low viscosity nonsolvent to suspend the polymer phase. Water is the most common suspending phase. Table 13.6 characterizes a variety of reaction mechanisms in which water is the continuous phase.

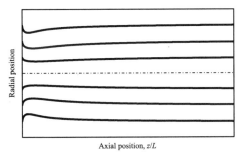

FIGURE 13.9 Curved streamlines resulting from a polycondensation in the laminar flow reactor of Example 13.10.

TABLE 13.6 Classification of Polymerization Mechanisms that Use Water as the Continuous, Suspending Phase

Process parameter	Polymerization type Emulsion → Dispersion → Suspension
Stabilizer type	Soaps → inorganic salts
Agitation requirements	Low → high
Stability of dispersion	High → Low
Particle size	0.2–200 μm
Purity of polymer	Low → high

The reaction engineering aspects of these polymerizations are similar. Excellent heat transfer makes them suitable for vinyl addition polymerizations. Free radical catalysis is mostly used, but cationic catalysis is used for nonaqueous dispersion polymerization (e.g., of isobutene). High conversions are generally possible, and the resulting polymer, either as a latex or as beads, is directly suitable for some applications (e.g., paints, gel-permeation chromatography beads, expanded polystyrene). Most of these polymerizations are run in the batch mode, but continuous emulsion polymerization is common.

Emulsion Polymerization. Emulsion polymerization uses soaps and anionic surfactants to create two-phase systems that have having long-term stability. The key steps in a batch emulsion polymerization are the following:

- Water and emulsifier are charged to the reactor, and the emulsifier forms aggregates known as *micelles.*
- One or more sparingly soluble monomers are charged to the reactor to form suspended drops. The monomer quickly saturates the aqueous phase.
- A water-soluble, free-radical initiator is charged to the reactor and initiates polymerization in the water phase.
- A chain growing in water soon becomes insoluble. It separates from the aqueous phase and penetrates a micelle, forming a *seed.* It may also add to an existing polymer particle.
- Mass transfer of monomer from the suspended drops through the aqueous phase to the seeded particles continues throughout the polymerization.
- The suspended drops of monomer are eventually depleted and polymerization stops.

There are many variations on this theme. Fed-batch and continuous emulsion polymerizations are common. Continuous polymerization in a CSTR is dynamically unstable when free emulsifier is present. Oscillations with periods of several hours will result, but these can be avoided by feeding the CSTR with seed particles made in a batch or tubular reactor.

Dispersion Polymerization. The suspending phase plays a more passive role in this form of polymerization. Dispersion polymerizations are typically batch processes with little or no mass transfer through the suspending phase. The monomer and initiator are charged to the reactor and react in a batch mode. Nonaqueous solvents are frequently used as the suspending phase. A great variety of surfactants are used, including polymers and block copolymers. The particle size of the polymer beads is generally a few microns, intermediate between those of emulsion and suspension polymerization. Agitation requirements are also intermediate. Through control of the surfactant type and concentration, monodispersed particles and core-shell structures can be made.

Suspension Polymerization. Water is the suspending phase. Inorganic salts and vigorous agitation prevent coalescence and agglomeration. The reaction mode is batch. The largest use of suspension polymerization is for the manufacture of expandable polystyrene beads.

Gas-Phase Polymerization. The fluidized-bed processes for polyolefins use gaseous monomers as the suspending phase. Finely ground catalyst particles containing a transition metal catalyst on a support such as $MgCl_2$ are fed continuously to the reactor. Polymerization occurs on the surface of the catalyst particle and ultimately encapsulates it. A cyclone separates the solid particles from unreacted monomer. The monomer is cooled and recycled. Temperature control is achieved by using cold monomer feed, by cooling the recycled monomer, and by heat transfer to the vessel walls. Fluidized-bed reactors are often modeled as being well mixed with respect to both phases. A typical residence time for the gas phase will be 20 s. That for the polymer will be a few hours.

Precipitation and Slurry Polymerization. Many polymers are insoluble in the reaction mixture from which they are formed. Upon significant polymerization, they precipitate to form a separate phase. The chemistry and reaction cycle must be managed to avoid agglomeration. Assuming this is possible, heat transfer is the major issue. Boiling of the suspending phase is used (e.g., when liquid propylene suspends polypropylene) and so is sensible heat transfer. Loop reactors may have most of their volume in the form of shell-and-tube heat exchangers. Reactor models normally assume CSTR behavior, although more detailed models are possible for loop reactors.

13.6 SCALEUP CONSIDERATIONS

The basic issues of scaleup are the same for polymer reactors as for ordinary chemical reactors. The primary problem is that the capacity for heat and mass transfer increases less rapidly than the reactor volume and throughput. The remedies are also similar, but the high viscosities characteristic of polymers cause special problems. These problems are most acute when the polymer phase is continuous.

13.6.1 Binary Polycondensations

Endpoint control is needed for binary polycondensations going to high molecular weight. The reaction must be stopped at the desired point and the polymer chains endcapped to prevent further polymerization. An example is the use of methyl chloride to endcap sodium-terminated chains. Classic mixing time problems arise in large tanks. Similar problems of stopping the reaction exist for batch vinyl polymerizations. Using multiple injection points for the chain stopper is a possibility. Stopping the polymerization with a quench and dilution with cold solvent is another.

Tubular reactors are used for some polycondensations. Para-blocked phenols can be reacted with formalin to form linear oligomers. When the same reactor is used with ordinary phenol, plugging will occur if the tube diameter is above a critical size, even though the reaction stoichiometry is outside the region that causes gelation in a batch reactor. Polymer chains at the wall continue to receive formaldehyde by diffusion from the center of the tube and can crosslink. *Local stoichiometry* is not preserved when the reactants have different diffusion coefficients. See Section 2.8.

13.6.2 Self-Condensing Polycondensations

The removal of condensation by-products becomes increasingly difficult upon scaleup. Some commercial PET processes use CSTRs for the early stages of the reaction where most of the by-product ethylene glycol is removed. They use only the top, visible surface of the liquid for mass transfer and rely on jacket heating to supply the latent heat of vaporization. The surface area scales as $S^{2/3}$ and limits the production rate in some processes because the previous limit, a downstream finishing reactor, has been improved. A pump-around loop containing a heat exchanger with a flash into the top of the vessel is one possibility for increasing capacity.

The finishing reactors used for PET and other equilibrium-limited polymerizations pose a classic scaleup problem. Small amounts of the condensation product are removed using devolatilizers (rotating-disk reactors) that create surface area mechanically. They scale as $S^{2/3}$.

13.6.3 Living Addition Polymerizations

The main problem with a living polymer is maintaining the strict cleanliness that is demanded by the chemistry. This is a particularly severe problem for large-scale, batch polymerizations, but it is a problem more of economics than technology. Living polymerizations are usually run to near completion, so that end point control is not a problem. Most living polymerizations operate at low temperatures, $-40°C$ to $+40°C$, to avoid chain transfer reactions. Thus, temperature control is a significant scaleup problem. The usual approach is to use 85–95 w% solvent and to rely on sensible heat transfer to the vessel walls.

The sanitation requirements are easier to meet in continuous operation, but polymerizations in a tubular geometry raise a stability issue. Living polymer formed near the wall will continue to grow due to the outward diffusion of monomer. The fear is that this very-high-molecular-weight polymer will eventually plug the reactor, or at least cause a severe form of velocity profile elongation. Possibly, there is a maximum tube diameter for stable operation. Tubular reactors have apparently not been used for living polymerizations, even though the low polydispersities would be an advantage in some applications. CSTRs have been used, but the polydispersity advantage of living polymerizations is lost.

13.6.4 Vinyl Addition Polymerizations

Heat removal is the key problem in scaling up a vinyl polymerization. Section 5.3 discusses the general problem and ways of avoiding it. Processes where the polymer phase is suspended in a nonsolvent are relatively easy to scale up. If the pilot-plant reactor is sensitive to modest changes in agitator speed or reactant addition rates, there are likely to be mixing time or dispersion quality problems upon scaleup. In batch processes, endpoint control can also be a problem, but it is usually managed by reacting to near-completion and controlling the temperature and initiator concentrations to give the desired molecular weight. Tubular geometries often have a diameter limit beyond which scaleup must proceed in parallel or series. The mechanically stirred, tube-cooled reactors (also called *stirred tube reactors*) that are commonly used for styrenic polymerizations have such a limit as well. Modern designs use three or more reactors in series to increase single-train capacity. This is analogous to adding extra length to a high-pressure polyethylene reactor and is a form of scaling in series.

PROBLEMS

13.1. A binary polycondensation of AMA and BNB is to be performed in a batch reactor. A number average chain length of at least 100 is required. What minimum accuracy is required for weighing the two components?

13.2. Find some real chemistry where a binary polycondensation will give a homopolymer; i.e., AMA reacts with BMB to form poly-M.

13.3. The phenol–formaldehyde polycondensation readily forms gels. The generally accepted limits for avoiding them are $[Phenol]_0 < 0.75[Formaldehyde]_0$ when the catalyst is a strong acid or

[Formaldehyde]$_0$ > 1.5[Phenol]$_0$ when the catalyst is a strong base. Are these limits universal for a trifunctional monomer condensing with a difunctional monomer or are they specific to the phenol–formaldehyde reaction? Assume the three sites are equally reactive. [Practical experience shows the gelation point to be at $S_{AB} > 0.8$. This is presumably due to lower reactivity of the para position compared with the ortho positions.] A good answer to this question may require recourse to the literature. Becoming acquainted with percolation theory might be helpful in understanding the phenomenon.

13.4. Refer to Equations (13.24) governing the MWD for a binary polycondensation. Derive and explain each of the limiting cases: $S_{AB}=0$, $S_{AB}=1$, $X_A=0$, and $X_A=1$.

13.5. Use Equation (13.26) to find the moments μ_0 through μ_2 for the Flory distribution. Use your results to validate Equation (13.27).

13.6. Find the standard deviation of the Flory distribution as given by Equation (13.26) and relate it to the polydispersity. Extend the calculations in Problem 13.5 to μ_3. Find the kurtosis of the distribution in the limit of high conversion.

13.7. The Flory distribution gives a polydispersity of 2 in the limit of high conversion. Yet, a thought experiment suggests that a small batch of self-condensing molecules would eventually condense to form a single, cyclic molecule. Reconcile this apparent inconsistency.

13.8. Determine the copolymer composition for a styrene–acrylonitrile copolymer made at the azeotrope (62 mol% styrene). Assume $\bar{l}_N = 1000$. One approach is to use the Gaussian approximation to the binomial distribution. Another is to "synthesize" 100,000 or so molecules using a random number generator and to sort them by composition.

13.9. Find the analytical integral of Equation (13.43).

13.10. A continuous polymerization train consisting of two stirred tanks in series is used to copolymerize styrene, $r_X = 0.41$, and acrylonitrile, $r_Y = 0.04$. The flow rate to the first reactor is 3000 kg/h and a conversion of 40% is expected. Makeup styrene is fed to the second reactor and a conversion of 30% (based on the 3000 kg/h initial feed) is expected there. What should be the feed composition and how much styrene should be fed to the second reactor if a copolymer containing 58 wt% styrene is desired?

13.11. Derive the equivalent of Equation (13.41) when penultimate effects are considered.

13.12. Consider a laminar flow tubular polymerizer with cooling at the tube wall. At what radial position will a hotspot develop: at the tube wall, at the centerline, or at an intermediate radius? Justify your answer. Will the situation change with heating at the wall?

13.13. Use the computer program in Appendix 13.1 to explore thermal runaway. Feed pure styrene at 135°C and maintain a wall temperature of 300 K. At what tube diameter does the reactor run away? Repeat this varying T_{wall}. Is the case of $T_{wall} = 400$ K with a tube radius of 0.4 in predicted to be stable? Compare your results with those for the detailed model shown in Figure 13.7. Rationalize any differences.

13.14. Use the kinetic model in Appendix 13.1 to design a CSTR for the production of polystyrene. The entering feed is pure styrene. It is desired to produce 50% by weight of polystyrene with a number average molecular weight of 85,000. The feed flow rate is 25,000 kg/h. Determine the required operating temperature and reactor capacity (in mass units).

13.15. The reactor in Problem 13.14 is to be cooled by autorefrigeration. Determine the boilup rate in the reactor assuming that the condensate is returned to the reactor without subcooling.

13.16. Suppose the reactor sized in Problem 13.14 is converted to manufacture a styrene–acrylonitrile copolymer containing 36% acrylonitrile by weight. Assume 50% conversion as before. What is the required feed composition to the reactor and what is the composition of the unreacted monomer mixture?

REFERENCE

1. Ham, G. E., *Copolymerization,* Interscience, New York, 1964.

SUGGESTIONS FOR FURTHER READING

A general reference on polymerization reaction engineering is

Biesenberger, J. A. and Sebastian, D. H., *Principles of Polymerization Engineering,* Wiley, New York, 1983.

A comprehensive treatment of molecular weight distributions is given in

Peebles, L. H., Jr., *Molecular Weight Distributions in Polymers,* Interscience, New York, 1971.

A briefer survey of suitable mathematical techniques for determining molecular weight distributions is given in

Chappelear, D. C. and Simon, R. H., "Polymerization reaction engineering," *Advances in Chemistry,* Vol. 91, *Addition and Condensation Polymerization Processes,* Platzer, N., Ed., 1–24 (1969).

There are a large number of proprietary process models for the industrially important polymerizations. Public domain descriptions of these models are seldom complete enough to allow independent evaluation. Given below are a few general references and models published in the referred literature.

A sophisticated model of the high-pressure process for polyethylene is described in

Gupta, S. K., Kumar, A., and Krishnamurthy, M. V. G. "Simulation of tubular low-density polyethylene," *Polym. Eng. Sci.,* **25**, 37–47 (1985).

An example of a design and optimization study using a fairly sophisticated model for styrene polymerization is given in

Mallikarjun, R. and Nauman, E. B., "A staged multitubular process for crystal polystyrene," *Polym. Process Eng.*, **4**, 31–51 (1986).

A modern polystyrene process consists of a CSTR followed by several stirred tube reactors in series. A description of this typical process is given in

Chen, C.-C., "Continuous production of solid polystyrene in back-mixed and linear-flow reactors," *Polym. Eng. Sci.*, **40**, 441–464 (2000).

An overview of the various dispersed-phase polymerizations is given in

Arshady, R., "Suspension, emulsion, and dispersion polymerization: a methodological survey," *Colloid. Polym. Sci.*, **270**, 717–732 (1992).

APPENDIX 13.1: LUMPED PARAMETER MODEL OF A TUBULAR POLYMERIZER

The following is a complete program for solving the lumped parameter model of styrene polymerization in a tube. Example 13.8 describes the governing equations. The propagation rate and instantaneous chain lengths are calculated using the model of Hui and Hamielec.[A1] The viscosity correlation is due to Kim and Nauman.[A2] The program is structured as an Excel macro. Sample output is given at the end. It reasonably represents the performance of a multitubular finishing reactor used in some polystyrene processes.

```
DefDbl A-Z
Sub Append13_1()
'Physical Properties of Entering Fluid in SI Units
MWSty=104.15
Cp=1880#
HeatReact=670000#
ThermCond=0.126
CummLn=960
PD=2.1
CummLw=PD * CummLn
WPoly=0.7
Wdil=0.06
WSty=1 - WPoly - Wdil

'Design and Operating Variables in SI Units
'but Temperature in C
```

```
TubeDia = 0.027
TubeLength = 3.6
Qmass = 5/3600
TcIn = 170
TcWall = 240
Twall = TcWall + 273.15
T = TcIn + 273.15

'Initializarion
dtime = 0.25
z = 0
GoSub Caption 'Prints Excel Column Headers

'Main Loop
Do
  RhoSty = 924 - 0.918 * Tc
  RhoPoly = 1084 - 0.605 * Tc
  RhoMix = WPoly * RhoPoly + (1 - WPoly) * RhoSty
  Sty = RhoMix * WSty
  GoSub Kinetics

  dWSty = (-RP - RD - RT + RU) * dtime/RhoMix
  dWPoly = RP * dtime/RhoMix
  NumMoles = WPoly/CummLn + dWPoly/Ln
  CummLn = (WPoly + dWPoly)/NumMoles
  CummLw = (WPoly * CummLw + dWPoly * Lw)/(WPoly + dWPoly)

  WPoly = WPoly + dWPoly
  WSty = WSty + dWSty
  Tv = Twall
  GoSub Viscosity
  ViscWall = Visc
  Tv = T
  GoSub Viscosity
  Velocity = Qmass/RhoMix/(3.14159 * TubeDia * TubeDia/4)
  Re = RhoMix * TubeDia * Velocity/Visc
  Pr = Cp * Visc/ThermCond
  U = (Re * Pr * TubeDia/TubeLength) ^ (1/3)
+     * (Visc/ViscWall) ^ 0.14
  U = ThermCond/TubeDia * 1.86 * U
  dT = (HeatReact * RP + 4/TubeDia * (Twall - T) * U)/RhoMix
+     /Cp * dtime
  T = T + dT
  Tc = T - 273.15
```

```
    Time = Time + dtime
    z = z + Velocity * dtime

    If Int(z/0.3) >= IntPrint Then 'Controls output interval
      IntPrint = IntPrint + 1
      GoSub Output
    End If
Loop While z < TubeLength
GoSub Output

Exit Sub

Kinetics: 'Kinetics subroutine: all variables are
          'shared with the main program
  Kp = 100900# * Exp(-3557/T)
  A1 = 2.57 - 0.00505 * T
  A2 = 9.56 - 0.0176 * T
  A3 = -3.03 + 0.00785 * T
  Ft = Exp(-2 * (A1 * WPoly + A2 * WPoly ^ 2 + A3 * WPoly ^ 3))
  Kt = 12050000# * Exp(-844/T) * Ft
  KFM = 22180# * Exp(-6377/T)
  KI = 20.19 * Exp(-13810/T)
  B1 = -0.0985 + 0.000777 * T
+    - 0.00000204 * T * T + 0.00000000179 * T ^ 3
  CM = KFM/Kp + B1 * WPoly
  RP = Kp * (2 * KI * Sty ^ 3/Kt) ^ 0.5 * Sty
  Beta = Kt * RP/Kp/Kp/Sty ^ 2
  Ln = (CM + Beta + 1)/(CM + Beta/2)
  Lw = (2 * CM + 3 * Beta)/(CM + Beta) ^ 2
Return

Viscosity: 'Viscosity Subroutine: All variables
           'are shared with main program.
  V = (3.915 * WPolyt - 5.437 * WPoly * 2
+      + (0.623 + 1387/Tv) * WPoly ^ 3)
  V = (CummLw * MWSty) ^ 0.18 * V
  Visc = Exp(-13.04 + 2013/Tv + V)
Return

Output: 'Outputs Results to Excel Cells
  Range("B" & CStr(IntPrint + 1)).Select
  ActiveCell.FormulaR1C1 = z
  Range("C" & CStr(IntPrint + 1)).Select
  ActiveCell.FormulaR1C1 = WSty
  Range("D" & CStr(IntPrint + 1)).Select
```

```
    ActiveCell.FormulaR1C1 = WPoly
    Range("E"& CStr(IntPrint + 1)).Select
    ActiveCell.FormulaR1C1 = CummLn
    Range("F"& CStr(IntPrint + 1)).Select
    ActiveCell.FormulaR1C1 = CummLw
    Range("G"& CStr(IntPrint + 1)).Select
    ActiveCell.FormulaR1C1 = CummLw/CummLn
    Range("H"& CStr(IntPrint + 1)).Select
    ActiveCell.FormulaR1C1 = Tc
Return
Caption: 'Prints Column Headers
    Range("B"& CStr(1)).Select
    ActiveCell.FormulaR1C1 = "Length"
    Range("C"& CStr(1)).Select
    ActiveCell.FormulaR1C1 = "WSty"
    Range("D"& CStr(1)).Select
    ActiveCell.FormulaR1C1 = "WPoly"
    Range("E"& CStr(1)).Select
    ActiveCell.FormulaR1C1 = "CummLn"
    Range("F"& CStr(1)).Select
    ActiveCell.FormulaR1C1 = "CummLw"
    Range("G"& CStr(1)).Select
    ActiveCell.FormulaR1C1 = "PD"
    Range("H"& CStr(1)).Select
    ActiveCell.FormulaR1C1 = "Temp"
Return
End Sub
```

The following is the output from the preceding Excel macro:

Length	WSty	WPoly	CummLn	CummLw	PD	Temp
0.00	0.240	0.700	960	2016	2.10	170.0
0.30	0.229	0.711	954	2006	2.10	186.9
0.60	0.211	0.729	933	1981	2.12	202.2
0.90	0.186	0.754	888	1941	2.19	216.5
1.20	0.158	0.782	821	1891	2.30	229.5
1.50	0.131	0.809	750	1841	2.46	240.0
1.80	0.108	0.832	689	1800	2.61	247.6
2.10	0.090	0.850	643	1769	2.75	252.4
2.40	0.077	0.863	611	1746	2.86	255.0
2.70	0.067	0.873	589	1729	2.94	256.2
3.00	0.059	0.881	573	1717	3.00	256.4
3.30	0.053	0.887	562	1707	3.04	256.1
3.60	0.049	0.891	553	1700	3.07	255.3

APPENDIX 13.2: VARIABLE-VISCOSITY MODEL FOR A POLYCONDENSATION IN A TUBULAR REACTOR

The following is a program, complete except for output, for solving Example 13.10. Sample output is given at the end and in Figure 13.9.

```
DefDbl A-H, P-Z
DefLng I-O
Public aold(256)
Sub Append13_2()
Dim Vz(256)
Dim Vr(256)
Dim FlowInt(256) 'Right-hand side of Equation 13.52
Dim FlowInt0(256) 'Left-hand side of Equation 13.52
Dim Stream(256) 'Radial position of streamline
                'with initial position i*dr
Dim G1(256) 'Used to calculate Vz per Example 8.10

Dim anew(256)

Itotal = 64
dr = 1/Itotal
DA = 0.1
RateConst = 1
Order = 2 'Reaction order

For i = 0 To Itotal 'Initial conditions
  aold(i) = 1
  Vz(i) = 2 * (1 - (i * dr) ^ 2)
Next

'Calculate the the left-hand side of Equation 13.52
Sum = 0
For i = 1 To Itotal
  Sum = Sum + i * dr * Vz(i) * dr
  FlowInt0(i) = Sum
Next

Do 'Start of main loop

dz = dr * dr * Vz(Itotal - 1)/DA/2 'Stability criterion,
                                  'refer to Equation 8.29
If dz > 0.001 Then dz = 0.001 'Reduce dz if needed for
                              'accuracy or physical
                              'stability
```

```
  If z+dz > 1 Then dz=1 - z   'The last step goes exactly to
                              'the end of the reactor
  'Centerline concentration, refer to Equation 8.26
  A=DA * 4/dr/dr/Vz(0)
  B=-A
  anew(0) =A * dz * aold(1) + (1+B * dz) * aold(0)
  anew(0) =anew(0) - RateConst * aold(0) ^ Order
+           * dz/Vz(0)

  'Interior concentrations, refer to Equation 8.25
  For i=1 To Itotal - 1
  C1 =DA * 0.5/i/dr/dr/Vz(i)
  C2 =DA/dr/dr/Vz(i)
  A=C1+C2
  C=-C1+C2
  B=-A - C
  anew(i) =A * dz * aold(i+1) + (1+B * dz) * aold(i)
+          +C * dz * aold(i -1)
  anew(i) =anew(i) - RateConst * aold(i) ^ Order
+           * dz/Vz(i)
  'Radial convection terms, refer to Equation 13.50
  anew(i) =anew(i) - Vr(i)/Vz(i) * (aold(i+1)
+          - aold(i - 1))/2/dr * dz
Next

'Wall concentration, refer to Equation 8.24
anew(Itotal) = (4 * anew(Itotal - 1) - anew(Itotal - 2))/3

For i=0 To Itotal
  aold(i) =anew(i)
Next

'Calculate mixing cup average
avgC=0
avgV=0
For i=0 To Itotal
  avgC=avgC+anew(i) * Vz(i) * i * dr * dr
  avgV=avgV+Vz(i) * i * dr * dr
Next
avgC=avgC/avgV

'Calculate axial velocity component using
'the code of Example 8.8
```

```
G1(Itotal) = 0
  For ii = 1 To Itotal
  i = Itotal - ii
  G1(i) = G1(i+1) + dr ^ 2/2 * ((i+1)/visc(i+1)
+         + i/visc(i))
Next
G2 = 0
For i = 1 To Itotal - 1
  G2 = G2 + i * dr * G1(i) * dr
Next
G2 = G2 + Itotal * dr * G1(Itotal) * dr/2
G2 = G2 + dr * dr/8

'The new value for Vz is G1(i)/G2 /2 but calculate the
'radial component prior to updating Vz

G3 = 0
For i = 1 To Itotal
  G3 = G3 + i * dr * (G1(i)/G2/2 - Vz(i))/dz * dr
  'G3 is the integral in Equation 13.50
  Vr(i) = -G3/i/dr
Next i

'Update Vz
Vz(0) = G1(0)/G2/2
For i = 1 To Itotal
  Vz(i) = G1(i)/G2/2
Next i

'Calculate right-hand side of Equation 13.52
Sum = 0
For i = 1 To Itotal
  Sum = Sum + i * dr * Vz(i) * dr
  FlowInt(i) = Sum
Next

'Locate radial position of streamlines. Use linear
'interpolation to smooth the results for display.
For iii = 1 To 3
  ii = iii * Itotal/4
  For i = 1 To Itotal - 2
```

```
    If FlowInt0(ii) > FlowInt(i) Then
        del = (FlowInt(i+1) - FlowInt(i))
        del = (FlowInt0(ii) - FlowInt(i))/del
        Stream(ii) = (i+del) * dr
    End If
  Next i
Next iii

'Output results as a function of z here
  z = z + dz
Loop While z < 1

'Output end-of-reactor results here

End Sub

Function visc(ii)
  visc = 100 - 99 * aold(ii)
End Function
```

Systematic verification is needed for a program as complex as this. The following tests were made using Itotal = 32. A sequence like this helps in debugging and lends confidence to the final results, although they do not prove accuracy.

1. Turn off the reaction by setting RateConst = 0. The resulting final value for the mixing-cup average, avgC, is 1, confirming the material balance.
2. Set the initial Vz(i) to 1 and turn off the updating. This gives piston flow. The calculated result for avgC for a first-order reaction is 0.3677 versus 0.3679 in theory. The result for a second-order reaction is 0.4998 versus 0.5 in theory.
3. Set the viscosity to a constant. Then the downstream velocity profile keeps the initial parabolic form.
4. Set the viscosity to a constant, set DA = 0.00001 so that diffusion is negligible and set Order = 1. Then the final value for avgC is 0.4433 versus 0.4432 calculated in Example 8.3.
5. Delete the radial convection term but otherwise run the full simulation. This gives avgC = 0.5197. Now add the radial term to get 0.5347. The change is in the correct direction since velocity profile elongation hurts conversion.

6. Run the full simulation at various values for Itotal:

Itotal	avgC	Δ
32	0.534663	
64	0.538913	0.004250
128	0.539984	0.001071

The results are converging almost exactly $O(\Delta r^2)$. The extrapolation to $\Delta r \to \infty$ gives a correction of $(1/4 + 1/16 + 1/64 + \cdots = 0.33333)$ $\Delta_{last} = 0.000357$. Thus, $a_{out}/a_{in} = 0.5403$.

References

A1. Hui, A. W. and Hamielec, A. E., "Thermal polymerization of styrene at high conversions and temperatures: an experimental study," *J. Appl. Polym. Sci.*, **16**, 749–769 (1972).

A2. Kim, D.-M. and Nauman, E. B., "Solution viscosity of polystyrene at conditions applicable to commercial manufacturing processes," *J. Chem. Eng. Data.*, **37**, 427–432 (1992).

CHAPTER 14
UNSTEADY REACTORS

The general material balance of Section 1.1 contains an accumulation term that enables its use for unsteady-state reactors. This term is used to solve steady-state design problems by the method of false transients. We turn now to solving real transients. The great majority of chemical reactors are designed for steady-state operation. However, even steady-state reactors must occasionally start up and shut down. Also, an understanding of process dynamics is necessary to design the control systems needed to handle upsets and to enable operation at steady states that would otherwise be unstable.

Unsteady mass and energy balances consider three kinds of accumulation:

Total mass $\quad \dfrac{d(\hat{\rho}V)}{dt}$

Component moles $\quad \dfrac{d(\hat{a}V)}{dt}$

Enthalpy $\quad \dfrac{d(\hat{H}\hat{\rho}V)}{dt}$

These accumulation terms are added to the appropriate steady-state balances to convert them to unsteady balances. The circumflexes indicate averages over the volume of the system, e.g.,

$$\hat{\rho} = \iiint_V \rho \, dV$$

The three accumulation terms represent the change in the total mass inventory, the molar inventory of component A, and the heat content of the system. The circumflexes can be dropped for a stirred tank, and this is the most useful application of the theory.

14.1 UNSTEADY STIRRED TANKS

The steady-state balance for total mass is

$$0 = Q_{in}\rho_{in} - Q_{out}\rho_{out}$$

A well-mixed stirred tank (which we will continue to call a CSTR despite possibly discontinuous flow) has $\bar\rho = \rho_{out}$. The unsteady-state balance for total mass is obtained just by including the accumulation term:

$$\frac{d(\rho_{out}V)}{dt} = Q_{in}\rho_{in} - Q_{out}\rho_{out} \tag{14.1}$$

Liquid-phase systems with approximately constant density are common. Thus, the usual simplification of Equation (14.1) is

$$\frac{dV}{dt} = Q_{in} - Q_{out}$$

The component balance for the general case is

$$\frac{d(a_{out}V)}{dt} = Q_{in}a_{in} - Q_{out}a_{out} + V\mathcal{R}_A \tag{14.2}$$

The general case treats time-dependent volumes, flow rates, and inlet concentrations. The general case must be used to for most startup and shutdown transients, but some dynamic behavior can be effectively analyzed with the constant-volume, constant-flow rate version of Equation (14.2):

$$\bar{t}\frac{da_{out}}{dt} = a_{in} - a_{out} + \bar{t}\mathcal{R}_A$$

The case of $a_{in} = a_{in}(t)$ will force unsteady output as will sufficiently complex kinetics.

The enthalpy balance for a reasonably general situation is

$$\frac{d(H_{out}\rho_{out}V)}{dt} = Q_{in}H_{in}\rho_{in} - Q_{out}H_{out}\rho_{out}$$
$$- V\,\Delta H_R \mathcal{R}_A + UA_{ext}(T_{ext} - T_{out}) \tag{14.3}$$

A still more general case is discussed in Problem 14.15. Typical simplifications are constant volume and flow rate, constant density, and replacement of enthalpy with $C_P(T - T_{ref})$. This gives

$$\bar{t}\frac{dT_{out}}{dt} = T_{in} - T_{out} - \frac{\bar{t}\Delta H_R \mathcal{R}_A}{\rho C_P} + \frac{UA_{ext}}{V\rho C_P}(T_{ext} - T_{out})$$

Equations (14.1)–(14.3) are a set of simultaneous ODEs that govern the performance of an unsteady CSTR. The minimum set is just Equation (14.2), which governs the reaction of a single component with time-varying inlet concentration. The maximum set has separate ODEs for each of the variables $V, H_{out}, a_{out}, b_{out}, \ldots$. These are the *state variables*. The ODEs must be supplemented by a set of initial conditions and by any thermodynamic relations needed to determine dependent properties such as density and temperature.

UNSTEADY REACTORS

The maximum set will consist of Equations (14.1) and (14.3) and N versions of Equation (14.2), where N is the number of components in the system. The maximum dimensionality is thus $2 + N$. It can always be reduced to 2 plus the number of independent reactions by using the reaction coordinate method of Section 2.8. However, such reductions are unnecessary from a computational viewpoint and they disguise the physics of the problem.

14.1.1 Transients in Isothermal CSTRs

If the system is isothermal with $T_{out} = T_{in}$, Equation (14.3) is unnecessary. Unsteady behavior in an isothermal perfect mixer is governed by a maximum of $N + 1$ ordinary differential equations. Except for highly complicated reactions such as polymerizations (where N is theoretically infinite), solutions are usually straightforward. Numerical methods for unsteady CSTRs are similar to those used for steady-state PFRs, and analytical solutions are usually possible when the reaction is first order.

Example 14.1: Consider a first-order reaction occurring in a CSTR where the inlet concentration of reactant has been held constant at a_0 for $t < 0$. At time $t = 0$, the inlet concentration is changed to a_1. Find the outlet response for $t > 0$ assuming isothermal, constant-volume, constant-density operation.

Solution: The solution uses a simplified version of Equation (14.2).

$$\bar{t}\frac{da_{out}}{dt} = a_1 - a_{out} + \bar{t}\mathscr{R}_A \quad \text{for} \quad t > 0$$

A general solution for constant a_1 and $\mathscr{R}_A = -ka_{out}$ is

$$a_{out} = \frac{a_1}{1+k\bar{t}} + C\exp[-(1+k\bar{t})t/\bar{t}]$$

as may be verified by differentiation. The constant C is found from the value of a_{out} at $t = 0$. For the current problem, this initial condition is the steady-state output from the reactor given an input of a_0:

$$a_{out} = \frac{a_0}{1+k\bar{t}} \quad \text{at} \quad t = 0$$

Applying the initial condition gives

$$a_{out} = \frac{a_1 + (a_0 - a_1)\exp[-(1+k\bar{t})t/\bar{t}]}{1+k\bar{t}}$$

as the desired solution. Figure 14.1 illustrates the solution and also shows the effect of restoring a_{in} to its original value at some time $t > 0$.

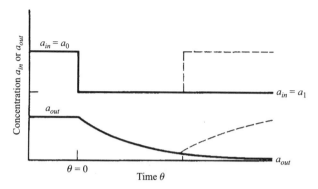

FIGURE 14.1 Dynamic response of a CSTR to changes in inlet concentration of a component reacting with first-order kinetics.

Stability. The first consideration is stability. Is there a stable steady state? The answer is usually yes for isothermal systems.

Example 14.1 shows how an isothermal CSTR with first-order reaction responds to an abrupt change in inlet concentration. The outlet concentration moves from an initial steady state to a final steady state in a gradual fashion. If the inlet concentration is returned to its original value, the outlet concentration returns to its original value. If the time period for an input disturbance is small, the outlet response is small. The magnitude of the outlet disturbance will never be larger than the magnitude of the inlet disturbance. The system is stable. Indeed, it is *open-loop stable*, which means that steady-state operation can be achieved without resort to a feedback control system. This is the usual but not inevitable case for isothermal reactors.

The steady-state design equations (i.e., Equations (14.1)–(14.3) with the accumulation terms zero) can be solved to find one or more steady states. However, the solution provides no direct information about stability. On the other hand, if a transient solution reaches a steady state, then that steady state is stable and physically achievable from the initial composition used in the calculations. If the same steady state is found for all possible initial compositions, then that steady state is unique and globally stable. This is the usual case for isothermal reactions in a CSTR. Example 14.2 and Problem 14.6 show that isothermal systems can have multiple steady states or may never achieve a steady state, but the chemistry of these examples is contrived. Multiple steady states are more common in nonisothermal reactors, although at least one steady state is usually stable. Systems with stable steady states may oscillate or be chaotic for some initial conditions. Example 14.9 gives an experimentally verified example.

Example 14.2: Suppose the rabbits and lynx of Section 2.5.4 become migratory. Model their behavior given a steady stream of rabbits and lynx entering a grassy plain. Ignore the depletion of grass.

Solution: The ODEs governing the population of rabbits and lynx are

$$\bar{t}\frac{dr_{out}}{dt} = r_{in} - r_{out} + \bar{t}(k_I g r_{out} - k_{II} l_{out} r_{out})$$

$$\bar{t}\frac{dl_{out}}{dt} = l_{in} - l_{out} + \bar{t}(k_{II} l_{out} r_{out} - k_{III} l_{out})$$

Figure 14.2 shows the numerical solution. Except for a continuous input of ten rabbits and one lynx per unit time, the parameter values and initial conditions are the same as used for Figure 2.6. The batch reactor has been converted to a CSTR. The oscillations in the CSTR are smaller and have a higher frequency than those in the batch reactor, but a steady state is not achieved.

Example 14.2 demonstrates that sustained oscillations are possible even in an isothermal flow system. This is hardly surprising since they are possible in a batch system provided there is an energy supply.

The rabbit and lynx problem does have stable steady states. A *stable steady state* is insensitive to small perturbations in the system parameters. Specifically, small changes in the initial conditions, inlet concentrations, flow rates, and rate constants lead to small changes in the observed response. It is usually possible to stabilize a reactor by using a control system. Controlling the input rate of lynx can stabilize the rabbit population. Section 14.1.2 considers the more realistic control problem of stabilizing a nonisothermal CSTR at an unstable steady state.

Startup and Shutdown Strategies. In addition to safe operation, the usual goal of a reactor startup is to minimize production of off-specification material. This can sometimes be accomplished perfectly.

Example 14.3: The initial portion of a reactor startup is usually fed-batch. Determine the fed-batch startup transient for an isothermal, constant-density

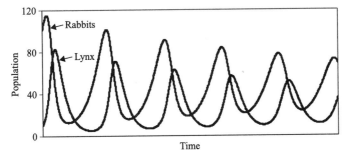

FIGURE 14.2 Population dynamics on a well-mixed grassy plain with constant migration of rabbits and lynx. Compare to the nonmigratory case in Figure 2.6.

stirred tank reactor. Suppose the tank is initially empty and is filled at a constant rate Q_0 with fluid having concentration a_{in}. A first-order reaction begins immediately. Find the concentration within the tank, a, as a function of time, $t < t_{full}$.

Solution: Equation (14.1) simplifies to

$$\frac{dV}{dt} = Q_{in}$$

so that $V = Q_{in}t$ throughout the filling period. Equation (14.2) becomes

$$V\frac{da}{dt} + a\frac{dV}{dt} = Q_{in}a_{in} - Vka$$

Note that $Q_{out} = 0$ during the filling period. Substituting $V = Q_{in}t$ and some algebra gives

$$t\frac{da}{dt} + (1 + kt)a = a_{in}$$

The initial condition is $a = a_{in}$ at $t = 0$. The solution is

$$a = \frac{a_{in}[1 - \exp(-kt)]}{kt} \tag{14.4}$$

This result applies until the tank is full at time $t_{full} = V_{full}/Q_{in}$. If the tank fills rapidly, $t_{full} \to 0$ and $a_{full} \to a_{in}$. If the tank fills slowly, $t_{full} \to \infty$ and $a_{full} \to 0$. By regulating Q_{in}, we regulate t_{full} and can achieve any desired concentration in the range from 0 to a_{in}.

The fed-batch scheme of Example 14.3 is one of many possible ways to start a CSTR. It is generally desired to begin continuous operation only when the vessel is full and when the concentration within the vessel has reached its steady-state value. This gives a *bumpless* startup. The results of Example 14.3 show that a bumpless startup is possible for an isothermal, first-order reaction. Some reasoning will convince you that it is possible for any single, isothermal reaction. It is not generally possible for multiple reactions.

A simpler (and faster) way to achieve a bumpless startup is to *fast fill and hold*. In the limiting case, the fill is instantaneous; and the reactor acts in a batch mode until the steady-state concentration is reached.

Example 14.4: Compare the fed-batch and fast-fill-and-hold methods for achieving a bumpless startup.

Solution: Steady-state operation will use the values V_{full} and Q_{full} and will give a steady-state outlet concentration of

$$a_{out} = \frac{a_{in}}{1 + k\bar{t}} = \frac{a_{in}Q_{full}}{Q_{full} + kV_{full}} \tag{14.5}$$

We want this concentration to be achieved at the end of the fed-batch interval when $t = t_{full} = V_{full}/Q_{in}$. Equate the concentrations in Equations (14.4) and (14.5) and solve for Q_{in}. The solution is numerical.

Suppose $V_{full} = 5\,\text{m}^3$, $Q_{full} = 2\,\text{m}^3/\text{h}$, $k = 3.5\,\text{h}^{-1}$, and $a_{in} = 15\,\text{mol/m}^3$. Then $a_{out} = 1.54\,\text{mol/m}^3$. Now assume values for Q_{in}, calculate t_{full}, and substitute into Equation (14.4) until this concentration is obtained. The result is $Q_{in} = 1.8\,\text{m}^3/\text{h}$ and $t_{full} = 2.78\,\text{h}$.

The fast-fill-and-hold method instantaneously achieves a full reactor, the concentration in which follows batch, first-order kinetics until the desired a_{out} is reached. Equation (14.5) is equated to $a_{in}\exp(-kt_{hold})$. An analytical solution is possible for this case:

$$t_{hold} = \frac{\ln(1 + k\bar{t})}{k} \tag{14.6}$$

The steady-state conversion is achieved at 0.65 h compared with 2.78 h for the fed-batch startup.

Obviously, the fast-fill-and-hold method is preferred from the viewpoint of elapsed time. More importantly, the fed-batch method requires an accurate process model that may not be available. The fast-fill-and-hold method can use a process model or it can use a real-time measurement of concentration.

Neither method will achieve a bumpless startup for complex kinetic schemes such as fermentations. There is a general method, known as *constant RTD control*, that can minimize the amount of off-specification material produced during the startup of a complex reaction (e.g., a fermentation or polymerization) in a CSTR. It does not require a process model or even a real-time analyzer. We first analyze shutdown strategies, to which it is also applicable.

Example 14.5: A CSTR is operating at steady state with a first-order reaction. It is desired to shut it down. Suppose this is done by setting $Q_{in} = 0$ while maintaining $Q_{out} = Q$ until the reactor is empty. Assume isothermal, constant-density operation with first-order reaction.

Solution: Stopping the input flow will cause the system to behave as a batch reactor even though the outlet flow continues. The initial concentration of the batch is the steady-state value $a_{in}/(1 + k\bar{t})$, and the concentration decreases exponentially as the vessel discharges:

$$a_{out} = \frac{a_{in}\exp(-kt)}{1 + k\bar{t}} \tag{14.7}$$

where we have assumed the shutdown transient to start at time $t = 0$. The transient lasts until the vessel is empty, $t_{empty} = \bar{t}$, assuming the discharge rate is held constant at its initial value.

Turning off the feed and letting the reactor empty itself passively is the most common method of shutting down a CSTR. The conversion increases during the discharge period, but this may not be a problem. Perhaps the reactor was already operating at such high conversion that the increase is of no consequence. For complex reactions, however, the increase in conversion may mean that the product is off-specification. So, we consider the following problem. The reactor is operating and full of good material. It is desired to shut it down while producing no material that is off-specification. One approach is to dump the entire contents quickly, but this is likely to cause downstream processing problems. Another approach allows a gradual discharge while maintaining constant product quality.

Example 14.6: Explore the consequences of the following shutdown strategy for an isothermal, constant-density CSTR that has been operating at steady state. At time zero, the discharge flow rate is increased by a factor of $1 + \delta$. Simultaneously, the inlet flow rate is made proportional to the fluid volume in the vessel. When does the vessel empty and what happens to the composition of the discharge stream during the shutdown interval?

Solution: The control strategy is to set the inlet flow rate proportional to the active volume in the vessel:

$$Q_{in} = \frac{V(t)}{\bar{t}} \qquad (14.8)$$

where $1/\bar{t}$ is the proportionality constant. This is the same proportionality constant that related flow rate to volume during the initial period of steady-state operation. The inlet flow rate gradually declines from its steady-state value of Q_0, and Equation (14.1) becomes

$$\frac{dV}{dt} = \frac{V}{\bar{t}} - (1+\delta)Q_0 \qquad (14.9)$$

The initial condition is $V = V_0$ at $t = 0$. Solving this ODE and setting $V = 0$ gives

$$t_{empty} = \bar{t} \ln\left[\frac{1+\delta}{\delta}\right] \qquad (14.10)$$

These shutdown times are moderate: $2.4\bar{t}$ for $\delta = 0.1$ and $3\bar{t}$ for $\delta = 0.05$. Equation (14.2) governs the outlet concentration during the shutdown interval. For this shutdown strategy, it becomes

$$a_{out}\frac{dV}{dt} + V\frac{da_{out}}{dt} = \frac{a_{in}V}{\bar{t}} - (1+\delta)a_{out}Q_0 + V\mathscr{R}_A$$

Substituting Equation (14.9) and simplifying gives

$$\bar{t}\frac{da_{out}}{dt} = a_{in} - a_{out} + \bar{t}\mathscr{R}_A$$

This equation applies for $t \geq 0$; and when t is exactly zero, a_{out} has its steady-state value, which is determined from the steady-state design equation:

$$0 = a_{in} - a_{out} + \bar{t}\mathscr{R}_A$$

Compare these results to see that $da_{out}/dt = 0$ so that the outlet concentration does not change during the shutdown transient.

Example 14.6 derives a rather remarkable result. Here is a way of gradually shutting down a CSTR while keeping a constant outlet composition. The derivation applies to an arbitrary $\mathscr{R}_A$ and can be extended to include multiple reactions and adiabatic reactions. It is been experimentally verified for a polymerization.[1] It can be generalized to shut down a train of CSTRs in series. The reason it works is that the material in the tank always experiences the same mean residence time and residence time distribution as existed during the original steady state. Hence, it is called constant RTD control. It will cease to work in a real vessel when the liquid level drops below the agitator.

Constant RTD control can be applied in reverse to startup a vessel while minimizing off-specification materials. For this form of startup, a near steady state is first achieved with a minimum level of material and thus with minimum throughput. When the product is satisfactory, the operating level is gradually increased by lowering the discharge flow while applying Equation (14.8) to the inlet flow. The vessel fills, the flow rate increases, but the residence time distribution is constant.

Product Transitions. A common practice in the manufacture of polymers and specialty chemicals is to use the same basic process for multiple products. Batch reactions obviously lend themselves to this practice, but continuous production lines are also switched from one product to another as dictated by market demand. This is routinely done at production rates of 50 t/h. There is strong economic incentive to minimize downtime and to minimize the production of off-specification product. A complete shutdown and restart might minimize the amount of off-specification product but may cause appreciable downtime. A running transition will maintain productivity but may generate a large amount of off-specification material. Combination strategies such as partially emptying a reactor before making a chemical change are sometimes used. When the reactor can be modeled as one or more CSTRs in series, Equations (14.1) through (14.3) provide the general framework for studying product transitions.

Example 14.7: A polymer manufacturer makes two products in a CSTR. Product I is made by the reaction

$$A \xrightarrow{k_I} P$$

Product II is a C-modified version of Product I in which a second reaction occurs:

$$P + C \xrightarrow{k_{II}} Q$$

The reactor operates at constant volume, constant density, constant flow rate, and isothermally. The only difference between the two products is the addition of component C to the feed when Product II is made.

Explore methods for making a running transition from Product I to Product II. There is no P or Q in the reactor feed.

Solution: One version of Equation (14.2) is written for each reactant:

$$\bar{t}\frac{da_{out}}{dt} = a_{in} - a_{out} - \bar{t}ka_{out}$$

$$\bar{t}\frac{dp_{out}}{dt} = -p_{out} + \bar{t}(k_I a_{out} - k_{II} p_{out} c_{out})$$

$$\bar{t}\frac{dq_{out}}{dt} = -q_{out} + \bar{t}_I k_{II} p_{out} c_{out}$$

$$\bar{t}\frac{dc_{out}}{dt} = c_{in} - c_{out} - \bar{t}k_{II} p_{out} c_{out}$$

The analysis from this point will be numerical. Suppose $a_{in} = 20\,\text{mol/m}^3$ for both products, $c_{in} = 9\,\text{mol/m}^3$ when Product II is being made at steady state, $\bar{t} = 1\,\text{h}$, $k_I = 4\,\text{h}^{-1}$, and $k_{II} = 1\,\text{h/(m}^3\cdot\text{mol)}$. This kinetic system allows only one steady state. It is stable and can be found by solving the governing ODEs starting from any initial condition. The steady-state response when making Product I is $a_{out} = 4\,\text{mol/m}^3$ and $p_{out} = 16\,\text{mol/m}^3$. When Product II is made, $a_{out} = 4\,\text{mol/m}^3$, $p_{out} = 8\,\text{mol/m}^3$, $q_{out} = 8\,\text{mol/m}^3$, and $c_{out} = 1\,\text{mol/m}^3$.

Consider a transition from Product I to Product II. The simplest case is just to add component C to the feed at the required steady-state concentration of $c_{in} = 9\,\text{mol/m}^3$. The governing ODEs are solved subject to the initial condition that the reactor initially contains the steady-state composition corre-sponding to Product I. Figure 14.3 shows the leisurely response toward the new steady state. The dotted lines represent the specification limits for Product II. They allow any Q concentration between 7 and $9\,\text{mol/m}^3$. The outlet composition enters the limits after 2.3 h. The specification for Product I allows $1\,\text{mol/m}^3$ of Q to be present, but the rapid initial increase in the concentration of Q means that the limit is quickly exceeded. The total transition time is about 2 h, during which some 100 t of off-specification material would be produced.

A far better control strategy is available. Figure 14.3 shows the response to a form of bang-bang control where C is charged as rapidly as possible to

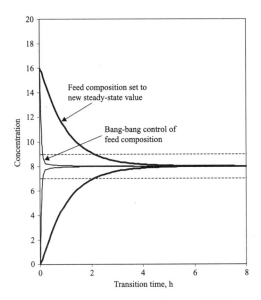

FIGURE 14.3 Transitions from Product I to Product II in Example 14.7.

quickly change the reactor concentration to $11\,\mathrm{mol/m^3}$. This is the first bang and it is assumed to be instantaneous. The second bang completely stops the feed of C for 12 min. This prevents the outlet concentration of Q from overshooting its steady-state value. After the 12 min duration of the second bang, the inlet concentration of Q is set to its steady-state value. The transition time is reduced to about 7 min and the amount of off-specification material to about 6 t. This is not yet the optimal response, which probably shows some overshoot in the outlet concentration of Q, but it is a reasonable start. Problems 14.9 and 14.10 pursue this problem.

14.1.2 Nonisothermal Stirred Tank Reactors

Nonisothermal stirred tanks are governed by an enthalpy balance that contains the heat of reaction as a significant term. If the heat of reaction is unimportant, so that a desired T_{out} can be imposed on the system regardless of the extent of reaction, then the reactor dynamics can be analyzed by the methods of the previous section.

This section focuses on situations where Equation (14.3) must be considered as part of the design. Even for these situations, it is usually possible to control a CSTR at a desired temperature. If temperature control can be achieved rapidly, the isothermal design techniques again become applicable. Rapid means on a time scale that is fast compared with reaction times and composition changes.

Example 14.8: The styrene polymerization example of Example 5.7 shows three steady states. The middle steady state with $a_{out} = 0.738$ and $T_{out} = 403$ K is unstable. Devise a control system that stabilizes operation near it.

Solution: There are several theoretical ways of stabilizing the reactor, but temperature control is the normal choice. The reactor in Example 5.7 was adiabatic. Some form of heat exchange must be added. Possibilities are to control the inlet temperature, to control the pressure in the vapor space thereby allowing reflux of styrene monomer at the desired temperature, or to control the jacket or external heat exchanger temperature. The following example regulates the jacket temperature. Refer to Example 5.7. The component balance on styrene is unchanged from Equation (5.29):

$$\frac{da_{out}}{d\tau} = a_{in} - a_{out} - 2 \times 10^{10} \exp(-10{,}000/T_{out}) a_{out}$$

A heat exchange term is added to the energy balance, Equation (5.30), to give

$$\frac{dT_{out}}{d\tau} = T_{in} - T_{out} + 8 \times 10^{12} \exp(-10{,}000/T_{out}) a_{out}/a_{in} + \frac{UA_{ext}\bar{t}}{V\rho C_p}(T_{ext} - T)$$

The heat transfer group, $UA_{ext}\bar{t}/V\rho C_p$, is dimensionless. Assume its value is 0.02. A controller is needed to regulate T_{ext}. The industrial choice would be a two-term controller, proportional plus reset. We skirt the formal control issues and use a simple controller of the form

$$T_{ext} = 375 + 20(T_{set} - T)$$

Suppose the reactor has been started using the fast-fill-and-hold method and has reached $a = 0.65$ at $T = 420$ K. Continuous flow is started with $a_{in} = 1$, $T_{in} = 375$ K, and $T_{set} = 404$ K. Figure 14.4 shows the response. The temperature response is very rapid, but the conversion increases slightly during the first seconds of operation. Without temperature control, the reaction would have run away. The concentration is slowly evolving to its eventual steady-state value of about 0.26. There is a small offset in the temperature because the controller has no reset term.

Example 14.9: This example cites a real study of a laboratory CSTR that exhibits complex dynamics and limit cycles in the absence of a feedback controller. We cite the work of Vermeulen, and Fortuin,[2] who studied the acid-catalyzed hydration of 2,3-epoxy-1-propanol to glycerol:

$$\underset{\substack{|\;\;|\;\;|\\ OH\;\;\;O}}{H-C-C-C-H} + H_2O \xrightarrow{H_2SO_4} \underset{\substack{|\;\;|\;\;|\\ OH\;OH\;OH}}{H-C-C-C-H}$$

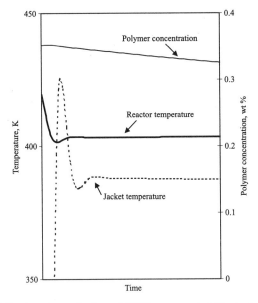

FIGURE 14.4 Stabilization of a nonisothermal CSTR near a metastable steady state.

The reactor has separate feed streams for an aqueous solution of the epoxy and for an aqueous solution of the acid. Startup begins with the vessel initially full of acid.

The chemistry seems fairly simple. The water concentration is high and approximately constant so that the reaction is pseudo-first-order with respect to the epoxy. The rate is also proportional to the hydrogen ion concentration h. Thus,

$$\mathcal{R} = k_0 \exp\left[\frac{-E}{R_g T}\right] eh$$

where e is the epoxy concentration. The sulfuric acid dissociates in two equilibrium steps:

$$H_2SO_4 \rightleftharpoons H^+ + HSO_4^- \qquad K_1 = \frac{[H^+][HSO_4^-]}{[H_2SO_4]}$$

$$HSO_4^- \rightleftharpoons H^+ + SO_4^{2-} \qquad K_2 = \frac{[H^+][SO_4^{2-}]}{[HSO_4^-]}$$

The hydrogen ion concentration can be found from

$$\frac{h_{out}^3}{K_1} + h_{out}^{2+} + (K_2 - s_{out})h_{out} - 2K_2 s_{out} = 0$$

where s is the total sulfate concentration.

There are three ODEs that govern the system. For sulfate, which is not consumed,

$$\bar{t}\frac{ds_{out}}{dt} = s_{in} - s_{out} \qquad s_{out} = s_0 \quad \text{at} \quad t = 0$$

For the epoxy,

$$\bar{t}\frac{de_{out}}{dt} = e_{in} - e_{out} - k_0 \bar{t} \exp\left[\frac{-E}{R_g T}\right] e_{out} h_{out} \qquad e_{out} = 0 \quad \text{at} \quad t = 0$$

For temperature,

$$(\rho V C_P + m_R C_R)\frac{dT_{out}}{dt} = \rho Q (C_P)_{in} T_{in} - \rho Q C_P T_{out} + UA_{ext}(T_{ext} - T_{out})$$

$$+ q - \Delta H_R \rho V k_0 \exp\left[\frac{-E}{R_g T}\right] e_{out} h_{out} \qquad T_{out} = T_0 \quad \text{at} \quad t = 0$$

This heat balance contains two terms not seen before: $m_R C_R$ represents the mass times specific heat of the agitator and vessel walls and q represents the energy input by the agitator. Although the model is nominally for constant physical properties, Vermeulen and Fortuin found a better fit to the experimental data when they used a slightly different specific heat for the inlet stream $(C_P)_{in}$.

Figure 14.5 shows a comparison between experimental results and the model. The startup transient has an initial overshoot followed by an apparent approach to steady state. Oscillations begin after a phenomenally long delay, $t > 10\bar{t}$, and the system goes into a limit cycle. The long delay before the occurrence of the oscillations is remarkable. So is the good agreement between model and experiment. Two facts are apparent: quite complex behavior is possible with a simple model, and one should wait a long time before reaching firm conclusions regarding stability. The conventional wisdom is that steady state is closely approached after 3–5 mean residence times.

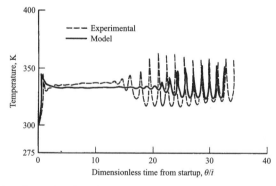

FIGURE 14.5 Experimental and model results on the acid-catalyzed hydration of 2,3-epoxypropanol to glycerol.

14.2 UNSTEADY PISTON FLOW

Dynamic analysis of piston flow reactors is fairly straightforward and rather unexciting for incompressible fluids. Piston flow causes the dynamic response of the system to be especially simple. The form of response is a limiting case of that found in real systems. We have seen that piston flow is usually a desirable regime from the viewpoint of reaction yields and selectivities. It turns out to be somewhat undesirable from a control viewpoint since there is no natural dampening of disturbances.

Unlike stirred tanks, piston flow reactors are distributed systems with one-dimensional gradients in composition and physical properties. Steady-state performance is governed by ordinary differential equations, and dynamic performance is governed by partial differential equations, albeit simple, first-order PDEs. Figure 14.6 illustrates a component balance for a differential volume element.

$$\mathscr{R}_A \Delta V + aQ - \left[aQ + \frac{\partial(aQ)}{\partial z}\Delta z\right] = \frac{\partial a}{\partial t}\Delta V$$

Input − Output = Accumulation

or

$$\frac{\partial a}{\partial t} + \frac{1}{A_c}\frac{\partial(aQ)}{\partial z} = \frac{\partial a}{\partial t} + \frac{1}{A_c}\frac{\partial(A_c \bar{u}a)}{\partial z} = \mathscr{R}_A \qquad (14.11)$$

where $A_c = \Delta V/\Delta z$ is the cross-sectional area of the tube. The tube has rigid walls and a fixed length so $\partial V/\partial t = 0$. Compare Equation (14.11) to Equation (3.5). All we have done is add an accumulation term. An overall mass balance gives

$$\frac{\partial \rho}{\partial t} + \frac{1}{A_c}\frac{\partial(\rho Q)}{\partial z} = 0 \qquad (14.12)$$

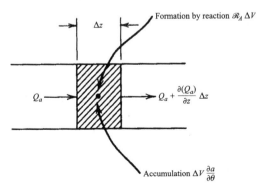

FIGURE 14.6 Differential volume element in an unsteady piston flow reactor.

If ρ is constant, Equation (14.12) shows Q to be constant as well. Then the component balance simplifies to

$$\frac{\partial a}{\partial t} + \bar{u}\frac{\partial a}{\partial z} = \mathscr{R}_A \qquad (14.13)$$

This result is valid for variable A_c but not for variable ρ. It governs a PFR with a time-dependent inlet concentration but with other properties constant. The final simplification supposes that A_c is constant so that $\bar{u}$ is constant. Then Equation (14.13) has a simple analytical solution:

$$\frac{z}{\bar{u}} = t = \int_{a_{in}(t-z/\bar{u})}^{a(t,z)} \frac{da'}{\mathscr{R}_A} \qquad (14.14)$$

Formal verification that this result actually satisfies Equation (14.13) is an exercise in partial differentiation, but a physical interpretation will confirm its validity. Consider a small group of molecules that are in the reactor at position z at time t. They entered the reactor at time $t' = t - (z/\bar{u})$ and had initial composition $a(t', z) = a_{in}(t') = a_{in}(t - z/\bar{u})$. Their composition has subsequently evolved according to batch reaction kinetics as indicated by the right-hand side of Equation (14.14). Molecules leaving the reactor at time t entered it at time $t - \bar{t}$. Thus,

$$\frac{L}{\bar{u}} = \bar{t} = \int_{a_{in}(t-\bar{t})}^{a_{out}(t)} \frac{da'}{\mathscr{R}_A} \qquad (14.15)$$

When a_{in} is constant, Equation (14.14) is a solution of Equation (3.1) evaluated at position z, and Equation (14.15) is a solution evaluated at the reactor outlet.

The temperature counterpart of Equation (14.11) is

$$\frac{\partial(\rho H)}{\partial t} + \rho Q\frac{\partial H}{\partial z} = \frac{\partial(\rho H)}{\partial t} + \rho\bar{u}A_c\frac{\partial H}{\partial z} = -\Delta H_R\mathscr{R}A_c + UA'_{ext}(T_{ext} - T) \qquad (14.16)$$

With constant physical properties and A_c this becomes

$$\frac{\partial T}{\partial t} + \bar{u}\frac{\partial T}{\partial z} = \frac{-\Delta H\mathscr{R}}{\rho C_p} + \frac{2U}{R\rho C_p}(T_{ext} - T) \qquad (14.17)$$

If the reactor is adiabatic, $U=0$ and Equation (14.17) has the following formal solution:

$$\frac{z}{\bar{u}} = t = \int_{T_{in}(t-z/\bar{u})}^{T(t,z)} \frac{\rho C_P}{-\Delta H_R\mathscr{R}}dT' \qquad (14.18)$$

This formal solution is not useful for finding $T(t, z)$ since the reaction rate will depend on composition. It does, however, show that the temperature at time t and position z is determined by inlet conditions at time $t - z/\bar{u}$. Temperature, like composition, progresses in a batch-like trajectory from its entering value to its exit value without regard for what is happening elsewhere in the tube. Including heat exchange to the environment, $U > 0$, does not change this fact provided T_{ext} is uncoupled to T. A solution for $a_{out}(t)$ and $T_{out}(t)$ can be found by solving the ordinary differential equations that govern steady-state piston flow:

$$\bar{u}\frac{da}{dz} = \mathscr{R}_A$$

$$\bar{u}\frac{dT}{dz} = \frac{-\Delta H \mathscr{R}}{\rho C_P} + \frac{2U}{R\rho C_P}(T_{ext} - T)$$

Solve these ODEs subject to the initial conditions that $a = a_{in}(t - \bar{t})$ and $T = T_{in}(t - \bar{t})$ at $z = 0$. Evaluate the solution at $z = L$ to obtain $a_{out}(t)$ and $T_{out}(t)$.

The most important fact about piston flow is that disturbances at the inlet are propagated down the tube with no dissipation due to mixing. They arrive at the outlet $\bar{t}$ seconds later. This pure time delay is known as *dead time*. Systems with substantial amounts of dead time oscillate when feedback control is attempted. This is caused by the controller responding to an output signal that may be completely different than that corresponding to the current input. *Feedforward* control represents a theoretically sound approach to controlling systems with appreciable dead time. Sensors are installed at the inlet to the reactor to measure fluctuating inputs. The appropriate responses to these inputs are calculated using a model. The model used for the calculations may be imperfect but it can be improved using feedback of actual responses. In *adaptive* control, this feedback of results is done automatically using a special error signal to correct the model.

Piston flow reactors lack any internal mechanisms for memory. There is no axial dispersion of heat or mass. What has happened previously has no effect on what is happening now. Given a set of inlet conditions $(a_{in}, T_{in}, T_{ext})$, only one output (a_{out}, T_{out}) is possible. A PFR cannot exhibit steady-state multiplicity unless there is some form of external feedback. External recycle of mass or heat can provide this feedback and may destabilize the system. Figure 14.7 shows an example of external feedback of heat that can lead to the same multiple steady states that are possible with a CSTR. Another example is when the vessel walls or packing have significant thermal capacity. Equation (14.16) no longer applies. Instead, the reactor must be treated as two phase with respect to temperature, even though it is single phase with respect to concentration.

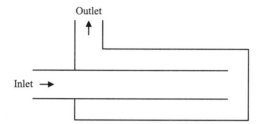

FIGURE 14.7 Piston flow reactor with feedback of heat.

14.3 UNSTEADY CONVECTIVE DIFFUSION

The unsteady version of the convective diffusion equation is obtained just by adding a time derivative to the steady version. Equation (8.32) for the convective diffusion of mass becomes

$$\frac{\partial a}{\partial t} + V_z(r)\frac{\partial a}{\partial z} = \mathscr{D}_A\left[\frac{\partial^2 a}{\partial z^2} + \frac{1}{r}\frac{\partial a}{\partial r} + \frac{\partial^2 a}{\partial r^2}\right] + \mathscr{R}_A \qquad (14.19)$$

The analogous equation for the convective flow of heat is

$$\frac{\partial T}{\partial t} + V_z(r)\frac{\partial T}{\partial z} = \alpha_T\left[\frac{1}{r}\frac{\partial T}{\partial r} + \frac{\partial^2 T}{\partial r^2} + \frac{\partial^2 T}{\partial z^2}\right] - \frac{\Delta H_R \mathscr{R}}{\rho C_P} \qquad (14.20)$$

These equations assume that the reactor is single phase and that the surroundings have negligible heat capacity. In principle, Equations (14.19) and (14.20) can be solved numerically using the simple methods of Chapters 8 and 9. The two-dimensional problem in r and z is solved for a fixed value of t. A step forward in t is taken, the two-dimensional problem is resolved at the new t, and so on.

The axial dispersion model discussed in Section 9.3 is a simplified version of Equation (14.19). Analytical solutions for unsteady axial dispersion are given in Chapter 15.

PROBLEMS

14.1. Determine the fractional filling rate Q_{fill}/Q that will fill an isothermal, constant-density, stirred tank reactor while simultaneously achieving the steady-state conversion corresponding to flow rate Q. Assume a second-order reaction with $a_{in}k\bar{t} = 1$ and $\bar{t} = 5\,\text{h}$ at the intended steady state.

14.2. Devise a fast-fill-and-hold startup strategy for the reaction of Problem 14.1.

14.3. Suppose the consecutive elementary reactions

$$2A \xrightarrow{k_I} B \xrightarrow{k_{II}} C$$

occur in an isothermal CSTR. Suppose $a_{in}k_I \bar{t} = 2$, and $k_{II} = 1$ with $b_{in} = c_{in} = 0$. Determine the steady-state outlet composition and explore system stability by using a variety of initial conditions a_{in} and b_{in}.

14.4. Find a nontrivial (meaning $r_{out} > 0$ and $l_{out} > 0$) steady state for the rabbit and lynx problem in Example 14.2. Test its stability by making small changes in the system parameters.

14.5. Suppose the following reactions are occurring in an isothermal perfect mixer:

$$A + B \rightarrow 2B \qquad \mathscr{R}_I = k_I a b$$

Suppose there is no B in the feed but that some B is charged to the reactor at startup. Can this form of startup lead to stable operation with $b_{in} = 0$ but $b_{out} > 0$?

14.6. Suppose the following reactions are occurring in an isothermal CSTR:

$$A + 2B \rightarrow 3B \qquad \mathscr{R}_I = k_I a b^2$$
$$B \rightarrow C \qquad \mathscr{R}_{II} = k_{II} b$$

Since the autocatalytic reaction is third order, a steady-state material balance gives a cubic in b_{out}. This means there are one or three steady states. Suppose $b_{in}/a_{in} = 1/15$ and explore the stability of the single or middle steady state for each of the following cases:

(a) $a_{in}^2 k_I \bar{t} = 190$, $k_{II} \bar{t} = 4.750$ (a small disturbance from the steady state gives damped oscillations)

(b) $a_{in}^2 k_I \bar{t} = 225$, $k_{II} \bar{t} = 5.625$ (a small disturbance from the steady state gives sustained oscillations)

(c) $a_{in}^2 k_I \bar{t} = 315$, $k_{II} \bar{t} = 7.875$ (a small disturbance from the steady state gives undamped oscillations and divergence to a new steady state)

See Gray and Scott[3] for a detailed analysis of this (hypothetical) reaction system.

14.7. Determine t_{full} for the fed-batch method of Examples 14.3 and 14.4 in the limiting cases as $k \rightarrow \infty$ and $k \rightarrow 0$.

Hint: the range is $\bar{t} < t_{full} < 2\bar{t}$. Determining one of these limits is an easy exercise using L'Hospital's rule.

14.8. Suppose there are two parallel, first-order reactions in a steady-state CSTR. Show that neither the fed-batch nor fast-fill-and-hold strategies can achieve a bumpless startup if the reactions have different rate constants. Is it possible to use a combination strategy to achieve a bumpless startup? A numerical example will be sufficient.

14.9. Improve the control strategy for the product transition in Example 14.7. Ignore mixing time constraints, flow rate limitations on the addition of component C, and any constraints on the allowable value for c_{out}. The concentration of Q can exceed its steady-state value of $8\,\text{mol/m}^3$ but must not be allowed to go outside the upper specification limit of $9\,\text{mol/m}^3$.

14.10. The transition control strategy in Example 14.7 quickly increases the concentration of C in the vessel from 0 to $11\,\text{mol/m}^3$. This means that $c_{out} = 11\,\text{mol/m}^3$, at least temporarily. Suppose the downstream recovery system is unable to handle more than $2\,\text{mol/m}^3$ of unreacted C. The obvious start to the transition is to quickly charge enough C to the reactor to get $c_{out} = 2\,\text{mol/m}^3$, but then what?

14.11. Use the inlet temperature rather than the jacket temperature to control the reactor in Example 14.8.

14.12. Suppose the reactor in Example 14.8 remains in batch mode after the fast-fill-and-hold startup. Will the temperature control system still work? A preliminary answer based on the approximate kinetics of Example 14.7 is sufficient, but see the next problem.

14.13. Use the more rigorous kinetic model of Appendix 13.1 to repeat the previous problem. Also consider how the viscosity increase might affect the heat transfer group. Use the viscosity correlation in Appendix 13.1.

14.14. Control systems can fail in many ways, and highly energetic reactions like the styrene polymerization in Examples 5.7 and 14.8 raise major safety concerns. The contents of the vessel are similar to napalm. Discuss ways of preventing accidents or of mitigating the effects of accidents. Is there one best method for avoiding a disastrous runaway?

14.15. Standard thermodynamic texts give a more general version of Equation (14.3). See Smith et al.[4] This more general version is

$$\frac{d}{dt}[\rho(H-PV)V] = [H_{in} + \bar{u}_{in}^2/2 + Z_{in}g_{in}]Q_{in}\rho_{in}$$
$$- [H_{out} + \bar{u}_{out}^2/2 + Z_{out}g_{out}]Q_{out}\rho_{out}$$
$$- V\Delta H_R \mathscr{R} + UA_{ext}(T_{ext} - T_{out})$$

Identify the added terms. When could they be important? When might other terms be important? Remember, this is a CSTR, not a spaceship, but note the extra terms included in Example 14.9.

14.16. Referring to Example 14.9, Vermeulen and Fortuin estimated all the parameters in their model from physical data. They then compared model predictions with experimental results and from this they made improved estimates using nonlinear regression. Their results were as follows:

Parameter	Estimate from physical data	Estimate from regression analysis	Units
ρQ	0.0019	0.001881615	kg/s
ρV	0.30	0.2998885	kg
$m_r C_R$	392	405.5976	J/K
e_{in}	8.55	8.532488	mol/kg
$(C_P)_{in}$	2650	2785.279	J/(kg·K)
C_P	2650	2517017	J/(kg·K)
s_{in}	0.15	0.1530875	mol/kg
k_0	8.5×10^{10}	8.534612×10^{10}	kg/(mol·s)
UA_{ext}	30	32.93344	J/(s·K)
$-\Delta H_R$	88200	87927.31	J/mol
q	30	32.62476	J/s
T_{in}	273.91	273.9100	K
T_{ext}	298.34	298.3410	K
E/R_g	8827	8815.440	K
K_1	1000	1000	mol/kg
K_2	0.012023	0.012023	mol/kg
T_0	300.605	300.605	K
e_0	0	0	mol/kg
s_0	0.894	0.894	mol/kg

(a) Show that, to a good approximation,

$$h_{out} = 0.5\left[s_{out} - K_2 + \sqrt{s_{out}^2 + 6s_{out}K_2 + K_2^2}\right]$$

(b) Using the parameter estimates obtained by regression analysis, confirm the qualitative behavior shown in Figure 14.5. A sophisticated integration routine may be necessary, but it is not necessary to match Vermeulen and Fortuin exactly.

(c) The parameter-fitting procedure used experimental data from a single run. Determine the sensitivity of the model by replacing the regression estimates with the physical estimates. What does this suggest about the reproducibility of the experiment?

(d) Devise a means for achieving steady operation at high conversion to glycerol. Undesirable side reactions may become significant at 423 K. At atmospheric pressure and complete conversion, the mixture boils at 378 K.

14.17. Upon entering engineering, freshmen tend to choose their curriculum based on job demand for the current graduating class. It is easy to change curriculum in the freshman year but it becomes difficult in subsequent years. Thus, we might model the engineering education process as a stirred tank with $\bar{t}_s = 1$ year followed by a piston flow reactor with

$\bar{t}_p = 3$ years. Does this model predict a good balance of supply and demand? What strategy would you suggest to a freshman whose sole concern was being in high demand upon graduation?

14.18. Blood vessels have elastic walls that expand or contract due to changes in pressure or the passage of corpuscles. How should Equations (14.11) and (14.12) be modified to reflect this behavior?

REFERENCES

1. Nauman, E. B. and Carter, K., "A control system for the minimization of substandard during product transitions," *I & EC Proc. Des. Dev.*, **13**, 275–279 (1974).
2. Vermeulen, D. P. and Fortuin, J. M. H., "Experimental verification of a model describing the transient behavior of a reaction system approaching a limit cycle or a runaway in a CSTR," *Chem. Eng. Sci.*, **41**, 1089–1095 (1986).
3. Gray, P. and Scott, S. K., "Autocatalytic reactions in the isothermal, continuous stirred tank reactor," *Chem. Eng. Sci.*, **39**, 1087–1097 (1984).
4. Smith, J. M., Van Ness, H. C., and Abbott, M. M., *Introduction to Chemical Engineering Thermodynamics*, 6th ed., McGraw-Hill, New York, 2001.

SUGGESTIONS FOR FURTHER READING

This chapter has presented time-domain solutions of unsteady material and energy balances. The more usual undergraduate treatment of dynamic systems is given in a course on control and relies heavily on Laplace transform techniques. One suitable reference is

Stephanopoulos, G., *Chemical Process Control: An Introduction to Theory and Practice*, Prentice-Hall, Englewood Cliffs, NJ, 1984.

A more recent book that stresses numerical solutions using Matlab® is

Bequette, B. W., *Process Dynamics: Modeling, Analysis and Simulation*, Prentice-Hall, Englewood Cliffs, NJ, 1998.

Unsteady reaction data are often an excellent means for estimating physical parameters that would be difficult or impossible to elucidate from steady-state measurements. However, the associated problems in nonlinear optimization can be formidable. A recent review and comparison of methods is given by

Biegler, L. T., Damiano, J. J., and Blau, G. E., "Nonlinear parameter estimation: a case study comparison," *AIChE J.*, **32**, 29–45 (1986).

CHAPTER 15
RESIDENCE TIME DISTRIBUTIONS

Reactor *design* usually begins in the laboratory with a kinetic study. Data are taken in small-scale, specially designed equipment that hopefully (but not inevitably) approximates an ideal, isothermal reactor: batch, perfectly mixed stirred tank, or piston flow. The laboratory data are fit to a kinetic model using the methods of Chapter 7. The kinetic model is then combined with a transport model to give the overall design.

Suppose now that a pilot-plant or full-scale reactor has been built and operated. How can its performance be used to confirm the kinetic and transport models and to improve future designs? Reactor *analysis* begins with an operating reactor and seeks to understand several interrelated aspects of actual performance: kinetics, flow patterns, mixing, mass transfer, and heat transfer. This chapter is concerned with the analysis of flow and mixing processes and their interactions with kinetics. It uses *residence time theory* as the major tool for the analysis.

In a batch reactor, all molecules enter and leave together. If the system is isothermal, reaction yields depend only on the elapsed time and on the initial composition. The situation in flow systems is more complicated but not impossibly so. The counterpart of the batch reaction time is the age of a molecule. Aging begins when a molecule enters the reactor and ceases when it leaves. The total time spent within the boundaries of the reactor is known as the exit age, or *residence time*, t. Except in batch and piston flow reactors, molecules leaving the system will have a variety of residence times. The distribution of residence times provides considerable information about homogeneous, isothermal reactions. For single, first-order reactions, knowledge of the *residence time distribution* allows the yield to be calculated exactly, even in flow systems of arbitrary complexity. For other reaction orders, it is usually possible to calculate tight limits, within which the yield must lie. Even if the system is nonisothermal and heterogeneous, knowledge of the residence time distribution provides substantial insight regarding the flow processes occurring within it.

15.1 RESIDENCE TIME THEORY

The time that a molecule spends in a reactive system will affect its probability of reacting; and the measurement, interpretation, and modeling of residence time distributions are important aspects of chemical reaction engineering. Part of the inspiration for residence time theory came from the *black box* analysis techniques used by electrical engineers to study circuits. These are *stimulus–response* or *input–output* methods where a system is disturbed and its response to the disturbance is measured. The measured response, when properly interpreted, is used to predict the response of the system to other inputs. For residence time measurements, an inert tracer is injected at the inlet to the reactor, and the tracer concentration is measured at the outlet. The injection is carried out in a standardized way to allow easy interpretation of the results, which can then be used to make predictions. Predictions include the dynamic response of the system to arbitrary tracer inputs. More important, however, are the predictions of the steady-state yield of reactions in continuous-flow systems. All this can be done without opening the black box.

15.1.1 Inert Tracer Experiments

Transient experiments with inert tracers are used to determine residence time distributions. In real systems, they will be actual experiments. In theoretical studies, the experiments are mathematical and are applied to a dynamic model of the system.

Negative Step Changes and the Washout Function. Suppose that an inert tracer has been fed to a CSTR for an extended period of time, giving $C_{in} = C_{out} = C_0$ for $t < 0$. At time $t = 0$, the tracer supply is suddenly stopped so that $C_{in} = 0$ for $t > 0$. Equation (14.2) governs the transient response of the system. For $t > 0$,

$$V \frac{dC_{out}}{dt} = -Q_{out} C_{out}$$

where constant-volume operation with $\mathscr{R}_C = 0$ has been assumed. The solution is

$$\frac{C_{out}(t)}{C_0} = \exp\left[-\frac{Q_{out} t}{V}\right] = \exp\left[-\frac{t}{\bar{t}}\right] \tag{15.1}$$

Tracer molecules originally in the system at time $t = 0$ gradually wash out. The exponential form of Equation (15.1) is specific to a CSTR, but the concept of *washout* applies to any flow system. Consider some time $t > 0$ when the fraction of molecules remaining in the system is $W(t) = C_{out}(t)/C_0$. These molecules must necessarily have entered the reactor before time $t = 0$ since no tracer was fed after that time. Thus, these molecules have residence times of t or longer. The residence time *washout function* is defined as

$$W(t) = \text{Fraction of molecules leaving the system that experienced a residence time greater than } t \quad (15.2)$$

It is apparent that $W(0) = 1$ since all molecules must have a residence time of zero or longer and that $W(\infty) = 0$ since all molecules will eventually leave the system. Also, the function $W(t)$ will be nonincreasing.

Washout experiments can be used to measure the residence time distribution in continuous-flow systems. A good step change must be made at the reactor inlet. The concentration of tracer molecules leaving the system must be accurately measured at the outlet. If the tracer has a background concentration, it is subtracted from the experimental measurements. The flow properties of the tracer molecules must be similar to those of the reactant molecules. It is usually possible to meet these requirements in practice. The major theoretical requirement is that the inlet and outlet streams have unidirectional flows so that molecules that once enter the system stay in until they exit, never to return. Systems with unidirectional inlet and outlet streams are *closed* in the sense of the axial dispersion model; i.e., $D_{in} = D_{out} = 0$. See Sections 9.3.1 and 15.2.2. Most systems of chemical engineering importance are closed to a reasonable approximation.

The use of inert tracer experiments to measure residence time distributions can be extended to systems with multiple inlets and outlets, multiple phases within the reactor, and species-dependent residence times. This discussion ignores these complications, but see "Suggestions for Further Reading."

Positive Step Changes and the Cumulative Distribution. Residence time distributions can also be measured by applying a positive step change to the inlet of the reactor: $C_{in} = C_{out} = 0$ for $t < 0$ and $C_{in} = C_0$ for $t > 0$. Then the outlet response, $F(t) = C_{out}(t)/C_0$, gives the *cumulative distribution function*:

$$F(t) = \text{Fraction of molecules leaving the system that experienced a residence time less than } t \quad (15.3)$$

Properties of the cumulative distribution function are $F(0) = 0$, $F(\infty) = 1$, and $F(t)$ is nondecreasing. It is related to the washout function by

$$F(t) = 1 - W(t) \quad (15.4)$$

Thus, measurement of one readily gives the other. The washout experiment is generally preferred since $W(\infty) = 0$ will be known a priori but $F(\infty) = C_0$ must usually be measured. The positive step change will also be subject to possible changes in C_0 during the course of the experiment. However, the positive step change experiment requires a smaller amount of tracer since the experiment will be terminated before the outlet concentration fully reaches C_0.

Impulse Response and the Differential Distribution. Suppose a small amount of tracer is instantaneously injected at time $t=0$ into the inlet of a reactor. All the tracer molecules enter together but leave at varying times. The tracer concentration at the outlet is measured and integrated with respect to time. The integral will be finite and proportional to the total quantity of tracer that was injected. The concentration measurement at the reactor outlet is normalized by this integral to obtain the *impulse response function*:

$$f(t) = \frac{C_{out}(t)}{\int_0^\infty C_{out}(t)dt} \tag{15.5}$$

This function has the physical interpretation as

$$f(t)dt = \text{Fraction of molecules leaving the system that experienced a residence time between } t \text{ and } t+dt \tag{15.6}$$

It is normally called the *differential distribution function* (of residence times). It is also known as the density function or frequency function. It is the analog for a continuous variable (e.g., residence time t) of the probability distribution for a discrete variable (e.g., chain length l). The "fraction" that appears in Equations (15.2), (15.3), and (15.6) can be interpreted as a probability, but now it is the probability that t will fall within a specified range rather than the probability that t will have some specific value. Compare Equations (13.8) and (15.5).

The differential distribution is related to the cumulative distribution and to the washout function by

$$f(t) = \frac{dF}{dt} = -\frac{dW}{dt}$$

$$F(t) = \int_0^t f(t')dt' \tag{15.7}$$

$$W(t) = \int_t^\infty f(t')dt'$$

Its properties are that $f(t) \geq 0$ and that

$$\int_0^\infty f(t)dt = 1 \tag{15.8}$$

Experimental determination of the density function requires rapid injection of tracer molecules at the inlet to the system. Ideally, a finite number of molecules will be injected in an infinitesimal period of time. Think of quick injection using a syringe.

Mathematically, $f(t)$ can be determined from $F(t)$ or $W(t)$ by differentiation according to Equation (15.7). This is the easiest method when working in the time domain. It can also be determined as the response of a dynamic model to a unit impulse or *Dirac delta function*. The delta function is a convenient mathematical artifact that is usually defined as

$$\delta(t) = 0, \quad t \neq 1$$
$$\int_{-\infty}^{\infty} \delta(t)dt = 1 \quad (15.9)$$

The delta function is everywhere zero except at the origin, where it has an infinite discontinuity, a discontinuity so large that the integral under it is unity. The limits of integration need only include the origin itself; Equation (15.9) can equally well be written as

$$\int_{0-}^{0+} \delta(t)dt = 1$$

The delta function has another integral of substantial use

$$\int_{-\infty}^{\infty} \phi(t)\delta(t - t_0)dt = \phi(t_0) \quad (15.10)$$

where $\phi(t)$ is any "ordinary" function. This suggests that $\delta(t)$ itself is not an ordinary function. Instead, it can be considered as the limit of an ordinary function. This is illustrated in Example 15.1 (see also Example 15.2, which shows how delta functions are used in connection with Laplace transforms).

15.1.2 Means and Moments

Residence time distributions can be described by any of the functions $W(t)$, $F(t)$, or $f(t)$. They can also be described using an infinite set of parameters known as *moments*:

$$\mu_n = \int_0^{\infty} t^n f(t)dt \quad (15.11)$$

where $n = 0, 1, 2, \ldots$. Compare Equation (13.9). These moments are also called *moments about the origin*. The zeroth moment is 1. A useful result for $n > 0$ is

$$\mu_n = n \int_0^{\infty} t^{n-1} W(t)dt \quad (15.12)$$

Equation (15.12) is preferred for the experimental determination of moments.

The first moment is the mean of the distribution or the mean residence time.

$$\bar{t} = \int_0^\infty t f(t) dt = \int_0^\infty W(t) dt \qquad (15.13)$$

Thus, $\bar{t}$ can be found from inert tracer experiments. It can also be found from measurements of the system inventory and throughput since

$$\bar{t} = \frac{\hat{\rho} V}{\rho_{out} Q_{out}}$$

Agreement of the $\bar{t}$ values calculated by these two methods provides a good check on experimental accuracy. Occasionally, Equation (15.13) is used to determine an unknown volume or an unknown density from inert tracer data.

Roughly speaking, the first moment, $\bar{t}$, measures the size of a residence time distribution, while higher moments measure its shape. The ability to characterize shape is enhanced by using *moments about the mean*:

$$\mu_n' = \int_0^\infty (t - \bar{t})^n f(t) dt \qquad (15.14)$$

Of these, the second is the most interesting and has a special name, the *variance*:

$$\sigma_t^2 = \mu_2' = \int_0^\infty (t - \bar{t})^2 f(t) dt \qquad (15.15)$$

Expanding the parenthetical term and integrating term-by-term gives

$$\sigma_t^2 = \mu_2' = \int_0^\infty (t - \bar{t})^2 f(t) dt = \int_0^\infty (t^2 - 2t\bar{t} + \bar{t}^2) f(t) dt$$
$$= \mu_2 - 2\bar{t}\mu_1 - \bar{t}^2 \mu_0 = \mu_2 - \bar{t}^2 \qquad (15.16)$$

This equation is normally used to calculate the variance from experimental data, μ^2 being calculated from Equations (15.11) or (15.12) using $n=2$ and $\mu_1 = \bar{t}$ being calculated using $n=1$. Note that either $W(t)$ or $f(t)$ can be used to calculate the moments. Use the one that was obtained directly from an experiment. If moments of the highest possible accuracy are desired, the experiment should be a negative step change to get $W(t)$ directly. Even so, accurate moments beyond the second are difficult to obtain under the best of circumstances. The weightings of t^n or t^{n-1} in Equations (15.11) or (15.12) place too much emphasis on the tail of the residence time distribution to allow accurate numerical results.

The subscript t on σ_t^2 denotes that this variance has units of time squared. The *dimensionless variance* measures the breadth of a distribution in a way that is independent of the magnitude of $\bar{t}$:

$$\sigma^2 = \frac{\sigma_t^2}{\bar{t}^2} = \frac{\mu_2}{\bar{t}^2} - 1 \tag{15.17}$$

The dimensionless variance has been used extensively, perhaps excessively, to characterize mixing. For piston flow, $\sigma^2 = 0$; and for a CSTR, $\sigma^2 = 1$. Most turbulent flow systems have dimensionless variances that lie between zero and 1, and σ^2 can then be used to fit a variety of residence time models as will be discussed in Section 15.2. The dimensionless variance is generally unsatisfactory for characterizing laminar flows where $\sigma^2 > 1$ is normal in liquid systems.

The entire residence time distribution can be made dimensionless. A *normalized distribution* has the residence time replaced by the *dimensionless residence time*, $\tau = t/\bar{t}$. The first moment of a normalized distribution is 1, and all the moments are dimensionless. Normalized distributions allow flow systems to be compared in a manner that is independent of their volume and throughput. For example, all CSTRs have the same normalized residence time distribution, $W(\tau) = \exp(-\tau)$. Similarly, all PFRs have $f(\tau) = \delta(\tau - 1)$.

15.2 RESIDENCE TIME MODELS

This section opens the black box in order to derive residence time models for common flow systems. The box is closed again in Section 15.3, where the predictions can be based on either models or measurements.

15.2.1 Ideal Reactors and Reactor Combinations

The ideal flow reactors are the CSTR and the PFR. (This chapter later introduces a third kind of ideal reactor, the segregated CSTR, but it has the same distribution of residence times as the regular, perfectly mixed CSTR.) Real reactors sometimes resemble these ideal types or they can be assembled from combinations of the ideal types.

The Single CSTR. The washout function for a CSTR is found from its response to a negative step change in tracer concentration; from Equation (15.1):

$$W(t) = e^{-t/\bar{t}} \tag{15.18}$$

A CSTR has an *exponential distribution* of residence times. The corresponding differential distribution can be found from Equation (15.7):

$$f(t) = (1/\bar{t})e^{-t/\bar{t}} \tag{15.19}$$

Example 15.1 shows how it can be determined in the time domain as the response to a delta function input.

Example 15.1: Apply a delta function input to a CSTR to determine $f(t)$.

Solution: This solution illustrates a possible definition of the delta function as the limit of an ordinary function. Disturb the reactor with a rectangular tracer pulse of duration Δt and height A/t so that A units of tracer are injected. The input signal is $C_{in} = 0, t < 0$; $C_{in} = A/\Delta t, 0 < t < \Delta t$; $C_{in} = 0$, and $t > \Delta t$. The outlet response is found from the dynamic model of a CSTR, Equation (14.2). The result is

$$C_{out} = 0 \qquad t < 0$$
$$C_{out} = (A/\Delta t)[1 - e^{-t/\bar{t}}] \qquad 0 < t < \Delta t$$
$$C_{out} = (A/\Delta t)[1 - e^{-\Delta t/\bar{t}}]e^{-t/\bar{t}} \qquad t > \Delta t$$

Now consider the limit as τ approaches zero. L'Hospital's rule shows that

$$\lim_{\Delta t \to 0} A/\Delta t[1 - e^{-\Delta t/\bar{t}}] = A/\bar{t}$$

The transient response to a pulse of infinitesimal duration is

$$C_{out} = 0 \qquad t < 0$$
$$C_{out} = A/\bar{t} \qquad t = 0$$
$$C_{out} = (A/\bar{t})e^{-t/\bar{t}} \qquad t > 0$$

The differential distribution is the response to a *unit impulse*. Setting $A = 1$ gives the expected result, Equation (15.19).

Pulse shapes other than rectangular can be used to obtain the same result. Triangular or Gaussian pulses could be used, for example. The limit must be taken as the pulse duration becomes infinitesimally short while the amount of injected tracer remains finite. Any of these limits will correspond to a delta function input.

The above example shows why it is mathematically more convenient to apply step changes rather than delta functions to a system model. This remark applies when working with dynamic models in their normal form; i.e., in the *time domain*. Transformation to the *Laplace domain* allows easy use of delta functions as system inputs.

Example 15.2: Use Laplace transform techniques to apply a delta function input to a CSTR to determine $f(t)$.

Solution: Define the Laplace transform of $C(t)$ with respect to the transform parameter s as

$$\mathcal{L}_s[C(t)] = \int_0^\infty C(t)e^{-st}dt$$

The governing ODE,

$$V\frac{dC_{out}}{dt} = QC_{in} - QC_{out}$$

transforms to

$$Vs\mathcal{L}_s[C_{out}(t)] - VC_{out}(0) = Q\mathcal{L}_s[C_{in}(t)] - Q\mathcal{L}_s[C_{out}(t)])$$

The $C_{out}(0)$ term is the initial condition for the concentration within the tank. It is zero when the input is a delta function. Such a system is said to be *initially relaxed*. The term $\mathcal{L}_s[C_{in}(t)]$ is the Laplace transform of the input signal, a delta function in this case. The Laplace transform of $\delta(t)$ is 1. Substituting and solving for $a_{out}(s)$ gives

$$\mathcal{L}_s[C_{out}(t)] = \frac{\mathcal{L}_s[C_{in}(t)]}{1+\bar{t}s} = \frac{1}{1+\bar{t}s} = \frac{(1/\bar{t})}{(1/\bar{t})+s} \tag{15.20}$$

Equation (15.20) is inverted to give the time-domain concentration, $f(t) = C_{out}(t)$. The result is Equation (15.19).

Example 15.3: Determine the first three moments about the origin and about the mean for the residence time distribution of a CSTR.

Solution: Use Equation (15.11) and $F(t) = (1/\bar{t})e^{-t/\bar{t}}$ to obtain the moments about the origin:

$$\mu_n = \int_0^\infty t^n f(t)dt = (1/\bar{t})\int_0^\infty t^n e^{-t/\bar{t}}dt$$

$$= \bar{t}^n \int_0^\infty \tau^n e^{-\tau}d\tau = \bar{t}^n \Gamma(n+1) = n!\bar{t}^n$$

where $\Gamma(n+1) = n!$ is the gamma function. Thus, for a CSTR, $\mu_1 = \bar{t}$, $\mu_2 = 2\bar{t}^2$, and $\mu_3 = 6\bar{t}^3$. To find the moments about the mean, the parenthetical term in Equation (15.14) is expanded and the resulting terms are evaluated as moments about the origin. Equation (15.16) gives the result for $n=2$. Proceeding in the same way for $n=3$ gives

$$\mu_3' = \int_0^\infty (t-\bar{t})^3 f(t)dt = \mu_3 - 3\bar{t}\mu_2 + 2\bar{t}^3 \tag{15.21}$$

Equations (15.17) and (15.21) apply to any residence time distribution. For the exponential distribution of a CSTR, $\mu'_2 = \sigma_t^2 = \bar{t}^2$ (so that $\sigma^2 = 1$) and $\mu'_3 = 2\bar{t}^3$. The general result for a CSTR is $\mu'_n = (n-1)\bar{t}^n$.

The Piston Flow Reactor. Any input signal of an inert tracer is transmitted through a PFR without distortion but with a time delay of $\bar{t}$ seconds. When the input is a negative step change, the output will be a delayed negative step change. Thus, for a PFR,

$$\begin{array}{ll} W(t) = 1 & t < \bar{t} \\ W(t) = 0 & t > \bar{t} \end{array} \quad (15.22)$$

The same logic can be used for a delta function input.

Example 15.4: The differential distribution can be defined as the outlet response of a system to a delta function input.

Solution: The dynamic model governing the flow of an inert tracer through an unsteady PFR is Equation (14.13) with $\mathscr{R}_C = 0$:

$$\frac{\partial C}{\partial t} + \bar{u}\frac{\partial C}{\partial z} = 0 \quad (15.23)$$

The solution has any input signal being transmitted without distortion:

$$C(t, z) = C_{in}(t - z/\bar{u})$$

Evaluating this solution at the reactor outlet gives

$$C_{out}(t) = C(t, L) = C_{in}(t - \bar{t})$$

The input to the reactor is a delta function, $\delta(t)$, so the output is as well, $\delta(t - \bar{t})$. Thus,

$$f(t) = \delta(t - \bar{t}) \quad (15.24)$$

for a piston flow reactor. In light of this result, the residence time distribution for piston flow is called a *delta distribution*.

Example 15.5: Determine the moments about the origin and about the mean for a PFR.

Solution: Equation (15.11) becomes

$$\mu_n = \int_0^\infty t^n f(t) dt = \int_0^\infty t^n \delta(t - \bar{t}) dt$$

Applying the integral property of the delta function, Equation (15.10), gives $\mu_n = \bar{t}^n$. The moments about the mean are all zero.

The Fractional Tubularity Model. Piston flow has $\sigma^2 = 0$. A CSTR has $\sigma^2 = 1$. Real reactors can have $0 < \sigma^2 < 1$, and a model that reflects this possibility consists of a stirred tank in series with a piston flow reactor as indicated in Figure 15.1(a). Other than the mean residence time itself, the model contains only one adjustable parameter. This parameter is called the *fractional tubularity*, τ_p, and is the fraction of the system volume that is occupied by the piston flow element. Figure 15.1(b) shows the washout function for the fractional tubularity model. Its equation is

$$W(t) = 1 \qquad\qquad t < \tau_p \bar{t}$$
$$W(t) = \exp\left[-\frac{(t - \tau_p \bar{t})}{\bar{t}(1 - \tau_p)}\right] \qquad t > \tau_p \bar{t} \qquad (15.25)$$

This equation can be fit to experimental data in several ways. The model exhibits a sharp *first appearance time*, $t_{first} = \tau_p \bar{t}$, which corresponds to the fastest material moving through the system. The mean residence time is found using Equation (15.13), and $\tau_p = t_{first}/\bar{t}$ is found by observing the time when the experimental washout function first drops below 1.0. It can also be fit from the slope of a plot of $\ln W$ versus t. This should give a straight line (for $t > t_{first}$) with slope $= 1/(\bar{t} - t_{first})$. Another approach is to calculate the dimensionless variance and then to obtain τ_p from

$$\tau_p = 1 - \sigma \qquad (15.26)$$

All these approaches have been used. However, the best method for the great majority of circumstances is nonlinear least squares as described in Section 7.1.1.

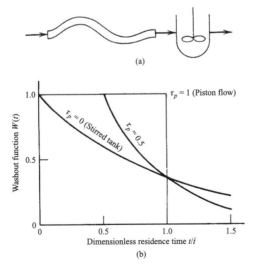

FIGURE 15.1 The fractional tubularity model: (a) physical representation; (b) washout function.

The fractional tubularity model has been used to fit residence time data in fluidized-bed reactors. It is also appropriate for modeling real stirred tank reactors that have small amounts of dead time, as would perhaps be caused by the inlet and outlet piping. It is not well suited to modeling systems that are nearly in piston flow since such systems rarely have sharp first appearance times.

The Tanks-in-Series Model. A simple model having fuzzy first appearance times is the tanks-in-series model illustrated in Figure 15.2. The washout function is

$$W(t) = e^{-Nt/\bar{t}} \sum_{i=0}^{N-1} \frac{N^i t^i}{i!\bar{t}^i} \qquad (15.27)$$

The corresponding differential distribution is

$$f(t) = \frac{N^N t^{N-1} e^{-Nt/\bar{t}}}{(N-1)!\bar{t}^{N-1}} \qquad (15.28)$$

where N (an integer) is the number of tanks in series. Each tank, individually, has volume V/N and mean residence time $\bar{t}/N$. This model reduces to the exponential distribution of a single stirred tank for $N=1$. It approaches the delta distribution of piston flow as $N \to \infty$. The model is well suited to modeling small deviations from piston flow. Physical systems that consist of N tanks (or compartments, or cells) in series are fairly common, and the model has obvious utility for these situations. The model is poorly suited for characterizing small

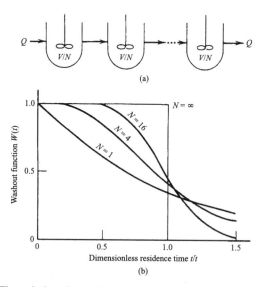

FIGURE 15.2 The tanks-in-series model: (a) physical representation; (b) washout function.

deviations from the exponential distribution of a single stirred tank because N takes only integer values. However, extensions to the basic tanks-in-series model allow N to take noninteger values and even $N < 1$. The *fractional tank extension* has a physical interpretation as N equal-sized tanks followed by one smaller tank. See Stokes and Nauman[1] or the references in "Suggestions for Further Reading." Another extension to the basic model extends the fact that

$$\sigma^2 = 1/N \tag{15.29}$$

when N is an integer. Using Equation (15.29) for noninteger N is possible but this destroys the physical basis for the model. The factorial in the denominator of Equation (15.28) must be interpreted as a gamma function. Thus, the model is called the *gamma function extension* to the tanks-in-series model.

Recycle Reactors. High rates of external recycle have the same effect on the residence time distribution as high rates of internal recycle in a stirred tank. The recycle reactor in Figure 4.2 can represent a physical design or it can be a model for a stirred tank. The model requires the once-through residence time distribution. In principle, this distribution can be measured by applying a step change or delta function at the reactor inlet, measuring the outlet response, and then destroying the tracer before it has a chance to recycle. In practice, theoretical models for the once-through distribution are used. The easiest way of generating the composite distribution is by simulation.

Example 15.6: Determine the washout function if a diffusion-free, laminar flow reactor is put in a recycle loop. Assume that 75% of the reactor effluent is recycled per pass.

Solution: Refer to Figure 4.2 and set $Q = Q_{in} = Q_{out} = 0.25 \, \text{m}^3/\text{s}$, $q = 0.75 \, \text{m}^3/\text{s}$, and $V = 1 \, \text{m}^3$. Then $\bar{t} = 4$ s for the overall system and 1 s for the once-through distribution. The differential distribution corresponding to laminar flow in a tube was found in Section 8.1.3. The corresponding washout function can be found using Equation (15.7). See also Section 15.2.2. The once-through washout function is

$$W(t) = \frac{1}{4t^2} \qquad t > \frac{1}{2}$$

Now select a few hundred thousand molecules. Twenty-five percent will leave after one pass through the reactor. For each of them, pick a random number, $0 < \text{Rnd} < 1$, and use the washout function to find a corresponding value for their residence time in the system, t. This requires a numerical solution when $W(t)$ is a complicated function, but for the case at hand

$$t = t_1 = \sqrt{\frac{1}{4W_1}} \qquad \text{where} \qquad W_1 = \text{Rnd}$$

Of the 75% that survive the first pass, 25% will leave after the second pass. Their residence time will be

$$t = t_1 + t_2 = \sqrt{\frac{1}{4W_1}} + \sqrt{\frac{1}{4W_2}}$$

where W_1 and W_2 are determined from independently selected random numbers. This procedure is repeated until nearly all the molecules have left. The various residence times are then sorted by duration, starting from the lowest value for t. The sorted results are counted as a function of t and the counts are divided by the original number of molecules. The result is the washout function for the system with recycle. Equation (15.13) provides a test for whether the original number of molecules was large enough. The integral of the tabulated washout function should exceed $0.999\bar{t}$ for reasonable accuracy. Results are shown in Figure 15.3.

The methodology of Example 15.6 works for any once-through residence time distribution. The calculations will require a very large number of original molecules if the recycle ratio is large. The data in Figure 15.3 came from a starting population of $2^{18} = 262{,}144$ molecules, and the recycle ratio Q/q was only $3:1$. The first appearance time for a reactor in a recycle loop is the first appearance time for the once-through distribution divided by $Q/q + 1$. It is thus 0.125 in Figure 15.3, and declines rather slowly as the recycle ratio is increased. However, even at $Q/q = 3$, the washout function is remarkably close to that for the exponential distribution.

Pathological Behavior. An important use of residence time measurements is to diagnose flow problems. As indicated previously, the first test is whether or not $\bar{t}$

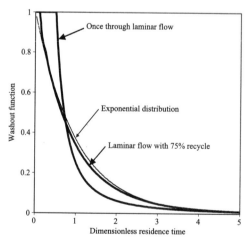

FIGURE 15.3 Effect of recycle on a laminar flow reactor.

has its expected value. A low value suggests fouling or stagnancy. A high value is more likely caused by experimental error.

The second test supposes that $\bar{t}$ is reasonable and compares the experimental washout curve with what would be expected for the physical design. The comparison is made with an ideal washout curve with the same value for $\bar{t}$. Suppose that the experimental curve is initially lower than the ideal curve. Then the system exhibits *bypassing*. On the other hand, suppose the tail of the distribution is higher than expected. Then the system exhibits *stagnancy*. Bypassing and stagnancy are easy to distinguish when the reactor is close to piston flow so that the experimental data can be compared with a step change. See Figure 15.4. They are harder to distinguish in stirred tanks because the comparison is made to an exponential curve. Figure 15.5(a) shows a design with poorly placed inlet and outlet connections that would cause bypassing. Figure 15.5(b) shows the two washout functions. Bypassing causes the washout curve initially to decline faster than the exponential distribution. However, the integral under the two curves must be the same since they have the same $\bar{t}$. See Equation (15.13). If the experimental washout function initially declines faster than expected, it must later decline more slowly.

When a stirred tank exhibits either bypassing or stagnancy, $\sigma^2 > 1$, so that the tanks-in-series model predicts $N < 1$. It is more common to model bypassing or stagnancy using vessels in parallel. A stirred tank might be modeled using large and small tanks in parallel. To model bypassing, the small tank would have a residence time lower than that of the large tank. To model stagnancy, the small tank would have the longer residence time. The *side capacity model* shown in Figure 15.6 can also be used and is physically more realistic than a parallel connection of two isolated tanks.

Example 15.7: Determine the washout function for the side capacity model given $Q = 8 \, \text{m}^3/\text{h}$, $q = 0.125 \, \text{m}^3/\text{h}$, $V_m = 7 \, \text{m}^3$, and $V_s = 1$.

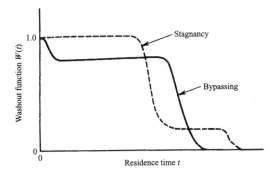

FIGURE 15.4 Bypassing and stagnancy in a system near piston flow.

CHEMICAL REACTOR DESIGN, OPTIMIZATION, AND SCALEUP

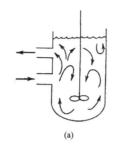

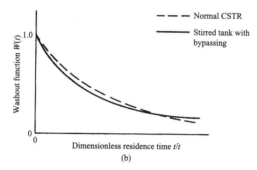

FIGURE 15.5 Pathological residence time behavior in a poorly designed stirred tank: (a) physical representation; (b) washout function.

Solution: Material balances for the two tanks are

$$V_m \frac{dC_{out}}{dt} = QC_{in} + qS_{out} - QC_{out} - qC_{out}$$

$$V_s \frac{dS_{out}}{dt} = qC_{out} - qS_{out}$$

These can be solved by classical methods (i.e., eliminate S_{out} to obtain a second-order ODE in C_{out}), by Laplace transformation techniques, or by numerical integration. The initial conditions for the washout experiment are that the entire system is full of tracer at unit concentration, $C_{out} = S_{out} = 1$. Figure 15.7 shows the result of a numerical simulation. The difference between the model curve and that for a normal CSTR is subtle, and would not normally be detected by a washout experiment. The semilog plot in Figure 15.8 clearly shows the two time constants for the system, but the second one emerges at such low values of $W(t)$ that it would be missed using experiments of ordinary accuracy.

The stagnant region can be detected if the mean residence time is known independently, i.e., from Equation (1.41). Suppose we know that $\bar{t} = 1\,\text{h}$ for this reactor and that we truncate the integration of Equation (15.13) after 5 h. If the tank were well mixed (i.e., if $W(t)$ had an exponential distribution), the integration of Equation (15.13) out to $5\bar{t}$ would give an observed $\bar{t}$ of

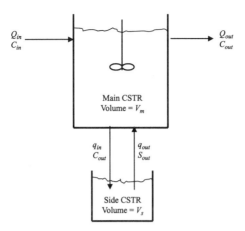

FIGURE 15.6 Side capacity model of stagnancy in a CSTR.

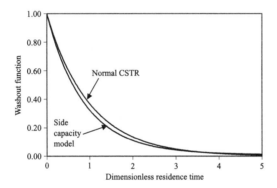

FIGURE 15.7 Effect of a stagnant zone in a stirred tank reactor according to the side capacity model.

0.993 h. Integration of the stagnancy model out to $5\bar{t}$ gives only 0.913 h. This difference is large enough to be detected and to initiate a search for the missing reactor volume.

15.2.2 Hydrodynamic Models

This section describes residence time models that are based on a hydrodynamic description of the process. The theory is simplified but the resulting models still have substantial utility as conceptual tools and for describing some real flow systems.

Laminar Flow without Diffusion. Section 8.1.3 anticipated the use of residence time distributions to predict the yield of isothermal, homogeneous reactions, and

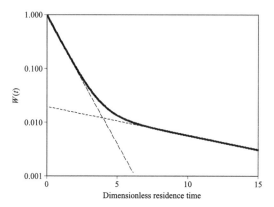

FIGURE 15.8 Semilog plot of washout function showing two slopes that correspond to the two time constants in the side capacity model.

Equation (8.11) gave the differential distribution function that corresponds to a parabolic velocity profile in a tube. This specific result is now derived in a more general way.

The velocity profile in a tube of length L is $V_z(r)$. The normal case is for $V_z(r)$ to have its maximum value at the centerline and to decline monotonically toward $V_z(r) = 0$ at $r = R$. The volumetric flow rate is Q. The fraction of that flow rate associated with the region from the centerline to radial position r is found from the following integral:

$$F(r) = \frac{1}{Q} \int_0^r 2\pi r' V_z(r') dr' \qquad (15.30)$$

Perform this integration to obtain a function of r that goes from 0 to 1 as r ranges from 0 to R. $F(r)$ gives the fraction of material leaving the reactor that flowed through it at a location of r or less. The residence time of material traveling along the streamlines at position r is

$$t = L/V_z(r) \qquad (15.31)$$

Material flowing at a position less than r has a residence time less than t because the velocity will be higher closer to the centerline. Thus, $F(r) = F(t)$ gives the fraction of material leaving the reactor with a residence time less that t where Equation (15.31) relates to r to t. $F(t)$ satisfies the definition, Equation (15.3), of a cumulative distribution function. Integrate Equation (15.30) to get $F(r)$. Then solve Equation (15.31) for r and substitute the result to replace r with t. When the velocity profile is parabolic, the equations become

$$F(t) = F(r) = \frac{2r^2 R^2 - r^4}{R^4}$$

$$t = \frac{\bar{t}}{2[1 - r^2/R^2]}$$

Elimination of r gives

$$F(t) = 1 - \frac{\bar{t}^2}{4t^2} \quad t > \frac{\bar{t}}{2} \quad (15.32)$$

Differentiating this result gives the differential distribution found in Equation (8.11). The washout function is

$$W(t) = 1 \quad t < \frac{\bar{t}}{2}$$
$$W(t) = \frac{\bar{t}^2}{4t^2} \quad t > \frac{\bar{t}}{2} \quad (15.33)$$

This function is shown in Figure 15.9. It has a sharp first appearance time at $t_{first} = \bar{t}/2$ and a slowly decreasing tail. When $t > 4.3\bar{t}$, the washout function for parabolic flow decreases more slowly than that for an exponential distribution. Long residence times are associated with material near the tube wall; $r/R = 0.94$ for $t = 4.3\bar{t}$. This material is relatively stagnant and causes a very broad distribution of residence times. In fact, the second moment and thus the variance of the residence time distribution would be infinite in the complete absence of diffusion.

The above derivation assumes straight streamlines and a monotonic velocity profile that depends on only one spatial variable, r. These assumptions substantially ease the derivation but are not necessary. Analytical expressions for the residence time distributions have been derived for noncircular ducts,

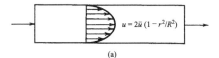

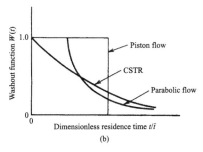

FIGURE 15.9 Residence time distribution for laminar flow in a circular tube: (a) physical representation; b) washout function.

non-Newtonian fluids, and helically coiled tubes. Computational fluid dynamics has been used for really complicated geometries such as motionless mixers.

In the absence of diffusion, all hydrodynamic models show infinite variances. This is a consequence of the *zero-slip* condition of hydrodynamics that forces $V_z = 0$ at the walls of a vessel. In real systems, molecular diffusion will ultimately remove molecules from the stagnant regions near walls. For real systems, $W(t)$ will asymptotically approach an exponential distribution and will have finite moments of all orders. However, molecular diffusivities are low for liquids, and σ^2 may be large indeed. This fact suggests the general inappropriateness of using σ^2 to characterize the residence time distribution in a laminar flow system. Turbulent flow is less of a problem due to eddy diffusion that typically results in an exponentially decreasing tail at fairly low multiples of the mean residence time.

Axial Dispersion. Rigorous models for residence time distributions require use of the convective diffusion equation, Equation (14.19). Such solutions, either analytical or numerical, are rather difficult. Example 15.4 solved the simplest possible version of the convective diffusion equation to determine the residence time distribution of a piston flow reactor. The derivation of $W(t)$ for parabolic flow was actually equivalent to solving

$$\frac{\partial C}{\partial t} + 2\bar{u}\left[1 - r^2/R^2\right]\frac{\partial C}{\partial z} = 0$$

subject to a negative step change of inert tracer. We go now to the simplest version of the convective diffusion equation that actually involves diffusion or a diffusion-like term. It is the axial dispersion model, written here in time-dependent form:

$$\frac{\partial C}{\partial t} + \bar{u}\frac{\partial C}{\partial z} = D\frac{\partial^2 C}{\partial z^2} \qquad (15.34)$$

The appropriate boundary conditions are the closed variety discussed in Section 9.3.1. The initial condition is a negative step change at the inlet. A full analytical solution is available but complex. For $\mathbf{Pe} = \bar{u}L/D > 16$, the following result is an excellent approximation:

$$W(\tau) = 1 - \int_0^\tau \frac{\mathbf{Pe}}{4\pi\theta^3}\exp\left[\frac{-\mathbf{Pe}(1-\theta)^2}{4\theta}\right]d\theta \qquad (15.35)$$

where $\tau = t/\bar{t}$ is the dimensionless residence time. Figure 15.10 shows the washout function for the axial dispersion model, including the exact solution for $\mathbf{Pe} = 1$. The model is defined for $0 < \mathbf{Pe} < \infty$, and the extreme values correspond to perfect mixing and piston flow, respectively. The axial dispersion model shows a fuzzy first appearance time. It is competitive with and generally preferable to the tanks-in-series model for modeling small deviations from piston flow.

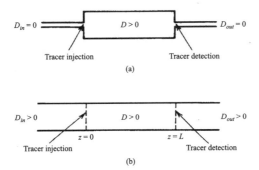

FIGURE 15.10 Transient response measurements for systems governed by the axial dispersion model: (a) closed system; (b) open system.

It should be used with caution for large deviations. As discussed in Chapter 9, predictions of the model at small **Pe** are likely to fail under close scrutiny.

Example 15.8: Find the relationship between **Pe** and σ^2.

Solution: The first step in the solution is to find a residence time function for the axial dispersion model. Either $W(t)$ or $f(t)$ would do. The function has **Pe** as a parameter. The methods of Section 15.1.2 could then be used to determine σ^2, which will give the desired relationship between **Pe** and σ^2.

We will begin by attempting to determine $f(t)$ for a closed system governed by the axial dispersion model.

Equation (15.34) is the system model. It is a linear PDE with constant coefficients and can be converted to an ODE by Laplace transformation. Define

$$\mathcal{L}_k[C(t,z)] = \int_0^\infty C(t,z)e^{-kt}dt$$

Note that the transform parameter is now being denoted as k. Equation (15.34) can be transformed term-by-term much like the transformation of an ODE. The result is

$$k\mathcal{L}_k[C(t,z)] - C(0,z) + \bar{u}\frac{d\mathcal{L}_k[C(t,z)]}{dz} = D\frac{d^2\mathcal{L}_k[C(t,z)]}{dz^2} \quad (15.36)$$

This is a second-order ODE with independent variable z and dependent variable $\mathcal{L}_k[C(t,z)]$, which is a function of z and of the transform parameter k. The term $C(t,0)$ is the initial condition and is zero for an initially relaxed system. There are two spatial boundary conditions. These are the Danckwerts conditions of Section 9.3.1. The form appropriate to the inlet of an unsteady system is a generalization of Equation (9.16) to include time dependency:

$$\bar{u}C_{in}(t) = \bar{u}C(t,0+) - D[\partial C(t,z)/\partial z]_{z=0+}$$

The outlet boundary condition for this unsteady but closed system is a generalization of Equation (9.17):

$$[\partial C(t,z)/\partial z]_{z=L} = 0$$

To use these boundary conditions with Equation (15.36), they must be transformed. The result for the inlet is

$$\bar{u}\mathcal{L}_k[C_{in}(t)] = \bar{u}\mathcal{L}_k[C(t,0+)] - D[d\mathcal{L}_k[C(t,z)]/dz]_{z=0+}$$

The inlet disturbance is applied through the inlet boundary condition with $C_{in}(t) = \delta(t)$ so that $\mathcal{L}_k[C_{in}(t)] = 1$. The outlet boundary condition is just $\partial \mathcal{L}_k[\partial C(t,z)/\partial z]_{z=L} = 0$.

The cumbersome notation of the Laplace transform can be simplified. $\mathcal{L}_s[C(t,z)]$ is a function of k and z. Denote it as $a = a(k,z)$ and set $C(0,z) = 0$. Also shift to a dimensionless length coordinate $\hat{z} = z/L$. Then Equation (15.36) and its associated (transformed) boundary conditions become

$$\frac{da}{d\hat{z}} = \frac{1}{\text{Pe}}\frac{d^2 a}{d\hat{z}^2} - k\bar{t}a$$

$$a_{in} = a(0+) - \frac{1}{\text{Pe}}\left[\frac{da}{d\hat{z}}\right]_{0+} \tag{15.37}$$

$$\left[\frac{da}{d\hat{z}}\right]_1 = 0$$

These equations are identical to Equations (9.15), (9.16), and (9.17) when we set $\mathcal{R}_A$ equal to $-ka$. The solution is necessarily identical as well. We evaluate that solution at $\hat{z} = 1$ to obtain

$$a_{out}(k) = \frac{4p\exp\left(\dfrac{\text{Pe}}{2}\right)}{(1+p^2)\exp\left(\dfrac{p\text{Pe}}{2}\right) - (1-p^2)\exp\left(\dfrac{-p\text{Pe}}{2}\right)} \tag{15.38}$$

where

$$p = \sqrt{1 + \frac{4k\bar{t}}{\text{Pe}}}$$

Equation (15.38) gives the Laplace transform of the outlet response to an inlet delta function; i.e., $a_{out}(k) = \mathcal{L}_k[f(t)]$. In principle, Equation (15.38) could be inverted to obtain $f(t)$ in the time domain. This daunting task is avoided by using properties of the Laplace transform to obtain

$$\mu_n = (-1)^n \lim_{k \to 0} \frac{d^n a(k)}{dk^n} \tag{15.39}$$

Equation (15.39) allows moments of a distribution to be calculated from the Laplace transform of the differential distribution function without need for finding $f(t)$. It works for any $f(t)$. The necessary algebra for the present case is formidable, but finally gives the desired relationship:

$$\sigma^2 = \frac{2}{\text{Pe}} - \frac{2}{\text{Pe}^2}[1 - \exp(-\text{Pe})] \tag{15.40}$$

To find **Pe** for a real system, perform a residence time experiment; calculate σ^2, and then use Equation (15.40) to calculate **Pe**.

The use of Equation (15.40) is limited to closed systems like that illustrated in Figure 15.10(a). Measurement problems arise whenever $D_{in} > 0$ or $D_{out} > 0$. See Figure 15.10(b) and suppose that an impulse is injected into the system at $z = 0$. If $D_{in} > 0$, some of the tracer may enter the reactor, then diffuse backward up the inlet stream, and ultimately reenter. If $D_{out} > 0$, some material leaving the reactor will diffuse back into the reactor to exit a second time. These molecules will be counted more than once by the tracer detection probes. The measured response function is not $f(t)$ but another function, $g(t)$, which has a larger mean:

$$\mu_{open} = \left[1 + \frac{D_{in} + D_{out}}{\bar{u}L}\right]\bar{t} = \left[1 + \frac{1}{\text{Pe}_{in}} + \frac{1}{\text{Pe}_{out}}\right]\bar{t} \tag{15.41}$$

If μ_{open} is erroneously interpreted as $\bar{t}$, results from an open system give significant errors when the inlet and outlet Peclet numbers are less than about 100. If the openness of the system cannot be avoided, the recommended approach is to rescale $g(t)$ so that it has the correct mean:

$$[g(t)]_{rescaled} = \frac{\bar{t}g(t)}{\int_0^\infty tg(t)dt} = \frac{\bar{t}g(t)}{\mu_{open}} \tag{15.42}$$

The rescaled function is still not $f(t)$, but should be a reasonable approximation to it.

15.3 REACTION YIELDS

The black box is closed again. This section assumes that the system is isothermal and homogeneous and that its residence time distribution is known. Reaction yields can be predicted exactly for first-order reactions. For other reactions, an exact prediction requires additional information about the state of mixing in the system, but knowledge of just the residence time distribution is usually sufficient to calculate close bounds on the reaction yield.

15.3.1 First-Order Reactions

For an isothermal, first-order reaction, the probability that a particular molecule reacts depends only on the time it has spent in the system:

$$P_R = 1 - e^{-kt}$$

To find the conversion for the reactor, we need the average reaction probability for a great many molecules that have flowed through the system. The averaging is done with respect to residence time since residence time is what determines the individual reaction probabilities:

$$X_A = \bar{P}_R = \int_0^\infty (1 - e^{-kt}) f(t) dt = 1 - \int_0^\infty e^{-kt} f(t) dt$$

Expressing this result in terms of the fraction unreacted gives a simpler form:

$$Y_A = \frac{a_{out}}{a_{in}} = \int_0^\infty e^{-kt} f(t) dt = 1 - k \int_0^\infty e^{-kt} W(t) dt \qquad (15.43)$$

For numerical integration, use whichever of $f(t)$ or $W(t)$ was determined experimentally. If a positive step change was used to determine $F(t)$, convert to $W(t)$ using Equation (15.4).

Example 15.9: Use residence time theory to predict the fraction unreacted for an isothermal, homogeneous, first-order reaction occurring in a CSTR and a PFR.

Solution: For the stirred tank, $W(t) = \exp(-t/\bar{t})$. Substitution into Equation (15.43) gives

$$\frac{a_{out}}{a_{in}} = 1 - k \int_0^\infty e^{-kt} e^{-t/\bar{t}} dt = \frac{1}{1 + k\bar{t}}$$

For the PFR, use $f(t) = \delta(t)$ and recall Equation (15.10) to obtain

$$\frac{a_{out}}{a_{in}} = \int_0^\infty e^{-kt} \delta(t - \bar{t}) dt = e^{-k\bar{t}}$$

Example 15.10: Use residence time theory to predict the fraction unreacted for a closed reactor governed by the axial dispersion model.

Solution: Equation (15.43) gives

$$\frac{a_{out}}{a_{in}} = \int_0^\infty e^{-kt} f(t) dt = \mathcal{L}_k[f(t)] \qquad (15.44)$$

Thus, the fraction unreacted is the Laplace transform with respect to the transform parameter k of the differential distribution function.

Example 15.8 used a delta function input in the Laplace domain to find $\mathcal{L}_k[f(t)]$. The result was Equation (15.38). Comparison with Equation (15.44) shows that $a_{out}(k)$ has already been normalized by a_{in} and is thus equal to the fraction unreacted, Y_A.

Given $\mathcal{L}_k[f(t)]$ for any reactor, you automatically have an expression for the fraction unreacted for a first-order reaction with rate constant k. Alternatively, given $a_{out}(k)$, you also know the Laplace transform of the differential distribution of residence time (e.g., $\mathcal{L}_k[f(t)] = \exp(-t/\bar{t})$ for a PFR). This fact resolves what was long a mystery in chemical engineering science. What is $f(t)$ for an open system governed by the axial dispersion model? Chapter 9 shows that the conversion in an open system is identical to that of a closed system. Thus, the residence time distributions must be the same. It cannot be directly measured in an open system because time spent outside the system boundaries does not count as residence but does affect the tracer measurements.

Equation (15.44) can be generalized to include operation with unsteady inlet concentrations where $a_{in} = a_{in}(t)$. The result is an unsteady output given by

$$a_{out}(t) = \int_0^\infty a_{in}(t-t') e^{-kt'} f(t') dt' = \int_{-\infty}^t a_{in}(t') e^{-k(t-t')} f(t-t') dt' \qquad (15.45)$$

This result allows the unsteady output to be calculated when component A reacts with first-order kinetics. The case $k=0$, corresponding to an inert tracer, is also of interest:

$$C_{out}(t) = \int_0^\infty C_{in}(t-t') f(t') dt' = \int_{-\infty}^t C_{in}(t') f(t-t') dt' \qquad (15.46)$$

Equation (15.46) is applicable to nonisothermal systems since there is no chemical reaction.

Example 15.11: Suppose the input of an inert tracer to a CSTR varies sinusoidally:

$$C_{in}(t) = C_0(1 + \beta \sin \omega t) \qquad \beta \leq 1$$

Find the outlet response. What is the maximum deviation in C_{out} from its midpoint C_0?

Solution: For a stirred tank, $f(t) = (1/\bar{t})\exp(-t/\bar{t})$. The second integral in Equation (15.46) can be used to calculate the outlet response:

$$C_{out}(t) = \int_{-\infty}^{t} C_{in}(t')f(t-t')dt' = \int_{-\infty}^{t} C_0(1+\beta\sin\omega t')(1/\bar{t})\exp\left[-\frac{(t-t')}{\bar{t}}\right]dt'$$

$$= C_0 \frac{1+\beta(\sin\omega t - \omega\bar{t}\cos\omega t)}{1+\omega^2\bar{t}^2}$$

The output tracer signal is attenuated and shows a phase shift, but there is no change in frequency. All solutions to Equations (15.45) and (15.46) have these characteristics. Differentiate $\sin\omega t - \omega\bar{t}\cos\omega t$ to show that the maximum deviation occurs when $\cot\omega t = -\omega\bar{t}$. Some trigonometry then shows that the maximum deviation is

$$|C_{out} - C_0|_{max} = \frac{\beta}{\sqrt{1+\omega^2\bar{t}^2}} \tag{15.47}$$

This result is useful in designing stirred tanks to damp out concentration fluctuations (e.g., as caused by a piston pump feeding the catalyst to a reactor). High-frequency noise is most easily dampened, and a single stirred tank is the most efficient means for such dampening. A PFR gives no dampening. Of course, if the reactor is a stirred tank, a preliminary dampening step may not be necessary.

Chapter 14 and Section 15.2 used a unsteady-state model of a system to calculate the output response to an inlet disturbance. Equations (15.45) and (15.46) show that a dynamic model is unnecessary if the entering compound is inert or disappears according to first-order kinetics. The only needed information is the residence time distribution, and it can be determined experimentally.

15.3.2 Other Reactions

For reaction other than first order, the reaction probability depends on the time that a molecule has been in the reactor and on the concentration of other molecules encountered during that time. The residence time distribution does not allow a unique estimate of the extent of reaction, but some limits can be found.

Complete Segregation. A perfect mixer has an exponential distribution of residence times: $W(t) = \exp(-t/\bar{t})$. Can any other continuous flow system have this distribution? Perhaps surprisingly, the answer to this question is a definite yes. To construct an example, suppose the feed to a reactor is encapsulated. The size of the capsules is not critical. They must be large enough to contain many molecules but must remain small compared with the dimensions of the reactor. Imagine them as small ping-pong balls as in Figure 15.11(a). The balls are agitated, gently enough not to break them but well enough to

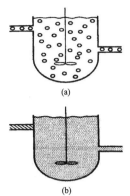

FIGURE 15.11 Extremes of micromixing in a stirred tank reactor: (a) ping-pong balls circulating in an agitated vessel—the completely segregated stirred tank reactor; (b) molecular homogeneity—the perfectly mixed CSTR.

randomize them within the vessel. In the limit of high agitation, the vessel can approach perfect mixing with respect to the ping-pong balls. A sample of balls collected from the outlet stream will have an exponential distribution of residence times:

$$W(t_b) = e^{-t_b/\bar{t}}$$

The molecules in the system are carried along by the balls and will also have an exponential distribution of residence time, but they are far from perfectly mixed. Molecules that entered together stay together, and the only time they mix with other molecules is at the reactor outlet. The composition within each ball evolves with time spent in the system as though the ball was a small batch reactor. The exit concentration within a ball is the same as that in a batch reactor after reaction time t_b.

We have just described a *completely segregated stirred tank reactor*. It is one of the ideal flow reactors discussed in Section 1.4. It has an exponential distribution of residence times but a reaction environment that is very different from that within a perfectly mixed stirred tank.

The completely segregated stirred tank can be modeled as a set of piston flow reactors in parallel, with the lengths of the individual piston flow elements being distributed exponentially. Any residence time distribution can be modeled as piston flow elements in parallel. Simply divide the flow evenly between the elements and then cut the tubes so that they match the shape of the washout function. See Figure 15.12. A reactor modeled in this way is said to be *completely segregated*. Its outlet concentration is found by averaging the concentrations of the individual PFRs:

$$a_{out} = \int_0^\infty a_{batch}(t) f(t) dt \qquad (15.48)$$

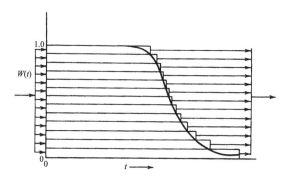

FIGURE 15.12 An arbitrary residence time distribution modeled as PFRs in parallel.

where $a_{batch}(t)$ is the concentration of a batch or piston flow reactor after reaction time t. The inlet concentration is the same for each batch and provides the initial condition for finding $a_{batch}(t)$.

Example 15.12: Find the outlet concentration from a completely segregated stirred tank for a first-order reaction. Repeat for a second-order reaction with $\mathscr{R}_A = -ka^2$.

Solution: The residence time distribution is exponential, $f(t) = (1/\bar{t}) \exp(-t/\bar{t})$. For first-order kinetics, $a_{batch}(t) = \exp(-kt)$, and Equation (15.48) gives

$$a_{out} = (1/\bar{t}) \int_0^\infty a_{in} e^{-kt} e^{-t/\bar{t}} dt = \frac{1}{1 + k\bar{t}}$$

which is the same as the outlet concentration of a normal CSTR. *The conversion of a first-order reaction is uniquely determined by the residence time distribution.*

For a second-order reaction, $a_{batch}(t) = a_{in}/(1 + a_{in}kt)$, and Equation (15.48) gives

$$\frac{a_{out}}{a_{in}} = \int_0^\infty \frac{e^{-t/\bar{t}} dt}{(1 + a_{in}kt)\bar{t}} = \frac{\exp[(a_{in}k\bar{t})^{-1}]}{a_{in}k\bar{t}} \int_{(a_{in}k\bar{t})^{-1}}^\infty \frac{e^{-x}}{x} dx$$

The integral can be evaluated using a tabulated function known as the exponential integral function, but numerical integration is easier. Figure 15.13 shows the performance of a segregated stirred tank and compares it with that of a normal, perfectly mixed CSTR (see Equation (1.51)). Segregation gives better performance, but a PFR will be still better. The hatched region in Figure 15.13 represents the conversion limits in normally designed reactors for a second-order reaction of the $2A \rightarrow P$ type

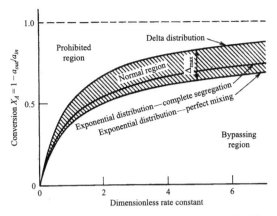

FIGURE 15.13 Conversion of a second-order reaction in the three ideal flow reactors.

with dimensionless rate constant $a_{in}k\bar{t}$, and for reactions of the type A + B → P with perfect initial stoichiometry. The region marked "prohibited" has better performance than a PFR and is impossible. The region marked "bypassing" has worse performance. It is possible to operate in this region, but it can be avoided through good engineering.

The largest difference in conversion between complete segregation and perfect mixing in a stirred tank reactor is 0.07, which occurs at $a_{in}k\bar{t} = 16$, complete segregation giving the higher conversions. The largest difference between piston flow and a normal CSTR is 0.192, which occurs at $a_{in}k\bar{t} = 4.9$. The differences shown in Figure 15.13 are not very large in absolute terms, but can have a profound effect on the reactor volume needed to achieve high conversion. In practice, single-phase, continuous-flow stirred tanks are similar to normal CSTRs with perfect internal mixing. In suspended phase systems, such as a continuous suspension polymerization, the system is physically segregated and Equation (15.48) applies.

Maximum Mixedness. For reactions other than first order, the conversion depends not only on the residence time distribution but also on mixing between molecules that have different ages. The age of a molecule is the time it has been in the reactor, and mixing between molecules with different ages is called *micromixing*. Completely segregated reactors have no mixing between molecules of different ages, and this zero level of micromixing is possible with any residence time distribution. At the opposite extreme, normal CSTRs have perfect mixing between molecules, but perfect mixing in a flow system implies an exponential distribution of residence times. Perfect micromixing is impossible except with the exponential distribution. Other residence time distributions have some maximum possible level of micromixing, which is known as *maximum mixedness*. Less micromixing than this is always possible. More would force a change in the residence time distribution.

A qualitative picture of micromixing is given in Figure 15.14. The x-axis, labeled "macromixing" measures the breadth of the residence time distribution. It is zero for piston flow, fairly broad for the exponential distribution of a stirred tank, and broader yet for situations involving bypassing or stagnancy. The y-axis is micromixing, which varies from none to complete. The y-axis also measures how important micromixing effects can be. They are unimportant for piston flow and have maximum importance for stirred tank reactors. Well-designed reactors will usually fall in the *normal region* bounded by the three apexes, which correspond to piston flow, a perfectly mixed CSTR, and a completely segregated CSTR. The line connecting the normal and segregated stirred tanks is vertical since these reactors have the same residence time distribution. Without even measuring the residence time distribution, we can determine limits on the performance of most real reactors just by calculating the performance at the three apexes of the normal region. The calculations require knowledge only of the rate constants and the mean residence time.

When the residence time distribution is known, the uncertainty about reactor performance is greatly reduced. A real system must lie somewhere along a vertical line in Figure 15.14. The upper point on this line corresponds to maximum mixedness and usually provides one bound limit on reactor performance. Whether it is an upper or lower bound depends on the reaction mechanism. The lower point on the line corresponds to complete segregation and provides the opposite bound on reactor performance. The complete segregation limit can be calculated from Equation (15.48). The maximum mixedness limit is found by solving *Zwietering's differential equation*:

$$\frac{da}{d\lambda} + \frac{f(\lambda)}{W(\lambda)}[a_{in} - a(\lambda)] + \mathscr{R}_A = 0 \qquad (15.49)$$

The solution does not use an initial value of a as a boundary condition. Instead, the usual boundary condition associated with Equation (15.50) is

$$\lim_{\lambda \to \infty} \frac{da}{d\lambda} = 0 \qquad (15.50)$$

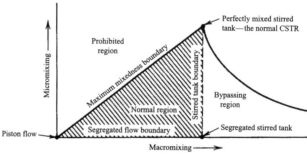

FIGURE 15.14 Macromixing versus micromixing—a schematic representation of mixing space.

which is another way of saying that a must have a finite value in the limit of large λ. The outlet concentration from a maximum mixedness reactor is found by evaluating the solution to Equation (15.49) at $\lambda = 0$ since $a_{out} = a(0)$.

Example 15.13: Solve Zwietering's differential equation for arbitrary reaction kinetics and an exponential residence time distribution.

Solution: The exponential distribution has $f(\lambda)/W(\lambda) = 1/\bar{t}$ so Equation (15.49) becomes

$$\frac{da}{d\lambda} + \frac{[a_{in} - a(\lambda)]}{\bar{t}} + \mathscr{R}_A = 0$$

Observe that the boundary condition will be satisfied if

$$\frac{[a_{in} - a(\lambda)]}{\bar{t}} + \mathscr{R}_A = 0$$

for all λ since this gives $da/d\lambda = 0$ for all λ. Set $\lambda = 0$ to obtain

$$a_{out} = a(\lambda = 0) = a_{in} + \bar{t}\mathscr{R}_A$$

Although this is an unusual solution to an ODE, it is the expected result since a stirred tank at maximum mixedness is a normal CSTR.

An analytical solution to Equation (15.49) can also be obtained for a first-order reaction. The solution is Equation (15.35). Beyond these cases, analytical solutions are difficult since the $\mathscr{R}_A$ is usually nonlinear. For numerical solutions, Equation (15.49) can be treated as though it were an initial value problem. Guess a value for $a_{out} = a(0)$. Integrate Equation (15.49). If $a(\lambda)$ remains finite at large λ, the correct $a(0)$ has been guessed. For any other $a(0)$, $a(\lambda)$ will tend toward $\pm\infty$ as $\lambda \to \infty$. This numerical approach is similar to the shooting methods of Section 9.5 even though the current ODE is only first order. The computed results are very sensitive to the guessed values for $a(0)$, and small changes will cause $a(\lambda)$ to range from $-\infty$ to $+\infty$. This sensitivity is beneficial since it allows $a(0)$ to be calculated with high precision.

Example 15.14: Solve Zwietering's differential equation for the residence time distribution corresponding to two stirred tanks in series. Use second-order kinetics with $a_{in}k\bar{t} = 5$.

Solution: Equations (15.27) and (15.28) give the residence time functions for the tanks-in-series model. For $N = 2$,

$$\frac{f(\lambda)}{W(\lambda)} = \frac{4\lambda}{\bar{t} + \lambda}$$

Set $a_{in} = 1$ so that a_{out} is the fraction unreacted. Then Zwietering's differential equation becomes

$$\frac{da}{d\tau} + \frac{4\tau}{1+2\tau}(1-a) - 5a^2 = 0$$

where $\tau = \lambda/\bar{t}$. An Euler's method solution with $\Delta\tau = 0.0625$ gives the following results:

$a(0)$	$a(\infty)$
0	$-\infty$
0.1	$-\infty$
0.2	$-\infty$
0.3	$+\infty$
0.25	$-\infty$
⋮	⋮
0.276	$-\infty$
0.277	$+\infty$

Obviously, $a_{out} = a(0)$ can be calculated with high precision. It happens that the precise estimate is not very accurate because of the large step size, but this can be overcome using a smaller $\Delta\tau$ or a more sophisticated integration technique. An accurate value is $a_{out} = 0.287$.

Example 15.15: Calculate limits on the fraction unreacted for a second-order reaction with $a_{in}k\bar{t} = 5$. Consider the following states of knowledge:

(a) You know the batch kinetics, the reactor volume and throughput, and the reactor operating temperature. It is from these values that you calculated $a_{in}k\bar{t} = 5$.
(b) You have measured the residence time distribution and know that it closely matches that for two stirred tanks in series.
(c) You know that the reactor physically consists of two stirred tanks in series.

Solution: The limits you can calculate under part (a) correspond to the three apexes in Figure 15.14. The limits are 0.167 for a PFR (Equation (1.47)), 0.358 for a CSTR (Equation (1.52)), and 0.299 for a completely segregated stirred tank. The last limit was obtained by integrating Equation (15.48) in the form

$$\frac{a_{out}}{a_{in}} = \int_0^\infty \frac{e^{-\tau}}{1+5\tau} d\tau$$

Thus, from part (a) we know that the fraction unreacted lies somewhere between 0.167 and 0.358.

The limits for part (b) are at the endpoints of a vertical line in Figure 15.14 that corresponds to the residence time distribution for two tanks in series. The maximum mixedness point on this line is 0.287 as calculated in Example 15.14. The complete segregation limit is 0.233 as calculated from Equation (15.48) using $f(t)$ for the tanks-in-series model with $N=2$:

$$\frac{a_{out}}{a_{in}} = \int_0^\infty \frac{4\tau e^{-2\tau}}{1+5\tau} d\tau$$

Thus, knowledge of the residence time distribution has narrowed the possible range on the fraction unreacted. It is now known to be between 0.233 and 0.287.

Part (c) considers the mixing extremes possible with the physical arrangement of two tanks in series. The two reactors could be completely segregated so one limit remains 0.233 as calculated in part (b). The other limit corresponds to two CSTRs in series. The first reactor has half the total volume so that $a_{in} k \bar{t}_1 = 2.5$. Its output is 0.463. The second reactor has $(a_{in})_2 k \bar{t}_2 = 1.16$, and its output is 0.275. This is a tighter bound than calculated in part (b). The fraction unreacted must lie between 0.233 and 0.275.

Part (c) in Example 15.15 illustrates an interesting point. It may not be possible to achieve maximum mixedness in a particular physical system. Two tanks in series—even though they are perfectly mixed individually—cannot achieve the maximum mixedness limit that is possible with the residence time distribution of two tanks in series. There exists a reactor (albeit semihypothetical) that has the same residence time distribution but that gives lower conversion for a second-order reaction than two perfectly mixed CSTRs in series. The next section describes such a reactor. When the physical configuration is known, as in part (c) above, it may provide a closer bound on conversion than provided by the maximum mixed reactor described in the next section.

The Bounding Theorem. The states of complete segregation and maximum mixedness represent limits on the extent of micromixing that is possible with a given residence time distribution. In complete segregation, molecules that enter together stay together. They are surrounded by molecules that have the same age, and they mix with molecules that have different ages only when they leave the reactor. This mixing situation can be represented by a parallel collection of piston flow elements as shown in Figure 15.12. It can also be represented as a single piston flow reactor with a large number of side exits. See Figure 15.15(a). The size and spacing of the side exits can be varied to duplicate any residence time distribution. Thus, piston flow with side exits is capable of modeling any residence time distribution. It is a completely segregated model

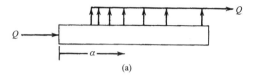

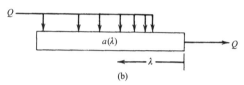

FIGURE 15.15 Extreme mixing models for an arbitrary residence time distribution: (a) complete segregation; (b) maximum mixedness.

since molecules in the reactor mix only with other molecules that have exactly the same age.

Another way of modeling an arbitrary residence time distribution is to use a single piston flow reactor with a large number of side entrances. See Figure 15.15(b). The size and spacing of the entrances can be adjusted to achieve a given residence time distribution. Thus, Figure 15.15 shows two ways of representing the same residence time distribution. The second way is quite different than the first. Molecules flow in through the side entrances and immediately mix with molecules that are already in the system. This is a maximum mixedness reactor, and there is substantial mixing between molecules that have different ages. Since there is only one exit, molecules that are mixed together will leave together, but they may have entered at different times. By way of contrast, there is only one entrance to the completely segregated reactor. Molecules that are mixed together in a completely segregated reactor must necessarily have entered together but they may leave separately.

Equation (15.48) governs the performance of the completely segregated reactor, and Equation (15.49) governs the maximum mixedness reactor. These reactors represent extremes in the kind of mixing that can occur between molecules that have different ages. Do they also represent extremes of performance as measured by conversion or selectivity? The *bounding theorem* provides a partial answer:

Suppose $\mathscr{R}_A$ is a function of a alone and that neither $d\mathscr{R}_A/da$ nor $d^2\mathscr{R}_A/da^2$ change sign over the range of concentrations encountered in the reactor. Then, for a system having a fixed residence time distribution, Equations (15.48) and (15.49) provide absolute bounds on the conversion of component A, the conversion in a real system necessarily falling within the bounds. If $d^2\mathscr{R}_A/da^2 > 0$, conversion is maximized by maximum mixedness and minimized by complete segregation. If $d^2\mathscr{R}_A/da^2 < 0$, the converse is true. If $d^2\mathscr{R}_A/da^2 = 0$, micromixing has no effect on conversion.

Example 15.16: Apply the bounding theory to an nth-order reaction, $\mathcal{R} = -ka^n$.

Solution:

$$d\mathcal{R}_A/da = -nka^{n-1}$$

and

$$d^2\mathcal{R}_A/da^2 = -n(n-1)ka^{n-2}$$

The first derivative is always negative. The second derivative is negative if $n > 1$, is zero if $n = 1$, and is positive if $n < 1$. Since it does not change sign for a fixed n, the bounding theorm applies. For $n > 1$ (e.g., second-order reactions), $d^2\mathcal{R}_A/da^2 < 0$ and conversion is highest in a completely segregated reactor. For $n = 1$, the reaction is first order, and micromixing does not affect conversion. For $n < 1$ (e.g., half-order), $d^2\mathcal{R}_A/da^2 > 0$ and maximum mixedness gives the highest possible conversion.

The bounding theory gives *sufficient* conditions for reactor performance to be bounded by complete segregation and maximum mixedness. These conditions are not *necessary*. In particular, the requirement that $d^2\mathcal{R}_A/da^2$ keep the same sign for $0 < a < a_{in}$ is not necessary. Some reactions show maximum rates so that the first derivative changes, yet the bounding theory still applies provided that $d^2\mathcal{R}_A/da^2$ does not change sign. If the second derivative does change sign, examples have been found that give a maximum conversion at an intermediate level of micromixing.

Micromixing Models. Hydrodynamic models have intrinsic levels of micromixing. Examples include laminar flow with or without diffusion and the axial dispersion model. Predictions from such models are used directly without explicit concern for micromixing. The residence time distribution corresponding to the models could be associated with a range of micromixing, but this would be inconsistent with the physical model.

Empirical models like fractional tubularity and tanks-in-series have a range of micromixing corresponding to their residence time distributions and sometimes a smaller range consistent with their physical configuration. For such models, it would seem desirable to have a micromixing model that, by varying some parameter, spans the possible range from maximum mixedness to complete segregation. It happens, however, that segregation is rarely observed in single-phase reactors.

The difference between complete segregation and maximum mixedness is largest when the reactor is a stirred tank and is zero when the reactor is a PFR. Even for the stirred tank case, it has been difficult to find experimental evidence of segregation for single-phase reactions. Real CSTRs approximate perfect mixing when observed on the time and distance scales appropriate to industrial reactions, provided that the feed is premixed. Even with unmixed feed, the experimental observation of segregation requires very fast reactions. The standard assumption of perfect mixing in a CSTR is usually justified. Worry when a highly reactive component is separately fed.

It is common to refer to the gross flow patterns in the reactor as *macromixing* and to molecular level mixing as *micromixing*. In this simplified view, the residence time distribution is determined by the macromixing, and micromixing is then imposed without substantially altering the residence time distribution. Some thought about laminar flow with diffusion will convince you that this idea is not rigorous in general, but it does work for the exponential distribution of a stirred tank. The *packet-diffusion model* supposes that the entering fluid is rapidly dispersed in small packets that are approximately the same size as the *Kolmogorov scale of turbulence*:

$$\eta = \left(\frac{\mu^3}{\rho^3 \nu}\right)^{1/4} \tag{15.51}$$

where ν is the power dissipation per unit mass of fluid. Following this rapid initial dispersion, the packets continue to evolve in size and shape but at a relatively slow rate. Molecular-level mixing occurs by diffusion between packets, and the rates of diffusion and of the consequent chemical reaction can be calculated. Early versions of the model assumed spherical packets of constant and uniform size. Variants now exist that allow the packet size and shape to evolve with time. Regardless of the details, these packets are so small that they typically equilibrate with their environment in less than a second. This is so fast compared with the usual reaction half-lives and with the mean residence time in the reactor that the vessel behaves as if it were perfectly mixed.

In laminar flow stirred tanks, the packet diffusion model is replaced by a *slab-diffusion model*. The diffusion and reaction calculations are similar to those for the turbulent flow case. Again, the conclusion is that perfect mixing is almost always a good approximation.

15.4 EXTENSIONS OF RESIDENCE TIME THEORY

The results in this chapter are restricted in large part to steady-state, homogeneous, isothermal systems. More general theories can be developed. The next few sections briefly outline some extensions of residence time theory.

15.4.1 Unsteady Flow Systems

The residence time distribution is normally considered a steady-state property of a flow system, but material leaving a reactor at some time θ will have a distribution of residence times regardless of whether the reactor is at steady state. The washout function for an unsteady reactor is defined as

$W(\theta, t)$ = Fraction of material leaving the reactor at time θ that remained in the system for a duration greater than t; i.e., that entered before time $\theta - t$.

A simple equation applies to a variable-volume CSTR:

$$W(\theta, t) = \exp\left[-\int_{\theta-t}^{\theta} \frac{Q_{in}}{V} d\theta'\right] \quad (15.52)$$

where $dV/d\theta = Q_{in} - Q_{out}$. The washout function can be used in the usual way to compute instantaneous values for the moments and reaction yields, including limits of complete segregation and maximum mixedness. The unsteady stirred tank is a maximum mixedness reactor when the tank is perfectly mixed. This is the usual case, and the reaction yield is more easily calculated using Equation (14.2) than by applying Zwietering's differential equation to the unsteady residence time distribution. Equation (15.48) applies to the complete segregation case appropriate to dispersed-phase reaction systems.

15.4.2 Contact Time Distributions

The yield of a gas–solid heterogeneous reaction depends not on the total time that molecules spend in the reactor but on the time that they spend on the catalyst surface. The *contact time distribution* provides a standardized measure of times spent in the absorbed state. A functional definition is provided by the following equation applicable to a first-order, heterogeneous reaction in an isothermal reactor:

$$\frac{a_{out}}{a_{in}} = \int_0^\infty e^{-kt_c} f_c(t_c) dt_c \quad (15.53)$$

where $f_c(t_c)$ is the differential distribution function for contact time. Equation (15.53) is directly analogous to Equation (15.44), and even provides a way of measuring $f_c(t_c)$. Vary the reaction temperature, and thus the rate constant k, over a wide range to measure $a_{out}(k)$ and then obtain $f_c(t_c)$ by inverse Laplace transformation. This approach has been used for a gas-fluidized bed, for which the assumption of isothermal operation is reasonable. The experiments detected bypassing as discussed in Section 11.3.1. Contact time distributions can also be measured using a combination of absorbable tracers. See Pustelnik and Nauman.[2]

15.4.3 Thermal Times

The analog of the residence time for a nonisothermal reaction is the thermal time:

$$t_T = \int_0^t \exp(-T_{act}/T) dt' \quad (15.54)$$

This is an integral along a molecule's path that weighs time and temperature in the manner appropriate to homogeneous but nonisothermal reactions. For a first-order reaction,

$$\frac{a_{out}}{a_{in}} = \int_0^\infty e^{-k_0} f_T(t_T) dt_T \tag{15.55}$$

where $f_T(t_T)$ is the differential distribution of thermal times and k_0 is the pre-exponential factor, the Arrhenius temperature dependence of the reaction having been incorporated in t_T. No measurement of $f_T(t_T)$ has been reported, but model-based calculations have been made for moving-wall devices such as extruders. The results show that a surprisingly uniform reaction environment is possible despite diffusion-free laminar flow and large point-to-point variations in temperatures. What happens is that the hot regions are associated with high velocities and low residence times. The integral in Equation (15.55) gives a similar value for t_T in the hot regions as in the cold regions that have long residence times.

15.5 SCALEUP CONSIDERATIONS

There are three situations where a residence time distribution can be scaled up with confidence.

1. The pilot reactor is turbulent and approximates piston flow. The larger reactor will have the same value of $\bar{t}$ and an aspect ratio, L/R, at least as great as that of the pilot reactor. These possibilities include the normal ways of scaling up a tubular reactor: in parallel, in series, by geometric similarity, and by constant pressure drop. The aspect ratio increases upon scaleup except when scaling in parallel or with geometric similarity, and the Reynolds number increases except when scaling in parallel. See Section 3.2 for the details. The worst case is scaling in parallel. The larger reactor will have the same residence time distribution as the small one. For the other forms of scaleup, the residence time distribution will more closely approach the delta distribution.

2. The pilot reactor is a tube in isothermal, laminar flow, and molecular diffusion is negligible. The larger reactor will have the same value for $\bar{t}$ and will remain in laminar flow. The residence time distribution will be unchanged by the scaleup. If diffusion in the small reactor did have an influence, it will lessen upon scaleup, and the residence time distribution will approach that for the diffusion-free case. This will hurt yield and selectivity.

3. The pilot reactor is a CSTR. The large reactor will be geometrically similar to the small one, and the scaleup will be done at constant power per unit volume. This form of scaleup exploits the fact that small vessels typically

use impeller speeds that are faster than necessary to achieve a close approach to the exponential distribution. Scaleup will eventually cause problems because of the ratio of circulation to throughput scales with the impeller speed, and this will decrease when scaling at constant power per unit volume. Correlations exist for the pumping capacity of the common impellers, so that the ratio of circulation to throughput can be calculated. If it is maintained at a reasonable value, say 8:1, the residence time distribution will remain close to exponential. Better, the impeller speed can be decreased in the small unit to test the water. Decrease it by a factor of $S^{2/9}$ where S is the intended scaleup factor for throughput. See Table 4.1. Scaling with constant power per unit volume will maintain the same Kolmogorov eddy size, so that micromixing should not become a problem.

PROBLEMS

15.1. A step change experiment of the turnoff variety gave the following results:

t	$C_{out}(t)/C_{out}(0)$
0	1.00
5	1.00
10	0.98
15	0.94
20	0.80
30	0.59
45	0.39
60	0.23
90	0.08
120	0.04

where t is in seconds. Estimate $\bar{t}$.

15.2. Determine the dimensionless variance of the residence time distribution in Problem 15.1. Then use Equation (15.40) to fit the axial dispersion model to this system. Is axial dispersion a reasonable model for this situation?

15.3. What, if anything can be said about the residence time distribution in a nonisothermal (i.e., $T_{in} \neq T_{out}$) CSTR with variable density (i.e., $\rho_{in} \neq \rho_{out}$ and $Q_{in} \neq Q_{out}$).

15.4. A washout experiment is performed on a CSTR to measure its mean residence time. What is the effect of starting the experiment before the outlet concentration has fully reached C_0? Assume that the normalized output response is based on the outlet concentration measured at $t=0$ so that the experimental washout function starts at 1.0.

15.5. A positive step change experiment is performed on a CSTR to determine $F(t)$ and, from it, the mean residence time. What is the effect of a variation in the inlet concentration C_0 during the course of the experiment? Consider a change from C_0 to $C_0 + \Delta C_0$ occurring at various times.

15.6. Compare Equation (15.11) to Equation (13.9). It is clear that $f(l)$ is the discrete analog of $f(t)$ and that summation over all possible chain lengths is equivalent to integration over the 0 to ∞ range on t. What is the analog of $W(t)$ for the chain length distribution?

15.7. What are the numerical values for the two time constants in the stagnancy model of Example 15.7? See Figure 15.8, but do not use a graphical method to determine them.

15.8. Apply the side capacity model of Figure 15.11 to bypassing. Calculate and plot $W(t)$ for the case where $Q = 8 \, \text{m}^3/\text{h}$, $q = 7 \, \text{m}^3/\text{h}$, $V_m = 1 \, \text{m}^3$, and $V_s = 8$.

15.9. Suppose that the tracer fed to the reactor in Example 15.11 is not inert but decomposes according to first-order kinetics. Show that

$$|a_{out} - a_0|_{max} = \frac{\beta}{\sqrt{(1+k\bar{t})^2 + \omega^2 \bar{t}^2}}$$

15.10. Suppose a piston pump operating at 100 strokes per minute is used to meter one component into a reactant stream. The concentration of this component should not vary by more than 0.1%. Devise a method for achieving this.

15.11. Experimental conditions prevented the application of a good step change at the inlet to the reactor, but it was possible to monitor both C_{in} and C_{out} as functions of time:

Time, s	C_{in}	C_{out}
0	0	0
3	0.072	0
6	0.078	0
9	0.081	0.008
15	0.080	0.017
20	0.075	0.020
30	0.065	0.027
40	0.057	0.035
60	0.062	0.043
80	0.068	0.051
100	0.068	0.057
120	0.068	0.062

The reactor is a gas-fluidized bed for which the fractional tubularity model is usually appropriate.

(a) Write the model as $f(t) = \alpha \exp[-\alpha(t - \tau)]$ and estimate the parameters α and τ.

(b) Use this estimate and the Equation (15.46) to predict $C_{out}(t)$ given the experimental values for $C_{in}(t)$. Can your estimates for α and τ be improved by this approach? A reasonable approximation to the input signal might be

t	C_{in}
0–20	0.078
20–40	0.066
40–60	0.060
60–80	0.065
80–100	0.068

15.12. Use the data in Problem 15.1 to estimate the conversion for an isothermal, first-order reaction with $k = 0.093 \text{ s}^{-1}$.

15.13. Apply the bounding theorem to the reversible, second-order reaction

$$A + B \underset{k_r}{\overset{k_f}{\rightleftharpoons}} C + D$$

Assume A, B, C, and D have similar diffusivities so that local stoichiometry is preserved. Under what circumstances will conversion be maximized by (a) complete segregation (b) by maximum mixedness?

15.14. Heterogeneous reactions are often modeled as if they were homogeneous. A frequently encountered rate expression is

$$\mathscr{R}_A = \frac{ka}{1 + K_A a}$$

Suppose $k = 2 \text{ s}^{-1}$ and $K_A = 0.8 \text{ m}^3/\text{mol}$. Determine bounds on the yield for a reactor having $\bar{t} = 3 \text{ s}$ and an inlet feed concentration of 2 mol/m^3.

15.15. Suppose the reactor in Problem 15.14 obeys the fractional tubularity model with $\tau_p = 0.5$. Use this information to calculate narrower bounds on the yield.

15.16. A typical power input for vigorous agitation is 10 hp per 1000 gal in systems with water-like physical properties.

(a) Calculate the Kolmogorov scale of turbulence.

(b) Assume that a spherical droplet with a diameter equal to the Kolmogorov size is placed in a large, homogeneous mass of fluid. How long will it take for concentrations inside the drop to closely approach those in the homogeneous fluid? Use $\mathscr{D} = 2 \times 10^{-9} \text{ m}^2/\text{s}$ and require a 95% response to the homogeneous phase concentration.

(c) Suppose a second-order reaction with unmixed feed streams is occurring in the agitated vessel. How large can the rate constant $a_{in}k$ be if mixing and diffusion times are to remain an order of magnitude smaller than reaction times?

REFERENCES

1. Stokes, R. L. and Nauman, E. B., "Residence time distribution functions for stirred tanks in series," *Can. J. Chem. Eng.*, **48**, 723–725 (1970).
2. Pustelnik, P. and Nauman, E. B., "Contact time distributions in a large fluidized bed," *AIChE J.*, **37**, 1589–1592 (1991).

SUGGESTIONS FOR FURTHER READING

The ideas explored in this chapter are discussed at length in

Nauman E. B. and Buffham, B. A. *Mixing in Continuous Flow Systems*, Wiley, New York, 1983.

Much of the material is also available in

Nauman, E. B., "Invited review: residence time distributions and micromixing," *Chem. Eng. Commun.*, **8**, 53–131 (1981).

INDEX

Absorption, 393–395
Activation, 438, 440
  energy, 151, 440
  temperature, 152
  volume, 184
Activity coefficient, 236
Addition polymers, 463, 467–470
  equilibrium, 469
  free-radical polymers, 482
  living polymers, 479
  polymerization kinetics, 479–492
  scaleup considerations, 504–505
  structure, 469
  thermal effects, 468
  transition metal catalysis, 487
  vinyl copolymerizations, 487
  (*see also* Chain-growth polymers)
Adiabatic reactors, 160, 174, 335
Adiabatic temperature change, 161
Adjustable constants
  in cosmological models, 209
  in kinetic equations, 209–210, 361, 439
  in nonisothermal axial dispersion, 319
  in residence time model, 549
Adsorption, 351–369, 437–441
Aerobic fermentation (*see* Fermentation)
Agitator power (*see also* Power in a stirred tank), 27
Air-lift, 403
Anaerobic fermentation (*see* Fermentation)
Arrhenius temperature dependence 151–154, 209, 440
  biochemical reactions, 154, 440
  sequence of fitting data, 217
Asymptotic property, 44
Autocatalytic, 54–58, 136
Autorefrigerated reactors, 137, 168, 174–175, 190
Avoiding scaleup problems, 174
Axial diffusion (*see* Diffusion, axial)
Axial dispersion coefficient, 329–330, 335
Axial dispersion model, 329–337, 339–341, 344–345

  multicomponent, 329
  nonisothermal, 336
  two-phase, 409
  use for scaleup calculations, 344–345
  utility of, 334–336

Backward difference, 311–316, 337
Batch reactor, 21, 35
Bimolecular reactions, 152–153, 360–362
Biochemical reactions, 435–459
Bounding theorem, 571–572
Breakthrough, 421
Bubbling fluidization, 416
Bypassing, 392, 553, 567–568

Cell culture, 446–459
  batch reactors, 447, 453–454, 455
  CSTRs with biomass recycle, 457
  CSTRs without biomass recycle, 454
  growth dynamics, 448–452
  growth limitations, 449
  immobilized cells, 459
  PFRs, 458
Central differencing, 273, 312–313, 337
Chain-growth polymers (*see also* Addition polymers), 463–464, 467–470, 478–492, 504–505
Chain length
  dynamic, 484, 495
  number and weight averages, 471, 474–475, 481, 484–485
Chain reaction, 51, 56
Chain transfer, 479, 481, 487, 505
Chaos, 76, 120, 173, 520
Chemical etching, 425
Chemical vapor deposition, 426
Chemostat (*see also* CSTRs), 443
Closed system, 331, 559–561, 563
Cold-flow models, 304, 430
Collision theory, 4, 5, 7, 152–153
Combustion
  branching chain reactions in, 56
  elementary reactions in, 8, 54, 247
  as a fast reaction, 25, 309

Combustion (*Cont.*)
  heats of, 232
  reactors for, 111, 418, 419
Competitive reactions, 35, 224
Complete segregation, 10, 269, 564–575
Complex reactions (*see also* Multiple reactions), 36, 153
Component balances
  axial dispersion, 329, 409
  batch, 11, 38, 58, 161
  catalyst pellet, 352–353, 368
  CSTR, 22, 118, 120, 167, 518
  differential, 19, 20, 83
  fed-batch, 65
  flow system, 19
  integral, 1, 3, 19, 22
  laminar flow reactors, 271
  laminar flow reactors with radial flow, 302
  laminar flow reactors with variable physical properties, 303
  matrix form for batch reactors, 68, 161
  matrix form for CSTRs, 118
  matrix form for PFRs, 85
  multiphase, 387, 401, 409
  packed-bed, 318
  piston flow, 17, 83–85
  transpired wall reactor, 111
  two-phase CSTRs, 387
  two-phase PFRs, 401
  unsteady CSTRs, 518
  unsteady laminar flow reactors, 534
  unsteady PFRs, 531
Computation scheme
  for axial velocity profiles, 299–301
  for gas-phase PFRs, 90
  for liquid-phase CSTRs, 125–126
  for liquid-phase PFRs, 96
Condensation polymers, 464
  average chain lengths, 474
  binary, 465–466, 504
  conversion, 473
  cyclization, 467
  equilibrium, 466
  kinetics, 473–478
  molecular weight distribution, 475
  random copolymers, 467
  self-condensing, 466, 504
  stoichiometry, 465
  (*see also* Step growth polymerization)
Confounded reactors, 224–226

Consecutive reactions, 35, 221–223
  batch, 47
  piston flow, 81–82
Contact time, 417, 430, 433, 575
Continuous-flow stirred tank reactors (*see* CSTRs)
Control volume, 1, 19
Convected-mean (*see also* Mixing-cup average), 266
Convective diffusion of heat (*see also* Energy balances), 291–295, 496
Convective diffusion of mass (*see also* Component balances), 269–271, 310–311
Convective diffusion, unsteady, 534, 558
Convergence
  of ODE solvers, 78–80
  of optimization techniques, 203, 206, 208
  of PDE solvers, 272–277
Convergence order
  Euler's method, 43–44
  finite difference approximations, 313–316
  radial velocity calculations, 500
  trapezoidal rule, 267–268, 277
Convergence testing
  Euler's method, 40–44, 90, 96
  explicit PDE solver, 296
  false transients, 126
  molecular weight distribution, 477
Conversion, 12, 29, 125
Copolymers
  azeotrope, 490
  composition distribution, 489–492
  condensation polymerization, 467
  effects on structure, 469–470
  living polymer polymerizations, 479
  vinyl addition, 487–492
Core turbulence, 328
Crank-Nicholson (*see also* Numerical methods), 316
Creeping flow, 298, 305
CSTRs, 22, 29, 117–133
  gas-phase, 127–131
  liquid-phase, 123–125
  scaleup, 131, 176–179
  two-phase, 382–397
  two-phase, operating modes, 388–395
Cumulative distribution of residence times, 541, 543, 556

Danckwerts boundary conditions, 330–332
Deactivation in enzyme catalysis, 440–442

INDEX

Deactivation in gas–solid catalysis, 357, 369–371, 414, 417, 421
Dead polymer, 483–485
Dead time, 533, 550
Debugging computer programs, 162, 515
Decoking, 369, 418–421
Decomposition reactions, 6, 10, 51–53
Delta distribution, 548, 550, 567, 576
Delta function, 543, 546, 548
Denominator terms in rate expressions, 210, 357, 361–362, 440, 449
Density function (*see also* Differential distribution), 542
Design equation (*see* Component balances)
Design variable, 132, 170–172, 497
Desorption, 351–360, 436
Differential distribution
  of chain lengths, 471
  of contact times, 575–576
  of exposure times, 410
  moments from Laplace transform, 561
  relation to Laplace transform, 561, 563, 575–576
  of residence time models, 545–546
  of residence times, 542
  of residence times in a CSTR, 423
  residence times for a parabolic velocity, 269
  residence times for tanks-in-series, 550–551
  as response to a delta input, 548
Differential reactor, 163, 212, 218
Diffusion
  axial, 270–271, 283–284, 303–304, 310–311
  in catalyst particles, 367–368
  in catalyst pores, 350, 363–366, 419
  changes in effects with reactor size, 107, 110, 576
  convective, 269–271, 310–311, 534, 558
  eddy, 328, 410
  Knudsen, 364–366
  molecular, 67, 160, 263–271, 328
  multicomponent, 272, 318
  when negligible, 264
  product layer, 419
  radial, 271, 284, 305, 310
  thermal or heat, 160, 263, 291–296
Diluents (*see also* Inerts), 174
Dilution rate, 443
Dimensionless rate constant, 13, 61
Dimensionless reaction time, 13
Dimensionless variables, 44–45, 61, 121–122, 282–284, 293–296

Dispersion polymerization, 502, 503
Dispersion, radial, 318
Distributed parameter system, 21

Economy of scale, 28, 145
Effective thermal conductivity, 319, 321
Effectiveness factor
  in enzyme reactions, 442–443
  general definition, 362
  isothermal model, 363–367
  nonisothermal model, 367–368
Elementary reactions, 4, 23, 35, 153, 209
  in gas–solid catalysis, 353
Elephants, 209, 217, 361
Eley-Rideal mechanism, 354, 377
Emulsion phase in bubbling fluidization, 416, 417
Emulsion polymerization, 173–174, 502–503
Endgroups, 465, 467, 473
Endothermic reactions, 155, 165, 174, 231
Energy balances, 151–186, 227–234, 518, 532
  for a batch reactor, 160
  for a batch reactor on a molar basis, 245
  for a CSTR 167
  enhancement factors, 411
  enthalpy, 158–159, 164, 227, 229, 244–245
  for a flow reactor, 158, 164
  for a laminar flow reactor, 291
  for a laminar flow reactor with radial velocities, 303
  for a laminar flow reactor with variable properties, 304
  for a packed bed, 319
  for two-phase CSTRs, 396
  for an unsteady CSTR, 518
  for an unsteady PFR, 532
Enzyme catalysis, 436–446
  reactor design, 443–446
Equal reactivity assumption, 473, 477–484, 494
Equilibrium compositions
  multiple reactions, 245–248
  single reactions, 240–245
Equilibrium constant
  kinetic, 235, 237–239
  multiple reactions, 245–246
  reconciliation of types, 237
  in reverse reaction rates, 237–239
  thermodynamic, 235–240
Ergun equation, 87, 372
Eulerian coordinates, 328

584 INDEX

Euler's method (*see* Numerical methods)
Exotherm, 56, 162
Exothermic reactions, 174, 231
Explicit differencing methods, 314
Exponential distribution, 545, 567–569
 of chain lengths, 494
 of exposure times, 410
 moments of, 547–548
 of residence times, 423–424
 scaleup of, 576–577
Extent of reaction (*see* Reaction coordinate)

False transients, 119–123, 240
Fast fluidization, 417
Fast-fill-and-hold, 522
Fed-batch, 21, 64–65
 biochemical reactors, 452
 polymerizers, 502
 startups, 521–523
Fermentation (*see also* Cell Culture), 64–65, 446–455
 aerobic, 447
 anaerobic, 447
 biomass production, 448
 cell death, 452
 growth phases in batch, 447
 maintenance coefficients, 450
 substrate consumption, 449
Fick's law, 310
Film mass transfer (*see* Mass transfer, across film resistance)
Finite difference approximations, 311–314
First appearance time, 549–557
First-order reactions, 6, 12, 46, 47, 85
 axial dispersion, 332
 batch, 59
 heterogeneous, 356–357
Flat-plate geometry, 285, 289–290
Flory distribution, 475, 477, 485
Fluidized-bed reactors, 414–418, 550, 575
 for heterogeneous catalysis, 350, 365, 370–371
 for polymers, 487, 493, 503
Flux
 convective, 84–85, 90, 96, 164
 diffusive, 84, 270, 310, 363
 zero flux boundary condition, 274, 290
Forward difference, 273–275, 312–314
Forward shooting, 337–338
Fraction unreacted, 12, 59, 85, 125

Fractional tank extension, 551
Fractional tubularity, 549–550
Free energy of reaction, 230–231, 235
Free-radical polymers
 copolymers, 487–492
 kinetics, 482–486
 (*see also* Addition polymers)
Free-radical reactions, 8, 51–54, 249, 426
Friction factor (Fanning), 87
Froude number, 133
Fugacity, 235–236

Gamma function extension, 551
Gas-fed slurry reactors, 413
Gas–liquid reactors, 381–411
Gas scrubbing, 393
Gel effect, 56, 484
Geometric similarity (*see also* Scaleup), 27
Gibbs free energy (*see* Free energy)
Graetz number, 179
Graetz problem, 294–296
Graphical or plotting methods, 152, 217, 219
Grass, 57, 74, 520

Half-integer kinetics, 53
Half-life, 12, 14, 25
Heat balance (*see* Energy balances)
Heat capacity, 159, 228–229
Heat-generation curve, 171–173
Heat of reaction, 231–234
 standard, 233
 summation convention for multiple reactions, 159
Heat-removal curve, 171–173
Heat transfer coefficient
 interfacial in two-phase reactors, 396
 tanks, 133, 176–177
 tubes, 179–180, 336
 packed-bed, 319–321
Heats of formation, 203–231, 232
Henry's law, 383–386, 402, 416
Heterogeneous rate expressions, 210, 361–362
Heterogeneous reactions, 65, 210, 318
Heterogeneous catalysis
 in biochemical systems, 435–459
 gas–solid, 349–376
 in polymerizations, 438, 487, 492–494
Holdup, 382
Homogeneous reaction, 4
Homopolymerization, 467

INDEX **585**

Hotspot, 324, 344
Hougen and Watson kinetics, 358, 438

Ideal gas law, 86
Ideal reactors, 10
   batch, 10, 35–116
   combinations of ideal reactors, 133–146
   CSTRs, 10, 117–133
   piston flow, 10, 81–116
   segregated CSTRs, 10, 565
   (*see also* individual reactor types)
Ill-conditioned computations, 338
Immobilized cells, 459
Immobilized enzymes, 441
Implicit numerical methods, 314–316, 394
Impulse, 542–543, 546, 561
Independent reactions, 36, 67–68, 248
Inerts, 165
Inhibition in enzyme catalysis, 440–441
Initial value problem, 82
Initiation, 51, 479–495
Integral reactor, 218
Interfacial area, 384
Interfacial mass transfer, 383–386
Intrinsic kinetics, 354–362, 362–371
   experimental measurement, 371–375
   recommended models, 361–362
Inventory
   mass, 2, 18, 26, 94, 125
   mass in an unsteady system, 517
   molar, 94
   scaleup factor (*see also* Scaleup factor), 26
   surface, 375–376

Kolomogorov scale of turbulence, 574, 577

Lagrangian coordinates, 328
Laminar flow reactors, 86, 177, 263–316
   axial dispersion approximation, 335–336
   scaleup at constant pressure drop, 108
   scaleup in series, 102, 104
   scaleup when nonisothermal, 305
   scaleup with geometric similarity, 106, 304–305
Langmuir 330–331
Langmuir and Hinshelwood, 361, 438
Laplace transforms, 546–547, 559–561, 563, 575
Least squares, 210–212, 255
L'Hospital's rule, 49, 274
Limit cycle, 172, 528, 530

Linear burn rate, 422
Liquid–liquid reactors, 381
Loop reactors (*see* Recycle reactors), 127
Lumped parameter systems, 22, 508
Lumping of chemical species, 501
Lynx, 57, 520

Macromixing, 568, 574
Maintenance coefficients for cell culture, 450
Marching (*see also* Numerical methods), 40
Marching backwards, 339
Mass balances, 1, 2
   overall or integral, 1, 82, 123
   (*see also* Material balances)
Mass transfer
   across a membrane, 386
   across film resistance, 351, 352, 366–367, 409
   limitation on reaction, 391
Mass transfer coefficients
   gas-side, 385
   liquid-side, 385
   nonreactive measurement, 397–398
   overall, 384
   overall for two-phase CSTRs, 395–396
   reactive measurement, 399–400
   typical values, 400
Material balances, 1
   closure, 216
   integral, 159–160
   recycle loop reactor, 150
   (*see also* Mass balances)
Maximum mixedness, 567–575
Mean residence time
   apparent in an open system, 561
   as a characteristic time in a flow system, 26, 44
   constant density system, 18
   in a CSTR, 124
   in a gas-phase tubular reactor, 92–95
   general expression, 18
   held constant upon scaleup, 27, 99, 304
   from inert tracer experiments, 544
   multiple of to achieve steady state, 530
   required to achieve desired reaction, 24
   in variable-density reactors, 124
Mechanism, 8, 36, 153
   of enzyme catalysis, 436–446
   of free-radical reactions, 51–54
   of heterogeneous catalysis, 351–354, 358, 361

Mechanism (*Cont.*)
   of heterogeneous catalysis, dual-site or bimolecular models, 361, 378, 438
   of polymerizations, 463
   of polymerizations, addition or chain growth, 467–470, 478–479, 482–484
   of polymerizations, condensation or step-growth, 464–467
   of polymerizations, suspended-phase, 501–503
   of polymerizations, transition metal catalysis, 487
   of polymerizations, vinyl copolymerizations, 487–492
Membrane reactors, 112–113, 386–387, 441–443
Metastable steady states, 168–173, 529
Method of moments, 480, 494
Michaelis constant, 438
Michaelis-Menten kinetics, 436–439, 444
Microelectronics, 110, 424–426
Micromixing, 565, 567–568, 571–574
Micromixing models, 573–574
Mixing-cup average, 265–268, 277
Mixing time, 25, 65–66, 133
Molecular weight distributions, 470–472, 475–478, 493
Moments
   from Laplace transforms, 561
   of molecular weight distribution, 470–471
   of residence time distribution, 543–545
Monod kinetics, 448
Monoliths, 326
Motionless mixers, 290–291, 336
Moving solids reactors, 413–418
Multiphase reactors, 381–430
Multiple reactions, 35–38, 154, 220, 245
   summation of heats of reaction, 159, 161
   batch reaction stoichiometry, 67–71
   CSTRs, 118–120
   piston flow, 82
Multiple steady states, 120, 169–173, 457, 520
Multitubular reactors (*see also* Shell-and-tube reactors), 100, 182, 326–327, 496–497

Nanotechnology, 424
Noncatalytic fluid–solid reactions, 418–423
Nonelementary reactions (*see also* Multiple reactions), 35

Nonisothermal effectiveness, 367
Nonisothermal reactors
   axial dispersion, 336
   batch, 160–163
   CSTRs, 167–173
   laminar, 291–306
   packed-bed, 318
   piston flow, 163–167
   scaleup, 173–183, 305
Non-Newtonian fluids, 287, 306, 397
nth-order reactions, 46
Numerical methods, 39, 49, 77–80, 205–208, 272–303
   binary search, 146
   Crank-Nicholson, 316
   Euler's method applied to ODEs, 40–43, 90, 96, 126
   Euler's method applied to PDEs, 275–277
   extrapolation of Euler's method, 78–79
   extrapolation of Runge-Kutta method, 79–80
   false transients, 119–123, 240
   implicit for PDEs, 314–316
   linear regression, 255
   method of lines, 272–275
   Newton's method, 119, 147–149
   optimization, 205–208
   Runge-Kutta integration, 44, 77
   shooting techniques, 337–344
   trapezoidal rule, 277

Objective function, 205
Occam's razor, 212
Open system, 333, 559, 561, 563
Operating variable, 122, 132, 170–172, 497
Optimization, 187
   constrained and unconstrained, 206
   functional, 207, 199
   by golden section, 207
   by gradient methods, 207
   parameter, 205, 208
   by a random search, 194, 206
   of reaction systems, 156, 187–209
   regression analysis, 210
   of temperature, 154–158, 296–297
Oscillations, 76
   in batch systems, 57–58
   in biological systems, 57, 457
   in isothermal CSTRs, 120, 520–522
   in nonisothermal CSTRs, 172–173, 530
   in polymerizations, 173, 502

INDEX

Packed-bed reactors
  component balance, 318
  isothermal axial dispersion, 330, 335
  scaleup at constant pressure drop, 109
  scaleup in series, 105
  scaleup with geometric similarity, 108
Parametric sensitivity, 325, 344
Pathological behavior, 136, 552–555
Peclet number
  for the axial dispersion model, 329, 330
  for the radial dispersion model, 320
Penetration theory, 410
Perfect mixer (*see also* CSTRs), 23
PFR (*see* Piston flow reactor)
Pipeline reactors, 334–335
Piston flow reactor, 17, 19, 21, 29, 81–116
  gas-phase, 86–95
  liquid-phase, 95–98
  two-phase, 401–406
  two-phase combination with a CSTR, 406
Polydispersity, 472
Polymerization, 59–62, 64, 463–505
Population balance methods, 400, 413, 422–423, 448
Pore diffusion, 353–354, 363–368, 419–421
Power in a stirred tank, 27, 133
Prandtl number, 179, 336
Pressure
  dependence of reaction rate, 63, 184
  dependence of thermodynamic properties, 227–235
  gradient in laminar flow, 86
  gradient in packed beds, 87
  gradient in turbulent flow, 87
Product transitions, 525
Propagation, 51, 479–488
Pseudo-first order, 9, 47
Pseudohomogeneous kinetics, 127, 318, 349, 352–356
  vs. heterogeneous kinetics, 374–375

Quasi-steady state hypothesis, 44, 49–54, 445

Rabbits, 57, 520
Radial dispersion coefficient, 318
Radial velocities, 301–303
Radioactive decay, 6, 47
Rate constant, 4, 6
  first-order, 48
  fitting to data, 152–153, 209–226
  identical or repeated, 48–49, 119
  pseudo-first-order, 9
  rate constant, temperature dependence, 151–154
  ratios of, 356–357, 361–362, 489
  second-order, 45
Rate-determining step, 357–361
Rate of formation (*see also* Reaction rate), 5
Reaction
  coordinate for batch reactors, 69, 76
  coordinate for CSTRs, 146
  equilibria, 234–250
  front, 420
  order, 8
  rate 5, 37, 237
Reactors in series and parallel, 134
Rectangular coordinates, 285–287, 289–290
Recycle reactors, 139, 175, 177
  for kinetic studies, 127, 355, 371
  as a model for a stirred tank, 131, 141, 177
  residence time distribution, 551
Regression analysis, 152
  linear, 211, 255
  nonlinear 152, 210–212
Residence time distributions, 268, 539–577
  models, 545–561
Residual standard deviation, 212
Residual sum-of-squares, 212
Reverse reaction rates, 237–240
Reverse shooting, 339
Reversible reactions, 6, 36, 210, 237, 358
Reynolds number
  impeller, 132, 133
  particle, 87
Runaway reactions, 168, 174, 277, 323–325, 496–499
Runge-Kutta (*see also* Numerical methods), 77

Scaleable heat transfer, 175
Scaleup
  avoiding problems during, 174–175
  of batch reactors, 65–66
  blind, 304
  with constant heat transfer, 182
  of CSTRs, 131–133, 176–179
  diplomatic, 175
  of exothermic packed beds, 326
  of gas–liquid reactors, 427
  of laminar flow reactors, 304–305
  of mixing times, 25
  model-based, 304–305
  of nonisothermal reactors, 173–183

588 INDEX

Scaleup (*Cont.*)
  in parallel, 100–101, 174, 326
  in series, 101–106
  of tubular reactors, 99–109, 179–183
Scaleup factor
  for inventory, 26
  for number of tubes, 99–100
  for pressure, 180
  for stirred tanks, 133
  for throughput, 26
  for tubular reactors, 99–109, 180
  for volume, 27
Scaling down, 109–110
Schmidt number, 320
Second-order reactions, 7, 13, 14
Segregation (*see also* Complete segregation), 400, 487, 496, 565
Selectivity, 16, 29, 45
Self-assembly, 427, 433
Semibatch (*see also* Fed-batch), 64
Shell-and-tube heat exchangers, use in loop reactors, 179, 388, 503
Shell-and-tube reactors
  feed distribution, 100, 174, 496
  use in scaleup, 99–101, 174, 190, 196, 326
  (*see also* Multitubular reactors)
Shooting method, 337–344, 394
Shutdown strategies, 521–525
Side capacity model, 553–556
Single-train process, 28
Site balance, 356
Site competition models, 349, 369
Slit flow, 285–287, 293
Slurry reactors, 413, 493, 503
Space time, 94
Spatial average, 3, 266, 306
Spouted-bed reactors, 417–418
Stability
  of isothermal CSTRs, 520
  by method of false transients, 120, 520
  of tubular reactors, 496, 504
Stability, numerical or discretization
  dimensionless form for laminar flow in a tube, 283–284
  for a flat profile in a slit, 288–289
  for a flat profile in a tube, 287
  for laminar flow in a tube, 276–277
  for packed-bed model, 321
  for slit flow, 286
  on thermal diffusivity, 292–294
Stability, physical, 277, 323, 496

Stagnancy, 553–555, 557
Standard deviation, 214
Startup strategies, 521–525
Static mixers (*see* Motionless mixers)
Step changes
  to measure mass transfer coefficient, 397
  to measure residence time distribution, 540–541
Step-growth polymerization, 463, 464–467, 473–478
  (*see also* Condensation polymers)
Stiff equations, 44, 49, 80, 272–274
Stirred tank reactors (*see* CSTRs)
Stoichiometric coefficients, 5, 37, 69, 159, 209
  matrix of, 67
Stoichiometry
  global, 67
  local, 67–68, 272, 504
  of multiple reactions, 67–71
  of single reactions, 66–67
Streamlines, 264–268, 303, 500–501, 556–557
Structured packing, 326
Substrate
  activation, 437–440
  in biochemical reactions, 436
  inhibition, 437–440
Sum-of-squares, 152, 211–225, 255–256, 489
Surface concentrations
  of adsorbed molecules, 318, 353
  in gas phase at external surface of particle, 352–353
Surface reaction, 353, 358
Surface renewal, 409
Suspended cells, 452
Suspending-phase polymerizations, 501–503

Tanks in series, 137–139, 173, 550–551
Termination reactions, 52, 479–488
Thermal conductivity in packed beds, 319, 321
Thermal Peclet number for the axial dispersion model, 336–337
Thermal time, 297, 575–576
Thermodynamics of chemical reactions, 226–250, 255
Thiele modulus
  isothermal, 364, 367
  nonisothermal, 368
Third-order reactions, 7–8
Three-phase reactors, 412–413, 452

Throughput scaleup factor (*see also* Scaleup factor), 26
Time to approach steady state, 530
Transpired wall reactors, 21, 111–113
Tray columns, 393–395
Trickle-bed reactors, 412
Tubular reactors
　isothermal, 92–95
　laminar flow, 263–306
　scaleup, 99–110
　stability problems, 496–501
　turbulent flow, 317–345
Turbulence, 328, 410, 574
Turbulent flow reactors
　isothermal axial dispersion, 334
　scaleup at constant pressure drop, 109
　scaleup in series, 102, 103
　scaleup of CSTRs, 133
　scaleup with geometric similarity, 107

Unimolecular, 6, 358–361
Unmixed feed streams, 321, 345
Unsteady reactors, 119–123, 517–534

Velocity profile, turbulent, 327–328
Variable cross-section, 21, 82–86, 303–304
Variable density, 21, 59, 123–131, 164, 303–304
Variable physical properties
　in nonisothermal reactors, 161–164
　in laminar flow reactors, 303–304
　rigorous example, 243–245
　two-phase systems, 387
Variable pressures, 21

Variable viscosity, 115, 297–301, 500, 512
Variable volume, 21, 58–65, 240
　mechanically determined, 63
　thermodynamically determined, 63
　in two-phase systems, 388
　unsteady systems, 518, 575
Variance, 544–545, 549, 557–558, 577
Velocity profile
　axial, flat, 287–289, 328–329, 321, 335
　axial, for turbulent flow in tubes, 328–329
　in laminar flow, 263–305, 496–499
　in a packed bed, 318
　radial, 301–303
　residence time distribution, 268–269, 555–557
Velocity, superficial, 87, 318
Void fraction, 87, 318–319, 335
　random variations, 100
　superficial vs. total, 372
Volumetric scaleup factor (*see also* Scaleup), 27

Washout, 137, 455, 457
Washout function, 540–542
　for unsteady stirred tank, 575
Working volume, 382

Yield
　mass, 16
　molar, 15
　theoretical, 16

Zero molecule, 5
Zero-order reactions, 46–47, 214, 358
Zwietering's differential equation, 568–570

ABOUT THE AUTHOR

E. Bruce Nauman, Ph.D., is one of the world's most respected authorities on chemical engineering. An international consultant with E.B. Nauman & Associates since 1982, he worked in industry for nearly 20 years with such firms as Xerox and Union Carbide, and is now Professor of Chemical Engineering at Rensselaer Polytechnic Institute. He is the author of four books and more than 100 journal articles, and he holds six U.S. patents, with numerous foreign counterparts. Past president of the North American Mixing Forum of the American Institute of Chemical Engineers, associate editor of two chemical engineering journals, chairman of two Engineering Foundation Conferences, and a Fellow of the American Institute of Chemical Engineers, he earned his doctorate in chemical engineering at the University of Leeds, in the United Kingdom. He lives in Schenectady, New York.

RETURN TO: CHEMISTRY LIBRARY
100 Hildebrand Hall

LOAN PERIOD	1	2	3
4		5	1-MONTH USE

ALL BOOKS MAY BE RECALLED AFTER 7 DAYS.
Renewable by telephone.

DUE AS STAMPED BELOW.

NON-CIRCULATING
UNTIL: DEC 3, 2001 2PM

FEB 2 0 2005

MAR 1 1 2005

DEC 2 0 2005

DEC 1 9 2006

FORM NO. DD 10
3M 3-00

UNIVERSITY OF CALIFORNIA, BERKELEY
Berkeley, California 94720–6000